INTERNATIONAL UNION OF CRYSTALLOGRAPHY
BOOK SERIES

IUCr BOOK SERIES COMMITTEE

T0202396

The Basics of Crystallography and Diffraction

Fourth Edition

Christopher Hammond

University of Leeds

INTERNATIONAL UNION OF CRYSTALLOGRAPHY

OXFORD

UNIVERSITY PRESS

OXFORD
UNIVERSITY PRESS

Great Clarendon Street, Oxford, OX2 6DP,
United Kingdom

Oxford University Press is a department of the University of Oxford.
It furthers the University's objective of excellence in research, scholarship,
and education by publishing worldwide. Oxford is a registered trade mark of
Oxford University Press in the UK and in certain other countries

© Christopher Hammond 2015

The moral rights of the author have been asserted

First Edition published in 1997
Second Edition published in 2001
Third Edition published in 2009
Fourth Edition published in 2015
Reprinted 2016 (with corrections)
Reprinted with corrections 2024

Published in the United States of America by Oxford University Press
198 Madison Avenue, New York, NY 10016, United States of America

British Library Cataloguing in Publication Data
Data available

Library of Congress Control Number: 2015930096

ISBN 978–0–19–873867–1 (hbk.)
ISBN 978–0–19–873868–8 (pbk.)

Printed and bound by
CPI Group (UK) Ltd, Croydon, CR0 4YY

Preface to the First Edition (1997)

This book has grown out of my earlier *Introduction to Crystallography* published in the Royal Microscopical Society's Microscopy Handbook Series (Oxford University Press 1990, revised edition 1992). My object then was to show that crystallography is not, as many students suppose, an abstruse and 'difficult' subject, but a subject that is essentially clear and simple and which does not require the assimilation and memorization of a large number of facts. Moreover, a knowledge of crystallography opens the door to a better and clearer understanding of so many other topics in physics and chemistry, earth, materials and textile sciences, and microscopy.

In doing so I tried to show that the ideas of symmetry, structures, lattices and the architecture of crystals should be approached by reference to everyday examples of the things we see around us, and that these ideas were not confined to the pages of textbooks or the models displayed in laboratories.

The subject of diffraction flows naturally from that of crystallography because by its means—and in most cases only by its means—are the structures of materials revealed. And this applies not only to the interpretation of diffraction patterns but also to the interpretation of images in microscopy. Indeed, diffraction patterns of objects ought to be thought of as being as 'real', and as simply understood, as the objects themselves. One is, to use the mathematical expression, simply the transform of the other. Hence, in discussing diffraction, I have tried to emphasize the common aspects of the phenomena with respect to light, X-rays and electrons.

In Chapter 1 (Crystals and crystal structures) I have concentrated on the simplest examples, emphasizing how they are related in terms of the occupancy of atomic sites and how the structures may be changed by faulting. Chapter 2 (Two-dimensional patterns, lattices and symmetry) has been considerably expanded, partly to provide a firm basis for understanding symmetry and lattices in three dimensions (Chapters 3 and 4) but also to address the interests of students involved in two-dimensional design. Similarly in Chapter 4, in discussing point group symmetry, I have emphasized its practical relevance in terms of the physical and optical properties of crystals.

The reciprocal lattice (Chapter 6) provides the key to our understanding of diffraction, but as a *concept* it stands alone. I have therefore introduced it separately from diffraction and hope that in doing so these topics will be more readily understood. In Chapter 7 (The diffraction of light) I have emphasized the geometrical analogy with electron diffraction and have avoided any quantitative analysis of the amplitudes and intensities of diffracted beams. In my experience the (sometimes lengthy) equations which are required cloud students' perceptions of the basic geometrical conditions for constructive and destructive interference—and which are also of far more practical importance with respect, say, to the resolving power of optical instruments.

Chapter 8 describes the historical development of the geometrical interpretation of X-ray diffraction patterns through the work of Laue, the Braggs and Ewald. The diffraction of X-rays and electrons from single crystals is covered in Chapter 9, but only in the case of X-ray diffraction are the intensties of the diffracted beams discussed.

This is largely because structure factors are important but also because the derivation of the interference conditions between the atoms in the motif can be represented as

nothing more than an extension of Bragg's law. Finally, the important X-ray and electron diffraction techniques from polycrystalline materials are covered in Chapter 10.

The Appendices cover material that, for ease of reference, is not covered in the text. Appendix 1 gives a list of items which are useful in making up crystal models and provides the names and addresses of suppliers. A rapidly increasing number of crystallography programs are becoming available for use in personal computers and in Appendix 2 I have listed those which involve, to a greater or lesser degree, some 'self learning' element. If it is the case that the computer program will replace the book, then one might expect that books on crystallography would be the first to go! That day, however, has yet to arrive. Appendix 3 gives brief biographical details of crystallographers and scientists whose names are asterisked in the text. Appendix 4 lists some useful geometrical relationships.

Throughout the book the mathematical level has been maintained at a very simple level and with few minor exceptions all the equations have been derived from first principles. In my view, students learn nothing from, and are invariably dismayed and perplexed by, phrases such as 'it can be shown that'—without any indication or guidance of *how* it can be shown. Appendix 5 sets out all the mathematics which are needed.

Finally, it is my belief that students appreciate a subject far more if it is presented to them not simply as a given body of knowledge but as one which has been gained by the exertions and insight of men and women perhaps not much older than themselves. This therefore shows that scientific discovery is an activity in which they, now or in the future, can participate. Hence the justification for the historical references, which, to return to my first point, also help to show that science progresses, not by being made more complicated, but by individuals piecing together facts and ideas, and seeing relationships where vagueness and uncertainty existed before.

Preface to the Fourth Edition (2015)

The successive editions of this book have incorporated a wider range of subject matter, some of which is of a rather more 'advanced' nature. However, I hope that I have adhered throughout to my original objective of showing that crystallography (and diffraction) is not an 'abstruse and difficult subject'. And as the book has expanded I hope that it will serve a further function, namely that of providing a reader with a greater confidence in tackling the more advanced and mathematically sophisticated texts included in Further Reading.

To this end I have used the simplest possible figures (both the drawings and photographs of crystal models) that a student is able to reproduce for himself or herself and which, I hope, will complement and aid an understanding of, the beautiful but invariably more complex computer-generated images of crystal structures that are now so widely available.

Each Chapter and Appendix is largely self-contained, and the Contents and Index have been made sufficiently detailed such that the reader should be able to locate, at a glance, those pages that contain the information which she or he requires.

Finally, there are it seems to me, two 'landmark' X-ray diffraction photographs: Laue's 1912 photograph of zinc blende and Franklin's 1952 photograph of DNA and in view of which I have placed these 'by way of symmetry' at the beginning of this book.

Acknowledgements

In the preparation of the successive editions of this book, I have greatly benefited from the advice and encouragement of present and former colleagues in the University of Leeds who have appraised and discussed draft chapters or who have materially assisted in the preparation of the figures. In particular, I wish to mention Dr Andrew Brown, Professor Rik Brydson, Dr Tim Comyn, Dr Andrew Scott, and Mr David Wright (Institute for Materials Research); Dr Jenny Cousens and Professor Michael Hann (School of Design); Dr Peter Evennett (formerly of the Department of Pure and Applied Biology); Dr John Lydon (School of Biological Sciences); Professor Anthony North (Emeritus Professor of Biophysics); the late Dr John Robertson (former Chairman of the IUCr, Book Series Committee), and the late Dr Roy Shuttleworth (formerly of the Department of Metallurgy).

Dr Pam Champness (formerly of the Department of Earth Sciences, University of Manchester) read and advised me on much of the early draft manuscript; Mrs Kate Crennell (formerly Education Officer of the BCA) prepared several of the figures in Chapter 2; Professor István Hargittai (of the Budapest University of Technology and Economics) advised me on the work, and sought out biographical material on A.I. Kitaigorodskii; Professor Amand Lucas (of the University of Namur and Belgian Royal Academy) allowed me to use his optical simulation of the structure of DNA; Dr Keith Rogers (of Cranfield University) advised me on the Rietveld method; Professor Shigeru Ohba (of Keio University, Japan) kindly drew my attention to a number of errors and misprints overlooked in earlier editions; Professor Hans Grimmer of the Paul Scherrer Institute, Switzerland, advised me on the work of Leonhard Weber and pointed out a number of errors and Dr David Watkin (University of Oxford) instructed me on direct and charge-flipping methods and Professor Michael Glazer (Universities of Oxford and Warwick) advised me on space groups and group theory.

Ms Hema Latha of Integra Software Services, Pondicherry, India, Ms Ania Wronski and Dr Sonke Adlung of the Academic Division, Oxford University Press, have guided me in the overall preparation and submission of the manuscript.

Many other colleagues at Leeds and elsewhere, have permitted me to reproduce figures from their own publications, as have the copyright holders of books and journals. Individual acknowledgements are given in the figure captions.

I would like to thank Miss Susan Toon and Miss Claire McConnell for word processing the manuscript and for attending to my constant modifications to it and to Mr David Horner and Dr Peter Evennett for their careful photographic work.

Finally, I recall with gratitude the great influence of my former teachers, in particular Dr P.M. Kelly and Dr N.F.M. Henry.

C.H.

Institute for Materials Research
University of Leeds
Leeds, LS2 9JT
July 2014

Contents

Contents

Contents

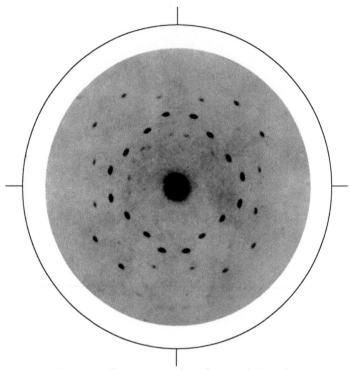

X-ray photograph of zinc blende

One of the eleven 'Laue Diagrams' in the paper submitted by Walter Friedrich, Paul Knipping and Max von Laue to the Bavarian Academy of Sciences and presented at its Meeting held on June 8th 1912—the paper which demonstrated the existence of internal atomic regularity in crystals and its relationship to the external symmetry.

 The X-ray beam (central black spot) is incident along one of the cubic crystal axes of the (face-centred cubic) ZnS structure and consequently the diffraction spots show the four-fold symmetry of the atomic arrangement about the axis. But notice also that the spots are not circular in shape—they are elliptical; the short axes of the ellipses all lying in radial directions. William Lawrence Bragg realized the great importance of this seemingly small observation: he had noticed that slightly divergent beams of light (of circular cross-section) reflected from mirrors also gave reflected spots of just these elliptical shapes. Hence he went on to formulate the Law of Reflection of X-Ray Beams which unlocked the door to the structural analysis of crystals.

X-ray photograph of deoxyribonucleic acid

The photograph of the 'B' form of DNA taken by Rosalind Franklin and Raymond Gosling in May 1952 and published, together with the two papers by J. D. Watson and F. H. C. Crick and M. H. F. Wilkins, A. R. Stokes and H. R. Wilson, in the 25 April issue of *Nature,* 1953, under the heading 'Molecular Structure of Nucleic Acids'.

The specimen is a fibre (axis vertical) containing millions of DNA strands roughly aligned parallel to the fibre axis and separated by the high water content of the fibre; this is the form adopted by the DNA in living cells. The X-ray beam is normal to the fibre and the diffraction pattern is characterised by four lozenges or diamond-shapes outlined by fuzzy diffraction haloes and separated by two rows or arms of spots radiating outwards from the centre. These two arms are characteristic of helical structures and the angle between them is a measure of the ratio between the width of the molecule and the repeat-distance of the helix. But notice also the sequence of spots along each arm; there is a void where the fourth spot should be and this 'missing fourth spot' not only indicates that there are two helices intertwined but also the separation of the helices along the chain. Finally, notice that there are faint diffraction spots in the two side lozenges, but not in those above and below, an observation which shows that the sugar-phosphate 'backbones' are on the outside, and the bases on the inside, of the molecule.

This photograph provided the crucial experimental evidence for the correctness of Watson and Crick's structural model of DNA—a model not just of a crystal structure but one which shows its inbuilt power of replication and which thus unlocked the door to an understanding of the mechanism of transmission of the gene and of the evolution of life itself.

1

Crystals and crystal structures

1.1 The nature of the crystalline state

The beautiful hexagonal patterns of snowflakes, the plane faces and hard faceted shapes of minerals and the bright cleavage fracture surfaces of brittle iron have long been recognized as external evidence of an internal order—evidence, that is, of the patterns or arrangements of the underlying building blocks. However, the nature of this internal order, or the form and scale of the building blocks, was unknown.

The first attempt to relate the external form or shape of a crystal to its underlying structure was made by Johannes Kepler* who, in 1611, wrote perhaps the first treatise on geometrical crystallography, with the delightful title, 'A New Year's Gift or the Six-Cornered Snowflake' (*Strena Seu de Nive Sexangula*).[1] In this he speculates on the question as to why snowflakes always have six corners, never five or seven. He suggests that snowflakes are composed of tiny spheres or globules of ice and shows, in consequence, how the close-packing of these spheres gives rise to a six-sided figure. It is indeed a simple experiment that children now do with pennies at school. Kepler was not able to solve the problem as to why the six corners extend and branch to give many patterns (a problem not fully resolved to this day), nor did he extend his ideas to other crystals. The first to do so, and to consider the structure of crystals as a general problem, was Robert Hooke* who, with remarkable insight, suggested that the different shapes of crystals which occur—rhombs, trapezia, hexagons, etc.—could arise from the packing together of spheres or globules. Figure 1.1 is 'Scheme VII' from his book *Micrographia,* first published in 1665. The upper part (Fig. 1) is his drawing, from the microscope, of 'Crystalline or Adamantine bodies' occurring on the surface of a cavity in a piece of broken flint and the lower part (Fig. 2) is of 'sand or gravel' crystallized out of urine, which consist of 'Slats or such-like plated Stones ... their sides shaped into **Rhombs, Rhomboeids** and sometimes into **Rectangles** and **Squares**'. He goes on to show how these various shapes can arise from the packing together of 'a company of bullets' as shown in the inset sketches A–L, which represent pictures of crystal structures which have been repeated in innumerable books, with very little variation, ever since. Also implicit in Hooke's sketches is the *Law of the Constancy of Interfacial Angles*; notice that the solid lines which outline the crystal faces are (except for the last sketch, L) all at 60° or 120° angles which clearly arise from the close-packing of the spheres. This law was first stated by Nicolaus Steno,* a near contemporary of Robert

* Denotes biographical notes available in Appendix 3.

[1] *The Six-Cornered Snowflake*, reprinted with English translation and commentary. Classic Texts in the Physical Science Oxford University Press, Oxford, 2014.

The Basics of Crystallography and Diffraction. Fourth Edition. Christopher Hammond.
© Christopher Hammond 2015. Published in 2015 by Oxford University Press.

Fig. 1.1. 'Scheme VII' (from Hooke's *Micrographia,* 1665), showing crystals in a piece of broken flint (Upper—Fig. 1), crystals from urine (Lower—Fig. 2) and hypothetical sketches of crystal structures A–L arising from the packing together of 'bullets'.

Hooke in 1669, from simple observations of the angles between the faces of quartz crystals, but was developed much more fully as a general law by Rome de L'Isle* in a treatise entitled *Cristallographie* in 1783. He measured the angles between the faces of carefully-made crystal models and proposed that each mineral species therefore had an underlying 'characteristic primitive form'. The notion that the packing of the underlying building blocks determines both the shapes of crystals and the angular relationships between the faces was extended by René Just Haüy.* In 1784 Haüy showed how the different forms (or habits) of dog-tooth spar (calcite) could be precisely described by the packing together of little rhombs which he called 'molécules intégrantes' (Fig. 1.2). Thus the connection between an internal order and an external symmetry was established. What was not realized at the time was that an internal order could exist even though there may appear to be no external evidence for it.

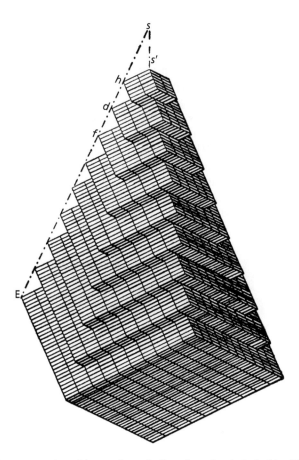

Fig. 1.2. Haüy's representation of dog-tooth spar built up from rhombohedral 'molécules integrantes' (from *Essai d'une théorie sur la structure des cristaux,* 1784).

* Denotes biographical notes available in Appendix 3.

It is only relatively recently, as a result primarily of X-ray and electron diffraction techniques, that it has been realized that most materials, including many biological materials, are crystalline or partly so. But the notion that a lack of external crystalline form implies a lack of internal regularity still persists. For example, when iron and steel become embrittled there is a marked change in the fracture surface from a rough, irregular 'grey' appearance to a bright faceted 'brittle' appearance. The change in properties from tough to brittle is sometimes vaguely thought to arise because the structure of the iron or steel has changed from some undefined amorphous or noncrystalline 'grey' state to a 'crystallized' state. In both situations, of course, the crystalline structure of iron is unchanged; it is simply the fracture processes that are different.

Given our more detailed knowledge of matter we can now interpret Hooke's spheres or 'bullets' as atoms or ions, and Fig. 1.1 indicates the ways in which some of the simplest crystal structures can be built up. This representation of atoms as spheres does not, and is not intended, to show anything about their physical or chemical nature. The diameters of the spheres merely express their nearest distances of approach. It is true that these will depend upon the ways in which the atoms are packed together, and whether or not the atoms are ionized, but these considerations do not invalidate the 'hard sphere' model, which is justified, not as a representation of the *structure* of atoms, but as a representation of the structures arising from the *packing together* of atoms.

Consider again Hooke's sketches A–L (Fig. 1.1). In all of these, except the last, L, the atoms are packed together *in the same way;* the differences in shape arise from the different crystal boundaries. The atoms are packed in a *close-packed hexagonal* or honeycomb arrangement—the most compact way which is possible. By contrast, in the *square* arrangement of L there are larger voids or gaps (properly called **interstices**) between the atoms. This difference is shown more clearly in Fig. 1.3, where the boundaries of the (two-dimensional) crystals have been left deliberately irregular to emphasize the

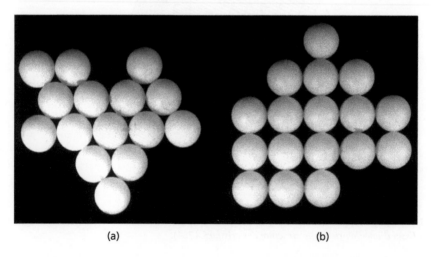

(a) (b)

Fig. 1.3. Layers of 'atoms' stacked (a) in hexagonal and (b) in square arrays.

point that is the internal regularity, hexagonal, or square, not the boundaries (or external faces) which defines the structure of a crystal.

Now we shall extend these ideas to three dimensions by considering not one, but many, layers of atoms, stacked one on top of the other. To understand better the figures which follow, it is very helpful to make models of these layers (Fig. 1.3) to construct the three-dimensional crystal models (see Appendix 1).

1.2 Constructing crystals from close-packed hexagonal layers of atoms

The simplest way of stacking the layers is to place the atom centres directly above one another. The resultant crystal structure is called the **simple hexagonal structure**. There are, in fact, no examples of elements with this structure because, as the model building shows, the atoms in the second layer tend to slip into the 'hollows' or interstices between the atoms in the layer below. This also accords with energy considerations: unless electron orbital considerations predominate, layers of atoms stacked in this 'close-packed' way generally have the lowest (free) energy and are therefore most stable.

When a third layer is placed upon the second we see that there are two possibilities: when the atoms in the third layer slip into the interstices of the second layer they may either end up directly above the atom centres in the first layer or directly above the unoccupied interstices between the atoms in the first layer.

The geometry may be understood from Fig. 1.4, which shows a plan view of atom layers. A is the first layer (with the circular outlines of the atoms drawn in) and B is the second layer (outlines of the atoms not shown for clarity). In the first case the third layer goes directly above the A layer, the fourth layer over the B layer, and so on; the stacking sequence then becomes ABABAB ... and is called the **hexagonal close-packed (hcp) structure.** The packing of idealized hard spheres predicts a ratio of interlayer atomic spacing to in-layer atomic spacing of $\sqrt{(2/3)}$ (see Exercise 1.1) and although interatomic forces cause deviations from this ratio, metals such as zinc, magnesium and the low-temperature form of titanium have the hcp structure.

In the second case, the third layer of atoms goes above the interstices marked C and the sequence only repeats at the fourth layer, which goes directly above the first layer.

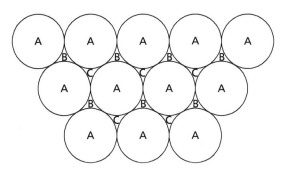

Fig. 1.4. Stacking sequences of close-packed layers of atoms. A—first layer (with outlines of atoms shown); B—second layer; C—third layer.

The stacking sequence is now ABCABC ... and is called the **cubic close-packed (ccp) structure.** Metals such as copper, aluminium, silver, gold and the high-temperature form of iron have this structure. You may ask the question: 'why is a structure which is made up of a three-layer stacking sequence of hexagonal layers called *cubic* close packed?' The answer lies in the shape and symmetry of the unit cell, which we shall meet below.

These labels for the layers A, B, C are, of course, arbitrary; they could be called OUP or RMS or any combination of three letters or figures. The important point is not the labelling of the layers but their **stacking sequence;** a two-layer repeat for hcp and a three-layer repeat for ccp. Another way of 'seeing the difference' is to notice that in the hcp structure there are open channels perpendicular to the layers running through the connecting interstices (labelled C in Fig. 1.4). In the ccp structure there are no such open channels—they are 'blocked' or obstructed because of the ABCABC ... stacking sequence.

Although the hcp and the ccp are the two most common stacking sequences of close-packed layers, some elements have crystal structures which are 'mixtures' of the two. For example, the actinide element americium and the lanthanide elements praseo-dymium, neodymium and samarium have the stacking sequence ABACABAC ... a four-layer repeat which is essentially a combination of an hcp and a ccp stacking sequence. Furthermore, in some elements with nominally ccp or hcp stacking sequences nature sometimes 'makes mistakes' in model building and faults occur during crystal growth or under conditions of stress or deformation. For example, in a (predomin-antly) ccp crystal (such as cobalt at room temperature), the ABCABC ... (ccp) type of stacking may be interrupted by layers with an ABABAB ... (hcp) type of stacking. The extent of occurrence of these **stacking faults** and the particular combinations of ABCABC ... and ABABAB ... sequences which may arise depend again on energy considerations, with which we are not concerned. What is of crystallographic import-ance is the fact that stacking faults show how one structure (ccp) may be transformed into another (hcp) and vice versa. They can also be used in the representation of more complicated crystal structures (i.e. of more than one kind of atom), as explained in Sections 1.6 and 1.9 below.

1.3 Unit cells of the hcp and ccp structures

A simple and economical method is now needed to represent the hcp and ccp (and of course other) crystal structures. Diagrams showing the stacked layers of atoms with irregular boundaries would obviously look very confused and complicated—the greater the number of atoms which have to be drawn, the more complicated the picture. The models need to be 'stripped down' to the fewest numbers of atoms which show the essential structure and symmetry. Such 'stripped-down' models are called the **unit cells** of the structures.

The unit cells of the simple hexagonal and hcp structures are shown in Fig. 1.5. The similarities and differences are clear: both structures consist of hexagonal close-packed layers; in the simple hexagonal structure these are stacked directly on top of each other,

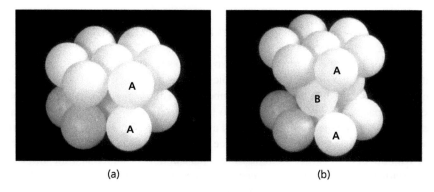

Fig. 1.5. Unit cells (a) of the simple hexagonal and (b) hcp structures.

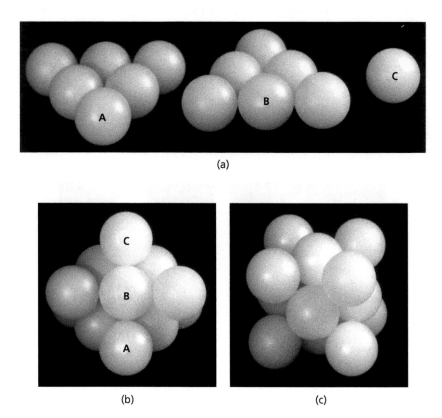

Fig. 1.6. Construction of the cubic unit cell of the ccp structure: (a) shows three close-packed layers A, B and C which are stacked in (b) in the 'ABC...' sequence from which emerges the cubic unit cell which is shown in (c) in the conventional orientation.

giving an AAA... type of sequence, and in the hcp structure there is an interleaving layer nestling in the interstices of the layers below and above, giving an ABAB... type of sequence.

The unit cell of the ccp structure is not so easy to see. There are, in fact, two possible unit cells which may be identified, a cubic cell described below (Fig. 1.6), which is almost invariably used, and a smaller rhombohedral cell (Fig. 1.7). Figure 1.6(a) shows three close-packed layers separately—two triangular layers of six atoms (identical to one of Hooke' s sketches in Fig. 1.1)—and a third layer stripped down to just one atom. If we stack these layers in an ABC sequence, the result is as shown in Fig. 1.6(b): it is a cube with the bottom corner atom missing. This can now be added and the unit cell of the ccp structure, with atoms at the corners and centres of the faces, emerges. The unit cell is usually drawn in the 'upright' position of Fig. 1.6(c), and this helps to illustrate a very important point which may have already been spotted whilst model building with the close-packed layers. The close-packed layers lie perpendicular to the body diagonal of the cube, but as there are four different body diagonal directions in a cube, there are therefore four different sets of close-packed layers—not just the one set with which

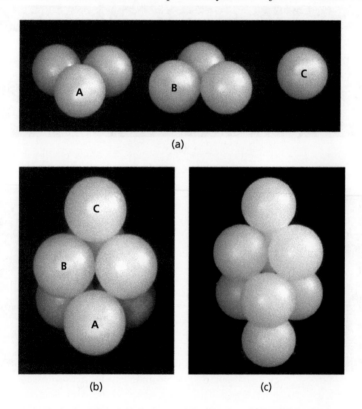

Fig. 1.7. Construction of the rhombohedral unit cell of the ccp structure: the close-packed layers (a) are again stacked (b) in the 'ABC...' sequence but the resulting rhombohedral cell (c) does not reveal the cubic symmetry.

we started. Thus three further close-packed layers have been automatically generated by the ABCABC... stacking sequence. This does not occur in the hcp structure—try it and see! The cubic unit cell, therefore, shows the **symmetry** of the ccp structure, a topic which will be covered in Chapter 4. The alternative rhombohedral unit cell of the ccp structure may be obtained by 'stripping away' atoms from the cubic cell such that there are only eight atoms left—one at each of the eight corners—or it may be built up by stacking triangular layers of only three atoms instead of six (Fig. 1.7). Unlike the larger cell, this does not obviously reveal the cubic symmetry of the structure and so is much less useful.

Although the atoms in the ccp and hcp structures are close-packed, there are (obviously) gaps or **interstices** between them; hence the volume occupied by the atoms is less than the total volume of the crystal. The proportion of the total volume occupied by the atoms is called the **packing density**, PD. Consider the ccp unit cell in Fig. 1.6(c). The volume of the unit cell $V = a^3$ where a is the distance along the cube-edge measured from atom-centre to atom-centre. The spheres', of radius r, are in contact along a face-diagonal, a distance $r + 2r + r = 4r = \sqrt{2}a$. Hence, $V = a^3 = 16\sqrt{2}r^3$. Now to calculate the number of atoms in the unit cell. There are eight at the corners, each 'shared' by eight surrounding cells $(8.\frac{1}{8} = 1)$ and six in the faces, each 'shared' by two adjacent cells $(6.\frac{1}{2} = 3)$, i.e. a total of four atoms per unit cell. The volume occupied by the atoms $V_a = 4(4/3)\pi r^3$. Hence, the packing density, $PD = \frac{\pi}{3\sqrt{2}} = 0.7405$. The same result is found for the hcp structure (Exercise 1.1).

1.4 Constructing crystals from square layers of atoms

It will be noticed that the atoms in the cube faces of the ccp structure lie in a square array like that in Fig. 1.3(b) and the ccp structure may be constructed by stacking these layers such that alternate layers lie in the square interstices marked X in Fig. 1.8(a). The models show how the four close-packed layers arise like the faces of a pyramid (Fig. 1.8(b)). If, on the other hand, the layers are all stacked directly on top of each other, a **simple cubic structure** is obtained (Fig. 1.8(c)). This is an uncommon structure for the same reason as the simple hexagonal one is uncommon. An example of an element with a simple cubic structure is α-polonium.

1.5 Constructing body-centred cubic crystals

The important and commonly occurring **body-centred cubic (bcc)** structure differs from those already discussed in that it cannot be constructed either from hexagonal close-packed or square-packed layers of atoms (Fig. 1.3). The unit cell of the bcc structure is shown in Fig. 1.9. Notice how the body-centring atom 'pushes' the corner ones apart so that, on the basis of the 'hard sphere' model of atoms discussed above, they are not 'in contact' along the edges (in comparison with the simple cubic structure of Fig. 1.8(c), where they are in contact). In the bcc structure the atoms are in contact only along the body-diagonal directions. The planes in which the atoms are most (not fully) closely packed is the face-diagonal plane, as shown in Fig. 1.9(a), and in plan

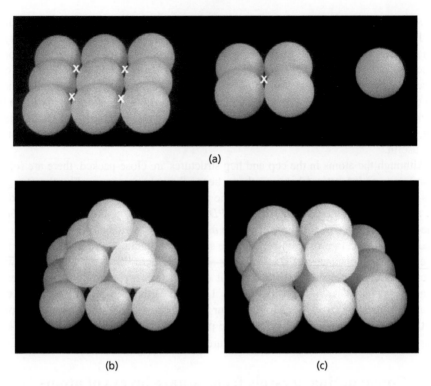

Fig. 1.8. (a) 'Square' layers of atoms with interstices marked X; (b) stacking the layers so that the atoms fall into these interstices, showing the development of the close-packed layers; (c) stacking the layers directly above one another, showing the development of the simple cubic structure.

view, showing more atoms, in Fig. 1.9(b). The atom centres in the next layer go over the interstices marked B, then the third layer goes over the first layer, and so on—an ABAB... type of stacking sequence. The interstices marked B have a slight 'saddle' configuration, and model building will suggest that the atoms in the second layer might slip a small distance to one side or the other (indicated by arrows), leading to a distortion in the cubic structure. Whether such a situation can arise in real crystals, even on a small scale, is still a matter of debate. Metals such as chromium, molybdenum, the high-temperature form of titanium and the low-temperature form of iron have the bcc structure.

Finally, notice the close similarity between the layers of atoms in Figs 1.3(a) and 1.9(b). With only small distortions, e.g. closing of the gaps in Fig. 1.9(b), the two layers become geometrically identical and some important bcc ⇌ ccp and bcc ⇌ hcp transformations are thought to occur as a result of distortions of this kind. For example, when iron is quenched from its high-temperature form (ccp above 910°C) to transform to its low-temperature (bcc) form, it is found that the set of the close-packed and closest-packed layers and close-packed directions are approximately parallel.

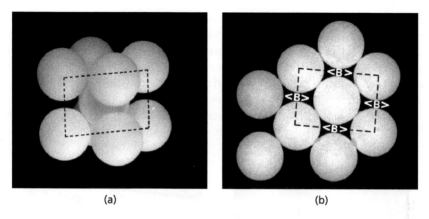

(a) (b)

Fig. 1.9. (a) Unit cell of the bcc structure, showing a face-diagonal plane in which the atoms are most closely packed; (b) a plan view of this 'closest-packed' plane of atoms; the positions of atoms in alternate layers are marked B. The arrows indicate possible slip directions from these positions.

The **packing density**, PD for the bcc structure is determined in the same way as for the ccp structure (Section 1.3 above). In this case, the spheres are in contact along a body-diagonal, distance $\sqrt{3}a = 4r$. Hence, the volume of the unit cell $V = a^3 = \left(\frac{4}{\sqrt{3}}\right)^3 r^3$. There are **two** atoms per unit cell, one at the corners as before, and one (unshared) in the centre. The volume occupied by the atoms $V_a = 2\left(\frac{4}{3}\right)\pi r^3$. Hence, the packing density, $PD = \pi\sqrt{3}/8 = 0.6802$, smaller than that for the fcc/hcp structures. A simple calculation for the simple cubic structure shows an even smaller $PD = \pi/6 = 0.5236$.

1.6 Interstitial structures

The different stacking sequences of one size of atom discussed in Sections 1.2 and 1.5 are not only useful in describing the crystal structures in many of the elements, where all the atoms are identical to one another, but can also be used to describe and explain the crystal structures of a wide range of compounds of two or more elements, where there are atoms of two or more different sizes. In particular, they can be applied to those compounds in which 'small' atoms or cations fit into the interstices between 'large' atoms or anions. The different structures of very many compounds arise from the different numbers and sizes of interstices which occur in the simple hexagonal, hcp, ccp, simple cubic and bcc structures and also from the ways in which the small atoms or cations distribute themselves among these interstices. These ideas can, perhaps, be best understood by considering the types, sizes and numbers of interstices which occur in the ccp and simple cubic structures.

In the ccp structure there are two types and sizes of interstice into which small atoms or cations may fit. They are best seen by fitting small spheres into the interstices between two-close-packed atom layers (Fig. 1.4). Consider an atom in a B layer which fits into the hollow or interstice between three A layer atoms: beneath the B atom is an interstice

which is surrounded at equal distances by four atoms—three in the A layer and one in the B layer. These four atoms surround or **'coordinate'** the interstice in the shape of a tetrahedron, hence the name **tetrahedral interstice** or **tetrahedral interstitial site,** i.e. where a small interstitial atom or ion may be situated. The position of one such site in the ccp unit cell is shown in Fig. 1.10(a) and diagrammatically in Fig. 1.10(b).

The other interstices between the A and B layers (Fig. 1.4) are surrounded or coordinated by six atoms, three in the A layer and three in the B layer. These six atoms surround the interstice in the shape of an octahedron; hence the name **octahedral**

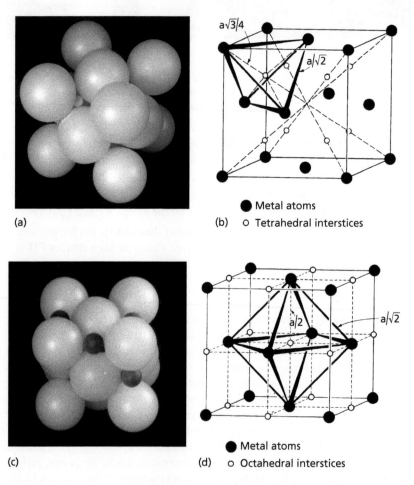

Fig. 1.10. (a) An atom in a tetrahedral interstitial site, $r_X/r_A = 0.225$ within the ccp unit cell and (b) geometry of a tetrahedral site, showing the dimensions of the tetrahedron in terms of the unit cell edge length a. (c) Atoms or ions in some of the octahedral interstitial sites, $r_X/r_A = 0.414$ within the ccp unit cell and (d) geometry of an octahedral site, showing the dimensions of the octahedron in terms of the unit cell edge length a. (From *The Structure of Metals,* 3rd edn, by C. S. Barrett and T. B. Massalski, Pergamon, 1980.)

interstice or **octahedral interstitial site.** The positions of several atoms or ions in octahedral sites in a ccp unit cell are shown in Fig. 1.10(c) and diagrammatically, showing one octahedral site, in Fig. 1.10(d).

Now the diameters, or radii, of atoms or ions which can just fit into these interstices may easily be calculated on the basis that atoms or ions are spheres of fixed diameter— the 'hard sphere' model. The results are usually expressed as a **radius ratio,** r_X/r_A; the ratio of the radius (or diameter) of the interstitial atoms, X, to that of the large atoms, A, with which they are in contact. In the ccp structure, r_X/r_A for the tetrahedral sites is 0.225 and for the octahedral sites it is 0.414. These radius ratios may be calculated with reference to Fig. 1.11. Figure 1.11(a) shows a tetrahedron, as in Fig. 1.10(b) outlined within a cube; clearly the centre of the cube is also the centre of the tetrahedron. The face-diagonal of the cube, or edge of the tetrahedron, along which the A atoms are in contact is of length $2r_A$. Hence the cube edge is of length $2r_A \cos 45 = \sqrt{2}r_A$ and the body-diagonal is of length $\sqrt{6}r_A$. The interstitial atom X lies at the mid-point of the body diagonal and is in contact with a corner atom A.

Hence $r_X + r_A = {}^1\!/_2\sqrt{6}r_A = 1.225r_A$; whence $r_X = 0.225r_A$.

Figure 1.11(b) shows a plan view of the square of four A atoms in an octahedron surrounding an interstitial atom X. The edge of the square, along which the A atoms are in contact, is of length $2r_A$ and the diagonal along which the X and A atoms are in contact is of length $2\sqrt{2}r_A$.

Hence $2r_X + 2r_A = 2\sqrt{2}r_A$; whence $r_X = 0.414r_A$.

In the simple cubic structure (Fig. 1.8(c)) there is an interstice at the centre of the unit cell which is surrounded by the eight atoms at the corners of the cube (Fig. 1.12(a)), hence the name **cubic interstitial site.** Caesium chloride, CsC1, has this structure, as shown diagrammatically in Fig. 1.12(b). The radius ratio for this site may be calculated in a similar way to that for the tetrahedral site in the ccp structure. In this case the atoms are in contact along the cube edge, which is of length $2r_A$ and the body diagonal, along which the atoms A and X are in contact, is of length $2\sqrt{3}r_A$.

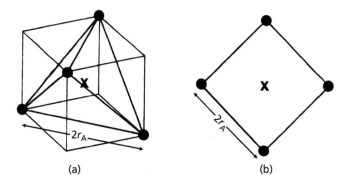

(a) (b)

Fig. 1.11. (a) A tetrahedral interstitial site X within a tetrahedron of four A atoms (shown as small filled circles for clarity) outlined within a cube, (b) A plan view of an octahedral interstitial site X with four surrounding A atoms (plus one above and one below X). The centres of the A atoms are shown as solid circles for clarity.

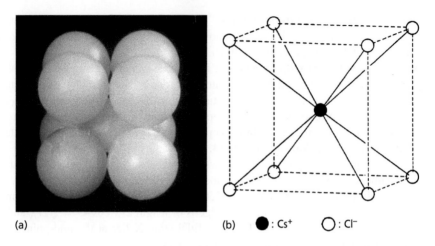

(a) (b) ● : Cs⁺ ○ : Cl⁻

Fig. 1.12. (a) Cubic interstitial site, $r_X/r_A = 0.732$, within the simple cube unit cell and (b) the CsC1 structure (ions not to scale).

Hence $r_X + r_A = \frac{1}{2}2\sqrt{3}r_A$; whence $r_X = 0.732 r_A$.

As well as being of different relative sizes, there are different numbers, or propor-tions, of these interstitial sites. For both the octahedral sites in ccp, and the cubic sites in the simple cubic structure, the proportion is one interstitial site to one (large) atom or ion, but for the tetrahedral interstitial sites in ccp the proportion is two sites to one atom. These proportions will be evident from model building or, if preferred, by geo-metrical reasoning. In the simple cubic structure (Fig. 1.12) there is one interstice per unit cell (at the centre) and eight atoms at each of the eight corners. As each corner atom or ion is 'shared' by seven other cells, there is therefore one atom per cell—a ratio of 1:1. In the unit cell of the ccp structure (Fig. 1.10(c)), the octahedral sites are situated at the midpoints of each edge and in the centre. As each edge is shared by three other cells there are four octahedral sites per cell, i.e. twelve edges divided by four (number shared), plus one (centre). There are also four atoms per cell, i.e. eight corners divided by eight (number shared), plus six faces divided by two (number shared), again giving a ratio of 1:1. The tetrahedral sites in the ccp structure (Fig. 1.10(a) and (b)) are situated between a corner and three face-centring atoms, i.e. eight tetrahedral sites per unit cell, giving a ratio of 1:2.

It is a useful exercise to determine also the types, sizes and proportions of intersti-tial sites in the hcp, bcc and simple hexagonal structures. The hcp structure presents no problem; for the 'hard sphere' model with an interlayer to in-layer atomic ratio of $\sqrt{(2/3)}$ (Section 1.2) the interstitial sites are identical to those in ccp. It is only the distribution or 'stacking sequence' of the sites, like that of the close-packed layers of atoms, which is different.

In the bcc structure there are octahedral sites at the centres of the faces and mid-points of the edges (Figs 1.13(a) and (b)) and tetrahedral sites situated between the centres of the faces and mid-points of each edge (Figs 1.13(c) and (d)). Note, however,

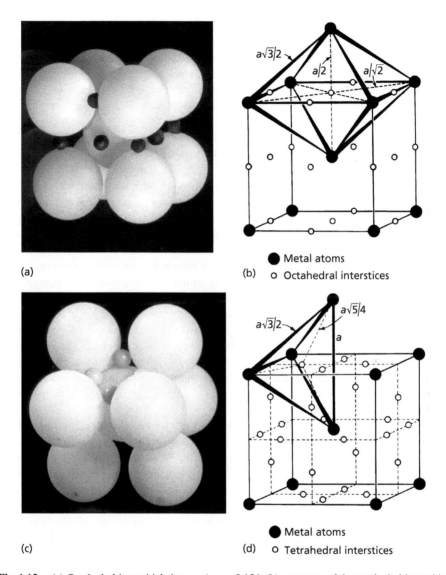

(a)

(b) ● Metal atoms
 ○ Octahedral interstices

(c)

(d) ● Metal atoms
 ○ Tetrahedral interstices

Fig. 1.13. (a) Octahedral interstitial sites, $r_X/r_A = 0.154$, (b) geometry of the octahedral interstitial sites, (c) tetrahedral interstitial sites $r_X/r_A = 0.291$, and (d) geometry of the tetrahedral interstitial sites in the bcc structure, (b) and (d) show the dimensions of the octahedron and tetrahedron in terms of the unit cell edge length a (from Barrett and Massalski, loc. cit.).

that both the octahedron and tetrahedron of the coordinating atoms do not have edges of equal length. The octahedron, for example, is 'squashed' in one direction and two of the coordinating atoms are closer to the centre of the interstice than are the other four.

It is very important to take this into account since the radius ratios are determined by the A atoms which are closer to the centre of the interstitial site and not those which

are further away. For the octahedral interstitial site the four A atoms which are further away lie in a square (Fig. 1.13(b)), just as the case for those surrounding the octahedral interstitial site in the ccp structure (Fig. 1.10(d)), but it is not these atoms, but the two atoms in the 'squashed' direction in Fig. 1.13(b) which determine the radius ratio. These are at a closer distance $a/2$ from the interstitial site where a is the cube edge length. Since in the bcc structure the atoms are in contact along the body diagonal, length $\sqrt{3}a$, then $4r_A = \sqrt{3}a$.

Hence $r_X + r_A = a/2 = 2r_A/\sqrt{3}$, whence $r_X = 0.154r_A$.

This is a very small site—smaller than the tetrahedral interstitial site (Fig. 1.13(c) and (d))—which has a radius ratio, $r_X/r_A = 0.291$ (see Exercise 1.2).

In the simple hexagonal structure the interstitial sites are coordinated by six atoms— three in the layer below and three in the layer above (Fig. 1.5(a)). It is the same coordination as for the octahedral interstitial sites in the ccp structure except that in this case the surrounding six atoms lie at the corners of a prism with a triangular base, rather than an octahedron, and the radius ratio is larger, $r_X/r_A = 0.528$ (see Exercise 1.3).

The radius ratios of interstitial sites and their proportions provide a very rough guide in interpreting the crystal structures of some simple, but important, compounds. The first problem, however, is that the 'radius' of an atom is not a fixed quantity but depends on its state of ionization (i.e. upon the nature of the chemical bonding in a compound) and coordination (the number and type of the surrounding atoms or ions). For example, the atomic radius of Li is about 156 pm but the ionic radius of the Li^+ cation is about 60 pm. The atomic radius of Fe in the ccp structure, where each atom is surrounded by twelve others, is about 258 pm but that in the bcc structure, where each atom is surrounded by eight others, is about 248 pm—a contraction in going from twelve- to eight-fold coordination of about 4 per cent.

Metal hydrides, nitrides, borides, carbides, etc., in which the radius ratio of the (small) non-metallic or metalloid atoms to the (large) metal atoms is small, provide good examples of interstitial compounds. However, in almost all of these compounds the interstitial atoms are 'oversize' (in terms of the exact radius ratios) and so, in effect, 'push apart' or separate the surrounding atoms such that they are no longer strictly close-packed although their pattern or distribution remains unchanged. For example, Fig. 1.14(a) shows the structure of TiN; the nitrogen atoms occupy all the octahedral interstitial sites and, because they are oversize, the titanium atoms are separated but still remain situated at the corners and face centres of the unit cell. This is described as a **face-centred cubic (fcc) array,** rather than a ccp array of titanium atoms, and TiN is described as a **face-centred cubic structure.** This description also applies to all compounds in which some of the atoms occur at the corners and face centres of the unit cell. The ccp structure may therefore be regarded as a special case of the fcc structure in which the atoms are in contact along the face diagonals.

In TiH_2 (Fig. 1.14(b)), the titanium atoms are also in an fcc array and the hydrogen atoms occupy all the tetrahedral sites, the ratio being of course 1:2. In TiH (Fig. 1.14(c)) the hydrogen atoms are again situated in the tetrahedral sites, but only half of these sites are occupied. Notice that in these interstitial compounds the fcc arrangement of the titanium atoms is *not* the same as their arrangement in the elemental form which is bcc (high temperature form) or hcp (low temperature form). In fact, interstitial *compounds*

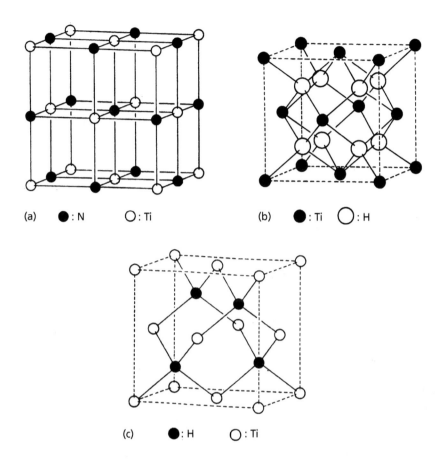

(a) ● : N ○ : Ti (b) ● : Ti ○ : H

(c) ● : H ○ : Ti

Fig. 1.14. (a) TiN structure (isomorphous with NaC1), (b) TiH_2 structure (isomorphous with CaF_2) and (c) TiH structure (isomorphous with sphalerite or zinc blende, ZnS or gallium arsenide, GaAs) (from *An Introduction to Crystal Chemistry*, 2nd edn, by R. C. Evans, Cambridge University Press, 1964).

based on a bcc packing of metal atoms are not known to exist; bcc metals such as vanadium, tungsten or iron (low temperature form) form interstitial compounds in which the metal atoms are arranged in an fcc pattern (e.g. VC), a simple hexagonal pattern (e.g. WC) or more complicated patterns (e.g. Fe_3C). Hence although as mentioned in Section 1.2, no elements have the simple hexagonal structure in which the close-packed layers are stacked in an AAA... sequence directly on top of one another (Fig. 1.5(a)), the metal atoms in some metal carbides, nitrides, borides, etc., are stacked in this way, the carbon, nitrogen, boron, etc., atoms being situated in some or all of the interstices between the metal atoms. The interstices are halfway between the close-packed (or nearly close-packed) layers and are surrounded or coordinated by six atoms—not, however, as described above in the form of an octahedron but in the form of a triangular prism. In the $A1B_2$ structure all these sites are occupied (Fig. 1.15(a)) and in the WC structure only half are occupied (Fig. 1.15(b)).

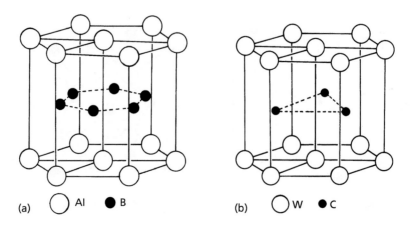

Fig. 1.15. (a) AlB$_2$ structure, (b) WC structure.

However, although there are no bcc interstitial *compounds* as such, interstitial elements can enter into the interstitial sites in the bcc structure to a limited extent to form what are known as *interstitial solid solutions*. One very important example in metallurgy is that of carbon in the distorted octahedral interstitial sites in iron, a structure called ferrite. The radius ratio of carbon to iron, r_C/r_{Fe}, is about 0.6, much greater than the radius ratio calculated above according to the 'hard sphere' model and the solubility of carbon is thus very small—1 carbon atom in about 200 iron atoms. The carbon atoms 'push apart' the two closest iron atoms and distort the bcc structure in a non-uniform or a *uniaxial* way—and it is this uniaxial distortion which is ultimately the origin of the hardness of steel. In contrast, a much greater amount of carbon can enter the (uniform) octahedral interstitial sites in the ccp (high temperature) form of iron (called austenite)—1 carbon atom in about 10 iron atoms. The carbon atoms are still oversize, but the distortion is a uniform expansion and the hardening effect is much less.

1.7 Some simple ionic and covalent structures

The ideas presented in Section 1.6 above can be used to describe and explain the crystal structures of many simple but important ionic and covalent compounds, in particular many metal halides, sulphides and oxides. Although the metal atoms or cations are smaller than the chlorine, oxygen, sulphur, etc. atoms or anions, radius ratio considerations are only one factor in determining the crystal structures of ionic and covalent compounds and they are not usually referred to as interstitial compounds even though the pattern or distribution of atoms in the unit cells may be exactly the same. For example, the TiN structure (Fig. 1.14(a)) is isomorphous with the NaCl structure. Similarly, the TiH$_2$ structure (Fig. 1.14(b)) is identical to the Li$_2$O structure and the TiH structure (Fig. 1.14(c)) is isomorphous with ZnS (zinc blende or sphalerite) and GaAs (gallium arsenide) structure.

Unlike the fcc NaCl or TiN structure, structures based on an hcp packing of ions or atoms with all the octahedral interstitial sites occupied only occur in a distorted form. The frequently given example is nickel arsenide (niccolite, NiAs). The arsenic atoms are stacked in the ABAB ... hcp sequence but with an interlayer spacing rather larger than that for close packing (see answer to Exercise 1.1) and the Ni atoms occupy all the (distorted) octahedral interstitial sites. These are all 'C' sites between the ABAB ... layers (see Fig. 1.4) and so the nickel atoms are stacked one above the other in a simple hexagonal packing sequence (Fig. 1.5(a))—they approach each other so closely that they are, in effect, nearest neighbours. Several sulphides (TeS, CoS, NiS, VS) all have the NiAs structure but there are no examples of oxides.

For a similar reason, structures based on the hcp packing of ions or atoms with all the tetrahedral sites occupied do not occur; there is no (known) such hcp interstitial structure corresponding to the fcc Li_2O structure. This is a consequence of the distribution of tetrahedral sites which occur in 'pairs' perpendicular to the close packed planes, above and below which are either A or B layer atoms. The separation of these sites is only one-quarter the hexagonal unit cell height (see Exercise 1.1) and both sites cannot be occupied by interstitial ions or atoms at the same time. However, half the interstitial sites can be occupied, one example of such a structure being wurtzite, the hexagonal form of ZnS, described below.

The differences in stacking discussed in Sections 1.2 and 1.6 also explain the different crystal structures or different crystalline polymorphs sometimes shown by one compound. As mentioned above, zinc blende has an fcc structure, the sulphur atoms being stacked in the ABCABC ... sequence. In wurtzite, the other crystal structure or polymorph of zinc sulphide, the sulphur atoms are stacked in the hexagonal ABABAB ... sequence, giving a hexagonal structure. In both cases the zinc atoms occupy half the tetrahedral interstitial sites between the sulphur atoms. As in the case of cobalt, stacking faults may arise during crystal growth or under conditions of deformation, giving rise to 'mixed' sphalerite-wurtzite structures.

Examples of ionic structures based on the simple cubic packing of anions are CsCl and CaF_2 (fluorite). In CsCl all the cubic interstitial sites are occupied by caesium cations (Fig. 1.12(b)) but in fluorite only half the sites are occupied by the calcium cations. The resulting unit cell is not just one simple cube of fluorine anions, but a larger cube with a cell side double that of the simple cube and containing therefore $2 \times 2 \times 2 = 8$ cubes, four of which contain calcium cations and four of which are empty.

The distribution of the small calcium cations in the cubic sites is such that they form an fcc array and the fluorite structure can be represented alternatively as an fcc array of calcium cations with all the tetrahedral sites occupied by fluorine anions. It is identical, in terms of the distribution of ionic sites, to the structure of TiH_2 or Li_2O (Fig. 1.14(b)), except that the positions of the cations and anions are reversed; hence Li_2O is said to have the *anti*fluorite structure. However, these differences, although in principle quite simple, may not be clear until we have some better method of representing the atom/ion positions in crystals other than the sketches (or clinographic projections) used in Figs. 1.10–1.15.

1.8 Representing crystals in projection: crystal plans

The more complicated the crystal structure and the larger the unit cell, the more difficult it is to visualize the atom or ion positions from diagrams or photographs of models—atoms or ions may be hidden behind others and therefore not seen. Another form of representation, the **crystal plan** or **crystal projection,** is needed, which shows precisely the atomic or ionic positions in the unit cell. The first step is to specify axes x, y and z from a common origin and along the sides of the cell (see Chapter 5). By convention the 'back left-hand corner' is chosen as the origin, the z-axis 'upwards', the y-axis to the right and the x-axis 'forwards', out of the page. The atomic/ionic positions or coordinates in the unit cell are specified as fractions of the cell edge lengths in the order x, y, z. Thus in the bcc structure the atomic/ionic coordinates are (000) (the atom/ion at the origin) and $\left(\frac{1}{2}\frac{1}{2}\frac{1}{2}\right)$ (the atom/ion at the centre of the cube). As all eight corners of the cube are *equivalent positions* (i.e. any of the eight corners can be chosen as the origin), there is no need to write down atomic/ionic coordinates (100), (110), etc.; (000) specifies *all* the corner atoms, and the two-coordinates (000) and $\left(\frac{1}{2}\frac{1}{2}\frac{1}{2}\right)$ represent the two atoms/ions in the bcc unit cell. In the fcc structure, with four atoms/ions per unit cell, the coordinates are (000), $\left(\frac{1}{2}\frac{1}{2}0\right)$, $\left(\frac{1}{2}0\frac{1}{2}\right)$, $\left(0\frac{1}{2}\frac{1}{2}\right)$.

Crystal projections or plans are usually drawn perpendicular to the z-axis, and Fig. 1.16(a) and (b) are plans of the bcc and fcc structure, respectively. Note that only the z coordinates are indicated in these diagrams; the x and y coordinates need not be written down because they are clear from the plan. Similarly, no z coordinates are indicated for all the corner atoms because all eight corners are equivalent positions in the structure, as mentioned above.

Figure 1.16(c) shows a projection of the antifluorite (Li_2O) structure; the oxygen anions in the fcc positions and the lithium cations in all the tetrahedral interstitial sites with z coordinates one-quarter and three-quarters between the oxygen anions are clearly shown. Notice that the lithium cations are in a simple cubic array, i.e. equivalent to the fluorine anions in the fluorite structure. The alternative fluorite unit cell, made up of eight simple cubes (see Section 1.7), is drawn by shifting the origin of the axes in Fig. 1.16(c) to the ion at the $\left(\frac{1}{4}\frac{1}{4}\frac{1}{4}\right)$ site and relabelling the coordinates. The result is shown in Fig. 1.16(d).

Sketching crystal plans helps you to understand the similarities and differences between structures; in fact, it is very difficult to understand them otherwise! For example, Fig. 1.17(a) and (b) show the same crystal structure (perovskite, $CaTiO_3$). They look different because the origins of the cells have been chosen to coincide with different atoms/ions.

1.9 Stacking faults and twins

As pointed out in Sections 1.2 and 1.7, the close packing of atoms (in metals and alloys) and anions (in ionic and covalent structures) may depart from the ABCABC... (ccp or fcc) or ABAB... (hcp or hexagonal) sequences: 'nature makes mistakes' and may do

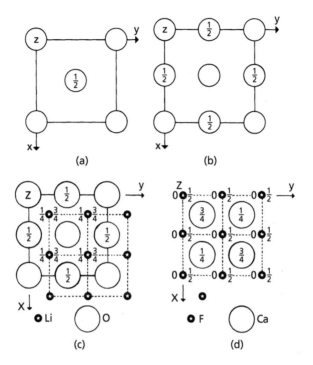

Fig. 1.16. Plans of (a) bcc structure, (b) ccp or fcc structure, (c) Li₂O (antifluorite) structure, (d) CaF₂ (fluorite) structure.

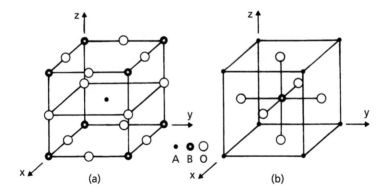

Fig. 1.17. Alternative unit cells of the perovskite ABO₃ structure.

so in a number of ways. First, stacking faults may occur during crystallization from the melt or magma: second, they may occur during the solid state processes or recrystallization, phase transitions and crystal growth (i.e. during the heating and cooling of metals and alloys, ceramics and rocks); and third, they may occur during deformation. The mechanisms of faulting have been most widely studied, and are probably most easily

understood in the simple case of metals in which there is no (interstitial) distribution of cations to complicate the picture. It is a study of considerable importance in metallurgy because of the effects of faulting on the mechanical and thermal properties of alloys—strength, work-hardening, softening temperatures and so on. However, this should not leave the impression that faulting is of lesser importance in other materials.

Consider first the ccp structure (or, better, have your close-packed raft models to hand). Three layers are stacked ABC (Fig. 1.4). Now the next layer should again by A; instead place it in the B layer position, where it fits equally well into the 'hollows' between the C layer atoms. This is the only alternative choice and the stacking sequence is now ABCB. Now, when we add the next layer we have two choices: either to place it in an A layer or in a C layer position. Now continue with our interrupted ABC... stacking sequence. In the first case we have the sequence ABCBABC... and in the second sequence ABCBCABC... In both cases it can be seen that there are layers which are in an hcp type of stacking sequence—but is there any difference between them, apart from the mere labelling of atom layers? Yes, there is a difference, which may be explained in two ways. If you examine the first sequence you will see that it is as if the mis-stacked B layer had been *inserted* into the ABCABC... sequence and this is called an *extrinsic* stacking fault, whereas in the second case it is as if an A layer had been *removed* from the ABCABC... sequence, and this is called an *intrinsic* stacking fault. However, this explanation, although it is the basis of the names *intrinsic* and *extrinsic,* is not very satisfactory. In order to understand better the distinction between stacking faults of different types (and indeed different stacking sequences in general), a completely different method of representing stacking sequences needs to be used.

You will recall (Section 1.2) that the labels for the layers are arbitrary and that it is the stacking *sequence* which is important; clearly then, some means of representing the sequence, rather than the layers themselves, is required. This requirement has been provided by F.C.Frank* and is named after him—the Frank notation. Frank proposed that each step in the stacking sequence A → B → C → A... should be represented by a little 'upright' triangle △, and that each step in the stacking sequence, C → B → A → C... should be represented by a little 'inverted' triangle ▽. Here are some examples, showing both the ABC... etc. type of notation for the layers and the Frank notation for the sequence of stacking of the layers. Note that the triangles come *between* the close packed layers, representing the stacking *sequence* between them.

A B C A B C A B C	C B A C B A C B A	Two fcc sequences
△ △ △ △ △ △ △ △	▽ ▽ ▽ ▽ ▽ ▽ ▽ ▽	
A B A B A B A B	C B C B C B C B	Two hcp sequences
△ ▽ △ ▽ △ ▽ △	▽ △ ▽ △ ▽ △ ▽	
A B C B A B C	A B C A C A B C	Extrinsic (left) and intrinsic (right) stacking faults
△ △ ▽ ▽ △ △	△ △ △ ▽ △ △ △	

* Denotes biographical notes available in Appendix 3.

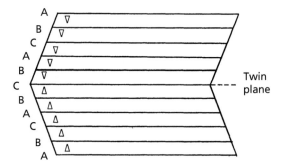

Fig. 1.18. Representation of the close-packed layers of a twinned fcc crystal indicating the atom layers 'edge on'. Notice that the stacking sequence reverses across the twin plane, such that the crystal on one side of the twin plane is a mirror reflection of the other.

Notice that the ccp (or fcc) stacking sequence is represented either by a sequence of upright triangles, or by a sequence of inverted ones—you could of course convert one to the other by simply turning your stack of close-packed layers upside down. However, the distinction is less arbitrary than this. An fcc crystal may grow, or be deformed (as explained below), such that the stacking sequence reverses, as indicated in Fig. 1.18 which shows the close-packed layers 'edge on' and both their ABC... etc. and Δ or ∇ labels. Such a crystal is said to be twinned and the twin plane is that at which the stacking sequence reverses. Note that the crystal on one side of the twin plane is a mirror reflection of the other, just like the pair of hands in Fig. 4.5(b).

The hcp stacking sequence is represented by alternate upright and inverted triangles—and the sequence is unchanged if the stack of close-packed layers is turned upside down. Hence twinning on the close-packed planes is not possible in the hcp structure—it is as if the backs and palms of your hands were identical, in which case, of course, your right hand would be indistinguishable from your left!

The Frank notation shows very well the distinction between extrinsic and intrinsic stacking faults; in the former case there are two inversions from the fcc stacking sequence, and in the latter case, one.

So far we have only considered stacking of close-packed layers of atoms and stacking faults in terms of the simple 'hard sphere' model. This model, given the criterion that the atoms should fit into the 'hollows' of the layers below (Fig. 1.4), would indicate that any stacking sequence is equally likely. We know that this is not the case—the fact that (except for the occurrence of stacking faults) the atoms, for example, of aluminium, gold, copper, etc., form the ccp structure, and zinc, magnesium etc., form the hcp structure, indicates that other factors have to be considered. These factors are concerned with the minimization of the energies of the nearest and second nearest neighbour configuration of atoms round an atom. It turns out that it is the configuration of the second nearest neighbours which determines whether the most stable structure is ccp or hcp. In one metal (cobalt) and in many alloys (e.g. α-brass) the energy differences between the two configurations is less marked and varies with temperature. Cobalt undergoes a

phase change ccp ⇋ hcp at 25°C, but the structure both above and below this temperature is characterized by many stacking faults. In α-brass the occurrence of stacking faults (and twins) increases with zinc content.

A detailed consideration of the stability of metal structures properly belongs to solid state physics. However, in practice we need to invoke some parameter which provides a measure of the occurrence of stacking faults and twins, and this is provided by the concept of the stacking fault energy (units mJ m^{-2}); it is simply the increased energy (per unit area) above that of the normal (unfaulted) stacking sequence. Hence the lower the stacking fault energy, the greater the occurrence of stacking faults. On this basis the energy of a twin boundary will be about half that of an intrinsic stacking fault, and the energy of an extrinsic stacking fault will be about double that of an intrinsic stacking fault.

As mentioned above, stacking faults, and the concept of stacking fault energy, play a very important role in the deformation of metals. During deformation—rolling, extrusion, forging and so on—the regular, crystalline arrangement of atoms is not destroyed. Metals do not, as was once supposed, become amorphous. Rather, the deformation is accomplished by the gliding or sliding of close-packed layers over each other. The overall gliding directions are those in which the rows of atoms are close-packed, but, as will also be evident from the models, the layers glide in a zig-zag path, from 'hollow to hollow' and passing across the 'saddle-points' between them. This is shown in Fig. 1.19, which is similar to Fig. 1.4, but re-drawn showing fewer atoms for simplicity. The overall glide direction of the B layer is along a close-packed direction, e.g. left to right, but the path from one B position to the next is over the saddle-points via a C position, as shown by the arrows. But the B layer may stop at a C position (partial slip), in which case we have an (intrinsic) stacking fault (Exercise 1.7). This partial slip is represented by the arrows or vectors B → C or C → B.

Extrinsic stacking faults, twins and the ccp → hcp transformation may be accomplished by mechanisms involving the partial slip of close-packed layers. The mechanism for the generation of a twinned crystal by deformation is illustrated sequentially in Fig. 1.20. Figure 1.20(a) shows the close packed layers of a single fcc crystal 'edge on'. Let us now slip an A layer (arrowed) into the sites of the B layer atoms (partial

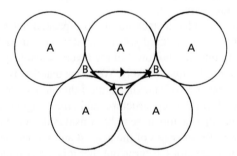

Fig. 1.19. On deformation, a B layer atom glides not in a straight line, e.g. B → B (left to right) but in two steps via a C site across the 'saddle points' between the underlying A layer atoms, i.e. B → C (first step), C → B (second step).

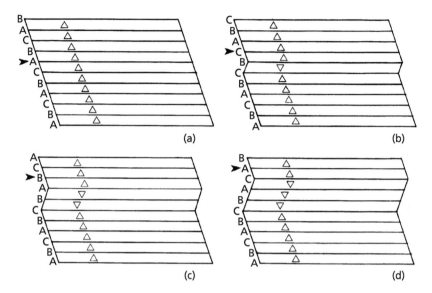

Fig. 1.20. (a)-(d) the sequence of faulting in a ccp (or fcc) crystal which leads to the formation of a twinned crystal as in Fig. 1.18. In Fig. 1.20(a) the close-packed layer to be faulted is an 'A' layer (arrowed) which in Fig. 1.20(b) becomes a 'B' layer and the layers above are relabelled accordingly; then the next layer above (the 'C' layer, arrowed) is faulted to become an 'A' layer (Fig. 1.20(c)) and so on.

slip), as shown in Fig. 1.20(b). In doing so all the layers above the A layer move too: B → C; C → A; A → B and so on. Figure 1.20(b) also shows these re-labelled layers of atoms. Now let us slip the next layer up (arrowed), C → A; and again A → B, B → C and so on (Fig. 1.20(c)). Again, slip the third layer up (arrowed) and re-label the layers—as shown in Fig. 1.20(d). As we can see we will ultimately generate the twinned crystal as shown in Fig. 1.18. This mechanism for deformation-twin formation may appear to be complicated; it occurs in practice (as in nearly all material deformation processes) through the movement of (partial) *dislocations,* a subject of great importance in materials science. The arrows in Fig. 1.19 represent the slip vectors for dislocations passing through the crystal; a slip vector such as B→B, in which the stacking sequence is unchanged, represents the passage of a *whole dislocation*; a slip vector such as B→C or C→B in which, as described above, the stacking sequence is changed, represents the passage of a *partial dislocation.* Such whole and partial dislocations also occur in bcc metal (and other) crystals. In the particular example above (Fig. 1.19) the partial dislocation vectors B → C and C → B (which lie in the close-packed slip plane) are called *Shockley partials,* after William Shockley* who first identified them.

The twinned crystal shown diagrammatically in Fig. 1.18 is just one particular example of very general phenomenon, which occurs in crystals with much more complex

* Denotes biographical notes available in Appendix 3.

structures and in which the two parts of the twinned crystal may not simply be related by reflection across the twin plane (reflection twins). In general a twinned crystal is one in which the two parts are related to each other by a rotation (usually, but not necessarily, 180°) about some particular direction called a *twin axis*. For example, we could create the twinned crystal in Fig. 1.18 by folding the figure 180° along the dashed line (the

(a)

(b)

(c)

(d)

0.1 mm

Fig. 1.21. Examples of twinned crystals: (a) rutile (TiO_2) twinned on a {101} plane (from Rutley's *Elements of Mineralogy,* 25th edn, revised by H. H. Read, George Allen and Unwin Ltd, 1962); (b) multiple twinning in rutile (from *Introduction to Crystallography,* 3rd edn, by F. C. Phillips, Longmans, 1963); (c) interpenetrant twin in mercurous chloride (HgC1), twin plane {101} (from F. C. Phillips loc. cit); (d) a photomicrograph of the etched surface of α-brass showing grain boundaries and (straight-sided) annealing twins, twin plane {111}.

twin axis) which lies in the twin plane; alternatively we could 'cut' the single crystal in Fig. 1.20(a) in two along the **c** layer and create the twinned crystal by rotating the upper half 180° about a vertical direction (another twin axis) which in this case is normal to the twin plane. In our crystal all these actions (reflection in the twin plane, rotation 180° about an axis in the twin plane, rotation 180° about an axis normal to the twin plane) are all equivalent, but this is not necessarily the case as we will discover when we have learned about the symmetry of crystals.

Some examples of twinned crystals from geology and metallurgy are shown in Fig. 1.21. The left–right-hand character of the twinned crystal in Fig. 1.21(a) is easily seen; in practice the twins may be interpenetrated and twinning may occur not just in one but in several different planes, as shown in Figs 1.21(b) and (c). In metals and alloys, the presence of twins may be seen in polished and etched surfaces (Fig. 1.21(d)). Old cast brass doorknobs provide a good and homely example. The stacking fault energy of α-brass (Cu \sim 30% Zn) is very low (30 mJ m^{-2}) and almost all the grains or crystals will contain twins. The corrosive contact of human hands reveals the irregular outlines or boundaries between the grains and also the perfectly straight lines or traces of the twin boundaries which terminate (unlike scratches) at the grain boundaries, as shown in Fig. 1.21(d).

Twinning is very common in minerals, frequently occurring as a result of phase transitions during cooling. They may be observed, like the brass, on the polished surfaces of minerals, where they may give rise to beautiful iridescent textures as a result of the diffraction and interference of light (Chapter 7). They may be more readily seen in petrographic thin sections under the polarizing light microscope. For the petrologist they constitute one of the most important means of mineral identification.

1.10 The crystal chemistry of inorganic compounds

The newcomer to crystallography is often dismayed by the daunting level of complexity, or immensity of all the different crystal structures—not only inorganic but also organic. The subject may seem to depend so much on rote-learning and the committing to memory of a vast number of facts. The subject is certainly immense, but it is not arbitrary, and the rules or criteria determining the simplest structures described in the preceding sections are the stepping-stones to an understanding of more complex structures.

We will consider the types of bonding which apply to inorganic crystals and the main differences to organic crystals (which are considered in more detail in Chapter 4). We have already seen how crystals may be represented as plans or projections (Section 1.8): now we will see how they can be represented in terms of coordination polyhedra, which lead naturally to a (brief) discussion of Pauling's rules which describe the conditions for the stability (and therefore the existence of) predominantly ionic crystal structures. Then, in Section 1.11, we describe some rather more complex crystal structures which either illustrate the principles we have already learned or for their own intrinsic interest and topicality.

1.10.1 Bonding in inorganic crystals

So far in this book we have described some simple crystal structures solely in geometrical terms—i.e. the various arrangements in which ions and atoms of different sizes can pack together. Now we must consider the forces which hold the ions and atoms together. This is of course a subject of profound importance in physical chemistry and we can only summarize, in a very simple way, the main features of the bonding mechanisms.

In inorganic crystals, the dominant bonding forces are *ionic* or *heteropolar bonds* and *covalent* or *homopolar bonds*, with lesser contributions from van der Waals and hydrogen bonds.

Ionic bonding dominates in inorganic crystals as a result of the high *electronegativity difference* between the atomic species—the (positive) metal cations on the one hand and the (negative) F^-, Cl^-, O^{2-} anions on the other. The simple compounds NaCl, CsCl and CaF_2 which we have already described are examples of almost 'pure' ionic bonding. Moreover, the ionic bond is non-directional, an important characteristic which it shares with the *metallic bond* which is the dominant cohesive force in metals with a high conductivity: the positive metal ions are held together by a 'sea' of conduction electrons which are wholly non-localized, i.e. the electrons are not bound to individual atoms.

As the electronegativity difference between the atomic species decreases, the covalent bond begins to dominate: the bonding electrons are 'shared' between the atoms instead of being transferred between the atomic orbitals. In sulphides, e.g. ZnS, the bonding is partly covalent. Unlike the ionic bond, the covalent bond is directional, the configuration around the atoms corresponds to, or arises from, the configuration of the atomic orbitals. An example of 'pure' covalent bonding is diamond (the hardest material known), described in more detail in Section 1.11.6. The carbon atoms in diamond are tetrahedrally coordinated, corresponding to the pattern of the sp^3 orbitals. The metallic bond, mentioned above, may be regarded as a 'special case' of a covalent bond—an atom shares electrons with its nearest neighbours and the empty orbitals permit the flow of conduction electrons. In many inorganic compounds, the bonding is a 'mixture' of (dominantly) ionic and (dominantly) covalent bonding. For example, in calcite, $CaCO_3$, the bonding between the Ca^{2+} and $(CO_3)^{2-}$ ions is dominantly ionic but the bonds which hold the $(CO_3)^{2-}$ groups together is dominantly covalent. The structure of calcite strongly resembles that of NaCl, except that the presence of the planar $(CO_3)^{2-}$ groups, all orientated the same way, parallel to the ABCABC... stacking sequence (see Fig. 1.6(b)), 'distorts' the cubic symmetry, or shape of the unit cell, to rhombohedral as shown in Fig. 1.22.

It is the high strength of ionic and covalent bonds which gives rise to the most characteristic mechanical property of inorganic compounds—their high hardness. However, many inorganic compounds are highly anisotropic in their mechanical properties, being 'strong' in some directions and 'weak' in others. Examples are the 'layer-structure' phyllosilicates, talc and mica (see Section 1.11.4) and graphite (see Section 1.11.6). Within the layers the ions/atoms are strongly ionically or covalently bonded, but the layers are only held together by much weaker *van der Waals* bonds which arise from the continuously changing dipole moments within atoms or molecules. An analogous situation applies to solid *organic* compounds; the atoms in the molecules themselves are

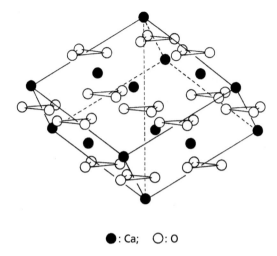

●: Ca; ○: O

Fig. 1.22. The rhombohedral structure of calcite, $CaCO_3$. The cell shown is not the smallest unit cell but corresponds to the cleavage rhombohedra which occur in natural crystals of calcite. The smallest unit cell may be outlined by linking together the eight innermost Ca^{2+} ions, following the pattern shown in Figure 1.7(c). (From *An Introduction to Crystal Chemistry*, 2nd Edn, by R. C. Evans, Cambridge University Press, 1964.)

strongly covalently bonded but the molecules are held together by weak van der Waals bonds and solid organic compounds are, in comparison, generally soft materials.

The concept of a molecule, introduced above, is of course fundamental in chemistry, but it can be an obstacle in understanding the arrangements and proportions of atoms and ions in inorganic compounds in the solid state. For example, it is not possible to identify a 'molecule' of NaCl in a crystal (Fig. 1.14(a)); each Na ion is bonded equally to six surrounding Cl ions and vice versa; the Na and Cl ions are not 'paired' to each other in any special way. W. L. Bragg records (in *The Development of X-ray Analysis*) that his chemical colleagues were loath to give up the idea of molecules in inorganic compounds and even as late as 1927, Henry E. Armstrong, an eminent chemist, published the following letter in *Nature*.

'Poor Common Salt'
"Some books are lies frae to end" says Burns. Scientific (save the mark) speculation would seem to be on the way to this state!... Prof. W.L. Bragg asserts that "In sodium chloride there appear to be no molecules represented by NaCl. The equality in number of sodium and chlorine atoms is arrived at by a chess-board pattern of these atoms; it is a result of geometry and not of a pairing-off of the atoms."

This statement is more than repugnant to common sense. It is absurd to the *n*..th degree, not chemical cricket. Chemistry is neither chess nor geometry, whatever X-ray physics may be. Such unjustified aspersion of the molecular character of our most necessary condiment must not be allowed any longer to pass unchallenged... It were time that chemists took charge of chemistry once more and protected neophytes against the worship of false gods; at least taught them to ask for something more than chess-board evidence.

However, perhaps the letter was written with a pinch of salt!

In addition, many inorganic compounds are *non-stoichiometric*. FeO, for example, which has the same structure as NaCl, rarely has the exact formula FeO, but a smaller proportion of Fe than in the ratio 1:1. This simply indicates the existence of vacant Fe^{2+} lattice sites and 2/3 the number of Fe^{3+} ions on other sites to ensure electrical neutrality and does not indicate a distinct chemical compound (see Section 1.11.3).

However, as mentioned above, the concept of a molecule is of vital importance in understanding the structures of solid *organic* compounds. The strong covalent bonds which bind together the carbon, nitrogen, oxygen, etc. atoms within the organic molecule, and which cause it to retain its identity when the compound is melted, dissolved, or even vaporized, are much stronger than the van der Waals bonds between the molecules in the solid state. The symmetries of the crystal structures which occur are, to a first approximation, determined by the overall shape or envelope of the molecules and the ways in which they pack together in the most efficient manner. However, we defer any further consideration of organic compounds until Chapter 4 when we have learned about symmetry, lattices and the methods needed to describe the geometry of crystals.

1.10.2　Representing crystals in terms of coordination polyhedra[2]

As we have seen in Sections 1.6 and 1.7, inorganic crystal structures may be described in terms of (small) atoms or cations, surrounded, or coordinated by, (large) atoms or anions. An alternative and very fruitful way of representing such structures is to concentrate attention on the coordinating polyhedra themselves rather than the individual atoms situated at their corners. We can then view the crystal structure as a pattern of linked polyhedra, without our view being obscured, as it were, by the large atoms or ions themselves.

The most important coordination polyhedra in inorganic crystals are the tetrahedron, $r_X/r_A = 0.225$; the octahedron, $r_X/r_A = 0.414$ (see Section 1.6) and the cubeoctahedron and hexagonal cubeoctahedron, $r_X/r_A = 1.000$, all of which occur in close-packed structures. The last two polyhedra represent the pattern of 12 atoms surrounding, or co-ordinating, an atom in face-centred and hexagonal close-packing respectively.

The polyhedra may be separate or linked together in various ways – corner-to-corner, edge-to-edge, face-to-face or in various combinations. Geometrically, the possibilities are almost endless, but in terms of the stability of inorganic crystals, they are not. The criteria which govern the stability, and hence possible structures, in (dominantly) ionic crystals, were set out by Linus Pauling* in a series of empirical rules. Essentially, these rules express the requirement for a charge balance between a cation and its surrounding polyhedron of anions and also between an anion and the cations that immediately surround it. The chemical law of valency is satisfied not by 'pairing off' individual anions and cations but by a 'sharing', or distribution of (say) the positive charge of a cation among the surrounding polyhedron of anions. This is further evidence of the

[2] See Appendix 2 – Polyhedra in crystallography.
* Denotes biographical notes available in Appendix 3.

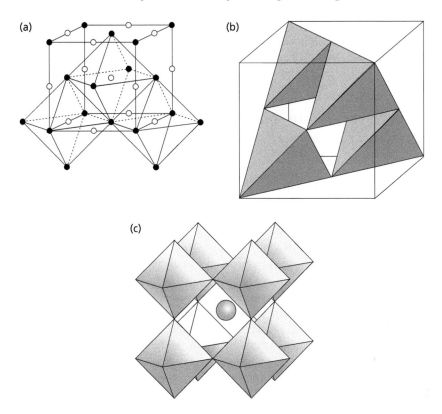

Fig. 1.23. Co-ordination polyhedra in three simple crystal structures: (a) edge-sharing octahedra in sodium chloride; (b) corner-sharing tetrahedra in zinc blende; and (c) corner-sharing octahedra and the (central) cubeoctahedron in perovskite. (From *An Introduction to Mineral Sciences* by Andrew Putnis, Cambridge University Press, 1992.)

non-existence of identifiable 'molecules' in inorganic ionic structures. Pauling's rules also determine, or limit, the ways in which the polyhedra can be linked together. The octahedra and cubeoctahedra surrounding larger and weaker cations may share corners, edges or even faces, but the tetrahedra surrounding the smaller and more highly positive cations tend only to share corners, such that the cations are as far apart as possible. We have already encountered this aspect of Pauling's rules in our discussion (Section 1.7) of the 'non-existence' of hcp structures in which all the tetrahedral sites are filled; the sites occur in pairs with the tetrahedra arranged face-to-face and the cations are too close to ensure stability. This also applies to the $(Si,Al)O_4$ tetrahedra which comprise the building blocks of the silicate minerals (see Section 1.11.4), the tetrahedra are isolated or share one, two, three or four corners, never edges or faces. However, this is not true of all synthesized ceramic materials. The new nitrido-silicates and sialons are based on the linkage of $(Si, Al)(N, O)_4$ (predominantly SiN_4) tetrahedra which share edges as well as corners and in addition, Si is octahedrally coordinated by N.

Figure 1.23 shows the coordination polyhedra in three simple crystal structures: (a) the edge-sharing octahedra in sodium chloride, NaCl; (b) the corner-sharing tetrahedra in zinc blende, ZnS; and (c) the corner-sharing octahedra and the central cubeoctahedron in perovskite, $CaTiO_3$. The unit cell shown is the same as that in Fig. 1.17(a).

1.11 Introduction to some more complex crystal structures

1.11.1 Perovskite $(CaTiO_3)$, barium titanate $(BaTiO_3)$ and related structures

Perovskite is an important 'type' mineral (in the same way as sodium chloride, NaCl) and is the basis of many technologically important synthetic ceramics in which the Ca is replaced by Ba, Pb, K, Sr, La or Co and the Ti by Sn, Fe, Zr, Ta, Ce or Mn. The general formula is ABO_3 (see Fig. 1.17), the A ion being in the large cubeoctahedral sites and the B ion being in the smaller octahedral sites (Fig. 1.23(c)). In perovskite itself, A is the divalent ion and B the tetravalent ion. This however is not a necessary restriction; trivalent ions can, for example, occupy both A and B sites; all that is needed is an aggregate valency of six to ensure electrical neutrality. It is, in short, a working out of Pauling's rules again. Of much greater importance are the sizes of these ions because they lead, separately or in combination, to different distortions of the cubic cell. In perovskite itself, the Ca cation is 'too small' for the large cubeoctahedral site and so the surrounding octahedra tilt, in opposite senses relative to one another, to reduce the size of the cubeoctohedral site. This is shown diagrammatically in Fig. 1.24(a). The unit cell is now larger, as outlined by the solid lines, the unit cell repeat distance now being between the similarly oriented octahedra (see Section 2.2). The symmetry is tetragonal, rather than cubic. The tilts are also slightly out of the plane of the projection which further reduces the symmetry to orthorhombic (see Chapter 3 for a description of these non-cubic structures).

In barium titanate, $BaTiO_3$, the Ti cation is 'too small' for the octahedral site and shifts slightly off-centre within the octahedron (Fig. 1.24(b)); the cubic unit cell is distorted to tetragonal. These shifts may occur along any of the three cube-edge directions such that a single crystal of $BaTiO_3$ may be divided into *domains*; within each domain, the shift is in the same direction (Fig. 1.25) (the idea of domains is discussed, with respect to ordered crystals, in Section 9.7). The consequence is that within a crystal (or within a domain) there is a net movement of charge, resulting in a structure with a dipole moment, and this spontaneous electrical polarization leads to the property of *ferroelectricity* which is of such great importance in many electronic devices and is the basis of much research on the titanates—some of which are anti-ferroelectric (e.g. $PbZrO_3$), ferromagnetic (e.g. $LaCo_{0.2}Mn_{0.8}O_3$) or antiferromagnetic (e.g. $LaFeO_3$). Barium titanate itself undergoes further transformations at lower temperatures (to orthorhombic and rhombohedral forms) which are, as with the tetragonal form, ferroelectric.

At higher temperatures (120° C for barium titanate), due to the increased thermal movement of the atoms, these structures revert to the cubic forms. This is an example

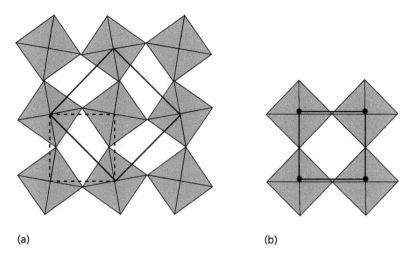

(a) (b)

Fig. 1.24. Two modifications to the perovskite structure: (a) octahedra tilted in opposite senses, the cubic cell is outlined by dashed lines and the tetragonal cell by full lines; (b) displacements of the B cations from the centres of the octahedra resulting in a distortion of the original cubic cell to a tetragonal unit cell (full lines). (From *An Introduction to Mineral Sciences* by Andrew Putnis, Cambridge University Press, 1992.)

Fig. 1.25. A polarized-light micrograph (crossed polars) of a single crystal of barium titanate which reveals the domains in each of which the tetragonal distortion is in a different cube-edge direction.

of a *displacive transformation*: no bonds are broken. Such displacive transformations also characterize the 'high temperature' (β) and 'low temperature' (α) forms of quartz, tridymite and cristobalite (see Section 1.11.5) and also relate the polymorphous forms of many *organic* compounds—the structures only differing in the way in which the molecules are packed together under the influence of the weak van der Waals forces.

1.11.2 Tetrahedral and octahedral structures—silicon carbide and alumina

Consider, for example, the crystal structures which consist of close-packed or nearly close-packed layers of large atoms or ions with the smaller atoms or ions occupying some or all of the tetrahedral interstitial sites. Zinc blende and wurtzite are two such structures (Section 1.7) which are re-drawn in Fig. 1.26 so as to emphasize the $\triangle\triangle\triangle\triangle$.... Stacking of the Zn (or S) atoms in zinc blende and the $\triangle\nabla\triangle\nabla$... stacking of the Zn (or S) atoms in wurtzite. Many carbides, including silicon carbide, SiC, also possess such close-packed structures with the carbon atoms occupying the tetrahedral sites and the metal or metalloid atoms stacked in various combinations of $\triangle\triangle\triangle$... and $\triangle\nabla\triangle\nabla$.... Figure 1.26 shows five structures or *polytypes* of silicon carbide (B3–B7) in which the silicon atoms are represented as solid spheres and the carbon atoms as small open spheres. The low-temperature form, β-SiC (B3) has the fcc structure, isomorphous with zinc blende. There are a number of high-temperature polytypes known collectively as α-SiC, the simplest structure of which, B4, is isomorphous with wurtzite. The other polytypes of α-SiC, three of which (B5, B6, B7) are shown in Fig. 1.26, have more complex stacking sequences, resulting in longer unit cell repeat distances.

B4 (2H)	B3 (3C)	B5 (4H)	B6 (6H)	B7 (15R)
Wurtzite	Zinc blende	Carborundum III	Carborundum II	Carborundum I

Fig. 1.26. Five common tetrahedral structure types. The bracketed symbols refer to the number of layers in the repeat sequence and the structure type: H (hexagonal), C (cubic) and R (rhombohedral) (after E. Parthe, *Crystal Chemistry of Tetrahedral Structures,* Gordon and Breach, New York, 1964, reproduced from *The Structure of Metals,* 3rd edn, C. S. Barrett and T. B. Massalski, Pergamon, 1980).

For example, the stacking sequence in B5 (Carborundum III) is (reading up) ABACABAC giving a four-layer repeat distance. In the Frank notation this is represented as $\triangle\triangledown\triangledown\triangle\triangle\triangledown\triangledown$... i.e. by inversions in the stacking sequence every two layers, rather than every layer as in B4, the wurtzite structure. Similarly, for B6 (Carborundum II) the stacking sequence is ABCACBA ... giving a six-layer repeat distance which in the Frank notation is $\triangle\triangle\triangle\triangledown\triangledown\triangledown$... i.e. an inversion every three layers.

Figure 1.27 shows an alternative representation of the polytypes of SiC showing (for clarity) only the close-packed layers of the metalloid (Si)atoms. The number below each polytype refers to the number of layers in the unit cell repeat distance and the letter refers to the type of unit cell (C—cubic; H—hexagonal; R—rhombohedral). All these polytypes of silicon carbide should not be thought of as distinct 'species', rather they should be regarded as interrelated as a result of different sequences of stacking faults and the transformations $\beta \rightleftharpoons \alpha$ SiC appear, from electron microscopy evidence, to occur

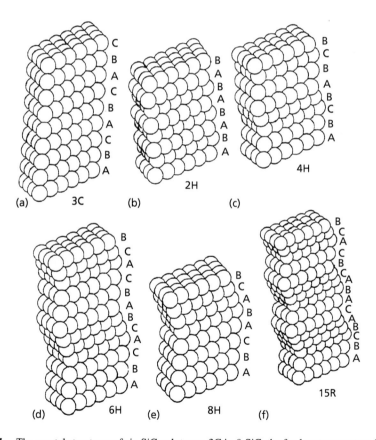

Fig. 1.27. The crystal structures of six SiC polytypes. 3C is β-SiC, the fcc low temperature form and the others with 2, 4, 6 and 8-layer repeat hexagonal cells or a 15-layer repeat rhombohedral cell are the α-SiC high temperature forms (from *Ceramic Microstructures* by W. E. Lee and W. M. Rainforth, Chapman & Hall, 1994).

as the result of the passage of partial dislocations across the close-packed planes in the same way as for the generation of a twinned crystal, as shown in Fig. 1.20. For example 6H, a common form of α-SiC, may be regarded as a 'microtwinned' form of 3C, the three-layer thick twins being generated by the passage of three partial dislocations on successive close-packed planes as shown in Fig. 1.20(d).

In aluminium oxide, Al_2O_3, the large oxygen anions occur in close-packed or nearly close packed layers with the aluminium cations occupying two-thirds of the octahedral interstitial sites. We now have an added complexity; not only may the oxygen anions be packed in different sequences but also the aluminium cations may be distributed differently throughout the interstitial sites—i.e. there may be different distributions of the one-third 'empty' sites. In the well-characterized form, α-Al_2O_3 (corundum-isomorphous with α-Fe_2O_3), the oxygen anions are stacked in the hcp ABAB ... stacking sequence and the aluminium cations between them are stacked in a rhombohedral sequence in exactly the same pattern as the carbon atoms in the rhombohedral form of graphite (see Fig. 1.37(b)). Hence the structure of α-Al_2O_3 is rhombohedral with a six-layer unit cell repeat distance of oxygen anions. The other polytypes of alumina, called the transition aluminas, are not so well characterized, particularly with respect to the distribution of the aluminium cations. A common form, γ-Al_2O_3 is based on an ABCABC... (fcc) stacking of oxygen anions with a distribution of aluminium cations which gives rise to a maghemite structure, described in Section 1.11.3.

1.11.3 The oxides and oxy-hydroxides of iron

Iron is remarkable for the range of oxides and hydroxides which can be formed and is one of the few elements which form compounds intermediate between these two—the oxy-hydroxides, the crystal structures of which are also of interest, and we will now describe them (as far as we are able) and trace their interrelationships.

We will begin with the simplest oxide, wustite, ferrous oxide, FeO and consider the structural changes which take place in the progressive oxidation of FeO to Fe_3O_4 (ferroso-ferric oxide) and Fe_2O_3 (ferric oxide). FeO has the NaCl structure—the Fe^{2+} ions are situated in the octahedral interstitial sites between the oxygen anions[3]. However, FeO is rarely stoichiometric but has vacant Fe^{2+} sites in the face-centred cubic structure, electrical neutrality being preserved by the presence of Fe^{3+} ions ($2Fe^{3+}$ replacing $3Fe^{2+}$, some of which are situated in the tetrahedral interstitial sites). The 'oxidation' of FeO to Fe_3O_4 proceeds, not by the 'addition' of oxygen atoms to the structure, but by the migration of Fe atoms to the surface (to combine with atmospheric oxygen there)—to a first approximation the close-packed oxygen atoms in the original FeO structure remain undisturbed. Within the structure the Fe^{2+} ions are progressively replaced by Fe^{3+} ions, half of which are situated in the tetrahedral interstitial sites.

[3] The terms 'anion' and 'cation' are used when we want to draw specific attention to the charge on the ionic species, otherwise the more general term 'atom' is used—it being understood that the term encompasses both neutral and charged species.

The structure of Fe_3O_4 (magnetite) thus formed is that of an *inverse spinel* with the general formula AB_2O_4 in which the 'A' tetrahedral sites are occupied by Fe^{3+} ions and the 'B' octahedral sites by equal numbers of Fe^{3+} and Fe^{2+} ions. Fe_3O_4 is therefore better represented by the formula $Fe^{3+}(Fe^{3+}Fe^{2+})O_4$. However, the occupancy of the interstitial sites is not random, but ordered such that the unit cell of Fe_3O_4 has twice the side-edge (eight times the volume) of the original FeO unit cell and contains 32 O^{2-} ions, 8 Fe^{2+} ions and 16 Fe^{3+} ions.

The distinction between an 'inverse' and a 'normal' spinel (both bad names!) is simple: spinel is the mineral $MgAl_2O_4$. The Mg^{2+} ions occupy the 'A' tetrahedral sites and the Al^{3+} ions occupy the 'B' octahedral sites—with regard to the 'A' sites only the inverse of Fe_3O_4. Again, we see that the occupancies of the interstitial sites are determined not simply by the valencies of the ions but also their sizes. Figure 1.28 shows a projection, or crystal plan of the spinel structure (see Section 1.8), split into two halves to make the atom/ion positions more clear.

As oxidation proceeds further the remaining eight Fe^{2+} ions in the unit cell are replaced by two-thirds of their number by Fe^{3+} ions (i.e. to maintain electrical neutrality) giving a total of $16 + 5^1/_3 = 21^1/_3$ Fe^{3+} ions, 32 O^{2-} ions and the overall composition Fe_2O_3. Except for a small reduction in volume (on account of the increased number of vacant lattice sites) the large cubic unit cell is unchanged. The structure is called maghemite, γ-Fe_2O_3 and is isostructural with the γ-Al_2O_3 structure (see Section 1.11.2). However, γ-Fe_2O_3 (like γ-Al_2O_3) is not the stable structural form and its conversion to α-Fe_2O_3 (haematite) occurs essentially by the rearrangement of the oxygen atoms from the fcc to the hcp stacking sequence.

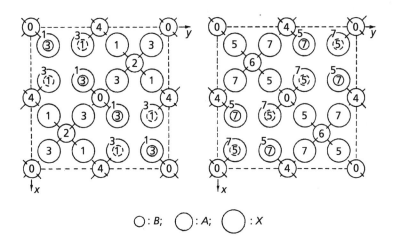

Fig. 1.28. Plan view of the unit cell of the cubic spinel structure AB_2X_4 projected on to a plane perpendicular to a cube axis. The heights of the atoms are indicated in units of one-eighth the cubic cell edge length. For clarity, the upper and lower halves of the cell are shown separately and only the tetrahedral coordination of the A ions is indicated. (From *Crystal Chemistry*, 2nd Edn, by R. C. Evans, Cambridge University Press, 1966.)

Similar relationships exist in the oxy-hydroxides in which iron is in the Fe^{3+} (ferric) form. The formulae are best written $FeO·OH$ (rather than $Fe_2O_3·H_2O$) to emphasize the replacement of oxygen ions by hydroxyl, $(OH)^-$ groups rather than the presence in the structure of discrete 'molecules' of water. In γ-$FeO·OH$ (lepidocrocite), the oxygen and hydroxyl ions are *approximately* cubic close-packed and in α-$FeO·OH$ (goethite), they are *approximately* hexagonal close-packed. The deviation from perfect close-packing arises from the formation of directed hydrogen bonds between the hydroxyl groups in different layers; the structures are not cubic or hexagonal but have orthorhombic symmetry (see Chapter 3). On heating, these oxy-hydroxide structures dehydrate to γ-Fe_2O_3 and α-Fe_2O_3, respectively.

Lepidocrocite and goethite are not the only oxy-hydroxides which occur. Of perhaps more biological or technological importance is ferrihydrite, a poorly crystallized mineral which is ubiquitous across the Earth's surface. It is a common product of weathering of iron-bearing minerals and of the microbial oxidation of ferrous ions and doubtless of the rusting of iron itself.

The determination of its crystal structure and composition is a matter of considerable difficulty because of the very small crystallite size, typically in the range 2–10 nm. However, it appears that the oxygen-hydroxyl groups are arranged in an hcp (2H, 4H or 6H) stacking sequence (see Fig. 1.27), i.e. closer to goethite than lepidocrocite and possibly isostructural with a natural alumina hydrate phase, akdaleite. The composition is commonly given as $Fe_5(OH)_8.4H_2O$ but the water content appears to depend on particle size.

Ferrihydrite is an example of a naturally occuring *nanocrystalline* material—upon which a whole new science and technology is now being built. Their importance consists in the simple fact that at these sizes a large proportion of the atoms or ions are at, or near, the surface and the conditions for the stability of the structures are found to be (as might be expected) quite different at the centres of the crystals and at their surfaces.

Such nano-sized crystals of ferrihydrite also appear to constitute the inorganic 'core' of the ferritin molecule—the iron storage molecule that occurs principally within the liver. The core is enclosed within a protein 'shell' (Fig. 1.29) which has a cubic symmetry rarely found in any inorganic crystals (see Chapter 4).

The mechanisms by which iron leaves and enters the ferritin molecule, and how it is taken up, and released, from the ferrihydrite, are matters of much research. But undoubtedly the crystallography—the valency and distribution of iron ions across the interstitial sites both at and below the crystal surfaces—will emerge as factors of great importance.

1.11.4 Silicate structures

Silicates—which constitute by far the most important minerals in the earth's crust—are based on the different ways in which SiO_4 tetrahedra[4] may be joined together, each tetrahedron being made up of four oxygen anions with the silicon cation in the

[4] The SiO_4 tetrahedra may be referred to more generally as Si–O tetrahedra since, as described in subsections (a)–(g), the silicon–oxygen ratio depends on how they are linked.

Fig. 1.29. The cubic (point group 432) structure of the ferritin protein shell viewed along a cube axis. (From *Mineralization in Ferritin: An Efficient Means of Iron Storage* by N Dennis Chasteen and Pauline M Harrison, *Journal of Structural Biology* **126,** 182–, 1999.)

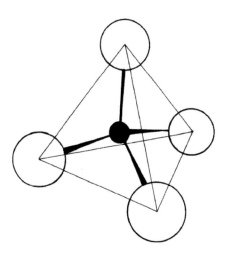

Fig. 1.30. A perspective of the SiO_4 tetrahedron. The oxygen anions at the corners of the tetrahedron are nearly close packed, as shown in the models, e.g. in Fig. 1.7 (the radius of the oxygen anion, $r_n = 0.132$ nm and that of the silicon cation, $r_X = 0.039$ nm, give a radius ratio, $r_X/r_n = 0.296$, slightly larger than the ideal value $r_X/r_n = 0.225$, Fig. 1.10(a) and (b)).

tetrahedral interstice in the centre (Fig. 1.30). Silicate chemistry is based on the linking of the SiO_4 tetrahedra, i.e. whether they occur separately, or whether they are linked by common oxygen anions to form chains, rings, sheets or complete frameworks. This provides the initial basis for the classification of silicate structures. If the Si–O bond is considered to be purely ionic, there are four positive charges associated with each silicon cation and two negative charges associated with each oxygen anion; hence there are four net negative charges associated with each SiO_4 tetrahedron. The 'charge balance' in silicates may be achieved in seven possible ways (see Fig. 1.31):

(a) Separate SiO_4 tetrahedra (*nesosilicates*): the charge balance (four net negative charges) is achieved with metal cations, e.g. Mg^{2+}, Fe^{2+}, which also link the tetrahedra together. Typical minerals are forsterite, $Mg_2(SiO_4)$, or fayalite, $Fe_2(SiO_4)$, the end-members of the olivine group, $MgFe(SiO_4)$.

(b) Two tetrahedra linked together sharing one oxygen anion (*sorosilicates*): the Si:O ratio is now Si_2O_7 giving six net negative charges which are balanced with metal cations. Typical minerals are melilite, $Ca_2Mg(Si_2O_7)$, or hemimorphite, $Zn_4(OH)_2H_2O(Si_2O_7)$.

(c) Three or more tetrahedra linked together to form rings, each tetrahedron sharing two oxygen anions (*cyclosilicates*): the Si:O ratio is Si_nO_{3n}, where n is the number of tetrahedra in the ring. A typical mineral is beryl, with a ring of six tetrahedra, $Be_3Al_2(Si_6O_{18})$.

(d) Many tetrahedra linked together to form single chains (*inosilicates*): each tetrahedron shares two oxygen anions, as in the ring structures above, and which therefore give rise to the same Si:O ratio. This is the basis of the group of minerals called the pyroxenes. Typical examples are enstatite, $Mg_2(Si_2O_6)$, or diopside, $CaMg(Si_2O_6)$.

(e) Tetrahedra linked together to form double chains (*inosilicates*), each tetrahedron sharing alternately two and three oxygen anions, giving the Si:O ratio Si_4O_{11}. This is the basis of the group of minerals called the amphiboles. Typical examples are anthophyllite, $Mg_7(OH)_2(Si_4O_{11})_2$, or tremolite, $Ca_2Mg_5(OH)_2(Si_4O_{11})_2$.

(f) Tetrahedra linked together to form sheets (*phyllosilicates*), each tetrahedron sharing three oxygen anions giving the Si: O ratio Si_2O_5. This is the basis of the micas, chlorites and the clay minerals.

(g) Tetrahedra linked together such that all the oxygen anions are shared giving a three-dimensional framework (*tectosilicates*). The Si:O ratio is now SiO_2 and there is an overall charge balance without the necessity of any linking cations.

Figure 1.31 shows the arrangements of SiO_4 tetrahedra in these seven silicate structures. However, having established the basic pattern there are very important complications both in the chemistry and the arrangements of the tetrahedra which must not be overlooked. In all the silicates, and in the chain, sheet and framework silicates in particular, the silicon in the centre of the tetrahedron can be substituted by aluminium—a trivalent rather than a tetravalent ion. For each such substitution an additional positive charge by way of a 'linking cation' is required. All sheet silicates or phyllosilicates show

Class	Arrangement of SiO$_4$ tetrahedra (central Si^{4+} not shown)	Unit composition	Mineral example
(a) Nesosilicates	Oxygen	$(SiO_4)^{4-}$	Olivine, $(Mg, Fe)_2 SiO_4$
(b) Sorosilicates		$(Si_2O_7)^{6-}$	Hemimorphite $Zn_4Si_2O_7(OH)_2 \cdot H_2O$
(c) Cyclosilicates		$(Si_6O_{18})^{12-}$	Beryl, $Be_3Al_2Si_6O_{18}$
(d) Inosilicates (single chain)		$(Si_2O_6)^{4-}$	Pyroxene e.g. Enstatite, $MgSiO_3$

Fig. 1.31. (*continued*)

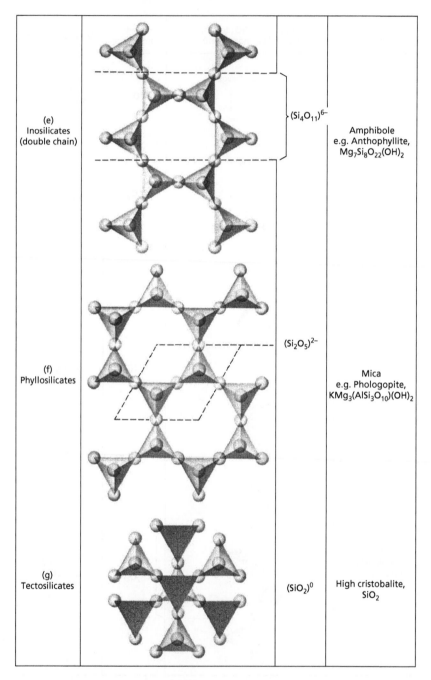

(e) Inosilicates (double chain) $(Si_4O_{11})^{6-}$ Amphibole e.g. Anthophyllite, $Mg_7Si_8O_{22}(OH)_2$

(f) Phyllosilicates $(Si_2O_5)^{2-}$ Mica e.g. Phologopite, $KMg_3(AlSi_3O_{10})(OH)_2$

(g) Tectosilicates $(SiO_2)^0$ High cristobalite, SiO_2

Fig. 1.31. The arrangements of the SiO_4 tetrahedra and Si:O ratios in the seven main types of silicate structures, with examples of typical minerals (from *Manual of Mineralogy*, 21st edn, by C. Klein and C. S. Hurlbut Jr., John Wiley & Sons Inc., 1993).

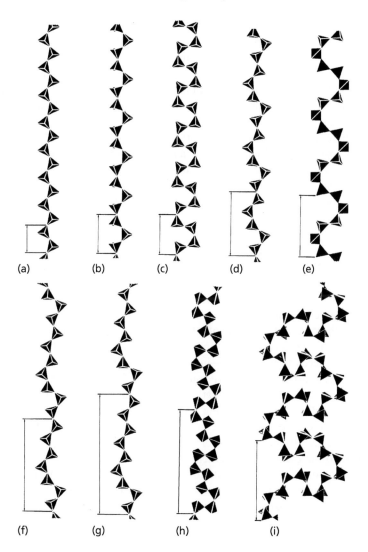

Fig. 1.32. The arrangements of the tetrahedra in the inosilicates. They are described (in German) according to the number of the tetrahedra in the repeat distance: a (zweier); b (dreier); c (vierer); d (fünfer); e (sechser); f (siebener); g (neuner); h (zwölfer); i (24er) (from *Structural Chemistry of Silicates* by F. Liebau, Springer-Verlag, 1985).

such substitution to a greater or lesser extent, for example phlogopite in which one silicon is replaced by one aluminium cation giving the formula $KMg_3(OH)_2(Si_3AlO_{10})$. In framework silicates (tectosilicates), substitution gives rise to an important new class of minerals—the feldspars, the most abundant minerals in the earth's crust, e.g. albite, $Na(Si_3AlO_8)$, or orthoclase, $K(Si_3AlO_8)$ (one silicon cation substituted), or anorthite, $Ca(Si_2Al_2O_8)$ (two silicon cations substituted).

With respect to the arrangements of the tetrahedra, one example—that of the inosilicates—will suffice to show the principles involved. Figure 1.32 shows nine possible patterns (a)–(i) or conformations of the (unbranched) single chains, giving rise to different repeat distances as indicated: (a) (the simplest—known as zweier single chains because there are two tetrahedra in the repeat distance) is that for diopside and enstatite; (b) (dreier single chains with three tetrahedra in the repeat distance) is that for wollastonite, $Ca_3(Si_3O_9)$, and so on.

Clearly, there are also many possible arrangements of the tetrahedra in the cyclosilicates, phyllosilicates and tectosilicates, and it is these which give rise (in part) to the many structural differences in silicate minerals. For example, in the tectosilicates the three different crystallographic forms of silica—quartz, tridymite and cristobalite—simply correspond to different ways in which the SiO_4 tetrahedra are linked together.

1.11.5 The structures of silica, ice and water

Of the three structural forms of silica—quartz, tridymite and cristobalite (not counting the high-pressure forms, coesite and stishovite)—quartz is by far the most common and is structurally stable at ambient temperatures, whereas tridymite is stable between 857 and 1470 °C and cristobalite is stable from 1470 °C to the melting point. At ambient temperatures, these latter two forms of silica are therefore metastable but they do not transform to quartz because in order to do so, a rearrangement of the linking of the SiO_4 tetrahedra needs to take place—in short, a *reconstructive* phase transformation must occur in contrast to a *displacive* transformation in which atomic bonds are not broken. However, displacive transformations occur in all three forms of silica by small rotations of the SiO_4 tetrahedra, giving rise to the 'more open, high temperature' β forms and the 'more closed, low temperature' α forms. This is illustrated, for quartz, in Fig. 1.33. Figure 1.33(b) is a plan view or projection of the hexagonal β-quartz structure perpendicular to the c-axis. For simplicity, only the Si atoms are indicated and their relative heights along the c-axis: white (0), grey (1/3) and black (2/3). Figure 1.33(a) is the corresponding projection for α-quartz; the structure is 'twisted' but no bonds are broken. The symmetry also changes from hexagonal to trigonal, as will be described in Section 3.3.

However, it is worth noticing at this stage one very important structural feature of quartz: the silicon atoms (and hence the SiO_4 tetrahedra) are arranged in a helical pattern along the c-axis. If we imagine a spiral staircase in the centre of a hexagon then the steps go: 0, 1/3, 2/3 ... in a clockwise fashion giving rise to a left-handed screw or helix. Now, we can interchange the positions of the grey and black atoms such that the steps go 0, 1/3, 2/3 ... in an anticlockwise fashion giving rise to a right-handed screw or helix. In short, quartz (both α and β) has two forms, one 'left-handed' and one 'right-handed' and is an example of an *enantiomorphous crystal structure* (see Table 3.1 and Section 4.5).

Quartz is the densest structural form (not counting the high-pressure form, coesite), (\sim2.66 g/cm^3); tridymite and cristobalite have much more 'open' structures (\sim2.33 g/cm^3). In their β forms, these two structures are similar to those of wurtzite and zinc blende (the two structural forms of zinc sulphide), respectively, the SiO_4 tetrahedra

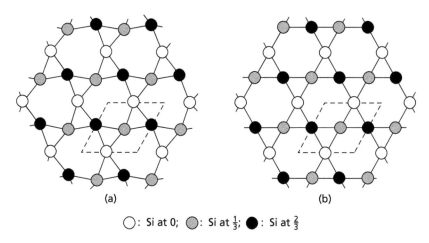

(a) (b)

◯: Si at 0; ◉: Si at $\frac{1}{3}$; ●: Si at $\frac{2}{3}$

Fig. 1.33. (a) Plan of the trigonal structure of α-quartz and (b) plan of the hexagonal structure of β-quartz. Only the silicon atoms are shown, the oxygen atoms are tetrahedrally arranged about those of silicon. The structures are clearly related by a displacive (shear) transformation. (From *Crystal Chemistry*, 2nd Edn, by R.C. Evans, Cambridge University Press, 1966.) See also Section 4.5 and Exercise 4.3.

being in the positions of the Zn and S atoms. More specifically, they correspond to the two structural forms of diamond shown in Fig. 1.36 (Section 11.1.6). Figure 1.36(a) shows the (common) face-centred cubic form of diamond; the positions of the carbon atoms correspond to the positions of the silicon atoms in β-cristobalite. If we now add the oxygen atoms half-way in between the silicon atoms then we generate the complete β-cristobalite structure. Similarly, Fig. 1.36(b) shows the (uncommon) hexagonal form of diamond (lonsdaleite); the carbon atoms correspond to the positions of the silicon atoms in β-tridymite and by adding the oxygen atoms half-way in between as before, we generate the complete β-tridymite structure. In the α forms of these structures there are small deviations from the cubic and hexagonal symmetries which give rise to lower-symmetry enantiomorphous crystal forms.

 Quartz is the best-known and most readily recognized mineral (as a visit to any 'rock-shop' will show). Its common name is rock-crystal and it was thought by the ancients to be a form of ice, frozen so hard that it did not melt. It was described in detail by Strabo (d. AD 24), the Greek geographer who travelled widely throughout the Roman empire, recording the geology and mining operations. He gave it the name Krystallos, which is Greek for ice, and from which our word crystal is derived[5]. The notion that rock-crystal was a form of ice has a long history and persisted as late as the seventeenth century, but what is very curious is that the structures of ice and water *do* have strong resemblances to those of silica as described below.

[5] The word 'crystallography', introduced in 1723 by the Swiss physician, Maurice Capeller, was first applied to the study of quartz.

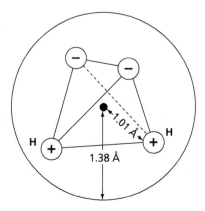

Fig. 1.34. The tetrahedral distribution of charge on the water molecule and its effective radius (1.38 Å). (From *Crystal Chemistry*, 2nd Edn, by R. C. Evans, Cambridge University Press, 1966.)

In structural terms, the water molecule may be regarded as a sphere, or rather a spherical envelope or radius 1.38 Å, with the large oxygen atom in the centre and two (small) hydrogen atoms 'embedded' within it (Fig. 1.34).

These hydrogen atoms may be supposed to occupy two corners of a tetrahedron, the other two corners being empty. However, it is not the location of the hydrogen atoms which is important, but rather that the corners of the tetrahedron are oppositely charged, as indicated in Fig. 1.34. The water molecule, although it is neutral, is nevertheless polar and the crystal structures of ice which arise are determined by the ways in which these spherical, polar molecules, pack together. They are *not* close-packed, rather they form tetrahedrally coordinated networks as in silica. In the common form of ice (ice-I_h) the pattern of H_2O molecules is similar to that of the SiO_4 tetrahedra in β-tridymite except that the hydrogen atoms are not disposed symmetrically between pairs of oxygen atoms as indicated in Fig. 1.35(a) (Fig. 1.35(b) shows a model of an ice-I_h snowflake crystal). When water is frozen at very low temperatures, a structure (ice-I_c) corresponding to β-cristobalite is formed (and there are, as with silica, high pressure forms as well).

In liquid water, from about 4 °C to 150 °C (water II), the molecules are of course in a state of flux, but at any moment small regions (10~100 molecules in size) are arranged in a β-quartz-like arrangement, denser, of course than the β-tridymite arrangement in ice. Only at very high temperatures (>150 °C) and pressures do the molecules approach random close-packing as in a liquid metal (water III).

Between 0 °C and 4 °C, as is known, the density of water increases. This is considered to arise from the slow breakdown of the less-dense tridymite-like arrangement (water I) to the more-dense quartz-like arrangement (water II). Above 4 °C, normal thermal expansion is dominant.

Water is unique. It owes its properties—high melting and boiling points, latent heat of boiling, etc.—to the divalency of oxygen, the polar nature of the molecules and the hydrogen bonding between them. The distribution of charge, and the positions of the hydrogen atoms (in both ice and water—Figs. 1.34 and 1.35(a)) is not fixed: the

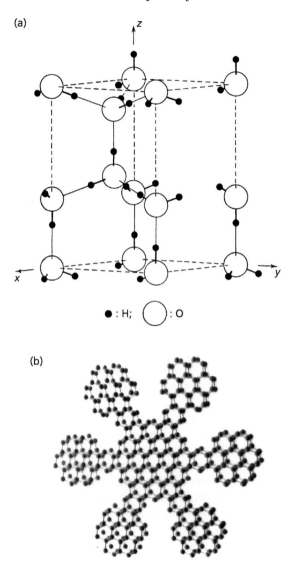

(a)

● : H; ◯ : O

(b)

Fig. 1.35. (a) The hexagonal structure of ice-I_h. The distribution of the hydrogen atoms is arbitrary. (b) A model of an ice (ice-I_h) crystal. The spheres represent water molecules tetrahedrally linked in the hexagonal β-tridymite-like structure.

hydrogen atom in any one bond may be associated with either of the two oxygen atoms each side of it to give the alternative configurations O-H···O or O···H-O. In other words, there is a multiplicity of ways in which a hydrogen atom can be arranged. In contrast, in H_2S the sulphur atom is insufficiently electronegative to form hydrogen bonds and H_2S has a low specific heat, a low boiling point ($-62°C$) and forms close-packed structures, like metals.

1.11.6 The structures of carbon

Of all the elements in the periodic table, only carbon and sulphur are capable of form-
ing elemental rings or chains whose stability is independent of length. However, the
chains formed by sulphur (with a higher valency than two), and those formed by
silicon, immediately below carbon in the periodic table, are thermally unstable and only
short-length chain compounds are possible. Hence only carbon is capable of forming
the complex ring- and long-chain compounds which are the prerequisite of life itself:
the 'sulphur man' and 'silicon man' must remain figments of the imagination!

We will briefly survey only those structures of carbon itself, partly for their own
intrinsic interest and partly because they serve to illustrate some of the basic ideas we
have already met. These are diamond, graphite, 'mesophase', the fullerenes or 'bucky
balls' and nanotubes.

In diamond the basic structural unit is a carbon atom linked to four equidistant
neighbours in a tetrahedral coordination, which arises, in chemical terms, from the sp^3
hybridization of the carbon atom. There are, however, two ways in which the tetrahedra
can be linked together. In by far the commonest form of diamond the tetrahedra are
linked together to generate a cubic structure, the pattern of carbon atoms being pre-
cisely the same as that in TiH (Fig. 1.14), or sphalerite or zinc blende, the cubic form
of ZnS. The analysis of this structure (by W. H. and W. L. Bragg in 1913) was indeed
one of the earliest triumphs of the X-ray diffraction technique. In the other, rare form
of diamond, called lonsdaleite, after Kathleen Lonsdale*, the tetrahedra are linked to-
gether to generate an hexagonal structure, the pattern of carbon atoms being precisely
the same as that in wurtzite, the hexagonal form of ZnS (Section 1.7). These structures
are shown in Fig. 1.36.

In graphite, the carbon atoms are linked together to form plane hexagonal nets or
graphene layers (sp^2 hybridization), the layers being stacked upon one another and held
together by weak (van der Waals) forces. As in the case of our close-packed layers of
metal atoms, the layers are not stacked immediately over each other (to give a simple
hexagonal structure) but are again displaced in either of two ways. In the commonest
form of graphite they are stacked in an ABAB sequence as in the hcp structure—carbon
atoms in the 'B' layers lying immediately above and below the hexagonal hollows of
the 'A' layers either side (Fig. 1.37(a)). This stacking pattern is precisely the same as
that of the cells in the two sides of a honeycomb—the corners of the cells on one side
corresponding with the centres of the cells on the other side.[6]

In the uncommon form of graphite the layers are stacked in an
ABCABC... sequence as in the ccp structure (Figs. 1.37(b)) and this is desig-
nated the rhombohedral form of graphite on the basis of the simplest unit cell which
can be drawn (see Fig. 1.7(c)). However, just as in the case of metals 'stacking faults'
doubtless occur particularly when we shear the graphene layers over each other as

* Denotes biographical notes available in Appendix 3.

[6] The dividing walls between the two sides of the honeycomb are not flat but are faceted at angles cor-
responding to those between the close-packed layers of atoms in the ccp structure (Section 1.3), and this
arrangement can be shown to result in the most economical use of beeswax.

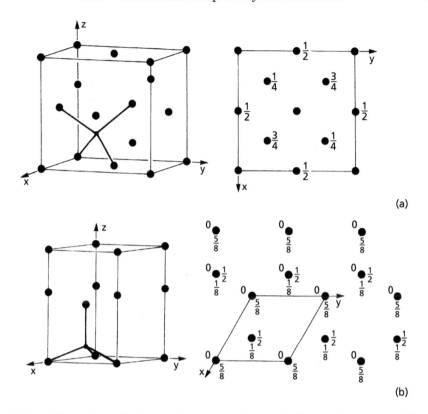

Fig. 1.36. (a) The cubic and (b) hexagonal structures of diamond in clinographic projection (left) and in plan view (right). In both cases the carbon atoms are linked to four others in tetrahedral coordination but the arrangement of the tetrahedra differs. The *pattern* of atom sites is precisely the same as in the cubic (sphalerite) and hexagonal (wurtzite) forms of ZnS. (From *Crystalline Solids* by D. McKie and C. McKie, Nelson, 1975.)

we do when we write with a pencil! It is also possible for the layers to be stacked in parallel layers but in a random orientation (with no correlation between the atoms in each layer) and this so-called turbostratic form appears to arise as an intermediate stage in the graphitization of pitch or some polymer precursors.

Plan views of the stacking of the layers in the hexagonal and rhombohedral graphite structures are shown in Fig. 1.37. These structures are related to the hexagonal and cubic diamond structures respectively as shown in Exercise 1.10.

Perhaps the most remarkable of the structures of carbon—remarkable, that is, for having been discovered so late—are the fullerenes or colloquially 'bucky balls'. These were postulated as an act of scientific imagination by D. E. H. Jones (who wrote under the pen-name Daedalus) in 1966, but their actual existence and structures were not established until 1985 by H. W. Kroto, R. W. Smalley, and their co-workers. If a hexagon is replaced by a pentagon in a graphite-like net then, as model-building shows in a most immediate way, the net assumes a dome-like shape (Fig. 1.38). This principle was

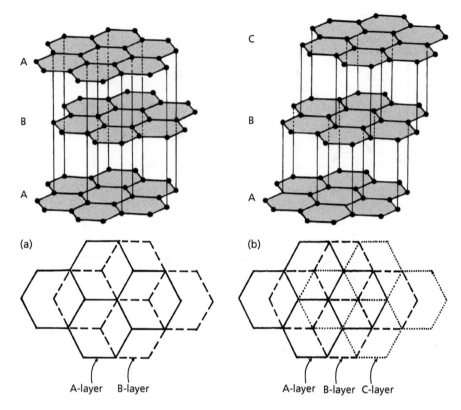

Fig. 1.37. (a) The hexagonal (layer sequence ABAB ...) and (b) the rhombohedral (layer sequence ABCABC ...) structures of graphite. These are shown (below) in plan view with the layers shown 'offset' for clarity. A-layer—full lines; B-layer—dashed lines and C-layer—dotted lines.

applied in engineering design by the American architect and polymath R. Buckminster Fuller*—the distribution of pentagons and hexagons in the structure determining, of course, the overall shape and stability of what he called a geodesic dome[7]. In order to create a complete closed sphere precisely 12 pentagons are needed and the smallest, simplest sphere consists of 12 pentagons only, without any hexagons. The shape or polyhedron which is formed is called a pentagonal dodecahedron (Fig. 1.39(a)) and is one of the five 'perfect' polyhedra, or Platonic solids since all the 12 pentagonal faces and all the 20 vertices or 'corners' where the carbon atoms are situated are identical. [The other Platonic solids are the cube (6 square faces, 8 vertices, Fig. 4.2(a)), the

* Denotes biographical notes available in Appendix 3.

[7] A geodesic, meaning 'earth dividing', is the shortest distance between two points across the surface of a sphere—just as a straight line is the shortest distance between two points in a plane. A geodesic is part of the circumference of what is called a great circle (see Chapter 12), a circle which has its centre at the centre of the sphere and which divides the sphere into two equal halves. In the design of geodesic domes, the constructional struts, the 'geodesics', follow the lines of great circles such that the stresses are distributed evenly. The edges of the Platonic, and many other polyhedra, also follow the lines of great circles (see Appendix 2).

(a)

(b)

Fig. 1.38. Model-building with simple linked straws or sticks, (a) The plane graphite-like or graphene net formed by linking hexagons and (b) the dome-like net formed when a hexagon is replaced by a pentagon.

octahedron (8 triangular faces, 6 vertices, Fig. 4.2(b)), the tetrahedron (4 triangular faces, 4 vertices, Fig. 4.2(c)), and the icosahedron, Fig. 1.39(b) (20 triangular faces, 12 vertices)]. Hence, the simplest and smallest of the 'bucky-balls' consists of 20 carbon atoms. It is called dodecahedrene, but it has not been isolated and is doubtless thermally unstable.

As an increasing number of hexagons are included in the structure a family of fullerenes of increasing thermal stability and an increasing number of carbon atoms is developed (Fig. 1.40). The most interesting of these is Buckminsterfullerene itself with 60 carbon atoms (C_{60}) consisting of 12 pentagons separated from each other by 20 hexagons (Fig. 1.40 (a)) precisely in the same way as a football (or US soccer ball). This polyhedron is called a truncated icosahedron and is one of the thirteen 'semi-regular' or Archimedean polyhedra in which every face is a regular polygon, though not all faces are of the same kind. In crystals of C_{60} the spheres arrange themselves in an fcc

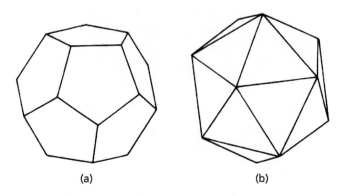

(a) (b)

Fig. 1.39. (a) the pentagonal dodecahedron (12 pentagonal faces) and (b) the icosahedron (20 triangular faces)—two of the five 'perfect' or 'Platonic' polyhedra (the others being the cube, the octahedron and the tetrahedron—see Fig. 4.2 and Appendix 2).

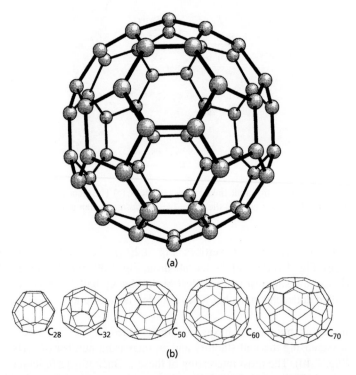

(a)

C_{28} C_{32} C_{50} C_{60} C_{70}

(b)

Fig. 1.40. (a) Buckminsterfullerene, C_{60}, consisting of 12 pentagonal and 20 hexagonal faces (from *Perfect Symmetry* by Jim Baggott, Oxford University Press, 1994) and (b) fullerenes C_{28}–C_{70} all of which have 12 pentagonal faces and an increasing number of hexagonal faces (from *Science* **242**, 1142, 1988: reproduced by courtesy of Prof. Sir Harry Kroto).

Fig. 1.41. The silicaceous skeleton of *Aulonia hexagona,* a deep sea-water radiolarian, drawn by E. Haeckel (from *'On Growth and Form'* by D'Arcy Wentworth Thompson, Cambridge University Press, 1942).

pattern but doubtless faulting can also occur. The next fullerene, C_{70}, with 70 carbon atoms, has the shape of a Rugby ball (US football) and fullerenes with much larger numbers of carbon atoms also occur. These structures have their counterparts in the structures of the spherical skeletons of the radiolaria (deep sea-water creatures) where pentagons (and octagons and heptagons) occur in a predominantly hexagonal network (Fig. 1.41). Here we see nature obeying the dictates of geometry both on an atomic and on a biological scale!

Finally, the graphene layers may not lie flat as in graphite, but may 'roll up' with the edges joined together to form carbon 'nanotubes'. These may consist of a single rolled-up layer (single walled nanotubes) or many such tubes, one inside the other (multi-walled nanotubes) the spacing between the graphene layers being rather larger than that for graphite. These structural forms of carbon were first identified by Iijima in 1991 and Bethune et al. in 1993, but again were suggested as a possible form of carbon by D.E.H. Jones in 1972. The tubes may also be 'capped' with fullerene hemispheres, to form close-ended nanotubes.

Carbon nanotubes also exhibit different geometries or patterns depending upon the orientations of the carbon hexagons with respect to the tube axis. Figure 1.42 shows nanotube models made from sheets depicting graphene layers rolled up in different ways and overlapped such that the hexagons match along the join. In Fig. 1.42(a) the carbon atoms show an 'armchair' pattern along the axis of the tube; in Fig. 1.42(b) a 'zig-zag' pattern and (perhaps of most interest) in Fig. 1.42(c) the carbon hexagons form a helical pattern along the axis of the tube. Clearly, there are many more helices possible, of different pitches, both right and left handed. These different patterns are not

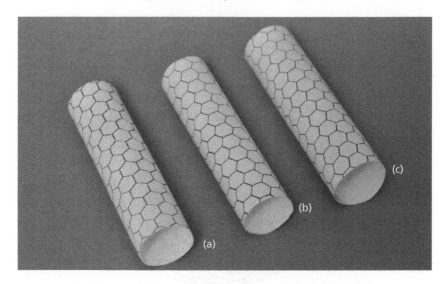

Fig. 1.42. Models of single-walled carbon nanotubes showing different orientations of the graphene cells along the tube axis (a) 'armchair', (b) 'zig-zag,' and (c) helical conformations.

just of crystallographic model-making interest but give rise to entirely different thermal, conductivity and strength properties and which make carbon nanotubes of such potential importance in nanotechnology.

A note on boron nitride—boron and nitrogen, which occur each side of carbon in the periodic table, form BN compounds with striking similarities to those of carbon described above. The boron and nitrogen atoms are linked alternately in a planar (sp^2) conformation, giving rise to graphite-like structures, or in a tetrahedral (sp^3) conformation giving diamond-like structures—the latter possess very high hardness and melting-points and are of great importance in materials engineering. It is not known whether BN also exhibits Buckminster fullerene structures—this may be unlikely in view of the necessity for adjacent boron or nitrogen atoms in the 5-ring components. However, open-ended 'nanotube' structures, which do not require 5-ring components, may occur.

Exercises

1.1 With reference to Fig. 1.5(b) or Fig. 1.11(a), determine the c/a ratio for the hcp structure. (*Hint: a* is equal to d, the atomic diameter and edge length of the tetrahedron, and c is twice the height of the tetrahedron. Determine also the height of the tetrahedron above the triangular base and the packing density for the hcp structure.

1.2 Determine the radius ratio, r_X/r_A, for the tetrahedral interstitial sites in the bcc structure.

1.3 Determine the radius ratio, r_X/r_A, for the six-fold coordinated interstitial site in the simple hexagonal structure.

1.4 Examine your crystal models and find:
 (a) the number of different (non-parallel) close-packed planes and close-packed directions in the ccp and hcp structures; and
 (b) the number of closest-packed planes and close-packed directions in the bcc structure.

1.5 In the deformation of ccp and bcc metals, slip generally occurs on the close- or closest-packed planes and in close-packed directions. Each combination of slip plane and direction is called a **slip system.** How many slip systems are there in these metals?

1.6 Draw a plan or crystal projection of the hcp structure perpendicular to the z-or c-axis. Assign axes x and y (at $120°$ to each other) and outline a primitive hexagonal cell (one atom at each corner and one within the cell). What are the atomic coordinates of the atoms in the cell?

1.7 Referring to Figs 1.26 or 1.27, express the stacking sequences in the α-SiC structures B6(6H) and B7(15R) in the Frank notation.

1.8 With reference to Fig. 1.20, show that an hcp structure may be generated from a ccp structure by faulting *alternate* close-packed planes.

1.9 Draw crystal plans of the perovskite structure shown in Fig. 1.17, relocate the origin of one cell, relabel the ionic coordinates and show that these two cells do represent the same crystal structure.

1.10 Make ball-and-stick models of the cubic-diamond (Fig. 1.36(a)) and rhombohedral-graphite (Fig. 1.37(b)) structures. Notice that in the diamond structure the carbon atoms lie in 'puckered' layers perpendicular to the body-diagonal directions of the cube. If we (physically or in imagination) make the atoms in these layers coplanar and also lengthen the bonds between them, then we have the rhombohedral form of graphite. A similar relationship exists between the hexagonal-diamond (Fig. 1.36(b)) and hexagonal-graphite (Fig. 1.37(a)) structures.

1.11 Express the stacking sequence of the close-packed layers in the element americium (Section 1.2) in terms of the Frank notation.

1.12 Make a graphene layer (print a pattern of hexagons on transparency film or a piece of paper) and roll it up in the ways shown in Fig. 1.42 to make different crystallographic forms of carbon nanotubes.

1.13 Identify the crystal structure(s) represented by the models in Fig 1.43(a) and (b). (Exercise by courtesy of Dr John Leake)

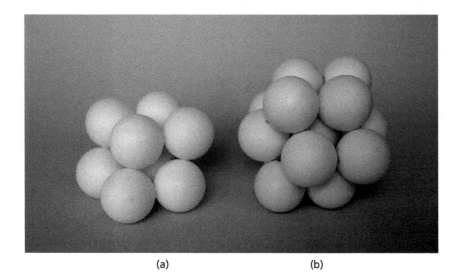

(a) (b)

Fig. 1.43.

2

Two-dimensional patterns, lattices and symmetry

2.1 Approaches to the study of crystal structures

In Chapter 1 we developed an understanding of simple crystal structures by first considering the ways in which atoms or ions could pack together and then introducing smaller atoms or ions into the interstices between the larger ones. This is a pragmatic approach as it not only provides us with an immediate and straightforward understanding of the atomic/ionic arrangements in some simple compounds, but also suggests the ways in which more complicated compounds can be built up.

However, it is not a systematic and rigorous approach, as all the possibilities of atomic arrangements in all crystal structures are not explored. The rigorous, and essentially mathematical approach is to analyse and classify the geometrical characteristics of quite general two-dimensional patterns and then to extend the analysis to three dimensions to arrive at a completely general description of all the patterns to which atoms or molecules or groups of atoms or molecules might conform in the crystalline state.

These two distinct approaches—or strands of crystallographic thought—are apparent in the literature of the nineteenth and early twentieth centuries. In general, it was the metallurgists and chemists, such as Tammann* and Pope,* who were the pragmatists, and the theoreticians and geometers, such as Fedorov* and Schönflies,* who were the analysts. It might be thought that the analytical is necessarily superior to the pragmatic approach because its generality and comprehensiveness provides a much more powerful starting point for progress to be made in the discovery and interpretation of the crystal structures of more and more complex substances. But this is not so. It was, after all, the simple models of sodium chloride and zinc blende of Pope (such as we also constructed in Chapter 1) that helped to provide the Braggs* with the necessary insight into crystal structures to enable them to make their great advances in the interpretation of X-ray diffraction photographs. In the same way, 40 years later, the discovery of the structure of DNA by Watson and Crick was based as much upon structural and chemical knowledge and intuition, together with model building, as upon formal crystallographic theory.

However, a more general appreciation of the different patterns into which atoms and molecules may be arranged is essential, because it leads to an understanding of the important concepts of symmetry, motifs and lattices. The topic need not be pursued rigorously—in fact it is unwise to do so because we might quickly 'lose sight of the wood for the trees!' The essential ideas can be appreciated in two dimensions, the

* Denotes biographical notes available in Appendix 3.

The Basics of Crystallography and Diffraction. Fourth Edition. Christopher Hammond.
© Christopher Hammond 2015. Published in 2015 by Oxford University Press.

subject of this chapter. The extension to three dimensions (Chapters 3 and 4) which relates to 'real crystal structures', should then present no conceptual difficulties.

2.2 Two-dimensional patterns and lattices

Consider the pattern of Fig. 2.1 (a), which is made up of the letter **R** repeated indefinitely. What does **R** represent? Anything you like—a 'two-dimensional molecule', a cluster of atoms or whatever. Representing the 'molecule' as an **R**, an *asymmetric* shape, is in effect representing an *asymmetric* molecule. We shall discuss the different types or elements of symmetry in detail in Section 2.3 below, but for the moment our general everyday knowledge is enough. For example, consider the symmetry of the letters **R M S**. **R** is asymmetrical. **M** consists of two equal sides, each of which is a reflection or mirror image of the other, there is a **mirror line** of symmetry down the centre indicated by the letter m, thus **M**$_{\leftarrow m}$. There is no mirror line in the **S**, but if it is rotated 180° about a point in its centre, an identical **S** appears; there is a **two-fold rotation axis** usually called a **diad axis** at the centre of the **S**. This is represented by a little lens-shape at ✦ the axis of rotation: **S** .

In Fig. 2.1(a) **R**, the repeating 'unit of pattern' is called the **motif**. These motifs may be considered to be situated at or near the intersections of an (imaginary) grid. The grid is called the **lattice** and the intersections are called **lattice points**.

Let us now draw this underlying lattice in Fig. 2.1(a). First we have to decide where to place each lattice point in relation to each motif: anywhere will do—above, below, to one side, in the 'middle' of the motif—the only requirement is that the *same* position with respect to the motif is chosen every time. We shall choose a position a little below the motif, as shown in Fig. 2.1(b). Now there are an infinite number of ways in which the lattice points may be 'joined up' (i.e. an infinite number of ways of drawing a lattice

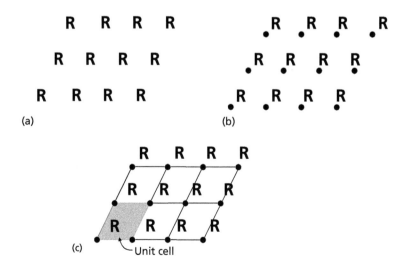

(a) (b)

(c) ⌐ Unit cell

Fig. 2.1. (a) A pattern with the motif **R**, (b) with the lattice points indicated and (c) the lattice and a unit cell outlined. (Drawn by K. M. Crennell.)

or grid of lines through lattice points). In practice, a grid is usually chosen which 'joins up' adjacent lattice points to give the lattice as shown in Fig. 2.1(c), and a unit cell of the lattice may also be outlined. Clearly, if we know (1) the size and shape of the unit cell and (2) the motif which each lattice point represents, including its orientation with respect to the lattice point, we can draw the whole pattern or build up the whole structure indefinitely. The unit cell of the lattice and the motif therefore define the whole pattern or structure. This is very simple: but observe an important consequence. Each motif is identical and, for an infinitely extended pattern, the environment (i.e. the spatial distribution of the surrounding motifs, and their orientation) around each motif is identical. This provides us with the definition of a lattice (which applies equally in two and three dimensions): *a lattice is an array of points in space in which the environment of each point is identical.* Again it should be stressed that by environment we mean the spatial distribution and orientation of the surrounding points.

Like all simple definitions (and indeed ideas), this definition of a lattice is often not fully appreciated; there is, to use a colloquial expression, 'more to it than meets the eye!' This is particularly the case when we come to three-dimensional lattices (Chapter 4), but, for the two-dimensional case, consider the patterns of points in Fig. 2.2 (which should be thought of as extending infinitely). Of these only (a) and (d) constitute a lattice; in (b) and (c) the points are certainly in a *regular* array, but the surroundings of each point are *not* all identical.

Figures 2.2(a) and (d) represent two two-dimensional lattice types, named **oblique** and **rectangular**, respectively, in view of the shapes of their unit cells. But what is the distinction between the oblique and rectangular lattices? Surely the rectangular lattice is just a special case of the oblique, i.e. with a 90° angle?

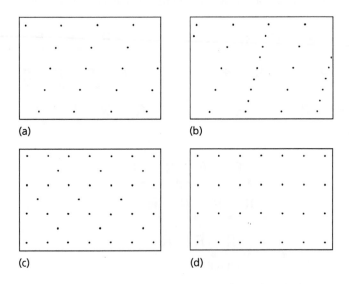

(a)

(b)

(c)

(d)

Fig. 2.2. Patterns of points. Only (a) and (d) constitute lattices.

The distinction arises from different symmetries of the two lattices, and requires us to extend our everyday notions of symmetry and to classify a series of symmetry elements. This precise knowledge of symmetry can then be applied to both the motif and the lattice and will show that there are a limited number of patterns with different symmetries (only seventeen) and a limited number of two-dimensional lattices (only five).

2.3 Two-dimensional symmetry elements

The clearest way of developing the concept of symmetry is to begin with an asymmetrical 'object'—say the **R** of Fig. 2.1—then to add successively mirror lines and axes of symmetry and to see how the **R** is repeated to form different patterns or groups. The different patterns or groups of **R**s which are produced correspond, of course, to objects or projections of molecules (i.e. 'two-dimensional molecules') with different symmetries which are not possessed by the **R** alone.

The patterns or groups which arise and which as explained below are of concern in crystallography are shown in Fig. 2.3. On the left are the patterns of **R**s, in the centre are decorative motifs with the same symmetry, and on the right are projections of molecules. Figure 2.3(1) shows the **R** 'on its own' and, as an example, the asymmetrical projection of the CHFCClBr molecule. Figure 2.3(2) shows 'right-' and 'left'-handed **R**s reflected in the 'vertical' mirror line between them. This pair of **R**s has the same mirror symmetry as the projection of the *cis*-difluoroethene molecule. Now add another 'horizontal' mirror line as in Fig. 2.3(3). A group of four **R**s (two right- and two left-handed) is produced. This group has the same symmetry as the projection of the ethene molecule.

The **R** may be repeated with a diad (two-fold rotation) axis, as in Fig. 2.3(4). The two **R**s (both right handed) have the same symmetry as the *trans*-difluoroethene molecule. Now look back to the group of **R**s in Fig. 2.3(3); notice that they also are related by a diad (two-fold rotation axis) at the intersection of the mirror lines: the action of reflecting the **R**s across two perpendicular mirror lines 'automatically' generates the two-fold symmetry as well. This effect, where the action of two symmetry elements generates another, is quite general as we shall see below.

Mirror lines and diad axes of symmetry are just two of the symmetry elements that occur in two dimensions. In addition there are three-fold rotation or **triad** (3) axes (represented by a little triangle, ▲, four-fold rotation or **tetrad** (4) axes (represented by a little square, ■), and six-fold (6) or **hexad** axes (represented by a little hexagon, ●). Asymmetrical objects are represented as having a one-fold or **monad** (1) axis of symmetry (for which there is no little symbol)—which means in effect that one 360° rotation brings the object into coincidence with itself.

Figure 2.3(5) shows the **R** related by a triad (three-fold) axis. The projection of the trifluoroalkylammonia molecule also has this same symmetry. Now add a 'vertical' mirror line as in Fig. 2.3(6). Three more left-handed **R**s are generated, and at the same time the **R**s are mirror related not just in the vertical mirror line but also in two lines inclined at 60° as shown; another example of additional symmetry elements (in this case mirror lines) being automatically generated.

This procedure (of generating groups of **R**s which represent motifs with different symmetries) may be repeated for tetrad (four-fold) axes (Fig. 2.3(7)); plus mirror lines (Fig. 2.3(8)); for hexad (six-fold) axes (Fig. 2.3(9)); plus mirror lines (Fig. 2.3(10)). Notice that not only do these axes of symmetry 'automatically' generate mirror lines at 90° (for tetrads) and 60° (for hexads) but also 'interleaving' mirror lines at 45° and 30° as well.

The ten arrangements of **R**s (and the corresponding two-dimensional motifs or projections of molecules) are called the **ten two-dimensional crystallographic or plane point groups**, so called because all the symmetry elements—axes (perpendicular to the page) and mirror lines (in the page)—pass through a point. The ten plane point groups are labelled with 'shorthand' symbols which indicate, as shown in Fig. 2.3, the

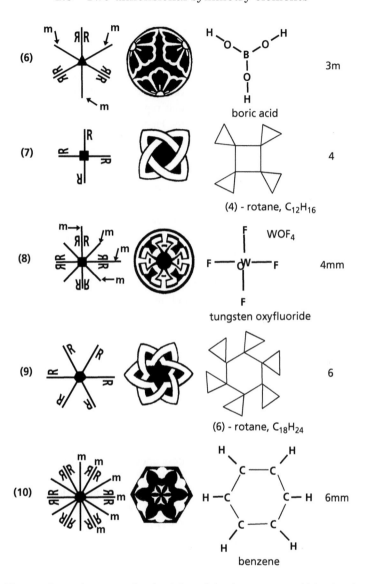

Fig. 2.3. The ten plane point groups showing left to right, the symmetry which arises based on an asymmetrical object **R**; examples of motifs; examples of molecules and ions (drawn as projections) and the point group symbols. (Drawn by K. M. Crennell.)

symmetry elements present: 1 for a monad (no symmetry), m for one mirror line, mm (or $2mm$) for two mirror lines (plus diad), 2 for a diad, 3 for a triad, $3m$ for a triad plus three mirror lines, 4 for a tetrad, $4mm$ for a tetrad plus four mirror lines, 6 for a hexad and $6mm$ for a hexad plus six mirror lines (the extra 'm' in the symbols referring to the 'interleaving' mirror lines).

Now, in deriving these ten plane point groups we have ignored groups of **R**s with fivefold (**pentad**), seven-fold (**heptad**) etc. axes of symmetry with and without mirror lines. Such plane point groups are certainly possible and are widely represented in nature—the pentagonal symmetry of a starfish for example. However, what makes the ten plane point groups in Fig. 2.3 special or distinctive is that *only* these combinations of axes and mirror lines can occur in regular repeating patterns in two dimensions as is explained in Sections 2.4 and 2.5 below. Hence they are properly called the two-dimensional *crystallographic* point groups as indicated above. Patterns with pentagonal symmetry are necessarily non-repeating, non-periodic or 'incommensurate' and consequently have in the past been rather overlooked by crystallographers. However, with the realization that groups of atoms (or viruses) can form 'quasicrystals' with five-fold symmetry elements (see Section 4.9), the study of non-periodic two-dimensional patterns has become of increasing interest and importance (see Section 2.9). A simple way at this stage of 'seeing the difference' is to compare, for example, the arrangement of six lattice points equally spaced around a central lattice point (hexagons) with the arrangement of five 'lattice' points equally spaced around a central point (pentagons). In the former case the arrangement of points can be put together to form a lattice (a pattern or *tiling* of hexagons with 'no gaps' and 'no overlaps'). In the latter case the points cannot be put together to form a lattice—there are always 'gaps' or 'overlaps' between the tiling of pentagons. Try it and see!

2.4 The five plane lattices

Having examined the symmetries which a two-dimensional motif may possess we can now determine how many two-dimensional or plane lattices there are. We will do this by building up patterns from the ten motifs in Fig. 2.3 with the important condition that the symmetry elements possessed by the single motif must also extend throughout the whole pattern. This condition is best understood by way of a few examples. Consider the asymmetrical motif **R** (Fig. 2.3(1)); there are no symmetry elements to be considered and the **R** may be repeated in a pattern with an oblique unit cell (i.e. the most asymmetrical) arrangement of lattice points. Now consider the motif which possesses one 'vertical' mirror line of symmetry (Fig. 2.3(2)). This mirror symmetry must extend throughout the whole pattern from motif to motif which means that the lattice must be rectangular. There are two possible arrangements of lattice points which fulfil this requirement: a simple rectangular lattice and a centred rectangular lattice as shown in Fig. 2.4(a).

These rectangular lattices also possess 'horizontal' mirror lines of symmetry corresponding to the motif with the two sets of mirror lines as shown in Fig. 2.3(3). Now consider the motifs with tetrad (four-fold) symmetry (Figs 2.3(7) and (8)). This four-fold symmetry must extend to the surrounding motifs which means that they must be arranged in a square pattern giving rise to a square lattice (Fig. 2.4(a)).

Altogether, five two-dimensional or plane lattices may be worked out, as shown in Fig. 2.4(a). They are described by the shapes of the unit cells which are drawn between lattice points—oblique *p*, rectangular *p*, rectangular *c* (which is distinguished

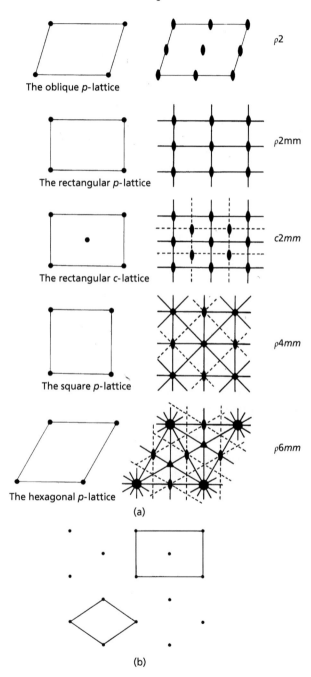

Fig. 2.4. (a) Unit cells of the five plane lattices, showing the symmetry elements present (heavy solid lines indicate mirror lines, dashed lines indicate glide lines) and their plane group symbols (from *Essentials of Crystallography*, by D. McKie and C. McKie, Blackwell, 1986). (b) The rectangular *c* lattice, showing the alternative primitive (rhombic *p* or diamond *p*) unit cell.

from rectangular p by having an additional lattice point in the centre of the cell), square p and hexagonal p. Notice again that additional symmetry elements are generated 'in between' the lattice points as shown in Fig. 2.4(a) (right). For example, in the square lattice there is a tetrad at the centre of the cell, diads halfway along the edges and vertical, horizontal and diagonal mirror lines as well as the tetrads situated at the lattice points.

All two-dimensional patterns must be based upon one of these five plane lattices; no others are possible. This may seem very surprising—surely other shapes of unit cells are possible? The answer is 'yes', a large number of unit cell shapes are possible, but the pattern of lattice points which they describe will always be one of the five of Fig. 2.4(a). For example, the rectangular c lattice may also be described as a rhombic p or diamond p lattice, depending upon which unit cell is chosen to 'join up' the lattice points (Fig. 2.4(b)). These are just two alternative descriptions of the *same* arrangement of lattice points. So the choice of unit cell is arbitrary: *any* four lattice points which outline a parallelogram can be joined up to form a unit cell. In practice we take a sensible course and mostly choose a unit cell that is as small as possible—or 'primitive' (symbol p)—which does not contain other lattice points within it. Sometimes a larger cell is more useful because the axes joining up the sides are at 90°. Examples are the rhombic or diamond lattice which is identical to the rectangular centred lattice described above and, to take an important three-dimensional case, the cubic cell (Fig. 1.6(c)) which is used to describe the ccp structure in preference to the primitive rhombohedral cell (Fig. 1.7(c)).

Now we combine the ten plane point group symmetries (Fig. 2.3) with the appropriate plane lattices (Fig. 2.4) in order to work out the total number of two-dimensional patterns. For example, plane point group symmetries 4 and $4mm$, combined with the square lattice, give two patterns $p4$ and $p4mm$ (Fig. 2.6). Notice that in these (and other) patterns additional mirror lines and axes of symmetry are 'automatically' generated within the unit cell (Fig. 2.6(b)). Notice also that there are two possible combinations of plane point symmetry $3m$ with the plane hexagonal lattice: the triad axes generated within the unit cell either have mirror lines passing through them ($p3m1$) or not ($p31m$). Continuing in this way we generate thirteen two-dimensional patterns known as the **symmorphic plane groups**. However, there is a complication: the combination of a point group symmetry with a lattice can give rise to an additional symmetry element called a **glide line**. Consider the two patterns in Fig. 2.5, both of which have a rectangular lattice. In Fig. 2.5(a) the motif has mirror symmetry as in Fig. 2.3(2); it consists of a pair of right- and left-handed **R**s. In Fig. 2.5(b) there is still a reflection—still pairs of right- and left-handed **R**s—but one set of **R**s has been translated, or glided half a lattice spacing. This symmetry is called a **reflection glide** or simply a **glide line of symmetry**. Notice that glide lines also arise automatically in the centre of the unit cell of Fig. 2.5(b) as do mirror lines in Fig. 2.5(a). Glide lines are, of course, as familiar to us as mirror lines; they represent the pattern of our footprints in the snow when we walk in a straight line!

The presence of the glide lines also has important consequences regarding the symmetry of the motif. In Fig. 2.5(a) the motif has mirror symmetry but in Fig. 2.5(b) it does not: the pair of right- and left-handed **R**s is asymmetric. It is the repetition of the

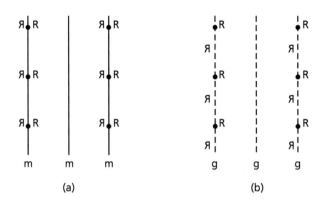

Fig. 2.5. Patterns with (a) reflection symmetry and (b) glide-reflection symmetry. The mirror lines (m) and glide lines (g) are indicated.

translational symmetry elements—the glide lines—that determines the overall rectangular symmetry of the pattern. The glide lines which are present in the five plane lattices are shown (in addition to the axes and mirror lines of symmetry) in Fig. 2.4(a).

2.5 The seventeen plane groups

Glide lines give rise to four more two-dimensional patterns (*pg, p2mg, p2gg*, and *p4gm*—Fig. 2.6)—the **non-symmorphic plane groups** giving seventeen in all—the seventeen **plane groups**. On a macroscopic scale the glide symmetry in a crystal would appear as simple mirror symmetry—the shift between the mirror-related parts of the motif would only be observable in an electron microscope which was able to resolve the individual mirror-related parts of the motif, i.e. distances of the order of 0.5–2 Å (50–200 pm).

The seventeen plane groups are shown in Fig. 2.6(a). They are labelled by 'shorthand' symbols which indicate the type of lattice (*p* for primitive, *c* for centred) and the symmetry elements present, *m* for mirror lines, *g* for glide lines, 4 for tetrads and so on. The symmetry elements within a unit cell are shown in Fig. 2.6(b). It is a good exercise in recognizing the symmetry elements present in the 17 plane groups to lay a sheet of tracing paper over Fig. 2.6(a), to indicate the positions of the axes, mirror and glide lines of symmetry in an (arbitrary) unit cell and then to compare your 'answers' with those shown in Fig. 2.6(b).

It is essential to practice recognizing the motifs, symmetry elements and lattice types in two-dimensional patterns and therefore to find to which of the seventeen plane groups they belong. Any regular patterned object will do—wallpapers, fabric designs, or the examples at the end of this chapter. Figure 2.7 indicates the procedure you should follow. Cover up Fig. 2.7(b) and examine only Fig. 2.7(a); it is a projection of molecules of $C_6H_2(CH_3)_4$. You should recognize that the molecules or groups of atoms are *not* identical in this two-dimensional projection. The motif is a *pair* of such molecules and this is the 'unit of pattern' that is repeated. Now look for symmetry elements and (using a piece of tracing paper) indicate the positions of all of these on the pattern. Compare

your pattern of symmetry elements with those shown in Fig. 2.7(b). If you did not obtain the same result you have not been looking carefully enough! Finally, insert the lattice points—one for each motif. Anywhere will do, but it is convenient to have them coincide with a symmetry element, as has been done in Fig. 2.7(b). The lattice is clearly oblique and the plane group is *p*2 (see Fig. 2.6).

Another systematic way of identifying a plane pattern is to follow the 'flow diagram' shown in Fig. 2.8. The first step is to identify the highest order of rotation symmetry present, then to determine the presence or absence of reflection symmetry and so on through a series of 'yes' and 'no' answers, finally identifying one of the seventeen plane patterns whose plane group symbols are indicated 'in boxes', corresponding to those given in Fig. 2.6.

2.6 One-dimensional symmetry: border or frieze patterns

Identifying the number of one-dimensional patterns provides us with a good exercise in applying our more general knowledge of plane patterns. It is also a useful exercise in that it tells us about the different types of patterns that can be designed for the borders of wallpapers, edges of dress fabrics, friezes and cornices in buildings, and so on.

In plane patterns the symmetry operations and symmetry elements are (clearly) repeated in a plane; in one-dimensional patterns they can only be repeated in or along a line—i.e. the line or long direction of the border or frieze. This restriction immediately rules out all rotational symmetry elements with the exception of diads: two-fold symmetry alone can be repeated in a line: three-, four-, and six-fold symmetry elements require the repetition of a motif in directions other than the line of the border. For the same reason glide-reflection lines of symmetry, other than that along the line of the border, are ruled out. Mirror lines of symmetry are restricted to those along, and perpendicular to, the line of the border.

These restrictions result in seven one-dimensional groups, shown in Fig. 2.9. It is a good and satisfying exercise for you to derive these from first principles as outlined above. It is also useful to compare Fig. 2.9 with Fig. 2.6; the bracketed symbols in Fig. 2.9 indicate from which plane pattern the one-dimensional pattern may be derived. Notice that in one case two one-dimensional patterns—these with 'horizontal' and 'vertical' mirror planes—are derived from one plane pattern (*pm*). This is because the mirror lines in the plane group *pm* can be oriented either along, or perpendicular to, the line of the one-dimensional pattern.

Figure 2.24 (see Exercise 2.6) also shows examples of some of the border patterns. You can practice recognizing such patterns either by overlaying the pattern with a piece of tracing paper, and indicating the positions of the diads, mirror and glide lines as described above for plane patterns or by following the flow diagram (Fig. 2.10).

2.7 Symmetry in art and design: counterchange patterns

We have a rich inheritance of plane and border patterns in printed and woven textiles, wallpapers, bricks and tiles which have been designed and made by countless craftsmen

Fig. 2.6. (*continued*)

(b)

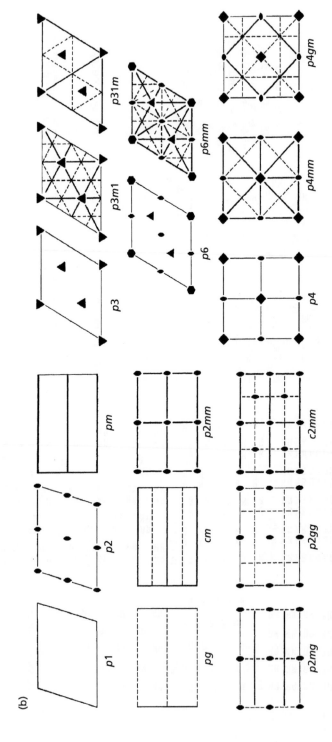

Fig. 2.6. (a) The seventeen plane groups (from *Point and Plane Groups* by K. M. Crennell). The numbering 1–17 is that which is arbitrarily assigned in the International Tables. Note that the 'shorthand' symbols do not necessarily indicate all the symmetry elements which are present in the patterns, (b) The symmetry elements outlined within (conventional) unit cells of the seventeen plane groups, heavy solid lines and dashed lines represent mirror and glide lines respectively (from *Manual of Mineralogy* 21st edn, by C. Klein and C. S. Hurlbut, Jr., John Wiley, 1999).

 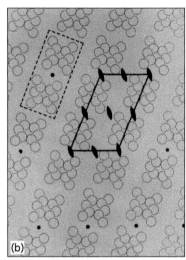

Fig. 2.7. Projection (a) of the structure of $C_6H_2(CH_3)_4$ (from *Contemporary Crystallography*, by M. J. Buerger, McGraw-Hill, 1970), with (b) the motif, lattice and symmetry elements indicated.

and artisans in the past 'without benefit of crystallography'. The question we may now ask is: 'Have all the seventeen plane groups and seven one-dimensional groups been utilized in pattern design or are some patterns and some symmetries more evident than others? If so, is there any relationship between the preponderance or absence of certain types of symmetry elements in patterns and the civilization or culture which produced them?'

Questions such as these have exercised the minds of archaeologists, anthropologists and historians of art and design. They are, to be sure, questions more of cultural than crystallographic significance, but patterns play such a large part in our everyday experience that a crystallographer can hardly fail to be absorbed by them, just as he or she is absorbed by the three-dimensional patterns of crystals.

The study of plane and one-dimensional patterns (and indeed three-dimensional (space) patterns) is complicated by the question of colour—'real' colours in the case of plane and one-dimensional patterns, or colours representing some property, such as electron spin direction or magnetic moment, in space patterns (Chapter 4). Colour changes may also be analysed in terms of symmetry elements in which colours are alternated in a systematic way. Clearly, the greater the number of colours, the greater the complexity. The simplest cases to consider are two-colour (e.g. black and white) patterns. Figure 2.11 shows the generation of plane motifs through the operation of what are called counter-change or colour symmetry elements,[1] which are distinguished from ordinary (rotation) axes and mirror lines of symmetry by a prime superscript. For example, the operation of a $2'$ axis is a twice repeated rotation of an asymmetric

[1] These are a special case of what are sometimes known as *anti-symmetry* elements, which relate the symmetry of opposites—black/white in this case.

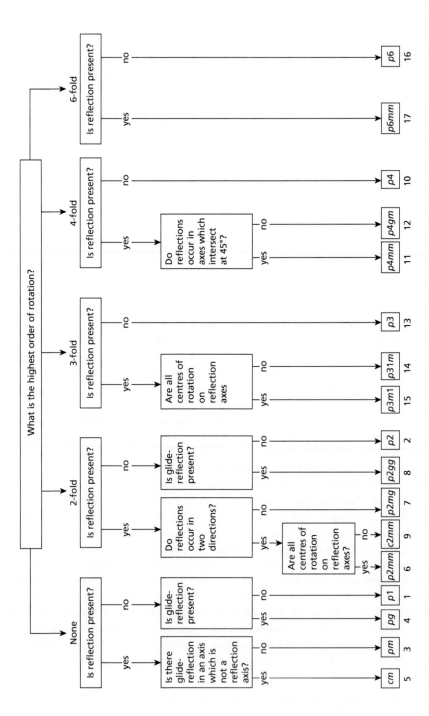

Fig. 2.8. Flow diagram for identifying one of the seventeen plane patterns (redrawn from *The Geometry of Regular Repeating Patterns* by M. A. Hann and G. M. Thomson, the Textile Institute, Manchester, 1992). The numbering is that which is arbitrarily assigned in the International Tables (see Fig 2.6).

Fig. 2.9. (*Left*) the seven one-dimensional groups or classes of border or frieze patterns (drawn by K. M. Crennell); (solid lines indicate mirror lines, dashed lines (symbol a) indicate glide lines and ● symbols indicate diads); (*centre*) their symmetry symbols and (*bracketed*) the plane groups from which they are derived; and (*right*) examples of Hungarian needlework border patterns (from *Symmetry Through the Eyes of a Chemist* 3rd edn. by M. and I. Hargittai, Springer, New York and London 2008).

object by 180° plus a colour change at each rotation; the operation of an m′ mirror line is a reflection plus colour change. Altogether there are eleven counterchange point groups (Fig. 2.11) compared with the ten plane point groups (Fig. 2.3). Note that there are **no** counterchange point groups corresponding to the plane point groups with only odd-numbered axes of symmetry (the monad and the triad), but that there are in each case two possible counterchange point groups corresponding to the plane point groups with symmetry 2*mm*, 4*mm* and 6*mm*.

Fig. 2.10. Flow diagram for identifying one of the seven border patterns (from *The Geometry of Regular Repeating Patterns*, in brackets the plane groups from which they are derived. loc. cit.).

The derivation of the counterchange one- and two-dimensional patterns also involves the operation of a g' glide line which involves a reflection plus a translation of half a lattice spacing **plus** a colour change and gives (to extend our footprint analogy) a sequence of black/white (i.e. right/left footprints).

Accounting for two-colour symmetry gives rise to a total of forty-six (rather than seventeen) plane patterns and seventeen (rather than seven) one-dimensional patterns. Figure 2.12 shows an example of plane group pattern *p2gg* (No. 8—see Fig. 2.6(a), (b)) and the two possible counterchange patterns (symbols *p2′ gg′* and *p2g′ g′*) which are based upon it.

Probably the most influential and pioneering study of patterns was *The Grammar of Ornament* by Owen Jones, first published in 1856.[2] Owen Jones attempted to categorize both plane and border patterns in terms of the different cultures that produced them, and although the symmetry aspects of patterns are touched on in the most fragmentary

[2] Owen Jones. *The Grammar of Ornament*, Day & Sons Ltd., London, reprinted by Studio Editions, London (1986).

Fig. 2.11. The eleven counterchange (black/white) point groups and (bracketed) the point group symbols for the plane point groups to which they correspond (see Fig. 2.3). The counterchange symmetry elements are denoted by prime superscripts. (Drawn by K. M. Crennell.)

(a) p2gg (b) p2'gg' (c) p2g'g'

Fig. 2.12. (a) Plane group $p2gg$ and (b) and (c) the two counterchange plane groups $p2'gg'$ and $p2g'g'$ respectively which are based upon it. (Drawn by K. M. Crennell.)

way, there is no doubt that the superb illustrations and encyclopaedic character of the book provided later writers with material which could be classified and analysed in crystallography terms. Perhaps the best known of these was M. C. Escher (1898–1971) who drew inspiration for his drawings of tessellated figures from visits to the Alhambra in the 1930s, and also presumably from Owen Jones' chapter on 'Moresque Ornament' in which he describes the Alhambra as 'the very summit of Moorish art, as the Parthenon

is of Greek art'. Escher's patterns encompass all the seventeen plane groups, eleven of which are represented in the Alhambra.

More recent work has identified clear preponderances of certain plane symmetry groups, and the absences of others.[3] For example, nearly 50% of traditional Javanese batik (wax-resist textile) patterns belong to plane group *p4mm* (Fig. 2.6), others, such as *p3*, *p3m1*, *p31m* and *p6* are wholly absent. In Jacquard-woven French silks of the last decade of the nineteenth century, nearly 80% of the patterns belong to plane group *pg*. In Japanese textile designs of the Edo period *all* plane groups are represented, with a marked preponderance for groups *p2mm* and *c2mm*. What these differences mean, or tell us about the cultures which gave rise to them, is, as the saying goes, 'another question'.

In X-ray crystallography crystal structures are frequently represented as two-dimensional projections (electron density maps—see Section 13.2). The beauty and variety of these patterns led Dr Helen Megaw*, a crystallographer at Birkbeck College, London, to suggest that they be made the basis for the design of wallpapers and fabrics in the same way that William Morris used flowers and birds in his pattern designs. Her suggestion eventually bore fruit in the work of the *Festival Pattern Group* of the 1951 Festival of Britain and the production of a remarkable variety of patterned wallpapers, carpets and fabrics based upon crystal structures as diverse as haemoglobin, insulin and apophyllite. These patterns, recently republished,[4] provide a rich source of material for plane group recognition.

2.8 Layer (two-sided) symmetry and examples in woven textiles

Woven textiles consist of interlacing warp 'north-south' threads and weft 'east-*weft*' threads. The various combinations of interlacings, which give rise to the different patterns of cloths, are very wide indeed, ranging from the simplest 'single cloth', plain weave fabric, where individual warp and weft threads pass over and under each time (Fig. 2.13), to more complex cloth structures. Common structures include twills (e.g. Fig. 2.14), herringbones, sateens, etc. Clearly, there are symmetry relationships between the 'face' and 'back' of such woven fabrics and the study of such relationships introduces us to what are known as *layer-symmetry groups* or classes.

We will not describe all the layer-symmetry groups or classes (of which there are a total of 80) but just some of the general principles of their construction. Readers who wish to follow this topic further should refer to the book by Shubnikov and Koptsik or the papers by Scivier and Hann.[5] However, we may note here that the 80 layer-symmetry

* Denotes biographical notes available in Appendix 3.

[3] M. A. Hann. *Symmetry of Regular Repeating Patterns: Case Studies from various cultural settings.* Journal of the Textile Institute (1992), Vol. 83, pp. 579–580.

[4] L. Jackson. *From Atoms to Patterns, Crystal Structure Designs from the 1951 Festival of Britain.* Richard Dennis Publications, Shepton Beauchamp, Somerset (2008).

[5] J.A. Scivier and M.A. Hann (2000) *The application of symmetry principles to the classification of fundamental simple weaves,* Ars Textrina **33**, 29 and (2000) *Layer symmetry in woven textiles,* Ars Textrina **34**, 81.

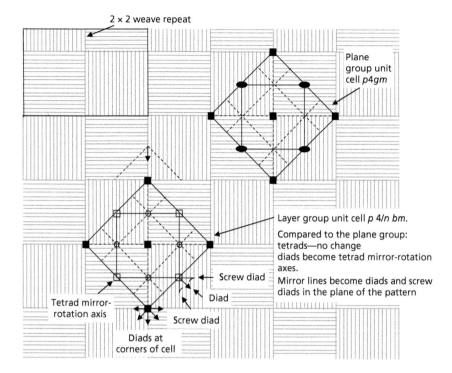

Fig. 2.13. Diagrammatic representation of a plain weave and (superimposed) the plane group unit cell *p4gm* (see also Fig. 2.6) and the layer-symmetry group unit cell *p4/nbm*. Note the differences between the symmetry elements in these unit cells: the diads and mirror lines in *p4gm* become tetrad mirror-rotation axes and screw diads respectively (in the plane) in *p 4/n bm*. (Drawn by C. McConnell.)

groups are sub-groups of the 230 space groups (Section 4.6) and that the 17 plane groups are, in turn, sub-groups of the 80 layer-symmetry groups.

Because of the structural restrictions imposed by the warp and weft character, the five plane groups with three or six fold symmetry are not applicable to woven fabrics and, correspondingly, neither are 16 of the 80 layer-symmetry groups.

As we have seen, in describing the 17 plane groups we are restricted to rotation axes perpendicular to the plane and reflection (mirror) and glide lines of symmetry within the plane. In describing layer-symmetry groups further symmetry elements or operations are required which relate the 'face' and 'back' of the fabric. These are (i) rotation-reflection (alternating) axes of symmetry perpendicular to the plane which consist of a rotation plus a reflection in the plane. These symmetry operations correspond to the black/white counterchange point groups (Fig. 2.11) in which the symbol **R** is now understood to have two sides—black on the face and white on the back. Again, because of the structural restrictions imposed by woven fabrics, only two such rotation-reflection axes are applicable—2′ and 4′ (Fig. 2.11). (ii) Diad axes lying *within* the

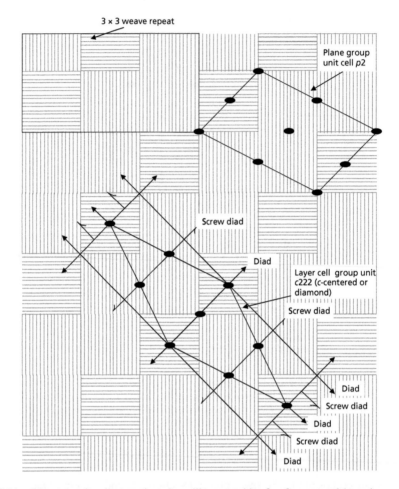

Fig. 2.14. Diagrammatic representation of a twill weave with a 3 × 3 repeat and (superimposed) the plane group unit cell *p*2 and the layer-symmetry group unit cell *c*222. Note the differences between these unit cells: *c*222 is a rectangular c-centred or 'diamond' unit cell (see Fig. 2.4) and contains diads and screw diads lying in the plane of the pattern. (Drawn by C. McConnell.)

plane—both the simple diad axes which we have already met and also *screw diad* axes which involve a rotation plus a translation (like glide lines) of half a lattice spacing (screw diad axes are but one example of screw axes which we shall meet in our description of three-dimensional symmetry and space groups). In both cases, because the axes lie in the plane, they 'turn over' the black face of the **R** to its white face. The operations of these in-plane axes are shown in Fig. 2.15. Notice that the operation of the diad is identical to that of the counterchange mirror line *m'* (Fig. 2.11). (iii) Planes (not lines) of symmetry coinciding with the plane; both reflection (mirror) planes (which are not

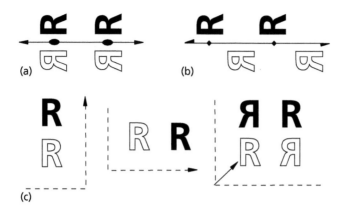

(a) (b)

(c)

Fig. 2.15. The additional symmetry operations for the 52 layer-symmetry groups applicable to woven fabrics (plus the counterchange symmetry operations $2'$ and $4'$ (Fig. 2.11). The 'face' and 'back' of the **R** symbols are shown here as black and white, respectively. (a) Operation of an in-plane diad axis (double arrow-head) (identical to counterchange symmetry element m'—see Fig. 2.11) and (b) an in-plane screw diad axis (single arrow-head). (c) Operation of in-plane glide planes (dashed lines) for three different orientations of the glide directions—along the axes and diagonally. (Drawn by K. M. Crennell.)

applicable to woven textiles because the back of the cloth is not identical to the front[6]) and glide-reflection planes (which are applicable to woven textiles). These symmetry operations are also shown in Fig. 2.15.

We will now apply these ideas to the plain weave and twill illustrated in Figs 2.13 and 2.14. The 'weave repeat' is the smallest number of warp and weft threads on which the weave interlacing can be represented; it is a 2×2 square for the plain weave (Fig. 2.13) and a 3×3 square for this example of a twill weave (Fig. 2.14). It is important to note that these weave repeat squares *do not* correspond with the unit cells of the plane patterns. These unit cells and the plane group symmetry elements are also shown in Figs 2.13 and 2.14. As can be seen, the plain weave has plane symmetry $p4gm$ and the twill plane symmetry $p2$ (see Fig. 2.6).

Figures 2.13 and 2.14 also show the unit cells and layer symmetry elements for these two weaves and the standard notation (which we will not describe in detail) which goes with them. Notice that for the plain weave that the unit cell is identical to that for the plane group symmetry but for the twill it is different—the primitive (p) lattice becomes a centred (c) or diamond lattice. Notice also that the layer-group symmetry of the plain weave is much more 'complicated' than that of the twill. It includes tetrad rotation-reflection axes perpendicular to the plane as well as diads and screw diads within the plane. The twill, by contrast, has no mirror lines of symmetry at all.

[6] This restriction further reduces (by 12) the number of layer-symmetry groups applicable to woven textiles, leaving a total of $80 - 16 - 12 = 52$. Still a lot!

Warp 'float' Weft facing Weft facing
 front of fabric back of fabric

Fig. 2.16. Operation of a tetrad rotation-reflection axis showing an **R** superimposed on a 'float' of the plain weave fabric (Fig. 2.13). The operations of reflection-rotation are of course repeated three times. (Drawn by K. M. Crennell.)

Finally, look carefully at the positions of the tetrad rotation–reflection axes in the centres of the plain weave warp and weft 'floats'. These axes help us to visualize the relation between the face and back of the fabric: for example, rotate a warp float 90° and we have a weft float, reflect it (black to white) and you have the weft float in the back face of the fabric as shown in Fig. 2.16.

2.9 Non-periodic patterns and tilings

Johannes Kepler was the first to show that pentagonal symmetry would give rise to a pattern which was non-repeating. Figure 2.17 is an illustration from perhaps his greatest work *Harmonices mundi* (1619) which shows in the figures captioned 'Aa' and 'z' a pattern or tiling of pentagons, pentagonal stars and 10 and 16-sided figures which radiate out in pentagonal symmetry from a central point. Grünbaum and Shephard[7] have shown how the tiling 'Aa' can be extended indefinitely giving long-range orientational order but the pattern does not repeat and cannot be identified with any of the seventeen plane groups (Fig. 2.6). A. L. Mackay[8] has shown how a regular, but non-periodic pattern, can be built up from regular pentagons in a plane with the triangular gaps covered by pieces cut from pentagons, which he describes with the title (echoing Kepler) *De nive quinquangula*—on the pentagonal snowflake.

These are but two examples of non-periodic or 'incommensurate' tilings, the mathematical basis of which was largely developed by Roger Penrose and are generally named after him. Figure 2.18 shows how a Penrose tiling may be constructed by linking together edge-to-edge 'wide' and 'narrow' rhombs or diamond-shaped tiles of equal edge lengths. The angles between the edges of the tiles (as shown in Fig. 2.18(a)) are not arbitrary but arise from pentagonal symmetry as shown in Fig. 2.18(b) (where the tiles are shown shaded in relation to a pentagon); nor are they linked together in an arbitrary fashion but according to local 'matching rules', shown in Fig. 2.18(a) by little triangular 'pegs' and 'sockets' along the tile edges. These are omitted in the resultant tiling (Fig. 2.18(c)), partly for clarity and partly because their work in constructing the

[7] B. Grünbaum and G. C. Shephard. *Tilings and Patterns: An Introduction.* W. H. Freeman, New York, 1989.
[8] A.L. Mackay (1976) *De nive quinquangula.* Physics Bulletin, p. 495.

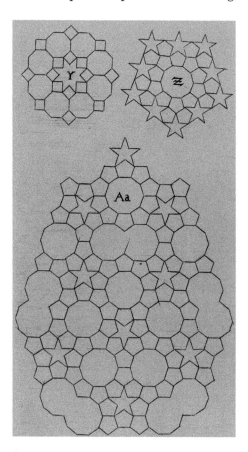

Fig. 2.17. Non-periodic tiling patterns 'z' and 'Aa' (from *Harmonices Mundi* by Johannes Kepler, 1619, reproduced from the copy in the Brotherton Library, University of Leeds, by courtesy of the Librarian).

pattern is done. (An alternative of showing how the tiles must be fitted together is to colour or shade them in three ways and then to match the colours, like the pegs and sockets, along the tile edges.) The tiling can be viewed as a linkage of little cubes where we see three cube faces; the 'front' and 'top' faces (represented by the 'wide' diamonds) and 'side' face (represented by the 'narrow' diamond).

However, building up a perfect Penrose tiling by adding wide and narrow tiles one-by-one is not a straightforward task. Even if we strictly adhere to the edge-matching rules, we soon find that 'blind alleys' are available at every step; nor is it easy to recognize the point at which a mistake has been made until further tiles have been added. In order to build a Penrose tiling without mistakes we need to invoke what are called **vertex matching rules**: these are discussed in C. Janot's book *Quasicrystals* (see Further Reading).

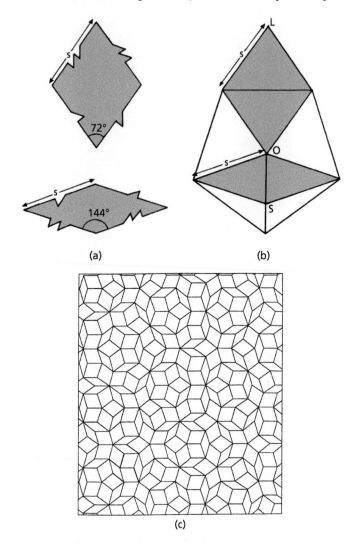

Fig. 2.18. (a) The two types ('wide' and 'narrow') tiles for the construction of a Penrose tiling. The triangular 'pegs' and 'sockets' along the tile edges indicate how they should be linked together edge-to-edge. (b) The geometry of the tiles in relation to a pentagon. The ratio OL/s (wide tile) = s/OS (narrow tile) = $(\sqrt{5}+1)/2 = 1.618\ldots$ (c) shows the resultant tiling (pegs and sockets omitted for clarity) (reproduced by courtesy of Prof. Sir Roger Penrose).

The mathematical analysis of non-repeating patterns is rather difficult (especially in three-dimensions—see Section 4.9), but we can perhaps understand their essential 'incommensurate' properties by way of a one-dimensional analogy or example. Consider a pattern made up of a row of arrows and a row of stars extending right and left from an origin O. If the spacings of the arrows and stars are in a ratio of whole numbers then,

depending on the values of these numbers, the pattern will repeat. Figure 2.19(a) shows
a simple case where the ratio of spacings is 3/2 and the pattern repeats (i.e. the ar-
rows and stars coincide) every third arrow or second star. If, however, the spacings of
the arrows and stars cannot be expressed as a ratio of whole numbers, in other words
if the ratio is an *irrational* number, then the pattern will never repeat—the arrows
and stars will never come into coincidence. Figure 2.19(b) shows an example where
the ratio of spacings is $\sqrt{2} = 1.414213\ldots$ an irrational number, like π, where there
is 'no end' to the number of decimal places and no cyclic repetition of the decimal
numbers.[9]

In Penrose five-fold or pentagonal tiling it turns out (Fig. 2.18(b)) that the ratio of
the diagonal OL to the edge length s (for the wide tile) and the ratio of the edge length
s to the diagonal OS (for the narrow tile) are also both equal to an irrational num-
ber $(\sqrt{5}+1)/2 = 1.618034\ldots$ called the *Golden Mean* or *Golden Ratio*. The Golden
Ratio also occurs as the convergence of the ratio of successive terms in the so-called
Fibonacci series of numbers where each term is the sum of the preceding two, i.e. 1,
1, 2, 3, 5, 8, 13, 21, 34, 55, …Not only is the Golden Ratio and Fibonacci series a
subject of mathematical interest but it also occurs in nature in, for example, the inter-
secting spirals of the florets traced in the head of a sunflower or in a pine-cone. In a
pine-cone there are, depending on species, 5/3, 8/5, 13/8 right-/left-handed intersect-
ing spirals and in a sunflower, depending on size, 55/34, 89/55 intersecting spirals—all
these numbers corresponding to successive terms in the Fibonacci series[10]. The Fi-
bonacci series may also describe the growth rate of molluscan shells (in which new

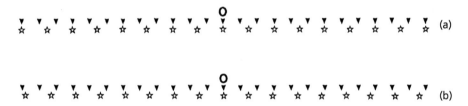

Fig. 2.19. One-dimensional examples of (a) a periodic pattern and (b) a non-periodic pattern. In
(a) the pattern repeats every third arrow and second star, in (b) the ratio of the spacings is $\sqrt{2}$ and
the pattern never repeats.

[9] The discovery that some numbers are irrational is one of the triumphs of Greek mathematics. The proof
that $\sqrt{2}$ is irrational, which is generally attributed to Pythagoras, may be expressed as follows. Suppose that
$\sqrt{2}$ *can* be expressed as a/b where a and b are whole numbers which have no common factor (if they had,
we could simply remove it). Hence $\sqrt{2}b = a$ and squaring $2b^2 = a^2$. Now $2b^2$ is an *even* number, hence a^2
is also an even number and, since the square of an even number is even, a is an even number. Now an even
number can be expressed as $2 \times$ (any number), i.e. $a = 2c$. Squaring again $a^2 = 4c^2 = 2b^2$, hence $2c^2 = b^2$
and, for the same reason as before, since $2c^2$ is an even number then b is an even number. So, both a and b
are even and have a common factor 2 which contradicts our initial hypothesis which therefore must be false.

[10] The configurations produced by the orderly arrangement of leaves and florets around a stem (a subject
known as phyllotaxis) and the growth and shapes of shells, horns, and teeth, are fully discussed in the classic
text *On Growth and Form* (see Further Reading).

material is added at the growing boundary) or the rate of population increase in animals. Suppose that in an interval of time a young animal (*S*) becomes an adult (*L*) and an adult (*L*) has one offspring (*S*). Then following the sequence $S \rightarrow L$ and $L \rightarrow LS$ we have: *S, L, LS, LSL, LSLLS, LSLLSLSL*, ... Notice that the numbers of *L* and *S* individuals in each term are equal to the numbers in the preceding and next-preceding terms: e.g. in the term *LSLLSLSL* there are 5*L* and 3*S*. As the series progresses the ratio *L/S* again converges to the Golden Ratio. The same is true of the ratio of wide and narrow tiles in a Penrose tiling. The Golden Ratio is also met in architectural proportion and design: a rectangle whose sides are in the (approximate) ratio 1.62:1 seems particularly restful to the eye—not too narrow and not too wide. It occurs in the shapes of window-panes, in the proportions of the facades of Greek temples or, to take a particular example, in the ratio of column spacing to height (5:8) in the Colosseum in Rome.

It is a simple exercise to show that if a square is cut off a Golden Rectangle, the rectangle which remains also has sides which are in the Golden Ratio and clearly the process can be continued indefinitely (Fig. 2.20). Finally, the rectangle can be used as the template for the construction of a **golden spiral**, a special case of an **equiangular spiral** (Fig. 2.20), so called because the tangent at any point on the spiral is at a con-stant angle to the corresponding radius-vector. This leads to the property of **continued similarity**: the shape remains the same irrespective of size. A large molluscan shell (e.g. of an ammonite or *nautilus*) is the same shape as a small one. Finally, we may note that a circle is also a special case of an equiangular spiral: all circles, large or small are the same shape.

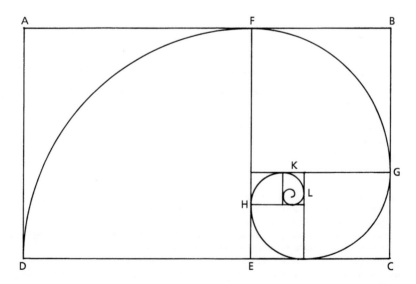

Fig. 2.20. A Golden Rectangle ABCD. Cutting off a square AFED leaves a smaller Golden Rectangle FBCE—and the process may be continued as shown. Notice the **continued similarity** of the inscribed equiangular spiral—e.g. the portion DFG has the same shape as HKL.

Exercises

2.1 Lay tracing paper over the plane patterns in Fig. 2.6. Outline a unit cell in each case and indicate the positions of all the symmetry elements within the unit cell. Notice in particular the differences in the distribution of the triad axes and mirror lines in the plane groups $p31m$ and $p3m1$.

2.2 Figure 2.21 is a design by M. C. Escher. Using a tracing paper overlay, indicate the positions of all the symmetry elements. With the help of the flow diagram (Fig. 2.8), determine the plane lattice type.

2.3 Figure 2.22 is a projection of the structure of FeS_2 (shaded atoms Fe, unshaded atoms S). Using a tracing paper overlay, indicate the positions of the symmetry elements, outline a unit cell and, with the help of the flow diagram in Fig. 2.8, determine the plane pattern type.

2.4 Figure 2.23 is a design by M. C. Escher. Can you see that the two sets of men are related by glide lines of symmetry? Draw in the positions of these glide lines, and determine the plane lattice type.

Fig. 2.21. A plane pattern (from *Symmetry Aspects of M. C. Escher's Periodic Drawings*, 2nd edn, by C. H. MacGillavry. Published for the International Union of Crystallography by Bohn, Scheltema and Holkema, Utrecht, 1976).

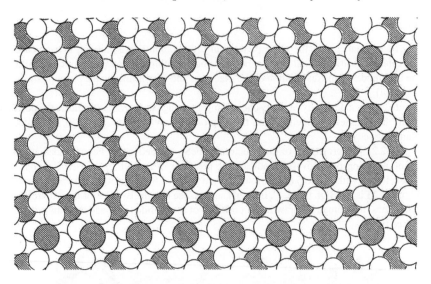

Fig. 2.22. A projection of the structure of marcasite, FeS_2 (from *Contemporary Crystallography* by M. J. Buerger, McGraw-Hill, New York, 1970).

Fig. 2.23. A plane pattern (from C. H. MacGillavry, loc. cit.).

2.5 Determine (with reference to Fig. 2.11) the counterchange (black–white) point group symmetry of a chessboard.

2.6 Figure 2.24 shows examples of border or frieze patterns from *The Grammar of Ornament* by Owen Jones. Using a tracing paper overlay, indicate the positions of the symmetry elements and, with the help of the flow diagram (Fig. 2.10), determine the one-dimensional lattice types.

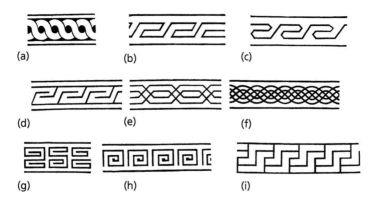

Fig. 2.24. Examples of border or frieze patterns (from *The Grammar of Ornament* by Owen Jones, Day & Son, London 1856, reprinted by Studio Editions, London, 1986). a, b, Greek; c, d, Arabian; e, Moresque; f, Celtic; g, h, Chinese; i, Mexican.

2.7 Figure 2.25(a) is a 'wood block floor' or 'herringbone' pattern with plane group symmetry *p2gg*. Using a tracing-paper overlay (and with the help of Fig. 2.6(b) and the flow chart, Fig. 2.8), locate the positions of the diad axes and glide lines. Now place your tracing paper over the counterchange pattern (Fig. 2.25(b)) and determine which of the symmetry elements become counterchange (2′ or *g*′) symmetry elements. To which of the counterchange patterns shown in Fig. 2.12 does this pattern belong?

2.8 The symmetry of border pattern *pma2* (*p2mg*) (Fig. 2.9) consists of a glide line *a* (or *g*) along the length of the border with vertical mirror lines and diad axes in between. Derive the two-colour (black and white) counterchange patterns based upon *pma2* by replacing, in turn, the glide lines, mirror lines and diad axes by the counterchange symmetry elements *g*′, *m*′ and 2′.

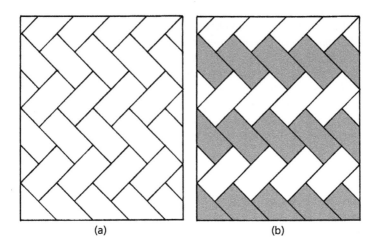

Fig. 2.25. 'Wood block floor' or 'herringbone brickwork' patterns (a) with all blocks the same colour, and (b) with alternating black and white blocks.

3

Bravais lattices and crystal systems

3.1 Introduction

The definitions of the motif, the repeating 'unit of pattern', and the lattice, an array of points in space in which each point has an identical environment, hold in three dimensions exactly as they do in two dimensions. However, in three dimensions there are additional symmetry elements that need to be considered: both **point symmetry elements** to describe the symmetry of the three-dimensional motif (or indeed any crystal or three-dimensional object) and also **translational symmetry elements,** which are required (like glide lines in the two-dimensional case) to describe all the possible patterns which arise by combining motifs of different symmetries with their appropriate lattices. Clearly, these considerations suggest that the subject is going to be rather more complicated and 'difficult'; it is obvious that there are going to be many more three-dimensional patterns (or space groups) than the seventeen two-dimensional patterns (or plane groups or the eighty two-sided patterns—Chapter 2), and to work through all of these systematically would take up many pages! However, it is not necessary to do so; all that is required is an understanding of the principles involved (Chapter 2), the operation and significance of the additional symmetry elements, and the main results. These main results may be stated straight away. The additional point symmetry elements required are centres of symmetry, mirror planes (instead of lines) and inversion axes; the additional translational symmetry elements are glide planes (instead of lines) and screw axes. The application and permutation of all symmetry elements to patterns in space give rise to 230 **space groups** (instead of seventeen plane groups) distributed among fourteen space lattices (instead of five plane lattices) and thirty-two point group symmetries (instead of ten plane point group symmetries).

In this chapter the concept of space (or Bravais) lattices and their symmetries is discussed and, deriving from this, the classification of crystals into seven systems.

3.2 The fourteen space (Bravais) lattices

The systematic work of describing and enumerating the space lattices was done initially by Frankenheim* who, in 1835, proposed that there were fifteen in all. Unfortunately for Frankenheim, two of his lattices were identical, a fact first pointed out by Bravais* in 1848. It was, to take a two-dimensional analogy, as if Frankenheim had failed to notice

* Denotes biographical notes available in Appendix 3.

The Basics of Crystallography and Diffraction. Fourth Edition. Christopher Hammond.
© Christopher Hammond 2015. Published in 2015 by Oxford University Press.

(see Fig. 2.4(b)) that the rhombic or diamond and the rectangular centred plane lattices were identical! Hence, to this day, the fourteen space lattices are usually, and perhaps unfairly, called Bravais lattices.

The unit cells of the Bravais lattices are shown in Fig. 3.1. The different shapes and sizes of these cells may be described in terms of three cell edge lengths or axial

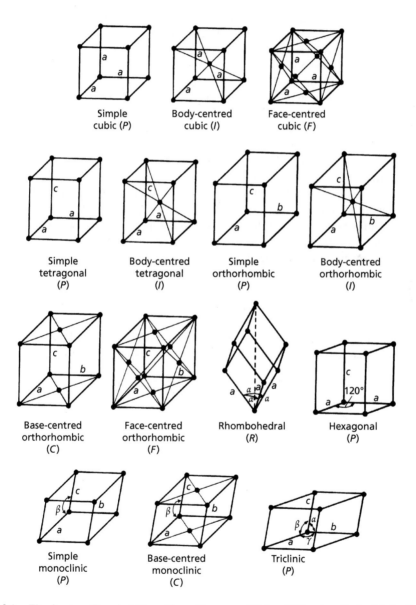

Fig. 3.1. The fourteen Bravais lattices (from *Elements of X-Ray Diffraction*, (2nd edn), by B. D. Cullity, Addison-Wesley, 1978).

distances, *a, b, c,* or lattice vectors **a, b, c** and the angles between them, α, β, γ, where α is the angle between **b** and **c**, β the angle between **a** and **c**, and γ the angle between **a** and **b**. The axial distances and angles are measured from one corner to the cell, i.e. a common origin. It does not matter where we take the origin—any corner will do—but, as pointed out in Chapter 1, it is a useful convention (and helps to avoid confusion) if the origin is taken as the 'back left-hand corner' of the cell, the *a*-axis pointing forward (out of the page), the *b*-axis towards the right and the *c*-axis upwards. This convention also gives a right-handed axial system. If any one of the axes is reversed (e.g. the *b*-axis towards the left instead of the right), then a left-handed axial system results. The distinction between them is that, like left and right hands, they are mirror images of one another and cannot be brought into coincidence by rotation.

The drawings of the unit cells of the Bravais lattices in Fig. 3.1 can be misleading because, as shown in Chapter 2, it is the *pattern* of lattice points which distinguishes the lattices. The unit cells simply represent arbitrary, though convenient, ways of 'joining up' the lattice points. Consider, for example, the three cubic lattices; cubic *P* (for *P*rimitive, one lattice point per cell, i.e. lattice points only at the corners of the cell), cubic *I* (for '*I*nnenzentrierte', which is German for 'body-centred', an additional lattice point at the centre of the cell, giving two lattice points per cell) and cubic *F* (for *F*ace-centred, with additional lattice points at the centres of each face of the cell, giving four lattice points per cell). It is possible to outline alternative primitive cells (i.e. lattice points only at the corners) for the cubic *I* and cubic *F* lattices, as is shown in Fig. 3.2. As mentioned in Chapter 1, these primitive cells are not often used (1) because the inter-axial angles are not the convenient 90° (i.e. they are not orthogonal) and (2) because they do not reveal very clearly the cubic symmetry of the cubic *I* and cubic *F* lattices. (The symmetry of the Bravais lattices, or rather the point group symmetries of their unit cells, will be described in Section 3.3.)

Similar arguments concerning the use of primitive cells apply to all the other centred lattices. Notice that the unit cells of two of the lattices are centred on the 'top' and

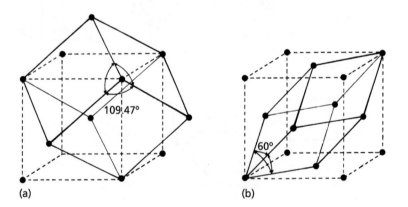

(a) (b)

Fig. 3.2. (a) the cubic *I*, **(b)** the cubic *F* lattice unit cells (dashed lines), and the corresponding primitive rhombohedral unit cells (full lines) with their inter-axial angles indicated.

'bottom' faces. These are called base-centred or *C*-centred because these faces are intersected by the *c*-axis.

The Bravais lattices may be thought of as being built up by stacking 'layers' of the five plane lattices, one on top of another. The cubic and tetragonal lattices are based on the stacking of square lattice layers; the orthorhombic *P* and *I* lattices on the stacking of rectangular layers; the orthorhombic *C* and *F* lattices on the stacking of rectangular centred layers; the rhombohedral and hexagonal lattice on the stacking of hexagonal layers and the monoclinic and triclinic lattices on the stacking of oblique layers. These relationships between the plane and the Bravais lattices are easy to see, except perhaps for the rhombohedral lattice. The rhombohedral unit cell has axes of equal length and with equal angles (α) between them. Notice that the layers of lattice points, perpendicular to the 'vertical' direction (shown dotted in Fig. 3.1) form triangular, or equivalently, hexagonal layers. The hexagonal and rhombohedral lattices differ in the ways in which the hexagonal layers are stacked. In the hexagonal lattice they are stacked directly one on top of the other (Fig. 3.3(a)) and in the rhombohedral lattice they are stacked such that the next two layers of points lie above the triangular 'hollows' or interstices of the layer below, giving a three layer repeat (Fig. 3.3(b)). These hexagonal and rhombohedral stacking sequences have been met before in the stacking of close-packed layers (Chapter 1); the hexagonal lattice corresponds to the simple hexagonal AAA... sequence and the rhombohedral lattice corresponds to the fcc ABCABC... sequence.

Now observant readers will notice that the rhombohedral and cubic lattices are therefore related. The primitive cells of the cubic *I* and cubic *F* lattices (Fig. 3.2) are rhombohedral—the axes are of equal length and the angles (α) between them are equal. As in the two-dimensional cases, what distinguishes the cubic lattices from the rhombohedral is their symmetry. When the angle α is 90° we have a cubic *P* lattice, when it is 60° we have a cubic *F* lattice and when it is 109.47° we have a cubic *I* lattice (Fig. 3.2). Or, alternatively, when the hexagonal layers of lattice points in the rhombohedral lattice are spaced apart in such a way that the angle α is 90°, 60° or 109.47°, then cubic symmetry results.

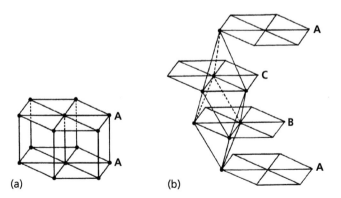

Fig. 3.3. Stacking of hexagonal layers of lattice points in (a) the hexagonal lattice and (b) the rhombohedral lattice.

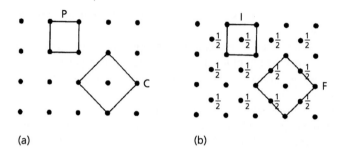

Fig. 3.4. Plans of tetragonal lattices showing (a) the tetragonal $P = C$ lattice and (b) the tetragonal $I = F$ lattice.

Finally, compare the orthorhombic lattices (all sides of the unit cell of different lengths) with the tetragonal lattices (two sides of the cell of equal length). Why are there four orthorhombic lattices, P, C, I and F, and only two tetragonal lattices, P and I? Why are there not tetragonal C and F lattices as well? The answer is that there *are* tetragonal C and F lattices, but by redrawing or outlining different unit cells, as shown in Fig. 3.4, it will be seen that they are identical to the tetragonal P and I lattices, respectively. In short, they represent no new arrangements of lattice points.

3.3 The symmetry of the fourteen Bravais lattices: crystal systems

The unit cells of the Bravais lattices may be thought of as the 'building blocks' of crystals, precisely as Haüy envisaged (Fig. 1.2). Hence it follows that the habit or external shape, or the observed symmetry of crystals, will be based upon the shapes and symmetry of the Bravais lattices, and we now have to describe the point symmetry of the unit cells of the Bravais lattices just as we described the point symmetry of plane patterns and lattices. The subject is far more readily understood if simple models are used (Appendix 1).

First, mirror lines of symmetry become mirror planes in three dimensions. Second, axes of symmetry (diads, triads, tetrads and hexads) also apply to three dimensions. The additional complication is that, whereas a plane motif or object can only have one such axis (perpendicular to its plane), a three-dimensional object can have several axes running in different directions (but always through a point in the centre of the object).

Consider, for example, a cubic unit cell (Fig. 3.5(a)). It contains a total of nine mirror planes, three parallel to the cube faces and six parallel to the face diagonals. There are three tetrad (four-fold) axes perpendicular to the three sets of cube faces, four triad (three-fold) axes running between opposite cube corners, and six diad (two-fold) axes running between the centres of opposite edges. This 'collection' of symmetry elements is called the **point group symmetry** of the cube because all the elements—planes and axes—pass through a point in the centre.

Why should there be these particular numbers of mirror planes and axes? It is because all the various symmetry elements operating at or around the point must be

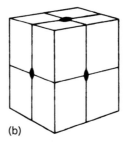

(a) (b)

Fig. 3.5. The point symmetry elements in (a) a cube (cubic unit cell) and (b) an orthorhombic unit cell.

consistent with one another. Self-consistency is a fundamental principle, underlying all the two-dimensional plane groups, all the three-dimensional point groups and all the space groups to be discussed in Chapter 4. If there are only two diad axes, for example, then they *have* to be mutually orthogonal, otherwise chaos would result; by the same token they also *must* generate a third diad perpendicular to both of them. It is the necessity for self-consistency which governs the construction of every one of the different combinations of symmetry, controlling the nature of each combination; it is this, also, which limits the total numbers of possible combinations to quite definite numbers such as thirty-two, in the case of the crystallographic point groups (the crystal classes), the fourteen Bravais lattices, and so on.

The cubic unit cell has more symmetry elements than any other: its very simplicity makes its symmetry difficult to grasp. More easy to follow is the symmetry of an orthorhombic cell. Figure 3.5(b) shows the point group symmetry of an orthorhombic unit cell. It contains, like the cube, three mirror planes parallel to the faces of the cell but no more—mirror planes do not exist parallel to the face diagonals. The only axes of symmetry are three diads perpendicular to the three faces of the unit cell.

In both cases it can be seen that the point group symmetry of these unit cells (Figs 3.5(a) and 3.5(b)) is independent of whether the cells are centred or not. All three cubic lattices, *P*, *I* and *F*, have the same point group symmetry; all four orthorhombic lattices, *P*, *I*, *F* and *C*, have the same point group symmetry and so on. This simple observation leads to an important conclusion: it is not possible, from the observed symmetry of a crystal, to tell whether the underlying Bravais lattice is centred or not. Therefore, in terms of their point group symmetries, the Bravais lattices are grouped, according to the shapes of their unit cells, into seven **crystal systems**. For example, crystals with cubic *P*, *I* or *F* lattices belong to the **cubic system**, crystals with orthorhombic *P*, *I*, *F* or *C* lattices belong to the **orthorhombic system**, and so on. However, a complication arises in the case of crystals with a hexagonal lattice. One might expect that all crystals with a hexagonal lattice should belong to the hexagonal system, but, as shown in Chapter 4, the external symmetry of crystals may not be identical (and usually is not identical) to the symmetry of the underlying Bravais lattice. Some crystals with a hexagonal lattice, e.g. α-quartz, do not show hexagonal (hexad) symmetry but have triad symmetry. (see Fig. 1.33a, Section 1.11.5) Such crystals are assigned to the **trigonal system** rather than to the hexagonal system. Hence the trigonal

system includes crystals with both hexagonal and rhombohedral Bravais lattices. There is yet another problem which is particularly associated with the trigonal system, which is that the rhombohedral unit cell outlined in Figs 3.1 and 3.3 is not always used—a larger (non-primitive) unit cell of three times the size is sometimes more convenient. The problem of transforming axes from one unit cell to another is addressed in Chapter 5.

The crystal systems and their corresponding Bravais lattices are shown in Table 3.1. Notice that there are no axes or planes of symmetry in the **triclinic system.** The only symmetry that the triclinic lattice possesses (and which is possessed by all the other lattices) is a **centre of symmetry.** This point symmetry element and inversion axes of symmetry are explained in Chapter 4.

3.4 The coordination or environments of Bravais lattice points: space-filling polyhedra

So far we have considered lattices as patterns of points in space in which each lattice point has the same environment in the same orientation. This approach is complete and sufficient, but it fails to stress, or even make clear, the fact that each of these environments is distinct and characteristic of the lattices themselves.

We need therefore a method of clearly and unambiguously defining what we mean by 'the environment' of a lattice point. One approach (which we have used already in working out the sizes of interstitial sites) is to state this in terms of 'coordination'—the numbers and distances of nearest neighbours. For example, in the simple cubic (cubic P) lattice each lattice point is surrounded by six other equidistant lattice points; in the bcc (cubic I) lattice each lattice point is surrounded by eight equidistant lattice points—and so on. This is satisfactory, but an alternative and much more fruitful approach is to consider the environment or domain of each lattice point in terms of a polyhedron whose faces, edges and vertices are equidistant between each lattice point and its nearest neighbours. The construction of such a polyhedron is illustrated in two dimensions for simplicity in Fig. 3.6. This is a plan view of a simple monoclinic (monoclinic P) lattice with the b axis perpendicular to the page. The line labelled ① represents the edge or trace of a plane perpendicular to the page and half way between the central lattice point 0 and its neighbour 1. All points lying in this plane (both in the plane of the paper and above and below) are therefore equidistant between the two lattice points 0 and 1. We now repeat the process for the other lattice points 2, 3, 4, etc., surrounding the central lattice point. The planes ①, ②, ③ etc. form the six 'vertical' faces of the polyhedron and in three dimensions, considering the lattice points 'above' and 'below' the central lattice point 0, the polyhedron for the monoclinic P lattice is a closed prism, shown shaded in plan in Fig. 3.6. Each lattice point is surrounded by an identical polyhedron and they all fit together to completely fill space with no gaps in between.

In this example (of a monoclinic P lattice) the edges of the polyhedron are where the faces intersect and represent points which are equidistant between the central lattice point and *two* other surrounding lattice points. Similarly, the vertices of the polyhedron represent points which are equidistant between the central lattice point and *three* other surrounding points. However, for lattices of higher symmetry this correspondence does

Table 3.1 The seven crystal systems, their corresponding Bravais lattices and symmetries

System	Bravais lattices	Axial lengths and angles	Characteristic (minimum) symmetry	Non-centrosymmetric point groups		Centrosymmetric point groups[a]
				Enantiomorphous[b]	Non-enantiomorphous[c]	Non-enantiomorphous[c]
Cubic	PIF	$a = b = c$ $\alpha = \beta = \gamma = 90°$	4 triads equally inclined at 109.47°	23, 432	$\bar{4}3m$	$m\bar{3}$, $m\bar{3}m$
Tetragonal	PI	$a = b \neq c$ $\alpha = \beta = \gamma = 90°$	1 rotation tetrad or inversion Tetrad	4^P, 422	$\bar{4}$, $4mm^P$, $\bar{4}2m$	$4/m$, $4/mmm$
Orthorhombic	PICF	$a \neq b \neq c$ $\alpha = \beta = \gamma = 90°$	3 diads equally inclined at 90°	222	$mm2^P$	mmm
Trigonal	R	$a = b = c$ $\alpha = \beta = \gamma \neq 90°$	1 rotation triad or inversion triad	3^P, 32	$3m^P$	$\bar{3}$, $\bar{3}m$
Trigonal	P	$a = b \neq c$ $\alpha = \beta = 90°$, $\gamma = 120°$	(= triad + centre of symmetry)			
Hexagonal	P	$a = b \neq c$ $\alpha = \beta = 90°$, $\gamma = 120°$	1 rotation hexad or inversion hexad (= triad + perp. mirror plane)	6^P, 622	$\bar{6}$, $6mm^P$, $\bar{6}m2$	$6/m$, $6/mmm$
Monoclinic	PC	$a \neq b \neq c$ $\alpha = \gamma = 90°, \beta \geq 90°$	1 rotation diad or inversion diad (= perp. mirror plane)	2^P	m^P	$2/m$
Triclinic	P	$a \neq b \neq c$ $\alpha \neq \beta \neq \gamma \neq 90°$	None	1^P		$\bar{1}$

(a) All the crystals which possess a centre of symmetry and/or a mirror plane are non-enantiomorphous.

(b) The eleven enantiomorphous point groups are those which do not possess a plane or a centre of symmetry. Hence enantiomorphous crystals can exist in right- or left-handed forms.

(c) Eleven of the twenty-one non-enantiomorphous point groups are centrosymmetric. Crystals which have a centre of symmetry do not exhibit certain properties, e.g. the piezoelectric effect.

The ten polar point (non-centrosymmetric) groups (indicated by a superscript P) possess a unique axis not related by symmetry. They are equally divided between the enantiomorphous point groups (1, 2, 3, 4, 6) and non-enantiomorphous point groups (m, mm2, 3m, 4mm, 6mm).

Trigonal crystals are divided into those which are represented by the hexagonal P lattice and those which are represented by the rhombohedral R lattice.

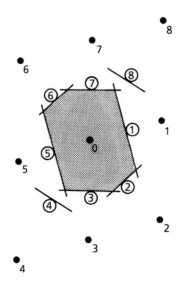

Fig. 3.6. The Voronoi polyhedron (Dirichlet region or Wigner-Seitz cell) for a monoclinic *P* lattice (plan view, *b* axis perpendicular to the page). The lines ①, ② etc. represent the edges or traces of planes which are equidistant between the central lattice point 0 and the surrounding or coordinating lattice points 1, 2, etc. The resulting Voronoi polyhedron is outlined in this two-dimensional section by the shaded area.

not hold. If, for example, we consider a cubic *P* lattice, square in plan, and follow the procedure outlined above, we find that the polyhedron is (as expected) a cube, but the edges of which are equidistant between the central lattice point and *three* surrounding lattice points and the vertices of which are equidistant between the central lattice point and *seven* surrounding lattice points.

The polyhedra constructed in this way and which represent the domains around each lattice point have various names: Dirichlet regions or Wigner-Seitz cells or Voronoi* polyhedra. There are altogether 24 such space-filling polyhedra corresponding to the 14 Bravais lattices; it is not a simple one-to-one correspondence in all cases because the shape of the polyhedron may depend upon the ratios between the axial lengths and angles and whether the Bravais lattice is centred or not. For example, Fig. 3.7(a) and (b) shows the two polyhedra for the tetragonal *I* lattice; Fig. 3.7(a) for the case where the axial ratio, *c/a*, is less than one and Fig. 3.7(b) for the case where it is greater than one.

The space-filling polyhedra for the cubic *P*, *I* and *F* lattices are particularly interesting. For the cubic *P* lattice it is simply a cube of edge-length equal to the spacing between nearest lattice points (Fig. 3.7(c)). For the cubic *I* lattice it is a truncated octahedron (Fig. 3.7(d), the eight hexagonal faces corresponding to the eight nearest neighbours at the corners of the cube and the six square faces corresponding to the six next-nearest neighbours at the centres of the surrounding cubes. For the cubic *F*

* Denotes biographical notes available in Appendix 3.

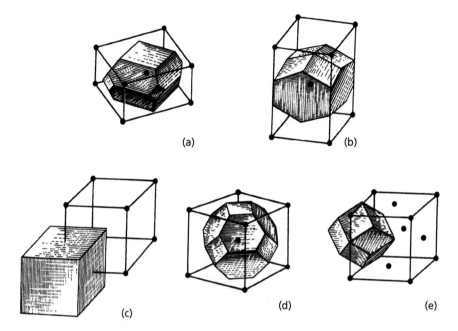

Fig. 3.7. Examples of domains or Voronoi polyhedra outlined around single lattice points (a) tetragonal I lattice, $c/a < 1$; (b) tetragonal I lattice, $c/a > 1$; (c) cubic P lattice; (d) cubic I lattice and (e) cubic F lattice (from *Modern Crystallography* by B. K. Vainshtein, Academic Press, 1981).

lattice it is a rhombic dodecahedron (Fig. 3.7(e)), the twelve diamond-shaped faces corresponding to the twelve nearest neighbours. (see Appendix 2).

It is of interest to compare the space-filling polyhedra for the fcc (cubic F) and hcp close-packing. These are shown in Fig. 3.8(a) and (b) respectively with the positions of the ABC and ABA atom layers indicated. If the 'central' atom is considered to be a B-layer then the 'bottom' three diamond-shaped faces correspond to the coordination of the three A-layer atoms below, the six 'vertical' diamond-shaped faces correspond to the coordination of the six surrounding B-layer atoms and the 'top' three diamond-shaped faces correspond to the coordination of the C-layer atoms for cubic close-packing (Fig. 3.8(a)) or the A-layer atoms for hexagonal close-packing (Fig. 3.8(b)). The polyhedron in Fig. 3.8(a) is a rhombic dodecahedron (as in Fig. 3.7(e)) and in Fig. 3.8(b) it is a trapezorhombic dodecahedron (see also Appendix 2).

The truncated octahedron (the Voronoi polyhedron for the cubic I lattice) also known as a **tetrakaidecahedron**, is of particular interest and is also an Archimedean polyhedron (see Appendix 2). It represents the 'special case' polyhedron for the tetragonal I lattice when the c/a ratio changes from <1 to >1 (compare Figs 3.7(a), (b) and (d)). More importantly, it is the space-filling solid with *plane* faces which has the largest volume-to-surface-area ratio and therefore *approximates* to the shapes of grains in annealed polycrystalline metals or ceramics or the cells in soap-bubble foams (Fig. 3.9). However, the angles between the faces and edges do not satisfy the

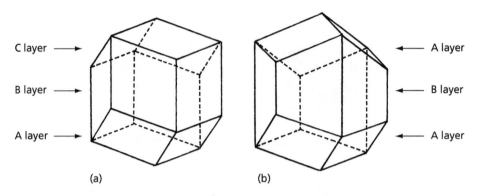

Fig. 3.8. Space filling polyhedra (a) for cubic close-packing (rhombic dodecahedron) and (b) for hexagonal close-packing (trapezorhombic dodecahedron).

Fig. 3.9. Space-filling by an assembly of truncated octahedra or tetrakaidecahedra (edges of equal length). These polyhedra have 14 faces (6 square plus 8 hexagonal) and are arranged at the points of a cubic *I* lattice. (From *Symmetry* by Hermann Weyl, Princeton University Press, 1952.)

equilibrium requirements for grain boundary energy (e.g. in two-dimensions the grain boundaries must meet at 120°). If, following Lord Kelvin, we (partly) accommodate these requirements by allowing the surfaces and edges to bow in or out, we obtain a (space-filling) solid with curved surfaces and edges called an β-tetrakaidecahedron. This, however, does not represent the 'last word' in the geometry of grain boundaries. If we relax the condition that all the polyhedra have an equal number of faces, then a space-filling structure with a slightly larger volume to surface area ratio can be built up consisting of pentagonal dodecahedra and 14-sided polygons consisting of 12 pentagonal faces and 2 hexagonal faces[1] ('C$_{24}$' in fullerene notation). However,

[1] D. Weaire (ed.) *The Kelvin Problem: Foam Structures of Minimal Surface Area.* Taylor & Francis, London and Bristol, Pa (1996).

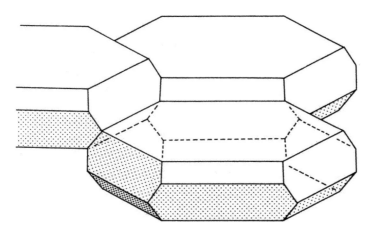

Fig. 3.10. Epidermal cells in mammalian skin which have the shapes of flattened tetrakaidecahedra arranged in vertical columns (compare with Fig. 3.9). (Illustration by courtesy of Professor Honda, Hyogo University, Japan.)

in practice, grains and the cells of soap-films are irregular in shape and size, although they do have on average about fourteen faces, like tetrakaidecahedra. In biological structures, the cells in the epidermis (the outer layer) of mammalian skin have also been shown to have the shape of intersecting flattened tetrakaidecahedra arranged in neat vertical columns (Fig. 3.10). In this case the edges are no longer equal in length; two of the eight hexagonal faces (parallel to the surface of the skin) are large and all the other faces are small and elongated. These epidermal cells are of course space-filling but have much smaller volume-to-surface area ratios.

The Voronoi approach to the partitioning of space may also be applied to the analysis of crystal *structures,* in which one alternative is to draw the planes equidistant between the outer radii of atoms or ions and not their centres—the sizes of the polyhedra being a measure of the relative sizes of the atoms or ions. All the polyhedra (now of different sizes and shapes) are space-filling. It may also be used in entirely non-crystallographic situations to determine, for example, the catchment areas for an irregular distribution of schools; pupils whose homes are on the dividing lines between the irregular polyhedra being equidistant from two schools and those whose homes are at the vertices being equidistant from three schools.

Exercises

3.1 The drawings in Fig. 3.11 show patterns of points distributed in orthorhombic-shaped unit cells. Identify to which (if any) of the orthorhombic Bravais lattices, P, C, I or F, each pattern of points corresponds.

 (Hint: It is helpful to sketch plans of several unit cells, which will show more clearly the patterns of points, and then to outline (if possible) a P, C, I or F unit cell.)

3.2 The unit cells of several **orthorhombic** structures are described below. Draw *plans* of each and identify the Bravais lattice, P, C, I or F, in each case.

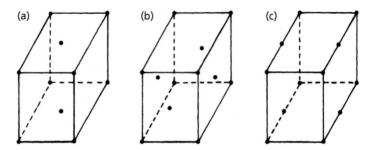

Fig. 3.11. Patterns of points in orthorhombic unit cells.

(a) One atom per unit cell located at $(x'y'z')$.

(b) Two atoms per unit cell of the *same* type located at $(0\frac{1}{2}0)$ and $\left(\frac{1}{2}0\frac{1}{2}\right)$.

(c) Two atoms per unit cell, one type located at $(00z')$ and $\left(\frac{1}{2}\frac{1}{2}z'\right)$ and the other type at $\left(00(\frac{1}{2}+z')\right)$ and $\left(\frac{1}{2}\frac{1}{2}\left(\frac{1}{2}+z'\right)\right)$.

(*Hint*: Draw plans of several unit cells and relocate the origin of the axes, x', y', z' should be taken as small (non-integral) fractions of the cell edge lengths.)

3.3 What are the shapes of the Voronoi polyhedra which correspond to the rhombohedral Bravais lattice?

(*Hint*: recall that the three cubic lattices are 'special cases' of the rhombohedral lattice in which the inter-axial angle α is 90° (cubic P), 60° (cubic F) or 109.47° (cubic I).)

3.4 Calculate the ratio between the 'long' and 'short' diagonals of the diamond-shaped faces in the primitive rhombohedral unit cells of the cubic I and cubic F lattices (see Fig. 3.2).

4

Crystal symmetry: point groups, space groups, symmetry-related properties and quasiperiodic crystals

4.1 Symmetry and crystal habit

As indicated in Chapter 3, the system to which a crystal belongs may be identified from its observed or external symmetry. Sometimes this is a very simple procedure. For example, crystals which are found to grow or form as cubes obviously belong to the cubic system: the external point symmetry of the crystal and that of the underlying unit cell are identical. However, a crystal from the cubic system may not grow or form with the external shape of a cube; the unit cells may stack up to form, say, an octahedron, or a tetrahedron, as shown in the models constructed from sugar-cube unit cells (Fig. 4.1). These are just two examples of a very general phenomenon throughout all the crystal systems: only very occasionally do crystals grow with the same shape as that of the underlying unit cell. The different shapes or **habits** adopted by crystals are determined by chemical and physical factors which do not, at the moment, concern us; what does concern us as crystallographers is to know how to recognize to which system a crystal belongs even though its habit may be quite different from, and therefore conceal, the shape of the underlying unit cell.

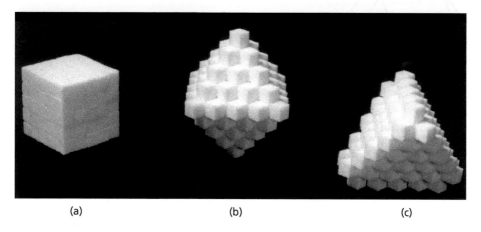

(a) (b) (c)

Fig. 4.1. Stacking of 'sugar-cube' unit cells to form (a) a cube, (b) an octahedron and (c) a tetrahedron. Note that the cubic cells in all three models are in the same orientation.

The Basics of Crystallography and Diffraction. Fourth Edition. Christopher Hammond.
© Christopher Hammond 2015. Published in 2015 by Oxford University Press.

The clue to the answer lies in the point group symmetry of the crystal. Consider, for example, the symmetry of the cubic crystals which have the shape or habit of a cube, an octahedron or a tetrahedron (Figs 4.1 and 4.2) or construct models of them (Appendix 1). The cube and octahedron, although they are different shapes, possess the same point group symmetry. The tetrahedron, however, has less symmetry: only six mirror planes instead of nine: only three diads running between opposite edges (i.e. along the directions perpendicular to the cube faces in the underlying cubes) and, as before, four triads running through each corner. The common, unchanged symmetry elements are the four (equally inclined) triads, and it is the presence of these four triads which

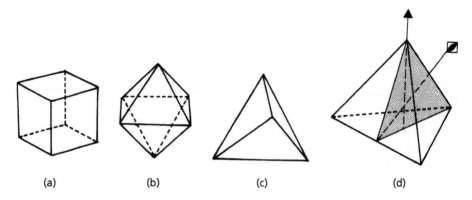

(a) (b) (c) (d)

Fig. 4.2. (a) A cube, (b) an octahedron and (c) a tetrahedron drawn in the same orientation as the models in Fig. 4.1. (d) A tetrahedron showing the positions of one variant of the point symmetry elements: mirror plane (shaded) ($\times 6$), triad ($\times 4$) and inversion tetrad (which includes a diad) ($\times 3$).

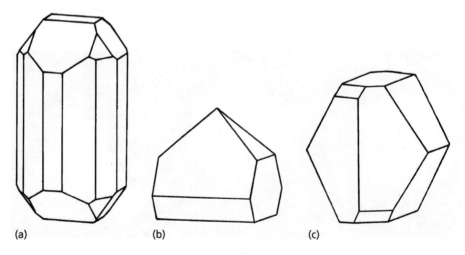

(a) (b) (c)

Fig. 4.3. Orthorhombic crystals (a) anglesite $PbSO_4$ (*mmm*), (b) struvite $NH_4MgPO_4 \cdot 6H_2O$ (*mm2*), (c) asparagine $C_4H_4O_3(NH_2)_2$ (222) (from *Introduction to Crystallography*, 3rd edn, by F. C. Phillips, Longmans 1963).

characterizes crystals belonging to the cubic system. Cubic crystals usually possess additional symmetry elements—the most symmetrical cubic crystals being those with the full point group symmetry of the underlying unit cell. But it is the four triads—not the three tetrads or the nine mirror planes—which are the 'hallmark' of a cubic crystal.

Similar considerations apply to all the other crystal systems. For example, Fig. 4.3 shows three orthorhombic crystals. Figure 4.3(a) shows a crystal with the full symmetry of the underlying unit cell—three perpendicular mirror planes and three perpendicular diads. Figure 4.3(b) shows a crystal with only two mirror planes and one diad along their line of intersection. Figure 4.3(c) shows a crystal with three perpendicular diads but no mirror planes.

4.2 The thirty-two crystal classes

The examples shown in Figs 4.1–4.3 are of crystals with different point group symmetries: they are said to belong to different **crystal classes.** Crystals in the same class have the same point group symmetry, so in effect the terms are synonymous. Notice that crystals in the same class do not necessarily have the same shape. For example, the cube and octahedron are obviously different shapes but belong to the same class because their point group symmetry is the same.

In two dimensions (Chapter 2) we found that there were ten plane point groups; in three dimensions there are thirty-two three-dimensional point groups. One of the great achievements of the science of mineralogy in the nineteenth century was the systematic description of the thirty-two point groups or crystal classes and their division into the seven crystal systems. Particular credit is due to J. F. C. Hessel,* whose contributions to the understanding of point group symmetry were unrecognized until after his death. The concept of seven different types or shapes of underlying unit cells then links up with the concept of the fourteen Bravais lattices; in other words, it establishes the connection between the external crystalline form or shape and the internal molecular or atomic arrangements.

It is not necessary to describe all the thirty-two point groups systematically; only the nomenclature for describing their important distinguishing features needs to be considered. This requires a knowledge of additional symmetry elements—centres and inversion axes.

Finally, we come to a 'practical' problem, which those of us who collect minerals or who grow crystals from solutions will immediately recognize. Our crystals are rarely uniformly developed, like those in Figs 4.2 or 4.3, but are irregular in appearance, with faces of different size and shape and from which it is almost impossible to recognize any point symmetry elements at all. Figure 4.4 shows examples of quartz crystals in which the corresponding faces are developed to different extents. It is this problem which hindered the development of crystallography until the discovery of the **Law of Constancy of Interfacial Angles** (Section 1.1) which enables us to focus on the underlying crystal symmetry rather than being diverted by the contingencies of crystal growth.

* Denotes biographical notes available in Appendix 3.

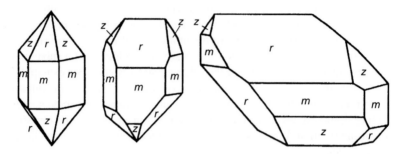

Fig. 4.4. Three quartz crystals with corresponding faces developed differently (from *Modern Crystallography* by B.K. Vainshtein, Springer-Verlag, 1981).

4.3 Centres and inversion axes of symmetry

If a crystal, or indeed any object, possesses a **centre of symmetry,** then any line passing through the centre of the crystal connects equivalent faces, or atoms, or molecules. A familiar example is a right hand and a left hand placed palm-to-palm but with the fingers pointing in opposite directions, as in Fig. 4.5(a). Lines joining thumb to thumb or fingertip to fingertip all pass through a centre of symmetry between the hands. When the hands are placed palm-to-palm but with the fingers pointing in the same direction, as in prayer, then there is no centre of symmetry but a **mirror (or reflection) plane of symmetry** instead, as in Fig. 4.5(b).

Notice the important relationship between these two symmetry elements: a centre of symmetry (Fig. 4.5(a)) *plus* a rotation of 180° (of one hand) is equivalent to a mirror plane of symmetry (Fig. 4.5(b)). Conversely, a mirror plane of symmetry *plus* a rotation of 180° (about an axis perpendicular to the mirror plane) is equivalent to a centre of symmetry. In short, centres and mirror planes of symmetry relate objects which (like hands) do not themselves possess these symmetry elements; conversely objects which themselves possess these symmetry elements do not occur in either right or left-handed forms (see Table 3.1).

In two dimensions a centre of symmetry is equivalent to diad symmetry. (See, for example, the motif and plane molecule shown in Fig. 2.3(4), which may be described as showing diad symmetry or a centre of symmetry.) In three dimensions this is not the case, as an inspection of Fig. 4.5(a) will show.

Inversion axes of symmetry are rather difficult to describe (and therefore difficult for the reader to understand) without the use of the stereographic projection—a method of representing a three-dimensional pattern of planes in a crystal on a two-dimensional plan. This topic is covered in Chapter 12 and the representation of symmetry elements in detail in Section 12.5.1. Geographers have the same problem when trying to represent the surface of the Earth on a two-dimensional map, and they too make use of the stereographic projection. In atlases, the circular maps of the world (usually with the north or south poles in the centre) are often stereographic projections.

(a)

(b)

Fig. 4.5. Right and left hands (a) disposed with a centre of symmetry between them and (b) disposed with a mirror plane between them.

Inversion axes are *compound* symmetry elements, consisting of a rotation followed by an inversion. For example, as described in Chapter 2, the operation of a tetrad (fourfold) rotation axis is to repeat a crystal face or pattern every 90° rotation, e.g. in two dimensions giving four repeating **R**s or the four-fold pattern of faces in a cube. The operation of an inversion tetrad, symbol ▨ or $\bar{4}$, is to repeat a crystal face or pattern every 90° rotation-plus-inversion through a centre. What results is a four-fold pattern of faces around the inversion axis, but with each alternate face inverted. Examples of a crystal and an object with inversion tetrad axes are shown in Figs 4.6(a) and (b). The tennis ball has, in fact, the same point group symmetry as the crystal. Notice that when it is rotated 90° about the axis indicated, the 'downwards' loop in the surface pattern is replaced by an 'upwards' loop. Another 90° rotation brings a 'downwards' loop and so on for the full 360° rotation. Notice also that the inversion tetrad includes a diad, as is indicated by the diad (lens) symbol in the inversion tetrad (open square) symbol, ▨ or $\bar{4}$.

Finally, compare the symmetry of the tetragonal crystal in Fig. 4.6(a) with that of the tetrahedron (Fig. 4.2(d)): the diad axes which we recognized passing through the centres of opposite edges in the tetrahedron are, in fact, inversion tetrad axes or, to

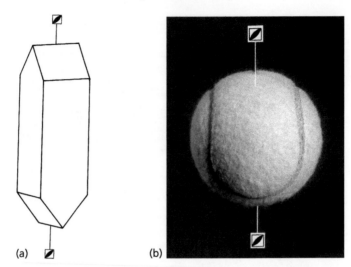

Fig. 4.6. Examples of a crystal and an object which have inversion tetrad axes (both point group $\bar{4}2m$). (a) Urea $CO(NH_2)_2$ and (b) a tennis ball.

develop one of the points made in Section 4.1, stacking the cubes into the form of a tetrahedron reduces the symmetry element along the cube axis directions from rotation to inversion tetrad.

There are also inversion axes corresponding to rotation monads diads, triads and hexads. The operation of an inversion hexad, for example, is a rotation of 60° plus an inversion, this compound operation being repeated a total of six times until we return to the beginning. However, for a beginner to the subject, these axes may perhaps be regarded as being of lesser importance than the inversion tetrad because they can be represented by combinations of other (better-understood) symmetry elements.

An inversion monad, symbol o or $\bar{1}$ is equivalent to a centre of symmetry.

An inversion diad, symbol $\bar{2}$ is equivalent to a perpendicular mirror plane.

An inversion triad, symbol \triangle or $\bar{3}$ is equivalent to a triad plus a centre of symmetry—which is the symmetry of a rhombohedral lattice (see Fig. 3.1). Notice that the 'top' three faces of the rhombohedron are related to the 'bottom' three faces by a centre of symmetry.

An inversion hexad, symbol ⬢ or $\bar{6}$ is equivalent to a triad plus a perpendicular mirror plane.

Again, these equivalences are best understood with the use of the stereographic projection (Chapter 12). The important point is that only inversion tetrads are unique (i.e. they cannot be represented by a combination of rotation axes, centres of symmetry or mirror planes) and therefore need to be considered separately.

The point group symmetries of the thirty-two classes are described by a 'short-hand' notation or **point group symbol** which lists the main (but not necessarily all) symmetry elements present. For example, the presence of centres of symmetry is not recorded

because they may arise 'automatically' from the presence of other symmetry elements, e.g. the presence of an inversion triad axis mentioned above. This notation for the thirty-two crystal classes or point groups, and their distribution among the seven crystal systems, is fully worked out in the *International Tables for Crystallography* published for the International Union of Crystallography and in F. C. Phillips' *Introduction to Crystallography*. Altogether there are five cubic classes, three orthorhombic classes, three monoclinic classes and so on. They are all listed in Table 3.1 (p. 93). The order in which the symmetry elements are written down in the point group symbol depends upon the crystal system.

In the cubic system the first place in the symbol refers to the axes parallel to, or planes of symmetry perpendicular to, the x-, y- and z-axes, the second refers to the four triads or inversion triads and the third the axes parallel to, or planes of symmetry perpendicular to, the face diagonal directions. Hence the point group symbol for the cube or the octahedron—the most symmetrical of the cubic crystals—is $4/m\bar{3}\,2/m$. This full point group symbol is usually (and rather unhelpfully) contracted to $m\bar{3}m$ because the operation of the four triads and nine mirror planes (three parallel to the cube faces and six parallel to the face diagonals) 'automatically' generates the three tetrads, six diads, and a centre of symmetry. The symbol for the tetrahedron is $\bar{4}3m$, the $\bar{4}$ referring to the three inversion tetrad axes along the x-, y- and z-axes together with the m referring to the face-diagonal mirror planes. The least symmetric cubic class has point group symbol 23, i.e. it only has diads along the x-, y- and z-axes and the characteristic four triads.

In the orthorhombic system the three places in the point group symbol refer to the symmetry elements associated with the x-, y- and z-axes. The most symmetrical class (Fig. 4.3(a)), which has the full point group symmetry of the underlying orthorhombic unit cell (Fig. 3.5), has the full point group symbol $2/m2/m2/m$, but this is usually abbreviated to mmm because the presence of the three mirror planes perpendicular to the x-, y- and z-axes 'automatically' generates the three perpendicular diads. The other two classes are $mm2$ (Fig. 4.3(b))—a diad along the intersection of two mirror planes—and 222 (Fig. 4.3(c))—three perpendicular diads.

In the monoclinic system the point group symbol simply refers to the symmetry elements associated with the y-axis. This may be a diad (class 2), an inversion diad (equivalent to a perpendicular mirror plane (class $\bar{2}$ or m)), or a diad plus a perpendicular mirror plane (class $2/m$).

In the tetragonal, hexagonal and trigonal systems, the first position in the point group symbol refers to the 'unique' z-axis. For example, the tetragonal crystals in Fig. 4.6 have point group symmetry $\bar{4}2m$; $\bar{4}$ referring to the inversion tetrad along the z-axis, 2 referring to the diads along the x- and y-axes and m to the mirror planes which bisect the x- and y-axes (which you will find by examining the model!). One of the trigonal classes has point group symbol 32 (not to be confused with cubic class 23!), i.e. a single triad along the z-axis and (three) perpendicular diads.

Not all classes are of equal importance; in 432 and $\bar{6} = 3/m$ there are very few examples of crystal structures (e.g. Ag_3AuTe_2 and $8Li_2O_2$ respectively). On the other hand, the two monoclinic classes m and $2/m$ contain about 50 per cent of all inorganic crystalline materials on a 'crystal counting' basis, including feldspar, the commonest

mineral in nature, and many other economically important minerals. As for the crystals of organic compounds, class 2/m is by far the most important, while crystals of biologically important substances which contain chiral (right- or left-handed enantiomorphic molecules) have a predilection for class 2. The commonest class in any system is the **holosymmetric** class, i.e. the class which shows the highest symmetry. The holosymmetric cubic class $m\bar{3}m$, the most symmetrical of all, contains only a few per cent of all crystals on this basis, but these also include many materials and ceramics of economic and commercial importance.

It is a great help in an understanding of point group symmetry simply to identify the symmetry elements of everyday objects such as clothes pegs, forks, pencils, tennis balls, pairs of scissors, etc. Or, one step further, you could make models showing the point group symmetries of all the 32 crystal classes as described in Appendix 1.

A note on alternating or rotation-reflection axes

These compound symmetry elements were used by Schönflies in his derivation of the 230 space groups. They are used in the description of layer-symmetry (Section 2.8) but are otherwise little used today. They consist of rotation plus a **reflection** in a plane perpendicular to the axis, rather than an **inversion**. Hence a monad alternating axis is equivalent to a perpendicular mirror plane (or inversion diad); a diad alternating axis is equivalent to a centre of symmetry (or inversion monad); a triad alternating axis is equivalent to an inversion hexad; a tetrad alternating axis is equivalent to an inversion tetrad and a hexad alternating axis is equivalent to an inversion triad.

4.4 Crystal symmetry and properties

The quantities which are used to describe the properties of materials are, as we know, simply represented as *coefficients,* i.e. as one measured (or measurable) quantity divided by another. For example, the property (coefficient) of electrical conductivity is given by the amount of electrical current flowing between two points (which may be measured in various ways) divided by the electrical potential gradient; the pyroelectric effect—the property of certain crystals of developing electrical polarization when the temperature is changed—is given by the polarization divided by the temperature change; the heat capacity is given by the quantity of heat absorbed or given out divided by the temperature change, and so on.

In many (in fact most) cases the measured quantities depend on direction and are called **vectors**.[1] In the examples above, electrical current flow, potential gradient and polarization are all vectors. The other quantities in the examples above, temperature change, quantity of heat, do not depend on direction and are called **scalars**. Finally, many quantities which are of importance in the description of the physical properties of crystals are described by **tensors**, a subject which is introduced in Chapter 14.

[1] See Appendix 5.

The important point is that, in those cases where one or more of the measured quantities vary with direction, so also do the crystal properties; they are said to be *anisotropic* (from the Greek tropos, direction or turn; (an)iso, (not the) same). Anisotropy clearly arises because the arrangements of atoms in crystals vary in different directions—you would intuitively *expect* crystals to be anisotropic, the only exceptions being those properties (the heat capacity) which are direction independent. You would also intuitively expect cubic crystals to be 'less anisotropic' than, say, monoclinic ones because of their greater symmetry, and this intuition would also be correct. For many properties, but not all, cubic crystals are isotropic—the property (and property coefficient) is direction independent. In the example given above, cubic crystals are isotropic with respect to electrical conductivity. They are also isotropic with regard to the pyroelectic effect, i.e. cubic crystals do not exhibit electrical polarization when the temperature is changed; the pyroelectric coefficient is zero. But cubic crystals are not isotropic with respect to all properties. For example, their elastic properties, which determine the mechanical properties of stiffness, shear and bulk moduli, are direction dependent and these are very important factors with respect to the properties of metals and alloys.

Hence, one major use of point groups is in relating crystal symmetry and properties; as the external symmetry of crystals arises from the symmetry of the internal molecular or atomic arrangements, so also do these in turn determine or influence crystal properties. Some examples have already been alluded to. For example, the pyroelectric effect cannot exist in a crystal possessing a centre of symmetry, and the pyroelectric polarization can only lie along a direction in a crystal that is *unique,* in the sense that it is not repeated by any symmetry element. There are only ten point groups or crystal classes which fulfil these conditions and they are called the **ten polar point groups:**

$$1 \quad 2 \quad\quad 3 \quad\quad 4 \quad\quad 6$$
$$m \quad mm2 \quad 3m \quad 4mm \quad 6mm.$$

Hence, pyroelectricity or the pyroelectric effect can only occur in these ten polar point groups or classes.

A very closely related property to pyroelectricity, and of great importance in electroceramics, is ferroelectricity. A ferroelectric crystal, like a pyroelectric crystal, can also show polarization, but in addition the direction of polarization may be reversed by the application of an electric field. Most ferroelectric crystals have a transition temperature (Curie point) above which their symmetry is non-polar and below which it is polar.

One such example is barium titanate, $BaTiO_3$, which has the perovskite structure (Fig. 1.17). Above the Curie temperature barium titanate has the fully symmetric cubic structure, point group $m\bar{3}m$, but below the Curie temperature, when the crystal becomes ferroelectric, distortions occur—a small expansion occurs along one cell edge and small contractions along the other two, changing the crystal system symmetry from cubic to tetragonal and the point group symmetry from $m\bar{3}m$ to $4mm$. As the temperature is further lowered below the Curie point, further distortions occur and the point group symmetry changes successively to $mm2$ and $3m$—all of them, of necessity, being polar point groups (see Section 1.11.1).

Another very important crystal property is piezoelectricity—the development of an electric dipole when a crystal is stressed, or conversely, the change of shape of a crystal when it is subjected to an electrical field. At equilibrium the applied stress will be centrosymmetrical, so if a crystal is to develop a dipole, i.e. develop charges of opposite sign at opposite ends of a line through its centre, it cannot have a centre of symmetry. There are twenty-one non-centrosymmetric point groups (Table 3.1), all of which, except one, point group 432, may exhibit piezoelectricity. It is the presence of the equally inclined triads, tetrads and diads in this cubic point group which in effect cancel out the development of a unidirectional dipole.

The optical properties of crystals—the variation of refractive index with the vibration and propagation direction of light (double refraction or birefringence), the variation of refractive index with wavelength or colour of the light (dispersion), or the associated variations of absorption of light (pleochroism)—are all symmetry dependent. The complexity of the optical properties increases as the symmetry decreases. Cubic crystals are optically isotropic—the propagation of light is the same in all directions and they have a single refractive index. Tetragonal, hexagonal and trigonal crystals are characterized by two refractive indices. For light travelling in a direction *perpendicular* to the principal (tetrad, hexad or triad) axis, such crystals exhibit two refractive indices—one for light vibrating along the principal axis, and another for light vibrating in a plane perpendicular to the principal axis. For light travelling *along* the principal axis (and therefore vibrating in the planes parallel to it), the crystal exhibits only one refractive index, and therefore behaves, for this direction only, as an optically isotropic crystal. Such crystals are called *uniaxial* with respect to their optical properties, and their principal symmetry axis is called the *optic axis*. Crystals belonging to the remaining crystal systems—orthorhombic, monoclinic and triclinic—are characterized by three refractive indices and two, not one, optic axes. Hence they are said to be biaxial since there are two, not one, directions for the direction of propagation of light in which they appear to be optically isotropic. It should be noted, however, that unlike uniaxial crystals, there is no simple relationship between the two optic axes of biaxial crystals and the principal symmetry elements; nor are they fixed, but vary as a result of dispersion, i.e. the variations in the values of the refractive indices with wavelength.

Finally, there is the phenomenon or property of optical activity or rotatory polarization, which should not be confused with double refraction. It is a phenomenon in which, in effect, the vibrational direction of light rotates such that it propagates through the crystal in a helical manner either to the right (dextrorotatory) or the left (laevorotatory). Now right-handed and left-handed helices are distinct in the same way as a right and left hand (Fig. 4.5) or the two parts of a twinned crystal (Fig. 1.18) and therefore optical activity would be expected to occur only in those crystals which occur in right-handed or left-handed forms, i.e. those which do not possess a mirror plane or a centre (or inversion axis) of symmetry. Such crystals are said to be enantiomorphous and there are altogether eleven enantiomorphous classes or point group symmetries (Table 3.1).

A famous example is sodium ammonium tartrate, a salt of tartaric acid (Fig. 4.7). In 1848 Louis Pasteur* first noticed these two forms 'hemihedral to the right' and 'hemihedral to the left' under the microscope and, having separated them, found that their solutions were optically active in opposite senses.

The study of enantiomorphism, or chirality, from the Greek word *chiros,* meaning hand, is becoming increasingly important. Louis Pasteur, as a result of his work on tartaric acid, was the first to suggest that the molecules themselves could be chiral—i.e. that they could exist in either right-handed or left-handed forms. The basic constituents of living things are chiral, including the amino acids[2] present in proteins, the nucleotides present in nucleic acids and the DNA double helix itself. But *only one* enantiomorph is ever found in nature—only L-amino acids are present in proteins and only D-nucleotides are present in nucleic acids (L stands for laevo—or left-rotating, and D stands for dextro—or right-rotating). Why this should be so is one of the mysteries surrounding the origin of life itself and for which many explanations or hypotheses have been offered. If, as many hypotheses suppose, it was the result of a chance event which was then consolidated by growth, then we might reasonably suppose that on another

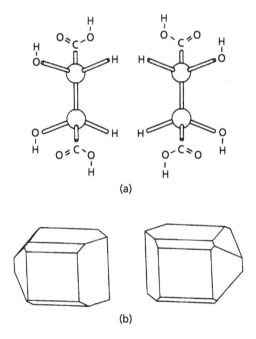

(a)

(b)

Fig. 4.7. (a) Left- and right-handed forms of tartaric acid molecules (from *Crystals: their Role in Nature and Science* by C. W. Bunn, Academic Press, New York, 1964); and (b) the left- and right-handed forms of tartaric acid crystals (from F. C. Phillips, loc. cit.).

[2] Except glycine, the simplest amino acid.

planet such as ours the *opposite* event might have occurred and that there exist living creatures, in every way like ourselves, but who are constituted of D-amino acids and L-nucleotides!

However, to return to earth, once our basic chirality has been established, the D- and L- (enantiomorphous) forms of many substances, including drugs in particular, may have very different chemical and therapeutic properties. For example, the molecule asparagine (Fig 4.3(c)) occurs in two enantiomorphous forms, one of which tastes bitter and the other sweet. Thalidomide also; the right-handed molecule of which acts as a sedative but the left-handed molecule of which gave rise to birth defects. Hence chiral separation, and the production of enantiomorphically pure substances, is of major importance.

4.5 Translational symmetry elements

The thirty-two point group symmetries (Table 3.1) may be applied to three-dimensional patterns just as the ten plane point group symmetries are applied to two-dimensional patterns (Chapter 2). As in two dimensions where translational symmetry elements or glide lines arise, so also in three dimensions do **glide planes** and also **screw axes** arise. It is only necessary to state the symmetry properties of patterns that are described by these translational symmetry elements. Glide planes are the three-dimensional analogues of glide lines; they define the symmetry in which mirror-related parts of the motif are shifted half a lattice spacing. In Fig. 2.5(b) the figures are related by glide lines, which can easily be visualized as glide-plane symmetry. Glide planes are symbolized as a, b, c (according to whether the translation is along the x-, y- or z-axes), n or d (diagonal or diamond glide—special cases involving translations along more than one axis).

Screw axes (for which there is no two-dimensional analogue—except for screw diads which arise in layer-symmetry patterns (see Section 2.8))—essentially describe helical patterns of atoms or molecules, or the helical symmetry of motifs. Several types of helices are possible and they are all based upon different combinations of rotation axes and translations. Figure 4.8 shows the possible screw axes (the direction of translation out of the plane of the page) with the heights **R** of the asymmetrical objects represented as fractions of the lattice repeat distance (compare to Fig. 2.3). Screw axes are represented in writing by the general symbol N_m, N representing the rotation (2, 3, 4, 6) and the subscript m representing the pitch in terms of the number of lattice translation or repeat distances for one complete rotation of the helix. m/N therefore represents the translation for each rotation around the axis. Thus the 4_1 screw axis represents a rotation of 90° followed by a translation of $\frac{1}{4}$ of the repeat distance, which repeated three times brings **R** to an identical position but displaced one lattice repeat distance; the 4_3 screw axis represents a rotation of 90° followed by a translation of $\frac{3}{4}$ of the lattice repeat distance, which repeated three times gives a helix with a pitch of three lattice repeat distances. This is equivalent to the 4_1 screw axis but of opposite sign: the 4_1 axis is a right-handed helix and 4_3 axis is a left-handed helix. In short they are enantiomorphs of each other. Similarly the 3_1 and 3_2 axes, the 6_1 and 6_5 axes, and the 6_2 and 6_4 axes are enantiomorphs of each other. In diagrams, screw axes are represented by the symbol for the

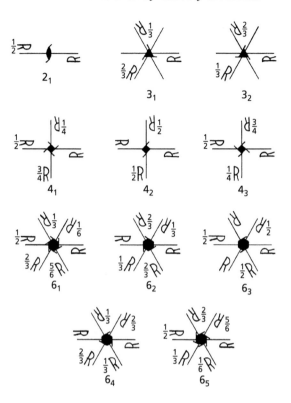

Fig. 4.8. The operation of screw axes on an asymmetrical motif, **R**. The fractions indicate the 'heights' of each motif as a fraction of the repeat distance.

rotation axis with little 'tails' indicating (admittedly not very satisfactorily) the pitch and sense of rotation (see Fig. 4.8). Screw axes have, of course, their counterparts in nature and design—the distribution of leaves around the stem of a plant, for example, or the pattern of steps in a spiral (strictly helical) staircase. Figure 4.9 shows two such examples. Figure 4.10 shows the 6_3 screw hexads which occur in the hcp structure; notice that they run parallel to the c-axis and are located in the 'unfilled' channels which occur in the hcp structure. They do not pass through the atom centres of either the A layer or the B layer atoms; these are the positions of the triad (not hexad) axes in the hcp structure.

Just as the external symmetry of crystals does not distinguish between primitive and centred Bravais lattices, so also it does not distinguish between glide and mirror planes, or screw and rotation axes. For example, the six faces of an hcp crystal show hexad, six-fold symmetry, whereas the underlying structure possesses only screw hexad, 6_3, symmetry.

In many crystals, optical activity arises as a result of the existence of enantiomorphous screw axes. For example, in α-quartz (enantiomorphous class 32), the SiO_2 structural units which are not themselves asymmetric, are arranged along the c-axis

(a) (b)

Fig. 4.9. (a) The 4_2 screw axis arrangement of leaves round a stem of pentstemon (after Walter Crane); and (b) a 6_1 screw axis spiral (helical) staircase (from *The Third Dimension in Chemistry* by A. F. Wells, Clarendon Press, Oxford, 1968).

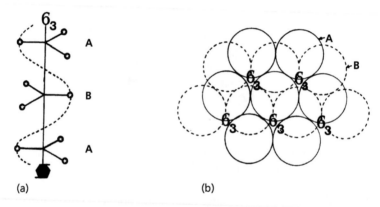

(a) (b)

Fig. 4.10. (a) A screw hexad (6_3) axis; and (b) location of these axes in the hcp structure. Notice that they pass through the 'unfilled channels' between the atoms in this structure.

(which is also the optic axis) in either a 3_1 or a 3_2 screw orientation (see Figure 1.33(a), Section 1.11.5). This gives rise to the two enantiomorphous crystal forms of quartz (class 32, Fig. 4.11). The plane of polarization of plane-polarized light propagating along the optic axis is rotated to right or left, the angle of rotation depending on the wavelength of the light and the thickness of the crystal. This is not, to repeat, the same phenomenon as birefringence; for the light travelling along the optic axis the crystal exhibits (by definition) one refractive index. If the 3_1 or 3_2 helical arrangement of the SiO_2 structural units in quartz is destroyed (e.g. if the crystal is melted and solidified as a glass), the optical activity will also be destroyed.

However, in other crystals such as tartaric acid (Fig. 4.7) and its derivatives, the optical activity arises from the asymmetry—the lack of a mirror plane or centre of symmetry—of the molecule itself (Fig. 4.7(a)). In such cases the optical activity is not destroyed if the crystal is melted or dissolved in a liquid. The left or right handedness of

Fig. 4.11. The enantiomorphic (right- and left-handed) forms of quartz. The optic axis is in the vertical (long) direction in each crystal (from F. C. Phillips, loc. cit).

the molecules, even though they are randomly orientated in a solution, is communicated at least in part, to the plane-polarized light passing through it. Unlike quartz, in which the optical activity depends on the direction of propagation of the light with respect to the optic axis, the optical activity of a solution such as tartaric acid is unaffected by the direction of propagation of the light. In summary, the optical activity of solutions arises from the asymmetry of the molecule itself; the optical activity which is shown in crystals, but not their solutions or melts, arises from the enantiomorphic screw symmetry of the arrangement of molecules in the crystal.

4.6 Space groups

In Section 2.4 it is shown how the seventeen possible two-dimensional patterns or plane groups (Fig. 2.6) can be described as a combination of the five plane lattices with the appropriate point and translational symmetry elements. Similarly, in three dimensions, it can be shown that there are 230 possible three-dimensional patterns or **space groups,** which arise when the fourteen Bravais lattices are combined with the appropriate point and translational symmetry elements. It is easy to see why there should be a substantially larger number of space groups than plane groups. There are fourteen space lattices compared with only five plane lattices, but more particularly there is a greater number of combinations of point and translational symmetry elements in three dimensions, particularly the presence of inversion axes (point) and screw axes (translational) which do not occur in two-dimensional patterns.

The first step in the derivation of 230 space groups was made by L. Sohncke* (who also first introduced the notion of screw axes described in Section 4.5). Essentially, Sohncke relaxed the restriction in the definition of a Bravais lattice—that the environment of each point is identical—by considering the possible arrays of points which have identical environments when viewed from *different* directions, rather than from the *same* direction as in the definition of a Bravais lattice. This is equivalent to

combining the fourteen Bravais lattices with the appropriate translational symmetry elements, and gives rise to a total of sixty-five space groups or Sohncke groups.

The second, final, step was to account for inversion axes and glide planes of symmetry which gives rise to a further 165 space groups. They were first worked out by Fedorov* in Moscow in 1890 and independently by Schönflies* in Göttingen in 1891 and Barlow* in London in 1894—an example of the frequently occurring phenomenon in science of progress being made almost at the same time by people approaching a problem entirely independently. (see Appendix 7)

The 230 space groups are systematically drawn and described in the *International Tables for Crystallography* Volume A, which is based upon the earlier *International Tables for X-ray Crystallography* Vol. 1 compiled by N. F. M. Henry* and Kathleen Lonsdale*—a work of great crystallographic scholarship. The space groups are arbitrarily numbered 1 to 230, beginning with triclinic crystals of lowest symmetry and ending with cubic crystals of highest symmetry. There are two space group symbols, one due to Schönflies* used in spectroscopy and the other, which is now generally adopted in crystallography, due to Hermann* and Mauguin.*

The Hermann–Mauguin space group symbol consists first of all of a letter *P, I, F, R, C, B or A* which describes the Bravais lattice type (Fig. 3.1) (the alternatives *C, B or A* being determined as to whether the unit cell axes are chosen such that the *C, B or A* faces are centred); then a statement, rather like a point group symbol, of the essential (not all) symmetry elements present. For example, the space group symbol *Pba*2 represents a space group which has a primitive (*P*) Bravais lattice and whose point group is *mm*2 (the *a* and *b* glide planes being simple mirror planes in point group symmetry). This is one of the point groups of the orthorhombic system (Fig. 4.3) and the lattice type is orthorhombic *P*. Similarly, space group *P*6₃/*mmc* has a primitive (*P*) (hexagonal) Bravais lattice with point group symmetry 6/*mmm*.

The space group itself is represented by means of two diagrams or projections, one showing the symmetry elements present and the other showing the operation of these symmetry elements on an asymmetric 'unit of pattern' represented by the circular symbol ○ and its mirror-image by ⊙: a circle with a comma inside. These symbols, which may represent an asymmetric molecule, a group of molecules, or indeed any asymmetrical structural unit, correspond to the **R** and **Я** of our two-dimensional patterns. The choice of a *circle* to represent an *asymmetric* object might be thought to be inadequate— surely a symbol such as **R** or, better still, a right hand would be more appropriate? In a sense it would, but there would then arise a serious problem of typography, of clearly and unambiguously representing the operation of all the symmetry elements in the projection. For example, in the case where a mirror plane lies *in* the plane of projection a right hand (palm-down) would be mirrored by a left hand (palm-up)—and the problem would be to represent clearly these two superimposed hands in a plan view. In the case of a circle this situation is easily represented by ◐ —a circle divided in the middle with the mirror-image indicated in one half. Similarly, the use of a symbol such as **R** would

* Denotes biographical notes available in Appendix 3.

lead to ambiguity. For example, a diad axis *in* the plane of projection would rotate an **R** 180° out of the plane of projection into an **Я**—which would be indistinguishable from an **R** reflected to an **Я** in a mirror plane perpendicular to the plane of projection—i.e. as for mirror-lines in the two-dimensional case. In the case of a circle there is no such ambiguity; in the former case we have

$$\circ \overset{2}{|} \circ \quad \text{(diad axis in plane of paper—no change of hand of motif)}$$

and in the latter case

$$\circ \overset{m}{|} \circ \quad \text{(mirror plane perpendicular to plane of paper—a change of hand of motif)}$$

The representation of space groups and some, but not all of the associated crystallographic information, is best described by means of four examples *Pba2* (No. 32), *P2₁/c* (No. 14), *P6₃/mmc* (No. 194) and *P4₁2₁2* (No. 92).

Figure 4.12, from the *International Tables for X-ray Crystallography*, shows space group No. 32, *Pba2* with the Hermann–Mauguin and Schoenflies symbols shown top left and the point group and crystal system top right. The two diagrams are projections in the $x - y$ plane, the right-hand one shows the symmetry elements present—the diads parallel to the z-axis at the corners, edges and centre of the unit cell and the a and b glide planes shown as dashed lines in between. It would be perfectly possible to draw the origin of the unit cell at an intersection of the glide planes—but to choose it, as shown, at a diad axis is more convenient, hence the note 'origin on 2'.

In the left-hand diagram the ◯ is placed at (small) fractions, x, y, z of the cell edge lengths away from the origin, the z parameter or 'height' being represented by a plus (+) sign. This is called a 'general position' because the ◯ does not lie on any of the symmetry elements present and the resulting pattern is known as the set of 'general equivalent positions'. The coordinates of these positions are listed below together with the total number of them, 4, the 'Wyckoff letter', c and the symbol 1 for a monad, indicating the asymmetry of the ◯ (and its glide plane image ⊙).

If the pattern unit ◯ were to be placed not in a general position but in a 'special position', on a diad axis in this example, then a simpler pattern results. The four asymmetric pattern units 'merge together' to give two units with diad symmetry and these are called 'special equivalent positions'. There are in fact two possibilities, denoted by the Wyckoff letters a and b and their co-ordinates are listed in the table on the left. The Wyckoff letters are purely arbitrary, like the numbering of the space groups themselves.

The table on the right lists the conditions (on the Laue indices *hkl*) limiting possible reflections, those not meeting these conditions being known as *systematic absences* in X-ray diffraction. These topics are covered in Chapter 9 and Appendix 6. Finally, the 'symmetry of special projections' shows the plane groups corresponding to the space group projected on different planes (just as in our projections of crystal structures in Chapter 2). For example, the projection on the (001) plane (which is that of the diagrams) corresponds to plane groups *pgg* (or *p2gg*—see Fig. 2.6).

$$\begin{array}{l} \boldsymbol{P\,b\,a\,2} \\ \boldsymbol{C_{2v}^8} \end{array}$$ No. 32 $P\,b\,a\,2$ $m\,m\,2$ Orthorhombic

Origin on 2

Number of positions, Wyckoff notation, and point symmetry	Co-ordinates of equivalent positions	Conditions limiting possible reflections

General:

4	c	1	$x,y,z;\quad \bar{x},\bar{y},z;\quad \tfrac{1}{2}-x,\tfrac{1}{2}+y,z;\quad \tfrac{1}{2}+x,\tfrac{1}{2}-y,z.$

hkl: No conditions
0kl: $k=2n$
h0l: $h=2n$
hk0: No conditions
h00: $(h=2n)$
0k0: $(k=2n)$
00l: No conditions

Special: as above, plus

2	b	2	$0,\tfrac{1}{2},z;\quad \tfrac{1}{2},0,z.$
2	a	2	$0,0,z;\quad \tfrac{1}{2},\tfrac{1}{2},z.$

$\left.\begin{array}{l} \\ \\ \end{array}\right\}$ *hkl*: $h+k=2n$

Symmetry of special projections

(001) *pgg*; $a'=a, b'=b$ (100) *pm*1; $b'=b/2, c'=c$ (010) *p*1*m*; $c'=c, a'=a/2$

Fig. 4.12. Space group *Pba*2 (No. 32) (from the *International Tables for X-ray Crystallography*).

Figure 4.13 is extracted from the entry for the most frequently occurring space group No. 14 ($P2_1/c$) in the *International Tables for Crystallography* in which two choices for the unique axis *b* or *c* (parallel to the (screw) diad axes) and three choices of unit cell are available. Figure 4.13(a) shows the usual choice of the *b* (or *y*-axis) parallel to the (screw) diad axes as indicated by the monoclinic point group symbol 2/*m* (see Section 4.3) and 'cell choice 1'. The pattern of general equivalent positions is shown in the lower right diagram and the symmetry elements (screw diads, centres of symmetry and glide planes) are shown in three different projections. The centres of symmetry are indicated by small circles, the glide planes by dashed or dotted lines depending as to whether the glide direction is in, or perpendicular to, the plane of the diagram, and similarly the screw diad axes normal to or in the plane of the projection are indicated by ⧊ or single-headed arrow symbols respectively. Figure 4.13(b) shows the three possible cell choices and the tables of the coordinates of the general and special equivalent positions (with their Wyckoff letters running from bottom to top) and the reflection conditions as before.

(a) *P 2₁/c* C_{2h}^5 2/m Monoclinic

No. 14 *P 1 2₁/c 1* Patterson symmetry *P 1 2/m 1*

UNIQUE AXIS *b*, CELL CHOICE 1

Origin at $\bar{1}$

Asymmetric unit $0\le x\le 1;\ 0\le y\le\frac{1}{4};\ 0\le z\le 1$

Symmetric operations

(1) 1 (2) $2(0,\frac{1}{2},0)\ 0,y,\frac{1}{4}$ (3) $\bar{1}\ 0,0,0$ (4) $c\ x,\frac{1}{4},z$

Fig. 4.13. (*continued*)

Figure 4.14 is extracted from the entry for space group *P6₃/mmc*, No. 194 in the *International Tables for X-ray Crystallography;* this is the space group for the hcp metals (Fig. 1.5(b)), graphite (Fig. 1.37(a)), lonsdaleite (Fig. 1.36(b)) and NiAs. Notice that there is a greater number of special equivalent positions (Wyckoff letters running from *a* to *k*) than in the two lower-symmetry space groups we have just looked at and that the coordinates of the pattern units are much reduced—from 24 for the general case to 2 for positions with Wyckoff letters *a, b, c, d*. In hcp metals the A and B layer atoms (Fig. 1.5(b)) are in the special equivalent positions denoted by Wyckoff letters *c* or *d*. Notice that if the origin of the unit cell is shifted so as to coincide with one of these atoms then their coordinates become (000), (2/3 1/3 1/2) *or* (000), (1/3 2/3 1/2) (see Exercise 1.6 and Section 9.2, Example 4). Finally, having studied Fig. 4.14 it is a good test of your powers of observation to turn back to Fig. 4.10(b) and fill in all the symmetry elements in addition to the 6₃ axes already indicated.

(b) $P\,2_1/c$ C_{2h}^5 $2/m$ Monoclinic

No. 14

UNIQUE AXIS b, DIFFERENT CELL CHOICES

$P\,12_1/c\,1$

UNIQUE AXIS b, CELL CHOICE 1

Origin at $\bar{1}$

Asymmetric unit $0\le x\le 1$; $0\le y\le\frac{1}{4}$; $0\le z\le 1$

Generators selected (1); $t(1,0,0)$; $t(0,1,0)$; $t(0,0,1)$; (2); (3)

Positions

Multiplicity, Wyckoff letter, Site symmetry	Coordinates				Reflection conditions
					General:
4 e 1	(1) x,y,z	(2) $\bar{x},y+\frac{1}{2},\bar{z}+\frac{1}{2}$	(3) \bar{x},\bar{y},\bar{z}	(4) $x,\bar{y}+\frac{1}{2},z+\frac{1}{2}$	$h0l:\ l=2n$ $0k0:\ k=2n$ $00l:\ l=2n$
					Special: as above, plus
2 d $\bar{1}$	$\frac{1}{2},0,\frac{1}{2}$	$\frac{1}{2},\frac{1}{2},0$			$hkl:\ k+l=2n$
2 c $\bar{1}$	$0,0,\frac{1}{2}$	$0,\frac{1}{2},0$			$hkl:\ k+l=2n$
2 b $\bar{1}$	$\frac{1}{2},0,\frac{1}{2}$	$\frac{1}{2},\frac{1}{2},\frac{1}{2}$			$hkl:\ k+l=2n$
2 a $\bar{1}$	$0,0,0$	$0,\frac{1}{2},\frac{1}{2}$			$hkl:\ k+l=2n$

Fig. 4.13. Space group $P2_1/c$ (No. 14) (from the *International Tables for Crystallography*), (a) unique axis b, cell choice 1, (b) unique axis b, different cell choices.

Figure 4.15 is extracted from the entry for space group $P4_12_12$ (No. 92) in the *International Tables for Crystallography*, Volume A. This space group contains principally 4_1 (screw tetrad) axes of symmetry but no glide, mirror planes or inversion axes of symmetry. It is enantiomorphous with space group $P4_32_12$ (No. 96). In the left-hand diagram two neighbouring cells are drawn to show clearly the operation of the 4_1 (right-handed) screw axes along the cell edges. To these two space groups belong the α (low-temperature) form of cristobalite (see Section 1.11.5) in which the distortion from the high temperature β (cubic) form gives rise to the enantiomorphous tetragonal forms.

P 6₃/m m c No. 194 P 6₃/m 2/m 2/c 6/m m m Hexagonal

D_{6h}^4

Origin at centre ($\bar{3}m1$)

Number of positions, Wyckoff notation, and point symmetry			Co-ordinates of equivalent positions	Conditions limiting possible reflections

General:

24	*l*	1	x,y,z; $\bar{y},x-y,z$; $y-x,\bar{x},z$; \bar{y},\bar{x},z; $x,x-y,z$; $y-x,y,z$;	*hkil*: No conditions
			\bar{x},\bar{y},\bar{z}; $y,y-x,\bar{z}$; $x-y,x,\bar{z}$; y,x,\bar{z}; $\bar{x},y-x,\bar{z}$; $x-y,\bar{y},\bar{z}$;	*hh2\bar{h}l*: *l*=2n
			$\bar{x},\bar{y},\frac{1}{2}+z$; $y,y-x,\frac{1}{2}+z$; $x-y,x,\frac{1}{2}+z$;	*h\bar{h}0l*: No conditions
			$x,y,\frac{1}{2}-z$; $\bar{y},x-y,\frac{1}{2}-z$; $y-x,\bar{x},\frac{1}{2}-z$;	
			$y,x,\frac{1}{2}+z$; $\bar{x},y-x,\frac{1}{2}+z$; $x-y,\bar{y},\frac{1}{2}+z$;	
			$\bar{y},\bar{x},\frac{1}{2}-z$; $x,x-y,\frac{1}{2}-z$; $y-x,y,\frac{1}{2}-z$.	

Special: as above, plus

12	*k*	*m*	$x,2x,z$; $2\bar{x},\bar{x},z$; x,\bar{x},z; $\bar{x},2\bar{x},\bar{z}$; $2x,x,z$; \bar{x},x,\bar{z};	⎫
			$\bar{x},2\bar{x},\frac{1}{2}+z$; $2x,x,\frac{1}{2}+z$ $\bar{x},x,\frac{1}{2}+z$;	⎪ no extra conditions
			$x,2x,\frac{1}{2}-z$; $2\bar{x},\bar{x},\frac{1}{2}-z$ $x,\bar{x},\frac{1}{2}-z$.	⎪
12	*j*	*m*	$x,y,\frac{1}{4}$; $\bar{y},x-y,\frac{1}{4}$; $y-x,\bar{x},\frac{1}{4}$; $\bar{y},\bar{x},\frac{1}{4}$; $x,x-y,\frac{1}{4}$; $y-x,y,\frac{1}{4}$;	⎪
			$\bar{x},\bar{y},\frac{3}{4}$; $y,y-x,\frac{3}{4}$; $x-y,x,\frac{3}{4}$; $y,x,\frac{3}{4}$; $\bar{x},x-y,\frac{3}{4}$; $x-y,\bar{y},\frac{3}{4}$.	⎭
12	*i*	2	$x,0,0$; $0,x,0$; $\bar{x},\bar{x},0$; $x,0,\frac{1}{2}$; $0,x,\frac{1}{2}$; $\bar{x},\bar{x},\frac{1}{2}$;	*hkil*: *l*=2n
			$\bar{x},0,0$; $0,\bar{x},0$; $x,x,0$; $\bar{x},0,\frac{1}{2}$; $0,\bar{x},\frac{1}{2}$; $x,x,\frac{1}{2}$.	
6	*h*	*mm*	$x,2x,\frac{1}{4}$; $2\bar{x},\bar{x},\frac{1}{4}$; $x,\bar{x},\frac{1}{4}$; $\bar{x},2\bar{x},\frac{3}{4}$; $2x,x,\frac{3}{4}$; $\bar{x},x,\frac{3}{4}$.	no extra conditions
6	*g*	2/m	$\frac{1}{2},0,0$; $0,\frac{1}{2},0$; $\frac{1}{2},\frac{1}{2},0$; $\frac{1}{2},0,\frac{1}{2}$; $0,\frac{1}{2},\frac{1}{2}$; $\frac{1}{2},\frac{1}{2},\frac{1}{2}$.	*hkil*: *l*=2n
4	*f*	3m	$\frac{1}{3},\frac{2}{3},z$; $\frac{2}{3},\frac{1}{3},\bar{z}$; $\frac{2}{3},\frac{1}{3},\frac{1}{2}+z$; $\frac{1}{3},\frac{2}{3},\frac{1}{2}-z$.	*hkil*: If *h–k*=3n, then *l*=2n
4	*e*	3m	$0,0,z$; $0,0,\bar{z}$; $0,0,\frac{1}{2}+z$; $0,0,\frac{1}{2}-z$;	*hkil*: *l*=2n
2	*d*	$\bar{6}m2$	$\frac{1}{3},\frac{2}{3},\frac{3}{4}$; $\frac{2}{3},\frac{1}{3},\frac{1}{4}$.	⎫ *hkil*: If *h–k*=3n, then *l*=2n
2	*c*	$\bar{6}m2$	$\frac{1}{3},\frac{2}{3},\frac{1}{4}$; $\frac{2}{3},\frac{1}{3},\frac{3}{4}$.	⎭
2	*b*	$\bar{6}m2$	$0,0,\frac{1}{4}$; $0,0,\frac{3}{4}$.	⎫ *hkil*: *l*=2n
2	*a*	$\bar{3}m$	$0,0,0$; $0,0,\frac{1}{2}$.	⎭

Fig. 4.14. Space group *P6₃/mmc* (No. 194) (from the *International Tables for X-ray Crystallography*).

$P\,4_1\,2_1\,2$ $\qquad\qquad$ D_4^4 $\qquad\qquad\qquad\qquad\qquad$ Tetragonal

No. 92 $\qquad\qquad\qquad\qquad$ $P\,4_1\,2_1\,2$ $\qquad\qquad\qquad\qquad$ Patterson symmetry $\;P\,4/m\,m\,m$

Origin on $2[1\,1\,0]$ at $2_1 1(1,2)$

Asymmetric unit $\quad 0\le x\le1;\quad 0\le y\le1;\quad 0\le z\le\tfrac{1}{8}$

Symmetry operations

(1) 1 $\qquad\qquad\qquad$ (2) $2(0,0,\tfrac{1}{2})$ $\;0,0,z$ \qquad (3) $4^+(0,0,\tfrac{3}{4})$ $\;0,\tfrac{1}{2},z$ \qquad (4) $4^-(0,0,\tfrac{1}{4})$ $\;\tfrac{1}{2},0,z$
(5) $2(0,\tfrac{1}{2},0)$ $\;\tfrac{1}{4},y,\tfrac{1}{8}$ \quad (6) $2(\tfrac{1}{2},0,0)$ $\;x,\tfrac{1}{4},\tfrac{3}{8}$ \qquad (7) $2\;\;x,x,0$ $\qquad\qquad$ (8) $2\;\;x,\bar{x},\tfrac{1}{4}$

Generators selected \quad (1); $\;t(1,0,0);\;\;t(0,1,0);\;\;t(0,0,1);\;\;(2);\;\;(3);\;\;(5)$

Positions

Multiplicity, $\qquad\qquad\qquad\qquad$ Coordinates $\qquad\qquad\qquad\qquad\qquad\qquad\qquad\qquad$ Reflection conditions
Wyckoff letter,
Site symmetry

$\qquad\qquad\qquad\qquad\qquad\qquad\qquad\qquad\qquad\qquad\qquad\qquad\qquad\qquad\qquad\qquad\qquad$ General:

8 $\quad b\quad$ 1 \quad (1) x,y,z $\qquad\quad$ (2) $\bar{x},\bar{y},z+\tfrac{1}{2}$ \qquad (3) $\bar{y}+\tfrac{1}{2},x+\tfrac{1}{2},z+\tfrac{1}{4}$ \quad (4) $y+\tfrac{1}{2},\bar{x}+\tfrac{1}{2},z+\tfrac{3}{4}$ \qquad $00l:\;l=4n$
$\qquad\qquad\qquad$ (5) $\bar{x}+\tfrac{1}{2},y+\tfrac{1}{2},\bar{z}+\tfrac{3}{4}$ $\;$ (6) $x+\tfrac{1}{2},\bar{y}+\tfrac{1}{2},\bar{z}+\tfrac{1}{4}$ $\;$ (7) y,x,\bar{z} $\qquad\qquad$ (8) $\bar{y},\bar{x},\bar{z}+\tfrac{1}{2}$ \qquad $h00:\;h=2n$

$\qquad\qquad\qquad\qquad\qquad\qquad\qquad\qquad\qquad\qquad\qquad\qquad\qquad\qquad\qquad\qquad$ Special: as above, plus

4 $\quad a\quad$..2 $\quad x,x,0\qquad \bar{x},\bar{x},\tfrac{1}{2}\qquad \bar{x}+\tfrac{1}{2},x+\tfrac{1}{2},\tfrac{1}{4}\quad x+\tfrac{1}{2},\bar{x}+\tfrac{1}{2},\tfrac{3}{4}$ $\qquad\qquad$ $0kl:\;l=2n+1$
$\qquad\qquad\qquad\qquad\qquad\qquad\qquad\qquad\qquad\qquad\qquad\qquad\qquad\qquad\qquad\qquad\qquad$ or $2k+l=4n$

Symmetry of special projections

Along [001] $\;p\,4g\,m$ $\qquad\qquad\qquad$ Along [100] $\;p\,2g\,g$ $\qquad\qquad\qquad$ Along [110] $\;p\,2g\,m$
$a'=a\quad b'=b$ $\qquad\qquad\qquad\qquad$ $a'=b\quad b'=c$ $\qquad\qquad\qquad$ $a'=\tfrac{1}{2}(-a+b)\quad b'=c$
Origin at $0,\tfrac{1}{2},z$ $\qquad\qquad\qquad\qquad$ Origin at $x,\tfrac{1}{4},\tfrac{1}{8}$ $\qquad\qquad\qquad$ Origin at $x,x,0$

Fig. 4.15. Space group $P4_12_12$ (No. 92) (from the *International Tables for Crystallography*, Volume A—partly redrawn and data relating to sub-groups omitted).

4.7 Bravais lattices, space groups and crystal structures

In the simple cubic, bcc and ccp structures of the elements, the three cubic lattices (Fig. 3.1) have exactly the same arrangement of lattice points as the atoms, i.e. in these examples the motif is just one atom. In more complex crystals the motif consists of more than one atom and, to determine the Bravais lattice of a crystal, it is necessary first to identify the motif and then to identify the arrangement of the motifs. In crystals consisting of two or more different types of atoms this procedure may be quite difficult, but fortunately simple examples best illustrate the procedure and the principles involved. For example, in NaCl (isomorphous with TiN; see Fig. 1.14(a)), the motif is one sodium and one chlorine ion and the motifs are arranged in an fcc array. Hence the Bravais lattice of NaCl and TiN is cubic F.

In Li_2O (isomorphous with TiH_2; see Fig. 1.14(b)) the motif is one oxygen and two lithium atoms; the motifs are arranged in an fcc array and the Bravais lattice of these compounds is cubic F. In ZnS (isomorphous with TiH; see Fig. 1.14(c)) the motif is one zinc and one sulphur atom; again, these are arranged in an fcc array and Bravais lattice of these compounds is cubic F. All the crystal structures illustrated in Fig. 1.14 have the cubic F Bravais lattice. They are called face-centred cubic structures not because the arrangements of atoms are the same—clearly they are not—but because they all have the cubic F lattice.

In CsCl (Fig. 1.12(b)), the motif is one caesium and one chlorine ion; the motifs are arranged in a simple cubic array and the Bravais lattice is cubic P. To be sure, the arrangement of ions in CsCl (and compounds isomorphous with it) is such that there is an ion or atom at the body-centre of the unit cell, but the Bravais lattice is *not* cubic I because the ions or atoms at the corners and centre of the unit cell are different. Nor, for the same reason, should CsCl and compounds isomorphous with it be described as having a body-centred cubic structure.

In the case of hexagonal structures the arrangements of lattice points in the hexagonal P lattice (Fig. 3.1) corresponds to the arrangement of atoms in the simple hexagonal structure (Fig. 1.5(a)) and *not* the hcp structure (Fig. 1.5(b)). In the simple hexagonal structure the environment of all the atoms is identical and the motif is just one atom. In the hcp structure the environment of the atoms in the A and B layers is different. The motif is a pair of atoms, i.e. an A layer and a B layer atom per lattice point. The environment of these pairs of atoms (as for the pairs of ions or atoms in the NaCl, or CsCl or ZnS structures) is identical and they are arranged on a simple hexagonal lattice. Notice that in these examples the motif is either asymmetric or has a mirror plane or centre of symmetry. These are further instances of the situation which we found in two-dimensional patterns (Section 2.5). It is the repetition of the motif by the lattice which generates the crystal structures.

The space groups of the simple cubic bcc and ccp structures of the elements are those of maximum symmetry, namely $Pm\bar{3}m$, $Im\bar{3}m$ and $Fm\bar{3}m$, and in which the atoms are all at the special positions 000 etc. Similarly, CsCl (Fig. 1.12(b)) and the cubic forms of perovskite $CaTiO_3$ (Fig. 1.17) or barium titanate $BaTiO_3$, in which all the atoms are in special positions, also belong to space group $Pm\bar{3}m$. All these space groups or structures have a centre of symmetry (at the origin) as indicated by the inversion triad axis, $\bar{3}$, symbol.

In all the examples above, the (special) atom positions are fixed or 'pinned down' by the symmetry elements. For example, in the CaF_2 (fluorite) or Li_2O (antifluorite) structures the space group, as with fcc metals, is of maximum symmetry $Fm\bar{3}m$ (point group $m\bar{3}m$) and the symmetry elements fix the positions of the atoms in their special positions precisely as shown in Figs 1.16(c) and (d). For example, an atom or ion must either be evenly bisected by a mirror plane or must be arranged in pairs equidistant each side of it: it cannot occupy an 'in between' position because the mirror symmetry would be violated. It is in crystals of lower symmetry that the positions of the atoms are not completely fixed. The 'classic' example is the structure of iron pyrites, FeS_2, which at first glance might be thought to have the same structure as CaF_2, the S atoms

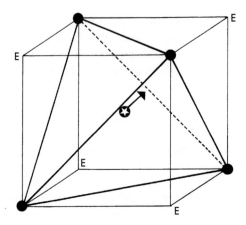

Fig. 4.16. A tetrahedral site in FeS$_2$ outlined within a cube. The Fe atoms are situated at four corners of the cube, the other four corners are 'empty' (denoted by E). The S atom (starred) is shifted from the centre of the tetrahedron/cube towards one of the four 'empty' corners as indicated by the arrow.

situated precisely at the centres of the tetrahedral sites co-ordinated by the Fe atoms. But this is not so: the S atoms do not lie in the centres of the tetrahedra but are shifted in a body-diagonal (triad axis) direction; in short, they lie in special positions in the structure. The geometry, just for one S atom, is shown in Figure 4.16. Within the whole unit cell the shifts of the S atoms are towards different 'empty' corners, preserving the cubic symmetry but reducing the space group symmetry from $Fm\bar{3}m$ to $Pa\bar{3}$ (point group symmetry reduced from $m\bar{3}m$ to $m\bar{3}$ (or $2/m\bar{3}$)).[3]

The 'amount' of shift of the S atoms in FeS$_2$, a single parameter, was deduced by W. L. Bragg in 1913 from the intensities of the X-ray reflections. It was the first structure to be analysed in which the atom positions are not fixed by the symmetry and provided Bragg, as he records long afterwards 'with the greatest thrill'.[4] Today of course the number of parameters required to be determined for the far more complex inorganic and especially organic crystals runs into the thousands and constitutes the major task in crystal structure determination.

Zinc blende, ZnS, and isomorphous structures such as TiH (Fig. 1.14(c)) and the technologically important gallium arsenide, GaAs, have the cubic F Bravais lattice, the atoms are again in special positions but the structure does not have a centre of symmetry; the space group in this case is $F\bar{4}3m$. This lack of a centre of symmetry, which is the origin, or crystallographic basis of the important electrical and physical properties in these structures, may be visualized with reference to Fig. 1.14(c). The TiH, ZnS or GaAs atoms are arranged in pairs in the body-diagonal directions of the cube (symbolized by $\langle 111 \rangle$—see Section 5.2) and the sequence of the atoms is either, e.g. ••• GaAs ••• GaAs ••• GaAs •••, or the reverse, i.e. ••• AsGa ••• AsGa ••• AsGa •••

[3] See Appendix 1 for illustrative models of the five cubic point groups.
[4] W. L. Bragg *The development of X-ray analysis*, Proc. Roy. Soc. **A262**, 145 (1961).

The body-diagonal directions are *polar* axes and the faces on he opposite sides of the crystal are terminated either by Ga or by As atoms.

In silicon, germanium and the common (cubic) form of diamond (see Section 1.11.6), the pattern of the atoms is the same as in ZnS or GaAs but of course all the atoms are of the same type (see Fig. 1.36). The body-diagonal directions are no longer polar because the sequence of pairs of atoms, e.g. ••• SiSi ••• SiSi ••• SiSi ••• , is obviously the same either way. These structures are centro-symmetric, the centres of symmetry lying half-way between the pairs of atoms. The space group in these cases is $Fd\bar{3}m$, the d referring to the special type of glide plane. Graphite and hcp metals, as mentioned above, belong to space group $P6_3/mmc$, as does also the common crystal structure of ice (see Section 1.11.5) in which the oxygen atoms lie in the same atomic positions as the carbon atoms in the hexagonal form of diamond (Fig. 1.36(b)) and in which the H atoms are between (but not equidistant between) neighbouring O atoms. Wurtzite, the hexagonal form of ZnS (Fig 1.26) has space group $P6_3mc$.

There are (Table 3.1) eleven enantiomorphous point groups (i.e. without a centre or mirror plane of symmetry) and upon which are based the 65 space groups first derived by L. Sohncke and in which there are eleven enantiomorphous pairs. We have already noticed the enantiomorphous pair for α-cristobalite ($P4_12_12$ and $P4_32_12$) based on the tetragonal point group 422. The others of particular interest are those for α-quartz ($P3_121$ and $P3_221$) based on the trigonal point group 32 and for β-quartz ($P6_222$ and $P6_422$) based on the hexagonal point group 622.

Not all the 230 space groups are of equal importance; for many of them there are no examples of real crystals at all. About 70% of the elements belong to the space groups $Fm\bar{3}m$, $Im\bar{3}m$ and $Fd\bar{3}m$ (all based on point group $m\bar{3}m$), $F\bar{4}3m$ (based on point group $\bar{4}3m$) and $6_3/mmc$ (based on point group $6/mmm$). Over 60% of organic and inorganic crystals belong to space groups $P2_1/c$, $C2/c$, $P2_1$, $P\bar{1}$, $Pbca$ and $P2_12_12_1$ and of these space group $P2_1/c$ (based on point group $2/m$, Fig. 4.12) is by far the commonest (see Table 4.1, p. 127).

4.8 The crystal structures and space groups of organic compounds

As mentioned in Section 1.10.1, the stability of inorganic molecules arises primarily from the strong, directed, covalent bonds which bind the atoms together. In comparison, the forces which bind organic molecules together are weak (in the liquid or solid states) or virtually non-existent (as in the gaseous state). The strongest of the intermolecular forces are hydrogen bonds, which link polar groups (as in water or ice, Section 1.11.5) or hydroxyl groups as in sugars. Indeed, organic crystals in which hydrogen bonds dominate are hard and rigid, like inorganic crystals. The remaining intermolecular forces are short-range and are generally described as van der Waals bonds. Apart from residual polarity, organic molecules are generally electrically neutral, and intermolecular ionic bonds, such as occur between atoms or groups of atoms in ionic crystals, do not exist.

The crystal structures which occur (if they occur at all) are largely determined by the ways in which the molecules pack together most efficiently: it is the 'organic equivalent'

of Robert Hooke's packing together of 'bullets' described at the very beginning of this book—except of course that organic molecules have far more complex shapes, or envelopes, than simple spheres.

As described below, it is from such packing considerations that the space groups of organic crystals can be predicted. However, it should be recognized at the outset that the determination of the space group provides little information on the positions of the atoms within the molecules themselves and which, particularly in macromolecules, are nearly all in general positions. The importance of crystallization (apart from its role in purification) lies in the fact that the structure of organic molecules may then be investigated by X-ray diffraction techniques: the space group determines the geometry of the pattern but it is the intensities of the X-ray reflections which ultimately determine the atom positions (see Chapters 6–10 and Chapter 13).

However, there is a further desideratum. Organic molecules which constitute living tissue—proteins, DNA, RNA—do not generally occur *in vivo* as crystals but are separated in an aqueous environment. The process of crystallization may not only reduce or eliminate the aqueous environment but may also distort the molecules away from their free-molecule geometry. An historically important example is the structure of DNA (see Section 10.5). Only when the parallel-orientated strands of DNA are examined in the wet or high-humidity condition (the B form) does the double-helical structure correspond to that which occurs *in vivo*. In the low-humidity or 'dry' condition (the A form) the repeat distance and conformation of the helices is changed—but at the same time giving rise to much sharper diffraction patterns. F.H.C. Crick realized that the transformation was in effect displacive rather than reconstructive (see Section 1.11.5) and that from the A form the double helical B form could be deduced.

4.8.1 *The close packing of organic molecules*

The first detailed analysis of the close (and closest) packing of organic molecules was made by A.I. Kitaigorodskii* who predicted the possible space groups arising from the close packing of 'molecules of arbitrary form'.[5] He proceeded on the principle that all the molecules were in contact, none interpenetrated, but rather that the 'protrusions' of one molecule fitted into the 'recesses' of a neighbouring molecule such that the amount of empty space was the least possible. He found, in summary, that the deviations from close-packing were small and that (as in the close-packing of spheres) a twelve-fold coordination was the general rule. No assumptions were made as to the nature of the intermolecular forces—the analysis is purely geometrical and must of course be modified when, for example, hydrogen bonding between molecules is taken into account.

The crystallographic interest of the analysis lies in its development from plane group symmetry (Section 2.5) to layer-group symmetry (Section 2.8) and then to space-group symmetry (Section 4.7). We shall follow these steps in outline (omitting the details of the analysis).

* Denotes biographical notes available in Appendix 3.
[5] A I Kitaigorodskii *Organic Chemical Crystallography*, USSR Academy of Sciences, Moscow, 1955; Eng. Trans (revised) Consultants Bureau Enterprises, New York, 1961.

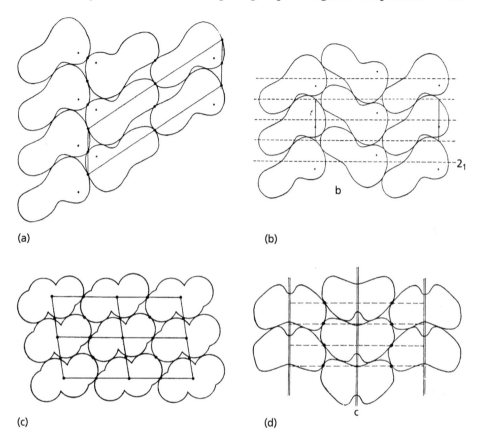

Fig. 4.17. Close-packing of two-dimensional motifs of 'arbitrary form' in oblique and rectangular unit cells. Motifs with point group symmetry 1 (a) and (b), 2 (c) and m (d) (from *Organic Chemical Crystallography* by A.I. Kitaigorodskii, Consultants Bureau, New York, 1961).

For plane molecules (or motifs) of arbitrary form having point group symmetry 1, 2 or *m* (see Fig. 2.3) it turns out that the requirement of close or closest-packing limits the plane groups to those with either oblique or rectangular unit cells (see Fig. 2.6). Figure 4.17 shows four examples to illustrate the motifs of 'arbitrary form' and the packing principles involved.

We now consider molecules or motifs which are three-dimensional, i.e. having 'top' and 'bottom' faces (as represented in Section 2.8, Fig. 2.15 by black and white **R**'s).

As in the two-dimensional case, such motifs can only be arranged with a minimum of empty space in layers in which the unit cells are oblique (total 7) or rectangular (total 41), i.e. a total of 48 out of the 80 possible layer symmetry groups (see Section 2.8). However, there are further restrictions. Layer symmetry groups with horizontal mirror planes are unsuitable for the close packing of such motifs since such planes would double the layers and cause protrusions to fall on protrusions and recesses on recesses.

Similarly, horizontal glide planes parallel to, or mirror planes perpendicular to, the axes of rectangular cells lead to four-fold, not six-fold coordination in the plane. Taking all these restrictions into account we are left with only ten layer symmetry groups which allow six-fold coordination close packing within the plane. These ten groups are shown in Fig. 4.18 where the black and white triangles indicate the 'top' and 'bottom' faces of the 'molecules of arbitrary form'.

Now we need to stack these layers upon each other to create a close-packed structure. Four of these layers are polar—the molecules all face the same way (all black triangles, Fig. 4.18 (a), (d), (f), (i)), represented diagrammatically in Fig. 4.19(a). The rest are non-polar, (Fig. 4.19(b)) and clearly only these non-polar layers can in principle give rise to close packing. Further, the presence of diad axes normal to the layers prohibit the close-packing of arbitrary shapes which just leaves us with layer-symmetry groups b, c, g and h (Fig. 4.18). Finally Kitaigorodskii concludes that close-packing can be achieved with molecules with monad symmetry (1) or a centre of symmetry ($\bar{1}$) but that for molecules with diad (2) or a single mirror plane (m) symmetry there is a reduction in full packing density; such structures he called 'limitingly close packed'.[5] Finally, he established those space groups which he termed 'permissible'. The space groups thus derived are listed in Table 4.1.

It is of interest to compare these predicted space groups with those of the molecular solids listed in the *Cambridge Structural Database* which (in 2015) had a total of 755,000 entries. Of this number, eight space groups account for 86% of all the

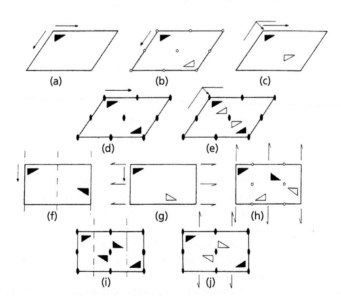

Fig. 4.18. Representation of the ten layer symmetry groups allowing coordination close packing of three-dimensional motifs in a plane. Single-headed arrows indicate in-plane screw diads, dashed lines indicate vertical glide planes (from *Macromolecular Physics*, Volume 1, by B. Wunderlich, Academic Press, New York and London, 1973).

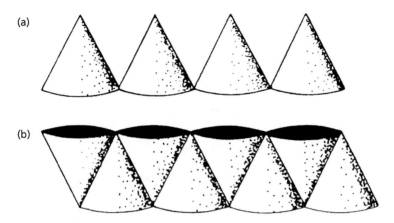

Fig. 4.19. Representation using a cone as a motif of the packing of (a) polar and (b) non-polar layers (from *Macromolecular Physics*, Volume 1, by B. Wunderlich, Academic Press, New York and London, 1973).

Table 4.1 Space groups for closest, limitingly and permissible close packing.

Motif symmetry	Closest packed	Limitingly close packed	Permissible
1	$P\bar{1}, P2_1, P2_1/c, Pca2_1,$ $Pna2_1, P2_12_12_1$	None	$P1,\ C_c,\ C2,\ P2_12_12,$ $Pbca$
$\bar{1}$	$P\bar{1}, P2_1/c, C2/c, Pbca$	None	$Pccn$
2	None	$C2/c, P2_12_12, Pbcn$	$C2, Aba2$
m	None	$Pmc2_1, Cmc2_1, Pnma$	$C_m,\ P2_1/m,\ Pmn2_1,$ $Abm2, Ima2, Pbcm$

entries, viz. $P2_1/c$ (34.6%), $P\bar{1}$ (24.4%), $P2_12_12_1$ (7.3%), $C2/c$ (8.3%), $P2_1$ (5.7%), $Pbca$ (3.4%), $Pnma$ (1.1%) and $Pna2_1$ (1.4%). All these space groups are included in Table 4.1—a remarkable predictive achievement when one considers how little chemistry was involved!

4.8.2 Long-chain polymer molecules

The crystal structures and space groups formed by long-chain polymer molecules are also in accord with the principles outlined above.

In the case of *atactic* polymers (i.e. those in which the side-groups are large and/or randomly distributed along the chain), crystal structures rarely occur—the side-groups keep the chains well apart—hence the name atactic. Crystal structures only occur in *tactic* polymers in which the side-groups are regularly distributed on one side of the chain (*isotactic*) or alternatively each side (*syndiotactic*). We shall consider just two polymers—polyethylene (polythene) and isotactic polypropylene (polypropene).

Fig. 4.20. Projections of polymer unit cells perpendicular to the chain axes. (a) Polyethylene, space group P*nam*, the centres of the carbon and hydrogen atoms in the planar zig-zag chains are shown by black and open circles respectively; the envelopes of the molecules show clearly the close packing (from *Macromolecular Physics*, Volume 1, by B. Wunderlich, Academic Press, New York and London, 1973). (b) Isotactic polypropylene, space group P$2_1/c$; the senses of the helices are indicated by the ↻ and ↺ symbols (from *Structure and properties of isotactic polypropylene* by G. Natta and P. Corradini, *Nuovo Cimento*, Suppl. to Vol **15** 1, 40, 1960).

Polyethylene $n(CH_2)$ is the simplest polymer, made up of a planar zig-zag chain of carbon atoms, each carbon tetrahedrally coordinated to two hydrogen atoms. Two crystal structures occur, polyethylene I (orthorhombic, space group P*nam*—equivalent to P*nma* by change of axes) and polyethylene II (monoclinic, space group C2/*m*). Polyethylene I is the common, stable form and the arrangement of the chains in the unit cell is shown in Fig. 4.20(a). The zig-zag planes of the chains are at 45° to the unit cell axes and are so arranged that the protrusions of one chain fit into the 'hollow' or recess formed by three neighbouring chains as is also shown in Fig. 4.20(a) by the outlines or the envelopes of the molecules. Screw diad axes of symmetry run in the directions of all three axes in the unit cell—principally along and through the centres of the chains.

In polypropylene (polypropene), $n(CH_2\text{-}CHCH_3)$, the CH_3 side-groups approach too closely for the backbone to remain planar and their efficient packing results in the backbone being twisted into a helical conformation, both right and left handed. In isotactic polypropylene the crystal structure is monoclinic and the space group is $P\ 2_1/c$—the commonest space group of all. Figure 4.20(b) shows a projection of the unit cell perpendicular to the chain axes. The packing together of the helices is dictated by the intermeshing of the CH_3 side-groups and this occurs most efficiently when the rows of helices along the *c*-axis are alternatively right and left handed as shown in Fig. 4.20(b). The packing is, in fact, very close to hexagonal, like a bundle of pencils, and an hexagonal unit cell may also occur. (Figure 10.11(b) shows a fibre photograph of isotactic polypropylene and Exercise 10.4 shows how the orientation of the chains in the unit cell may be determined.)

4.9 Quasicrystals (quasiperiodic crystals or crystalloids)

The 230 space groups represent all the possible combinations of symmetry elements, and therefore all the possible patterns which may be built up by the repetition, without any limit, of the structural units of atoms and molecules which constitute crystals. But real crystals are finite and the atoms or molecules at their surfaces obviously do not have the same environment as those inside. Moreover, crystals nucleate and grow not according to geometrical rules as such but according to the local requirements of atomic and molecular packing, chemical bonding and so on. The resulting repeating pattern or space group is the usual consequence of such requirements, but it is not a necessary one. We will now consider some such cases where 'crystals' nucleate and grow such that the resulting pattern of atoms or molecules is non periodic and does not conform to any of the 230 space groups—in short the three-dimensional analogy to the non-periodic patterns and tilings discussed in Section 2.9. But first we need to adopt a new name for such structures and, following Shechtman[6] can call them **quasiperiodic crystals** or **materials**, or following Mackay[7] call them **crystalloids** or simply, call them **quasicrystals**.

[6] D. Shechtman, I. Blech, D. Gratias, and J.W. Cahn (1984) *Metallic phase with long-range orientational order and no translational symmetry*. Phys. Rev. Lett. **53**, p. 1951.
[7] A.L. Mackay (1976) *De nive quinquangula*, Phys. Bull. p. 495.

(a) (b) (c)

Fig. 4.21. (a) The close packing of 12 spheres around a central sphere as in the ccp structure. The solid is a cubeoctahedron; note that the spheres are not *evenly* distributed round the central sphere, some of the interstices are square, some triangular, (b) The twelve spheres shifted to obtain an *even* distribution; note that the spheres are surrounded by, but not in contact with, five others. (c) The spheres brought together such that they are now in contact; the central sphere is now ~10% smaller. The solid now has the 20 triangular faces of an icosahedron (see also note on pp. 129–30).

We will start 'where we began' in Section 1.1 of this book by model-building with equal size closely packed spheres. In the ccp structure, as we have seen, each sphere is surrounded or coordinated by 12 others as shown in Fig. 4.21(a). The poly-hedron formed around the central sphere is a cubeoctahedron. It is one of the thirteen semi-regular or Archimedean solids (see Sections 3.4 and Appendix 2). However, even though all the spheres are close-packed, they are not all *evenly distributed* around the central sphere the interstices between them are different: some 'square', some 'triangular'.

Now, we can shift the spheres around the central sphere to obtain an even distribution and as we see (Fig. 4.21(b)), this occurs when each sphere is surrounded by (but not now touching) five others and with 'open' triangular interstices between them. (This operation is best carried out by making the central sphere out of soft modelling clay and using pegs sticking from the spheres into the clay to keep them in place.) Finally, we can squeeze the whole model (i.e. compress the central sphere) in our cupped hands to bring all the spheres into contact and make a close-packed shell of 12 spheres as shown in Fig. 4.21(c). We have created icosahedral packing because the solid has the 20 triangular faces of an icosahedron. The central sphere or interstice now has a radius some 10% smaller than the 12 surrounding spheres.

The icosahedron can be extended by adding a second icosahedral shell of 42 atoms, twice the size of the first, then a third shell of 92 atoms, as shown in Fig. 4.22. The atoms in the surfaces of the 20 triangular faces are clearly close-packed but they are not close-packed with the atoms in the underlying shells. As further shells are added, Mackay shows that the packing density converges to 0.688—a value only a little greater

than the packing density for a (monoatomic) bcc structure (see Section 1.5)—which is often the high temperature structure of metals and alloys.

Mackay[8] also shows that the structure of quasicrystals may be understood by an extension of Penrose tiling from two to three dimensions (see Section 2.9), using obtuse and acute rhombohedra (instead of 'wide' and 'narrow' tiles) and corresponding matching rules. Again, the Golden Ratio occurs as the limit of the ratio of the occurrence of these two types of rhombohedra. Icosahedral packing is not the densest packing (cubeoctahedral packing is the densest—packing density 0.7405—see Section 1.4), nor is it crystallographic packing—the non-repeating pattern of the shells of spheres constitutes a crystalloid with point group symmetry $\frac{2}{m}\bar{3}\bar{5}$ indicating the presence of 15 two-fold, 10 three-fold and 6 five-fold axes of symmetry. It is, however, an extremely stable structure (the spheres naturally 'lock' together during the squeezing operation) and it is the basis of Buckminster Fuller's construction of geodesic domes (Section 1.11.6) as well as being characteristic of many virus structures (e.g. the polio virus) which makes them so indestructible.

Icosahedral structures also occur in several metallic alloys, in particular those based on aluminium with copper, iron, ruthenium, manganese, etc. These quasicrystals were first recognized in an Al-25wt% $Mn(Al_6Mn)$ alloy by Dan Shechtman (see Section 11.4.3) from the ten-fold symmetry of their electron diffraction patterns (i.e. five-fold symmetry) plus a centre of symmetry resulting from diffraction—Friedel's

Fig. 4.22. Icosahedral packing of spheres showing close-packing on each of the 20 triangular faces (from 'A dense non-crystallographic packing of equal spheres', by A. L. Mackay (1962), Acta. Cryst. **15**, 916).

[8] A.L. Mackay (1982) *Crystallography and the Penrose Pattern*, Physica **114A**.

Fig. 4.23. A quasicrystal of a 63% Al, 25% Cu-11% Fe alloy showing pentagonal dodecahedral faces (from 'A stable quasicrystal in Al-Cu-Fe system' by An-Pang Tsai, Akihisa Inoue and Tsuyoshi Masumoto, *Jap. J. Appl. Phys.* **26**, 1505, 1987).

law—see Section 9.2). Many such quasicrystals, formed by rapid solidification from the melt, are metastable and revert to crystalline structures on heating, but stable quasicrystals as large as a few millimetres in size have been prepared. Figure 4.23 is a scanning electron micrograph of a 63% Al-25% Cu-11% Fe alloy quasicrystal showing the existence of beautiful pentagonal dodecahedral faces. The pentagonal arrangement of atoms in such a face can be revealed by scanning tunnelling microscopy of carefully prepared surfaces and Fig. 4.24 shows such a (Fourier filtered) image of the quasicrystalline alloy $Al_{70} Pd_{21} Mn_9$.

Icosahedral shells of atoms may also occur as the motif within crystal structures. For example, the alloy $MoAl_{12}$ consists of Mo atoms surrounded by icosahedral shells of 12 Al atoms, the icosahedra themselves being packed together in a bcc array. Similarly, the so-called complex alloy Frank–Caspar phases (with 162 atoms per unit cell) consist of clusters of atoms of icosahedral shape (or parts of icosahedra) which are assembled to form a periodic lattice.

Icosahedral groups of molecules also occur in a number of gas hydrates which can be crystallized in the form of highly hydrated solids, called clathrates. For example, chlorine hydrate, Cl_2nH_2O, has a body-centred cubic structure at the centre and corners of which the water molecules are arranged at the corners of pentagonal dodecahedra—an arrangement analogous to dodecahedrene (see Section 1.11.6). Further water molecules occupy the interstices between four such dodecahedra and the chlorine molecules are 'imprisoned' within this framework—hence the name clathrate, meaning latticed or screened.

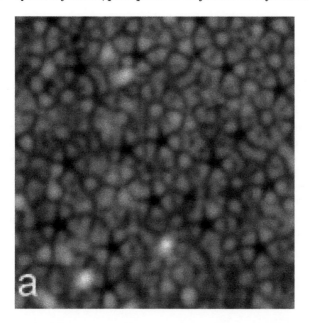

Fig. 4.24. Image (10 nm × 10 nm) of a surface of the alloy Al_{70} Pd_{21} Mn_9 showing the five-fold 'dark star' quasiperiodic pattern. (Photograph by courtesy of Prof. Ronan McGrath, University of Liverpool.)

It was the known existence of such phases which led some crystallographers, Linus Pauling in particular, to question the existence of quasicrystals. The arguments are complex but may be (partly) understood by the tetrahedron models shown in Fig. A1.4 (Appendix 1): a multiply twinned crystal may exhibit a symmetry very close to that of (true) pentagonal symmetry.

Quasicrystallography has an important role in the new science of nanomaterials. As particle size decreases, surface energy terms increasingly dominate and determine the relative stability of different crystal/quasicrystal structures. For example, W, Mo, and Nb, which are all bcc in the bulk, have fcc or hcp structures at particle sizes 5~10 nm. All particles assume polyhedral shapes and small (<5 nm) clusters may be quasicrystalline.

A note on the transformation from crystallographic to quasiperiodic atom packing

The transformation from the cubeoctahedron to the icosahedron (Fig 4.21 and the front cover illustration of this book) may be described in another way. In cubic close-packing the 'middle' ring of six atoms is *planar* with a group of three close-packed atoms above and three below. In icosahedral packing the corresponding ring of six atoms is *puckered* (resulting in a ~10% smaller central cavity) with again three close-packed atoms above and three below. Hence, the transformation may proceed by such small displacive atom movements.

Exercises

4.1 Draw the space group *Pba*2 with the pattern unit ◯ at the following positions:

(a) on the *b* glide plane, i.e. at $\left(\frac{1}{4}y'z'\right)$;

(b) at the intersections of the *a* and *b* glide planes, i.e. at $\left(\frac{1}{4}\frac{1}{4}z'\right)$;

(c) on a diad axis through the origin, i.e. at $(00z')$;

(d) on a diad axis through the mid-points of the cell edges, i.e. at $\left(\frac{1}{2}0z'\right)$.

Hence, show that only (c) and (d) constitute special positions.

4.2 Make and examine the crystal models of NaCl, CsCl, diamond, ZnS (sphalerite), ZnS (wurtzite), Li_2O or CaF_2 (fluorite), $CaTiO_3$ (perovskite). Identify the Bravais lattice and describe the motif of each structure.

4.3 Figure 1.33 shows the pattern of silicon atoms in one enantiomorphous form of α-quartz (trigonal) and the corresponding form in β-quartz. Identify the space groups of these two forms and describe the structural relationships between them.

5

Describing lattice planes and directions in crystals: Miller indices and zone axis symbols

5.1 Introduction

In previous chapters we have described the distributions of atoms in crystals, the symmetry of crystals, and the concept of Bravais lattices and unit cells. We now introduce what are essentially shorthand notations for describing directions and planes in crystals (whether or not they correspond to axes or planes of symmetry). The great advantages of these notations are that they are short, unambiguous and easily understood. For example, the direction (or zone axis) symbol for the 'corner-to-corner' (or triad axis) directions in a cube is simply ⟨111⟩. The plane index (or Miller index) for the faces of a cube is simply {100} or of an octahedron {111} (Fig. 4.1). The various faces and the directions of their intersections in crystals such as those illustrated in Fig. 4.3 can also be precisely described using these notations. Without them one would have to resort to carefully scaled drawings or projections.

Now direction symbols and plane indices are based upon the crystal axes or lattice vectors which outline or define the unit cell (see Section 3.2) and the only ambiguities which can arise occur in those cases in which different unit cells may be used. For example, crystals with the cubic F Bravais lattice may be described in terms of the 'conventional' face-centred cell (Fig. 1.6) or in terms of the primitive rhombohedral cell (Fig. 1.7). Because the axes are different, the direction symbols and plane indices will also be different. Hence it is important to know (1) which set of crystal axes, or which unit cell, is being used and (2) how to change or transform direction symbols and plane indices when the set of crystal axes or the unit cell is changed. This topic is covered in Section 5.8. It is a serious problem only in the case of the trigonal system for crystals with a rhombohedral lattice where there are two almost equally 'popular' unit cells—unlike, say, the rhombohedral cells for the cubic F and cubic I lattices which are rarely used. In addition, in the trigonal and hexagonal systems it is possible to introduce, because of symmetry considerations, a fourth axis, giving rise to 'Miller-Bravais' plane indices and 'Weber' direction symbols, each of which consist of four, rather than three, numbers. This topic is covered in Section 5.7, but first the concept of a zone and zone axis needs to be explained, a topic which is covered in more detail in Section 5.6.

A **zone** may be defined as 'a set of faces or planes in a crystal whose intersections are all parallel'. The common direction of the intersections is called the **zone axis.** All directions in crystals are zone axes, so the terms 'direction' and 'zone axis' are

The Basics of Crystallography and Diffraction. Fourth Edition. Christopher Hammond.

synonymous. So much for the definition. The concept of a zone is readily understood by examining an ordinary pencil. The six faces of a pencil all form or lie in a zone because they all intersect along one direction—the pencil lead direction—which is the zone axis for this set of faces. The number of faces in a zone is not restricted and the faces need not be crystallographically equivalent. For example, the edges of a pencil may be shaved flat to give a 12-sided pencil, i.e. an additional six faces in the zone. Or consider an orthorhombic crystal (Fig. 3.5(b)), or a matchbox. Each crystal axis is the zone axis for four faces, or two crystallographically equivalent pairs of faces. Each face lies in two zones; for example, the 'top' and 'bottom' faces of Fig. 3.5(b) lie in the zones which have the x- and y-directions as zone axes. In general, a face or plane in a crystal belongs to a whole 'family' of zones, the zone axes of which lie in or are parallel to the face.

Verbal definitions are frequently rather clumsy in relation to the simple concepts which they seek to express. Take a piece of paper and draw on it some parallel lines. Fold the paper along the lines—and there you have a zone!

5.2 Indexing lattice directions—zone axis symbols

First, the direction whose symbol is to be determined must pass through the origin of the unit cell. Consider the unit cell shown in Fig. 5.1, which has unit cell edge vectors **a, b, c** (which are not necessarily orthogonal or equal in length)[1]. The steps for determining the zone axis symbol for the direction OL are as follows. Write down the coordinates of a point—any point—in this direction—say P—in terms of fractions of the lengths a, b and c, respectively. The coordinates of P are $\frac{1}{2}$, 0, 1. Now express these fractions as the ratio of whole numbers and insert them into [square] brackets without commas; hence [102].

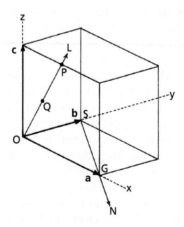

Fig. 5.1. Primitive unit cell of a lattice defined by unit cell vectors **a, b, c**. OL and SN are directions [102] and [1$\bar{1}$0] respectively.

[1] See Appendix 5 for a simple introduction to vectors.

This is then the direction or zone axis symbol for OL. Notice that if we had chosen a different point along OL—say Q—with coordinates $\frac{1}{4}$, 0, $\frac{1}{2}$, we would have obtained the same result.

Now consider the direction SN. To find its direction symbol the origin must be shifted from O to S. Proceeding as before (e.g. finding the coordinates of G with respect to the origin at S) gives the direction symbol [1$\bar{1}$0] (pronounced one bar-one oh), the bar or minus sign referring to a coordinate in the negative sense along the crystal axis.

Directions in crystals are, of course, vectors, which may be expressed in terms of components on the three unit cell edge or 'base' vectors **a, b** and **c**. In the above example the direction OL is written

$$\mathbf{r}_{102} = 1\mathbf{a} + 0\mathbf{b} + 2\mathbf{c}.$$

The general symbol for a direction is [uvw] or, written as a vector,

$$\mathbf{r}_{uvw} = u\mathbf{a} + v\mathbf{b} + w\mathbf{c}.$$

The direction symbols for the unit cell edge vectors **a, b** and **c** are [100], [010] and [001], and very often these symbols are used in preference to the terms x-axis, y-axis and z-axis.

5.3 Indexing lattice planes—Miller indices

First, for reasons which will be apparent shortly, the lattice plane whose index is to be determined must not pass through the origin of the unit cell, or rather the origin must be shifted to a corner of the cell which does not lie in the plane. Consider the unit cell in Fig. 5.2 (identical to Fig. 5.1 but drawn separately to avoid confusion). We shall determine the index of the lattice plane which is shaded and outlined by the letters RMS. It is important first to realize that this plane extends indefinitely through the crystal; the shaded area is simply that portion of the plane that lies within the unit cell of Fig. 5.2.

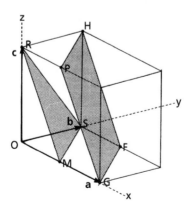

Fig. 5.2. Primitive unit cell (identical to Fig. 5.1), showing the first two planes, RMS and PGFH, in the family. These planes are shaded within the confines of the unit cell.

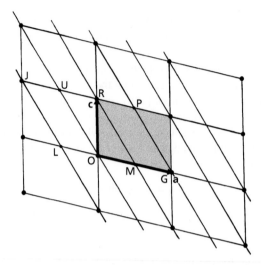

Fig. 5.3. Sketch of Fig. 5.2 with the **a** and **c** unit cell vectors in the plane of the paper (**b** into the plane of the paper) showing traces JL, UO, RM, PG of a family of planes.

It is also important to realize that we are not just considering *one* plane but a whole *family* of identical, parallel planes passing through the crystal. The next plane 'up' in the family is also shaded within the confines of the unit cell and is outlined by the letters PGFH. There is a whole succession of such planes, including one which passes through the origin of the unit cell and they all pass through successive 'layers' of lattice points, hence the name, lattice planes.

A two-dimensional sketch (Fig. 5.3) of Fig. 5.2, with the x- and z-axes in the plane of the paper, and showing the traces of these planes extending into neighbouring cells, will make this clear. Figure 5.3 also shows that all the planes in the family are identical in that they contain the same number or sequence of lattice points.

For RMS, the plane in the family nearest the origin, write down the *intercepts* of the plane on the axes or unit cell vectors **a, b, c,** respectively; they are $\frac{1}{2}a, 1b, 1c$. Expressed as fractions of the cell edge lengths we have $\frac{1}{2},1,1$. Now take the *reciprocals* of these fractions, and put the whole numbers into (round) brackets without commas; hence (211). This is the **Miller index** of the plane, so-called after the crystallographer, W. H. Miller,* who first devised the notation. Proceeding similarly for the next plane in the family, PGFH (and considering its extension beyond the confines of the unit cell), we have intercepts $1a,2b,2c$, which expressed as fractions becomes 1,2,2; the reciprocals of which are $1,\frac{1}{2},\frac{1}{2}$, which expressed as whole numbers again gives the Miller index (211). The plane through the origin, OU (Fig. 5.3), has intercepts 0, 0, 0 which gives an indeterminate 'Miller index' ($\infty\infty\infty$), but this is merely expressive of the fact that another corner of the unit cell must be selected as the origin. The plane JL (Fig. 5.3)

* Denotes biographical notes available in Appendix 3.

in the same family lies on the opposite side of the origin from RM and has intercepts –
$\frac{1}{2}a$,-$1b$,-$1c$, which gives the Miller index $(\bar{2}\bar{1}\bar{1})$ (pronounced bar-two, bar-one, bar-one),
the bar signs simply being expressive of the fact that the planes are recorded from the
opposite (negative axis) side of the origin.

The general index for a lattice plane is (hkl), i.e. (working backwards), the first plane
in the family from the origin makes intercepts a/h, b/k, c/l on the axes. This provides us
with an alternative method for determining the Miller index of a family of planes. Count
the number of planes intercepted in passing from one corner of the unit cell to the next.
For the family of planes (hkl) the first plane intercepts the x-axis at a distance a/h, the
second at $2a/h$, and so on; i.e. a total of h planes are intercepted in passing from one
corner of the unit cell to the next along the x-axis. Similarly, k planes are intercepted
along the y-axis and l planes along the z-axis.

Miller indices apply not only to lattice planes, as sketched in Figs. 5.2 and 5.3, but
also to the external faces of crystals, where the origin is conventionally taken to be
at the centre of the crystal. The intercepts of a crystal face will be many millions of
lattice spacings from the origin, depending of course on the size of the crystal but the
ratios of the fractional intercepts, and therefore the Miller indices, will be simple whole
numbers as before. This is sometimes expressed as 'The Law of Rational Indices', the
germ of which can be traced back to Haüy—see, for example, his representation of the
relationship between the crystal faces and the unit cell in dog-tooth spar (Fig. 1.2).

When a crystal plane lies parallel to an axis its intercept on that axis is infinity, the
reciprocal of which is zero. For example, the 'front' face of a crystal, i.e. the face which
intersects the x-axis only and is parallel to the y- and z-axes, has Miller index (100); the
'top' face, which intersects the z-axis is (001) and so on. It is useful to remember that
a zero Miller index means that the plane (or face) is parallel to the corresponding unit
cell axis.

Although Miller indices never include fractions they may, when used to describe lat-
tice planes, have common factors and this occurs when the unit cell is non-primitive.
Consider, for example, the lattice planes perpendicular to the x-axis in the cubic F unit
cell (Fig. 5.4). Because of the presence of the face-centring lattice points, lattice planes
are intersected every $\frac{1}{2}a$ distance along the x-axis. The first lattice plane in the family,
shown shaded in Fig. 5.4, makes fractional intercepts $\frac{1}{2}$,∞,∞, on the x-, y- and z-axes.
The Miller index of this family of planes is therefore (200). To refer to them as (100)
would be to ignore the 'interleaving' lattice planes within the unit cell[2]. This distinc-
tion does not apply to the Miller indices of the external crystal faces, which are many
millions of lattice planes from the origin.

The procedure described above for defining plane indices may seem rather odd—
why not simply express indices as fractional intercepts without taking reciprocals?
The Law of Rational Indices gives half a clue, but the full significance can only be
appreciated in terms of the reciprocal lattice (Chapter 6).

[2] In face-centred cells the Miller indices (hkl) describing lattice planes must be all odd or all even (see
Appendix 6). The common factor 2 (as in the example, Fig. 5.4) arises when they are all even.

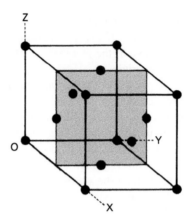

Fig. 5.4. A cubic F unit cell showing (shaded) the first plane from the origin of a family of planes perpendicular to the x-axis but with interplanar spacing $a/2$.

5.4 Miller indices and zone axis symbols in cubic crystals

Miller indices and zone axis symbols may be used to express the symmetry of crystals. This applies to crystals in all the seven systems, but the principles are best explained in relation to cubic crystals because of their high symmetry.

The positive and negative directions of the crystal axes x, y, z can be expressed by the direction symbols (Section 5.2) as [100], [$\bar{1}$00], [010], [0$\bar{1}$0], [001], [00$\bar{1}$]. Because, in the cubic system, the axes are crystallographically equivalent and interchangeable, so also are all these six direction symbols. They may be expressed collectively as ⟨100⟩, the (triangular) brackets implying all six permutations or variants of 1, 0, 0. Similarly, the triad axis corner-to-corner directions are expressed as ⟨111⟩, of which there are eight (four pairs) of variants, namely, [11$\bar{1}$], [$\bar{1}$$\bar{1}$1]; [1$\bar{1}$1], [$\bar{1}1\bar{1}$]; [$\bar{1}$11], [1$\bar{1}$$\bar{1}$]; [111], [$\bar{1}$$\bar{1}$$\bar{1}$]. The diad axis (edge-to-edge) directions are ⟨110⟩, of which there are twelve (six pairs) of variants. For the general direction ⟨uvw⟩ there are forty-eight (twenty-four pairs) of variants.

A similar concept can be applied to Miller indices. The six faces of a cube (with the origin at the centre) are (100), ($\bar{1}$00), (010), (0$\bar{1}$0), (001), (00$\bar{1}$). These are expressed collectively as planes 'of the form' {100}, i.e. in {curly} brackets. Again, for the general plane {hkl} there are forty-eight (twenty-four pairs) of variants.

In cubic crystals, directions are perpendicular to planes with the same numerical indices; for example, the direction [111] is perpendicular to the plane (111), or equivalently it is parallel to the normal to the plane (111). This parallelism between directions and normals to planes with the same numerical indices does not apply to crystals of lower symmetry except in special cases. This will be made clear by considering Figs 5.5(a) and (b), which show plans of unit cells perpendicular to the z-axis of a cubic and an orthorhombic crystal. The traces of the (110) plane and [110] direction are shown in each case. Clearly, the (110) plane and [110] direction are only perpendicular to each

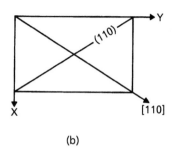

(a) (b)

Fig. 5.5. Plans of (a) cubic and (b) orthorhombic unit cells perpendicular to the z-axis, showing the relationships between planes and zone axes of the same numerical indices.

other in the cubic crystal. In the orthorhombic crystal it is only in the special cases, e.g. (100) planes and [100] directions, that the directions are perpendicular to planes of the same numerical indices.

5.5 Lattice plane spacings, Miller indices and Laue indices

The calculation for lattice plane spacings (also called interplanar spacings or d-spacings), d_{hkl}, is simple in the case of crystals with orthogonal axes. Consider Fig. 5.6, which shows the first plane away from the origin in a family of (hkl) planes. As there is another plane in the family passing through the origin, the lattice plane spacing is simply the length of the normal ON. Angle AON $= \alpha$ (angle between normal and x-axis) and angle ONA $= 90\,°$. Hence

$$\text{OA} \cos \alpha = \text{ON} \quad \text{or} \quad (a/h) \cos \alpha = d_{hkl} \quad \text{or} \quad \cos \alpha = \left(\frac{h}{a}\right) d_{hkl}.$$

Given that β and γ (not shown in Fig. 5.6) are the angles between ON and the y- and z-axes, respectively, then

$$\cos \beta = \left(\frac{k}{b}\right) d_{hkl} \quad \text{and} \quad \cos \gamma = \left(\frac{l}{c}\right) d_{hkl}.$$

For orthogonal axes $\cos^2 \alpha + \cos^2 \beta + \cos^2 \gamma = 1$ (Pythagoras' theorem), hence

$$\left(\frac{h}{a}\right)^2 d_{hkl}^2 + \left(\frac{k}{b}\right)^2 d_{hkl}^2 + \left(\frac{l}{c}\right)^2 d_{hkl}^2 = 1.$$

For a cubic crystal $a = b = c$, hence

$$\frac{1}{d_{hkl}^2} = \frac{h^2 + k^2 + l^2}{a^2}.$$

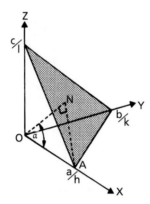

Fig. 5.6. Intercepts of a lattice plane (*hkl*) on the unit cell vectors **a, b, c**. ON $= d_{hkl}$, $=$ interplanar spacing.

Finally, since $\cos\alpha = (h/a)d_{hkl}$, $\cos\beta = (k/a)d_{hkl}$ and $\cos\gamma = (l/a)d_{hkl}$ the ratios of the direction cosines $\cos\alpha : \cos\beta : \cos\gamma$ for a cubic crystal equals the ratios of the Miller indices $h : k : l$.

The concept of lattice planes and interplanar spacings is the basis of the concept of the reciprocal lattice (Chapter 6) and also Bragg's law (Chapter 8)—an equation which every schoolboy (or schoolgirl) knows!

$$n\lambda = 2d_{hkl}\sin\theta$$

where n is the order of reflection, λ is the wavelength, d_{hkl} is the lattice plane spacing and θ is the angle of incidence/reflection to the lattice planes. However, it is very important, when using Bragg's law, to distinguish between lattice planes and **reflecting planes.**

Except in the cases of non-primitive cells discussed above (Section 5.3), indices for lattice planes do not have common factors. However, the indices for reflecting planes frequently do have common factors. They are sometimes called **Laue indices** or **Reflection indices** and are usually written without brackets. Their relationship with Miller indices for lattice planes is best illustrated by way of an example. Apply Bragg's law to the (111) lattice planes in a crystal:

$$\text{first-order reflection } (n = 1) : 1\lambda = 2d_{111}\sin\theta_1$$
$$\text{second-order reflection } (n = 2) : 2\lambda = 2d_{111}\sin\theta_2, \text{ etc.}$$

Now the order of reflection is written on the right-hand side, i.e. for the second-order reflection $(n = 2)$

$$1\lambda = 2\left(\frac{d_{111}}{2}\right)\sin\theta_2.$$

This suggests that second-order reflections from the (111) lattice planes of d-spacing d_{111} can be regarded as first-order reflections from planes of half the spacing, $d_{111}/2$.

Halving the intercepts implies doubling the indices, so these planes are called 222 (no brackets) of d-spacing $d_{222} = d_{111}/2$. These 222 planes are imaginary in the sense that only half of them pass through lattice points, but they are a useful fiction in the sense that the order of reflection, n in Bragg's law, can be omitted. Continuing the above example, third-order reflections from the (111) lattice planes can be regarded as first-order reflections from the 333 reflecting planes (only a third of which in a family pass through lattice points). As mentioned above, these unbracketed indices are sometimes called Laue indices or reflection indices.

However, it should be pointed out that, in practice, when analysing X-ray or electron diffraction patterns, crystallographers very often do not make this distinction between Miller indices and Laue indices, but simply refer, for example, to 333 reflections (no brackets) from (333) 'planes' (with brackets). This should not lead to any confusion, except perhaps in the case of centred lattices where lattice planes may have common factors. For example, the 200 reflecting planes in the cubic F lattice (Fig. 5.4) are also the (200) lattice planes, but the 200 reflecting planes in the cubic P lattice refer to second-order reflections from the (100) lattice planes.

5.6 Zones, zone axes and the zone law, the addition rule

The concept of a zone has been introduced in Section 5.1. In this section some useful geometrical relationships are listed, the proofs of which are given in Section 6.5, in which use is made of reciprocal lattice vectors. It is possible, of course, to prove the relationships given below without making use of the concept of the reciprocal lattice, but the proofs tend to be long, tedious and not very obvious.

5.6.1 The Weiss* zone law or zone equation

If a plane (hkl) lies in a zone $[uvw]$ (i.e. if the direction $[uvw]$ is parallel to the plane (hkl)), then

$$hu + kv + lw = 0.$$

5.6.2 Zone axis at the intersection of two planes

The line or direction of intersection of two planes in a zone $(h_1k_1l_1)$ and $(h_2k_2l_2)$ gives the zone axis, $[uvw]$, where

$$u = (k_1l_2 - k_2l_1); \quad v = (l_1h_2 - l_2h_1); \quad w = (h_1k_2 - h_2k_1).$$

To remember these relationships, use the following 'memogram':

* Denotes biographical notes available in Appendix 3.

$$
\begin{array}{c|c c c c|c}
h_1 & k_1 & & -\,l_1 & & -\,h_1 & & -\,k_1 & l_1 \\
h_2 & k_2 & \times & +\,l_2 & \times & +\,h_2 & \times & +\,k_2 & l_2 \\
\hline
& & u & & v & & w &
\end{array}
$$

Write down the indices twice and strike out the first and last pairs. Then u is given by cross-multiplying, i.e.

$$u = (\text{plus } k_1 l_2 \text{ minus } k_2 l_1),$$

and similarly for v and w.

5.6.3 Plane parallel to two directions

To find the plane lying parallel to two directions $[u_1 v_1 w_1]$ and $[u_2 v_2 w_2]$, write down the 'memogram'

$$
\begin{array}{c|c c c c|c}
u_1 & v_1 & & -\,w_1 & & -\,u_1 & & -\,v_1 & w_1 \\
u_2 & v_2 & \times & +\,w_2 & \times & +\,u_2 & \times & +\,v_2 & w_2 \\
\hline
& & h & & k & & l &
\end{array}
$$

and proceed as above, i.e. $h = (v_1 w_2 - v_2 w_1)$, etc.

5.6.4 The addition rule

Consider two planes $(h_1 k_1 l_1)$ and $(h_2 k_2 l_2)$ lying in zone. Then the index of another plane (HKL) in the zone lying between these planes is given by $H = (mh_1 + nh_2)$; $K = (mk_1 + nk_2)$; $L = (ml_1 + nl_2)$, where m and n are small whole numbers.

The rule enables us to work out the indices of 'in-between' planes in zones. Consider, for example, the plane marked P in Fig. 5.7, which lies both in the zone containing (100) and (011) and the zone containing (101) and (110). We need a choose values of m and n such that adding the pairs of indices as above gives the same result for (HKL). This is much more easily done than said! The plane is (211), i.e.

$$(101) + (110) = (211) \quad (m = 1, n = 1)$$

and

$$(011) + 2(100) = (211) \quad (m = 1, n = 2).$$

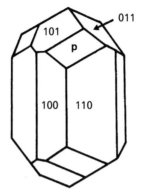

Fig. 5.7. A monoclinic crystal (class *m*) in which the face *P* lies in two zones, one containing (101) and (110), the other containing (011) and (100).

5.7 Indexing in the trigonal and hexagonal systems: Weber symbols and Miller-Bravais indices

As shown in Section 3.3, the rhombohedral and hexagonal lattices both consist of hexagonal layers of lattice points which in the rhombohedral lattice are 'stacked' in the ABCABC... sequence (Fig. 3.3(b)) and in the hexagonal lattice are 'stacked' in the AAA... sequence (Fig. 3.3(a)). In both cases, it is convenient to outline unit cells using the easily recognized hexagonal layers as the 'base' of the cells and with the z-axis (or **c** vector) perpendicular thereto. However, at least three choices of unit cell are commonly made and these are illustrated in the case of the hexagonal lattice in Fig. 5.8 (the corresponding unit cells of the rhombohedral lattice differ only insofar as they contain additional lattice points of the B and C layers at fractional distances of $\frac{1}{3}$ and $\frac{2}{3}$ of the unit edge length along the z-axis).

Figure 5.8(a) shows the primitive hexagonal unit cell, the smallest that can be chosen with the x- and y-axes at 120°. However, this is often inconvenient in that it does not reveal the hexagonal symmetry of the lattice. For example, all the (pencil) faces parallel to the z-axis are crystallographically equivalent or of the same form, but their indices differ in type as shown—some have indices of the type $(1\bar{1}0)$ and some have indices of the type (100). To overcome this problem a *fourth* axis—the *u*-axis—is inserted at 120° to both the x- and y-axes, as shown in Fig. 5.8(b). These are called Miller–Bravais axes x, y, u, z (or vectors **a, b, t, c**) and the indices of the lattice planes, called **Miller–Bravais indices,** now consist of four numbers (*hkil*). The indices for the pencil faces are shown in Fig. 5.8(b) and they are all of the same form $\{10\bar{1}0\}$. Notice that, in all cases the sum of the first three numbers is zero, i.e. $h+k+i=0$. As therefore $i=-(h+k)$, is sometimes simply represented as a dot, i.e. $\{hk.l\}$. However, this unnecessary abbreviation should be discouraged, as it defeats the object of using Miller–Bravais axes in the first place.

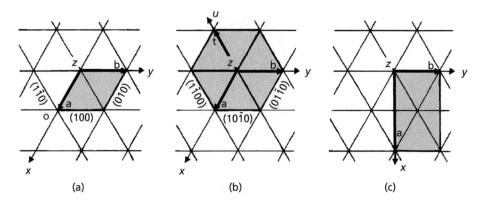

Fig. 5.8. Hexagonal net of the hexagonal P lattice showing (shaded) (a) primitive hexagonal unit cell with the traces of the six prism faces indexed $\{hkl\}$, (b) hexagonal (four-index) unit cell with the traces of the six prism faces indexed $\{hkil\}$, (c) orthohexagonal (*Base* or *C*-centred) unit cell. The z-axis is out of the plane of the page.

Similarly, zone axis symbols, sometimes called **Weber* symbols,** consist of four numbers $\langle UVTW\rangle$. However, determining these numbers (i.e. determining the components of a vector using four base vectors in three-dimensional space) is not straightforward. We cannot convert to 4-index Weber symbols $\langle UVTW\rangle$ from 3-index zone axis symbols $\langle uvw\rangle$ by 'adding' the third symbol T such that $T = -(u+v)$ as we did for the i index in Miller–Bravais indices. We need to derive the relationship between $\langle UVTW\rangle$ and $\langle uvw\rangle$ as follows. For a vector specified in both systems to be identical:

$$u\mathbf{a} + v\mathbf{b} + w\mathbf{c} = U\mathbf{a} + V\mathbf{b} + T\mathbf{t} + W\mathbf{c}.$$

Since the x, y and u axes are at $120°$ to each other, the vector sum, $\mathbf{a} + \mathbf{b} + \mathbf{t} = 0$. Substituting for $\mathbf{t} = -(\mathbf{a} + \mathbf{b})$,

$$u\mathbf{a} + v\mathbf{b} + w\mathbf{c} = U\mathbf{a} + V\mathbf{b} - T(\mathbf{a} + \mathbf{b}) + W\mathbf{c}, \text{ i.e.}$$
$$u\mathbf{a} + v\mathbf{b} + w\mathbf{c} = (U - T)\mathbf{a} + (V - T)\mathbf{b} + W\mathbf{c}.$$

Hence we obtain the identities:

$$u = (U - T); \quad v = (V - T); \quad w = W.$$

To obtain the identities for the reverse transformation we impose the condition that $U + V + T = 0$ (in the same way that $h + k + i = 0$ for Miller–Bravais indices). Eliminating T the identities become:

$$u = 2U + V; \quad v = U + 2V; \quad w = W$$

* Denotes biographical notes available in Appendix 3.

and by adding/subtracting the first two equations we obtain:

$$U = 1/3(2u - v); \quad V = 1/3(2v - u); \quad T = -1/3(u + v); \quad W = w.$$

On this basis the zone axis symbols for the x-, $y-$ and u-axes are $[2\bar{1}\bar{1}0]$, $[\bar{1}2\bar{1}0]$ and $[\bar{1}\bar{1}20]$ respectively.

When using Miller–Bravais indices and Weber symbols the zone law $hu+kv+lw = 0$ becomes (using the above identities)

$$h(U - T) + k(V - T) + lW = 0$$
$$\text{i.e. } hU + kV - (h + k)T + lW = 0$$

and since $i = -(h + k)$ the zone law for Miller–Bravais axes becomes:

$$hU + kV + iT + lW = 0.$$

In order to find the Weber symbol $[U\,V\,T\,W]$ for the zone axis along the intersection of two planes $(h_1\,k_1\,i_1\,l_1)$ and $(h_2\,k_2\,i_2\,l_2)$, first find $[u\,v\,w]$ using the 'memogram' in Section 5.6.2 and then find $[U\,V\,T\,W]$ using the above identities. Alternatively (and more conveniently), use the following 'memogram':

$(2h_1+k_1)$	(h_1+2k_1)	l_1	$(2h_1+k_1)$	(h_1+2k_1)	l_1
$(2h_2+k_2)$	(h_2+2k_2)	l_2	$(2h_2+k_2)$	(h_2+2k_2)	l_2
	U	V		W	

Write down the indices twice and strike out the first and last pairs. Then U is given by cross-multiplying, i.e.

$$U = (h_1 + 2k_1)l_2 - (h_2 + 2k_2)l_1$$

and similarly for V and W. T is simply $-(U + V)$. Conversely, the plane $(h\,k\,i\,l)$ lying parallel to two directions $[U_1\,V_1\,T_1\,W_1]$ and $[U_2\,V_2\,T_2\,W_2]$ is found by using a similar 'memogram', i.e.

$(2U_1+V_1)$	(U_1+2V_1)	W_1	$(2U_1+V_1)$	(U_1+2V_1)	W_1
$(2U_2+V_2)$	(U_2+2V_2)	W_2	$(2U_2+V_2)$	(U_2+2V_2)	W_2
	h	k		l	

Another unit cell—the orthohexagonal cell—is shown in Fig. 5.8(c). The ratio of the lengths of the edges, a/b is $\sqrt{3}$. As with the primitive hexagonal cell this does not reveal the hexagonal symmetry of the lattice. This base-centred cell has the advantage that the axes are orthogonal and is particularly useful in showing the relationships between

the crystals which have similar structures but where small distortions can change the symmetry from hexagonal to orthorhombic, i.e. in situations in which the ratio a/b is no longer precisely $\sqrt{3}$.

To summarize, great care must be taken in interpreting plane indices and zone axis symbols in the hexagonal and trigonal systems—an index such as (111) could refer to the primitive hexagonal or orthohexagonal unit cell, or it could even refer to the Miller–Bravais hexagonal unit cell and be the contracted form of $(11\bar{2}1)$, i.e. (11.1), in which the dot may have been omitted in printing!

5.8 Transforming Miller indices and zone axis symbols

As mentioned in Section 3.3, different choices of axes (i.e. different unit cells) are frequently encountered in trigonal crystals with the rhombohedral Bravais lattice and it is therefore important to know how to transform Miller indices and zone axis symbols from one axis system to the other. The appropriate matrix equations or **transformation matrices** are given in Section 5.9. However, the concept and derivation of transformation matrices are best understood by way of an easily visualized, in effect, 'two-dimensional' example (Fig. 5.9) in which one of the axes (out of the plane of the paper) is common to both cells.

Figure 5.9 shows a plan view of a lattice with two possible unit cells outlined by vectors **a, b, c** and **A, B, C**. The common vectors **c** = **C** lie out of the plane of the paper. The unit cell vectors **A, B, C** can be expressed as components of **a, b, c**:

$$\mathbf{A} = 1\mathbf{a} + 2\mathbf{b} + 0\mathbf{c}$$
$$\mathbf{B} = \bar{1}\mathbf{a} + 1\mathbf{b} + 0\mathbf{c}$$
$$\mathbf{C} = 0\mathbf{a} + 0\mathbf{b} + 1\mathbf{c}.$$

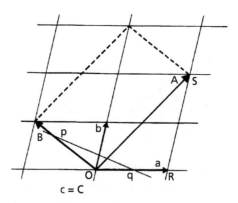

Fig. 5.9. Alternative unit cells in a lattice defined by unit cell vectors **a, b, c** and **A, B, C**.

Now there are two ways of writing these three equations in matrix form; the vectors may either by written as column matrices:

$$\begin{pmatrix} \mathbf{A} \\ \mathbf{B} \\ \mathbf{C} \end{pmatrix} = \begin{pmatrix} 1 & 2 & 0 \\ \bar{1} & 1 & 0 \\ 0 & 0 & 1 \end{pmatrix} \begin{pmatrix} \mathbf{a} \\ \mathbf{b} \\ \mathbf{c} \end{pmatrix}$$

or as row matrices:

$$(\mathbf{A}\,\mathbf{B}\,\mathbf{C}) = (\mathbf{a}\,\mathbf{b}\,\mathbf{c}) \begin{pmatrix} 1 & \bar{1} & 0 \\ 2 & 1 & 0 \\ 0 & 0 & 1 \end{pmatrix}.$$

Notice that the rows and columns of the 3×3 matrix are transposed.

Let (hkl) and $[uvw]$ be the Miller indices and zone axis symbols referred to the primitive unit cell $\mathbf{a}, \mathbf{b}, \mathbf{c}$ and (HKL) and $[UVW]$ be the Miller indices and zone axis symbols referred to the 'large' unit cell $\mathbf{A}, \mathbf{B}, \mathbf{C}$ (which has three lattice points per cell).

Consider first the transformation between the indices (hkl) and (HKL). Figure 5.9 shows the trace pq of the first plane from the origin in a family of planes in the lattice. By definition, the intercepts of this plane are a/h on the \mathbf{a} lattice vector, A/H on the \mathbf{A} lattice vector, b/k on the \mathbf{b} lattice vector and B/K on the \mathbf{B} lattice vector. Or, recall the equivalent definition that h is the number of planes intersected along the \mathbf{a} lattice vector, H is the number intersected along the \mathbf{A} lattice vector, and so on; h is the number intersected along \mathbf{a} (i.e. 0 to R) and $2k$ is the number intersected along $2\mathbf{b}$ (i.e. R to S). Hence, $h + 2k$ is the total number of planes intersected along $1\mathbf{a} + 2\mathbf{b}$ (i.e. O to S) and, as $\mathbf{A} = 1\mathbf{a} + 2\mathbf{b}$, then this number is the H index. Hence

$$H = 1h + 2k + 0l$$
$$K = \bar{1}h + 1k + 0l$$
$$L = 0h + 0k + 1l,$$

and the indices transform in the same way as the unit cell vectors, and again can be written in matrix form as column matrices or row matrices. We shall choose the *row* matrix form, viz.

$$(HKL) = (hkl) \begin{pmatrix} 1 & \bar{1} & 0 \\ 2 & 1 & 0 \\ 0 & 0 & 1 \end{pmatrix}.$$

The relationship between $[uvw]$ and $[UVW]$ may be derived as follows. A vector \mathbf{r} in the lattice is written in terms of its components on the two sets of lattice vectors, i.e.

$$\mathbf{r} = u\mathbf{a} + v\mathbf{b} + w\mathbf{c} = U\mathbf{A} + V\mathbf{B} + W\mathbf{C}.$$

Substituting for \mathbf{A}, \mathbf{B} and \mathbf{C}:

$$u\mathbf{a} + v\mathbf{b} + w\mathbf{c} = U(1\mathbf{a} + 2\mathbf{b} + 0\mathbf{c}) + V(\bar{1}\mathbf{a} + 1\mathbf{b} + 0\mathbf{c}) + W(0\mathbf{a} + 0\mathbf{b} + 1\mathbf{c}).$$

Collecting terms together for u, v and w gives

$$u = 1U + \bar{1}V + 0W$$
$$v = 2U + 1V + 0W$$
$$w = 0U + 0V + 1W.$$

Writing the zone axis symbols as *column* matrices gives

$$\begin{pmatrix} u \\ v \\ w \end{pmatrix} = \begin{pmatrix} 1 & \bar{1} & 0 \\ 2 & 1 & 0 \\ 0 & 0 & 1 \end{pmatrix} \begin{pmatrix} U \\ V \\ W \end{pmatrix}.$$

Hence the *same* 3 × 3 matrix relates plane indices written as *row* matrices from the primitive to 'large' unit cell and zone axis symbols written as *column* matrices from the 'large' to the primitive unit cell.

The reverse relationships are found from the inverse of the matrix. The procedures for inverting a matrix and for finding its determinant may be found in many A-level mathematics textbooks. A simple 'memogram' method for a 3 × 3 matrix is given in Section 5.10. For this example the inverse matrix is given by

$$\begin{pmatrix} 1 & \bar{1} & 0 \\ 2 & 1 & 0 \\ 0 & 0 & 1 \end{pmatrix}^{-1} = \frac{1}{3} \begin{pmatrix} 1 & 1 & 0 \\ \bar{2} & 1 & 0 \\ 0 & 0 & 3 \end{pmatrix};$$

hence

$$(hkl) = (HKL)\frac{1}{3} \begin{pmatrix} 1 & 1 & 0 \\ \bar{2} & 1 & 0 \\ 0 & 0 & 3 \end{pmatrix}$$

and

$$\begin{pmatrix} U \\ V \\ W \end{pmatrix} = \frac{1}{3} \begin{pmatrix} 1 & 1 & 0 \\ \bar{2} & 1 & 0 \\ 0 & 0 & 3 \end{pmatrix} \begin{pmatrix} u \\ v \\ w \end{pmatrix}.$$

Finally, note that the determinant of the matrix equals 3 (and its inverse equals $\frac{1}{3}$), the ratio of the volumes, or number of lattice points, in the unit cells.

In transforming indices and zone axis symbols from one set of crystal axes to another it is usually best to work from first principles as above because it is very easy to fall into error by using a transformation matrix without knowing which conventions for writing lattice planes and directions are being used.

5.9 Transformation matrices for trigonal crystals with rhombohedral lattices

It is often convenient to describe the rhombohedral R lattice in terms of a large **non-primitive** hexagonal unit cell with lattice vectors **A** and **B** at 120° to each other and **C** perpendicular thereto, rather than as a primitive cell with equi-inclined lattice vectors **a**, **b**, and **c**. The relationship between these two unit cells is shown in Fig. 5.10. Proceeding as described in Section 5.8,

$$(HKL) = (hkl) \begin{pmatrix} 1 & 0 & 1 \\ \bar{1} & 1 & 1 \\ 0 & \bar{1} & 1 \end{pmatrix}; \quad \begin{pmatrix} u \\ v \\ w \end{pmatrix} = \begin{pmatrix} 1 & 0 & 1 \\ \bar{1} & 1 & 1 \\ 0 & \bar{1} & 1 \end{pmatrix} \begin{pmatrix} U \\ V \\ W \end{pmatrix}$$

$$(hkl) = (HKL) \begin{pmatrix} \frac{2}{3} & \frac{\bar{1}}{3} & \frac{\bar{1}}{3} \\ \frac{1}{3} & \frac{1}{3} & \frac{\bar{2}}{3} \\ \frac{1}{3} & \frac{1}{3} & \frac{1}{3} \end{pmatrix}; \quad \begin{pmatrix} U \\ V \\ W \end{pmatrix} = \begin{pmatrix} \frac{2}{3} & \frac{\bar{1}}{3} & \frac{\bar{1}}{3} \\ \frac{1}{3} & \frac{1}{3} & \frac{\bar{2}}{3} \\ \frac{1}{3} & \frac{1}{3} & \frac{1}{3} \end{pmatrix} \begin{pmatrix} u \\ v \\ w \end{pmatrix},$$

where (HKL), $[UVW]$ and (hkl), $[uvw]$ refer to the Miller indices and zone axes symbols for the hexagonal and rhombohedral unit cells, respectively. Note that the determinants of these two matrices are 3 and $\frac{1}{3}$, respectively, i.e. equal to the ratios of the number of lattice points in the two cells. Hence the hexagonal unit cell is known as a triple hexagonal cell because it contains three lattice points per cell. Note also that it is possible to choose the hexagonal **A** and **B** axes differently (e.g. rotated 60° to those shown

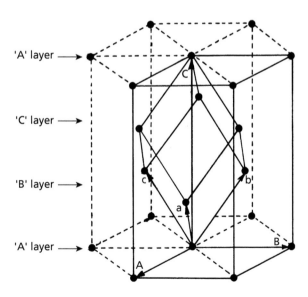

Fig. 5.10. The relationship between the rhombohedral R unit cell, lattice vectors **a, b, c**, and the (large, non-primitive) hexagonal unit cell, lattice vectors **A, B**, and **C**. The 'ABCA...' stacking of the layers of lattice points is indicated (see also Fig. 3.3b).

in Fig. 5.10); this will give rhombohedral unit cells that are mirror-related, sometimes known as 'obverse' and 'reverse' unit cells.

The corresponding Miller–Bravais indices and Weber symbols are determined as described in Section 5.7. For example the Miller–Bravais indices $(HKIL)$ are simply given by $H = (h - k)$; $K = (k - l)$; $I = (l - h)$, and $L = (h + k + l)$.

We showed (see Section 3.3 and Table 3.1) that the trigonal system includes crystals with **either** the rhombohedral R lattice **or** the hexagonal P lattice. When trigonal crystals are described in terms of a hexagonal unit cell, how are we to distinguish whether the underlying lattice is rhombohedral R or hexagonal P? We do so by considering the condition for the presence (or systematic absences) on the indices as described for the cubic lattices in Section A6.1 and Table A6.2. For trigonal crystals with the hexagonal P Bravais lattice, there are (as for cubic P) no conditions on the indices. For those with the rhombohedral R Bravais lattice (as in Fig. 5.10), we note that the sum $(-H + K + L)$ $= -(h - k) + (k - l) + (h + k + l) = 3k$, i.e. the sum is always a multiple of 3.

5.10 A simple method for inverting a 3 × 3 matrix

This method has the advantage that we do not need to memorize a sign convention. We will use the matrix derived in Section 5.8 as an example:

Step 1: Write out the matrix (top left) and repeat it three times as if it were the 'motif' of a pattern, thus:

$$
\begin{array}{cccccc}
1 & \bar{1} & 0 & 1 & \bar{1} & 0 \\
2 & 1 & 0 & 2 & 1 & 0 \\
0 & 0 & 1 & 0 & 0 & 1 \\
1 & \bar{1} & 0 & 1 & \bar{1} & 0 \\
2 & 1 & 0 & 2 & 1 & 0 \\
0 & 0 & 1 & 0 & 0 & 1 \\
\end{array}
$$

Step 2: Find the co-factor for each term. We will start with the term 1 in the first row and first column. 'Look down' towards the right (as indicated by the arrow) at the group of four terms (shown inside the dashed box) $\begin{array}{cc} 1 & 0 \\ 0 & 1 \end{array}$

Cross-multiply these terms following the 'memogram' procedure (Section 5.6),

i.e. $\begin{array}{cc} 1 & 0 \\ & \\ 0 & 1 \end{array}$

The result, 1, is the co-factor. Repeat this procedure for all the other terms—for each term 'look down' towards the right, identify the group of four terms and cross-multiply them. For example, for $\bar{1}$, the second term in the first row, the group of four terms is

$$\begin{array}{cc} 0 & 2 \\ 1 & 0 \end{array}$$ and the cofactor is $\bar{2}$.

Step 3: Write down, following the above procedure, the matrix of co-factors, which is:

$$\begin{array}{ccc} 1 & \bar{2} & 0 \\ 1 & 1 & 0 \\ 0 & 0 & 3 \end{array}$$

Step 4: The determinant, det, of the matrix is the sum of the terms in *any* row or *any* column multiplied by their co-factors. For example, for the first row:

$$\det = 1.1 + 1.2 + 0.0 = 3$$

and for the second row:

$$\det = 2.1 + 1.1 + 0.0 = 3$$

and for the second column:

$$\det = 1.2 + 1.1 + 0.0 = 3$$

(repeating in this way serves as a useful cross-check to see that you have not made any mistakes, particularly with signs).

Step 5: The inverse of the matrix is found by transposing the matrix of co-factors (i.e. interchanging rows and columns) and by dividing by the determinant, hence:

$$\frac{1}{3}\begin{pmatrix} 1 & 1 & 0 \\ \bar{2} & 1 & 0 \\ 0 & 0 & 3 \end{pmatrix} \quad \begin{array}{c} \text{which is} \\ \text{the inverse} \\ \text{of the matrix} \end{array} \quad \begin{pmatrix} 1 & \bar{1} & 0 \\ 2 & 1 & 0 \\ 0 & 0 & 1 \end{pmatrix}.$$

Exercises

5.1 Write down the Miller indices and zone axis symbols for the slip planes and slip directions in fcc and bcc crystals. (See also Exercise 1.5.)

5.2 Which of the directions [010], [432], [$\bar{2}$10], [23$\bar{1}$], if any, lie parallel to the plane (115)? Which of the planes (112), (321), ($\bar{9}$11$\bar{2}$), (1$\bar{1}$1), if any, lie parallel to the direction [1$\bar{1}\bar{1}$]?

5.3 Write down the planes whose normals are parallel to directions with the same numerical indices in the triclinic, monoclinic and tetragonal systems.

5.4 Using the 'memograms' on pp. 143–144, find the plane which lies parallel to the directions [131] and [0$\bar{1}$1]. Find the plane which lies parallel to the directions [102] and [$\bar{1}$11]. Find the direction which lies parallel to the intersection of the planes (342) and (10$\bar{3}$). Find

the direction which lies parallel to the intersection of the planes $(21\bar{3})$ and (110). Check your answers by using the zone law.

5.5 An orthorhombic crystal (cementite, Fe_3C) has unit cell vectors (or lattice parameters) of lengths $a = 452$ pm, $b = 508$ pm, $c = 674$ pm.

 (a) Find the d-spacings of the following families of planes: (101), (100), (111) and (202).

 (b) Find the angles α, β, γ (Fig. 5.6) between the normal to the plane (111) and the three crystal axes.

 (c) Find the angles p, q, r between the $[111]$ direction and the three crystal axes. Briefly explain why these angles are not the same as those in (b).

5.6 Figure 3.2 shows the cubic I and cubic F lattices and the corresponding primitive rhombohedral unit cells. Study these figures carefully, then close the book and redraw them for yourself—to ensure that you understand the geometries of the cells in each case. Assign unit cell vectors \mathbf{A}, \mathbf{B}, \mathbf{C} to the primitive rhombohedral cells and unit cell vectors \mathbf{a}, \mathbf{b}, \mathbf{c} to the cubic I and cubic F cells and derive the transformation matrices for indices $(hkl) \leftrightarrows (HKL)$ and direction symbols $[UVW] \leftrightarrows [uvw]$ in each case.

5.7 Draw the directions (zone axes) $[001]$, $[010]$, $[210]$, $[110]$ in a hexagonal unit cell, and determine their Weber zone axis symbols, $\langle UVTW \rangle$.

5.8 Using the 'memograms' on p. 147, find the direction (Weber symbol) $[U\,V\,T\,W]$ which lies parallel to the intersection of planes $(1\,2\,\bar{3}\,1)$ and $(0\,\bar{1}\,1\,2)$. Find the plane $(h\,k\,i\,l)$ which lies parallel to the directions $[3\,\bar{1}\,\bar{2}\,1]$ and $[1\,0\,\bar{1}\,\bar{2}]$. Check your answers by using the zone law for Miller–Bravais axes.

5.9 In an hcp crystal, sketch the traces of the (0002) and $\{10\bar{1}1\}$ planes in a $\langle \bar{1}2\bar{1}0 \rangle$ zone. Note the near-hexagonal arrangement of these planes around the zone axis. Determine the value of the axial ratio c/a, for which the arrangement of these planes is exactly hexagonal.

6

The reciprocal lattice

6.1 Introduction

The **reciprocal lattice** is often regarded by students of physics as a geometrical abstraction, comprehensible only in terms of vector algebra and difficult diffraction theory. It is, in fact, a very simple concept and therefore a very important one. It provides a simple geometrical basis for understanding not only the geometry of X-ray and electron diffraction patterns but also the behaviour of electrons in crystals—reciprocal space being essentially identical to 'k-space'.

The concept of the reciprocal lattice may be approached in two ways. First, reciprocal lattice unit cell vectors may be defined in terms of the (direct) lattice unit cell vectors **a**, **b**, **c**, and the geometrical properties of the reciprocal lattice developed therefrom. This is certainly an elegant approach, but it very often fails to provide the student with an immediate understanding of the relationships, for example, between the reciprocal lattice of a crystal and the diffraction pattern. The second approach, which is the one adopted in this chapter, is much less elegant. It develops the notion that families of planes in crystals can be represented simply by their normals, which are then specified as (reciprocal lattice) vectors and which can then be used to define a pattern of (reciprocal lattice) points, each (reciprocal lattice) point representing a family of planes. The advantage of this approach is that it accentuates the connections between families of planes in the crystals, Bragg's law and the directions of the diffracted or reflected beams.

The notion that crystal axes can be defined in terms of the *normals* to crystal faces belongs to Bravais* (1850) who called them 'polar axes' (not to be confused with current usage of the term which means axes or directions whose ends are not related by symmetry). However, the credit for the development of this idea to the concept of the reciprocal lattice, and the application of this concept to the analysis of X-ray diffraction patterns, belongs to P. P. Ewald,* whose story is told in Section 8.1 and Appendix 3.

A study of the reciprocal lattice does require an elementary knowledge of vectors, their symbolism and some simple rules governing their manipulation (vector algebra). If you are unfamiliar with these topics, or wish to refresh your memory, Appendix 5 provides all the information you will require in order to follow the rest of this chapter.

6.2 Reciprocal lattice vectors

Consider a family of planes in a crystal (for example, those in Figs 5.2 or 5.3). Geometrically, the planes can be specified by two quantities: (1) their orientation in the crystal and (2) their d-spacings. Now, the orientation of the planes is given or defined by the

* Denotes biographical notes available in Appendix 3.

The Basics of Crystallography and Diffraction. Fourth Edition. Christopher Hammond.
© Christopher Hammond 2015. Published in 2015 by Oxford University Press.

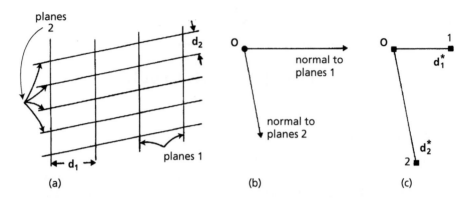

Fig. 6.1. (a) Traces of two families of planes 1 and 2 (perpendicular to the plane of the paper), (b) the normals to these families of planes drawn from a common origin and (c) definition of these planes in terms of the reciprocal (lattice) vectors \mathbf{d}_1^* and \mathbf{d}_2^*, where $|\mathbf{d}_1^*| = K/d_1$, $|\mathbf{d}_2^*| = K/d_2$, K being a constant.

direction of their normals. In Fig. 6.1(a) two families of planes, simply labelled 'planes 1' and 'planes 2', are sketched 'edge on'. Notice that the planes do not have the same d-spacings and are not at right angles to one another—in short we are starting with a completely general case. In Fig. 6.1(b) the normals to these planes are drawn from a common origin and these specify the orientation of the planes but as yet give no information about their d-spacings. Now the d-spacings, d_1 and d_2 need to be specified. An 'obvious' way of doing this might be to make the lengths of the normals directly proportional to the d-spacings, i.e. by expressing them as *vectors* with moduli equal to Kx (d-spacing) where K is a proportionality constant. However, this is *not* the way; instead, we make the lengths or the moduli of the vectors *inversely* proportional to the d-spacings, i.e. equal to K/d-spacing (where K is a proportionality constant, taken as unity or, in X-ray or electron diffraction, as λ, the X-ray or electron wavelength), i.e. a *longer* vector, indicating a *smaller* d-spacing. The reason for making the moduli of the vectors inversely proportional to the d-spacings will be apparent shortly (recall also the inversion of the intercepts in the definition of Miller indices). These vectors are called **reciprocal (lattice) vectors,** symbols \mathbf{d}_1^* and \mathbf{d}_2^*, and are shown in Fig. 6.1(c). The 'end points' of the vectors (called reciprocal lattice points) are labelled 1 and 2, corresponding to the planes which they represent, and are usually denoted by little squares (as in Fig. 6.1(c)) instead of arrow-heads. Reciprocal lattice vectors have dimensions of 1/length (for $K = 1$), i.e. reciprocal Ångstroms, Å$^{-1}$, or reciprocal picometres, pm^{-1}. For example, if $d_1 = 0.5$ Å, the length or modulus of \mathbf{d}_1^* (for $K = 1$) is

$$|\mathbf{d}_1^*| = \frac{1}{0.5\,\text{Å}} = 2\text{Å}^{-1}.$$

This completes the simple definition of reciprocal lattice vectors. But, like all simple concepts, it is capable of great development, which will be done for particular examples in the sections below. But first we need to show that the points do indeed form a grid

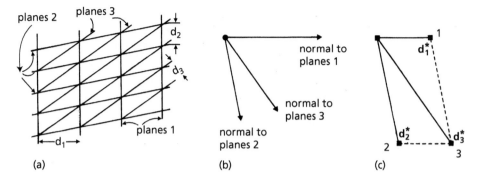

Fig. 6.2. As Fig. 6.1, showing in Fig. 6.2(a) a third set of intersecting planes (planes 3), their normals in Fig. 6.2(b) and their reciprocal lattice vectors in Fig. 6.2(c). Note that $\mathbf{d}_1^* + \mathbf{d}_2^* = \mathbf{d}_3^*$ and that the reciprocal lattice points do form a lattice.

or lattice—we have so far only two such points (plus an origin) in Fig. 6.1(c). Let us therefore add a third set of planes—'planes 3' to Fig. 6.1(a), as shown separately for clarity in Fig. 6.2(a). All the planes have common points of intersection. Figure 6.2(b) shows in addition the normal to planes 3 and Fig. 6.2(c) shows the reciprocal lattice vector \mathbf{d}_3^* and reciprocal lattice point 3. Without any calculations or measurement we can see intuitively that the reciprocal lattice points 1, 2 and 3 do indeed form a lattice as indicated by the dashed lines and that, by vector addition, $\mathbf{d}_1^* + \mathbf{d}_2^* = \mathbf{d}_3^*$. If you are not immediately convinced (and even if you are), it is worthwhile sketching some planes (of arbitrary spacings and orientations) as in Fig. 6.2(a) and then proceeding to draw their reciprocal lattice vectors as in Fig. 6.2(c). But first a practical note. In drawing crystals, and in sketching crystal planes, we have to use a 'map scale' in 'direct space', choosing a scale to fit our piece of paper, e.g. 1 Å equals 1 cm. So also when drawing reciprocal lattice vectors we have to use a quite separate 'map scale' in 'reciprocal space', e.g. 1 Å^{-1} equals 1 cm or 10 cm or 1 inch, as convenient. Strictly speaking we should add 'scale bars' to drawings of crystals and crystal planes as we do for micrographs, giving the scale of the drawing in terms of Å or nm or μ m, as appropriate, and quite different 'scale bars' to drawings of reciprocal lattices (as we ought to do for diffraction patterns) giving the scale of the reciprocal lattice/diffraction pattern in terms of reciprocal units; Å^{-1}, or nm^{-1} or μm^{-1} as appropriate.

The five plane lattices (Fig. 2.4(a)) also have their reciprocal lattice counterparts as is shown in Fig. 6.3—a comparison which further emphasizes the reciprocal relationships between them. The unit cells of the plane lattices are specified by the lattice vectors **a** and **b** and the unit cells of the reciprocal lattices are specified by the reciprocal lattice vectors **a*** and **b***, as explained in Section 6.3 below.

6.3 Reciprocal lattice unit cells

To avoid the pitfalls of making hasty assumptions about the relationships between reciprocal and (direct) lattice vectors, a monoclinic crystal will be used as an example.

<placeholder>
158 The reciprocal lattice
</placeholder>

ok

<render>
<header>
</header>
</render>

<p></p>

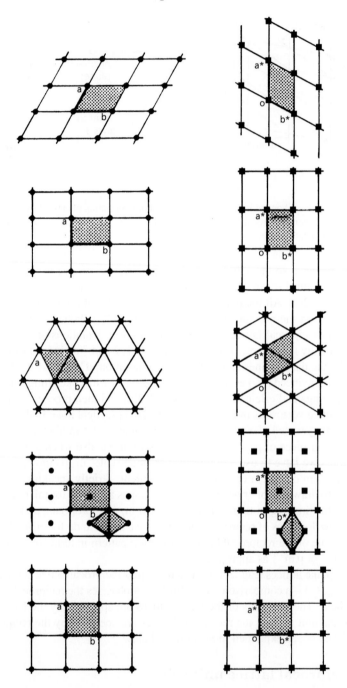

Fig. 6.3. The five plane lattices (left) and plane reciprocal lattices (right) indicating the corresponding unit cells, lattice vectors **a** and **b** and reciprocal lattice vectors **a*** and **b***. (The rectangular centred reciprocal lattice is drawn at half scale.)

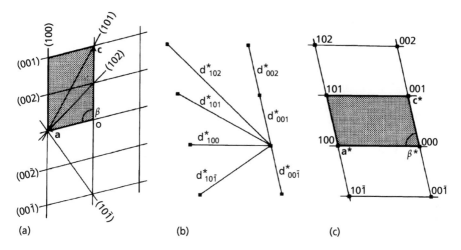

Fig. 6.4. (a) Plan of a monoclinic *P* unit cell perpendicular to the *y*-axis with the unit cell shaded. The traces of some planes of type {*h0l*} (i.e. parallel to the *y*-axis) are indicated, (b) the reciprocal (lattice) vectors, \mathbf{d}^*_{hkl} for these planes and (c) the reciprocal lattice defined by these vectors. Each reciprocal lattice point is labelled with the indices of the plane it represents and the unit cell is shaded. The angle β^* is the complement of β.

First, we shall draw the reciprocal lattice vectors in a section perpendicular to the *y*-axis (i.e. containing the **a** and **c** lattice vectors) from which we will find the reciprocal lattice unit cell vectors **a*** and **c***. This then enables us to express reciprocal lattice vectors in this section in terms of their components on these unit cell vectors. It is then a simple step to extend these ideas to three dimensions.

Figure 6.4(a) shows a section of a monoclinic *P* lattice through the origin and perpendicular to the *y*-axis or **b** unit cell vector. The section of the unit cell is outlined by the lattice vectors **a** and **c** at the obtuse angle β and also the traces of some planes of the type (*h0l*) (i.e. those planes parallel to the *y*-axis). In Fig. 6.4(a), (and Fig. 6.7(a)) no distinction is made between Miller indices and Laue indices—the indices of the planes are all put in brackets whether or not they all pass through lattice points.

Figures 6.4(b) and (c) show the reciprocal lattice vectors for these planes constructed in the same way as before (Section 6.2), with the reciprocal lattice points labelled with the index of the planes they represent. Note again the reciprocal relationships between the lengths of the vectors and the d_{hkl}-spacings, e.g. the (002) planes with *half* the *d*-spacing of the (001) planes are represented by a reciprocal lattice point 002, *twice* the distance from the origin as the reciprocal lattice point 001. Obviously, 003 will be three times the distance, and so on.

It can be seen that the reciprocal lattice points (the 'end points' of the reciprocal lattice vectors) do indeed form a grid or lattice—hence the name. This is emphasized in Fig. 6.4(c), which also shows the **reciprocal lattice unit cell** for this section outlined by reciprocal lattice unit cell vectors **a*** and **c***, where

$$\mathbf{a}^* = \mathbf{d}^*_{100} \quad \text{and} \quad |\mathbf{a}^*| = 1/d_{100}; \quad \mathbf{c}^* = \mathbf{d}^*_{001} \quad \text{and} \quad |\mathbf{c}^*| = 1/d_{001}.$$

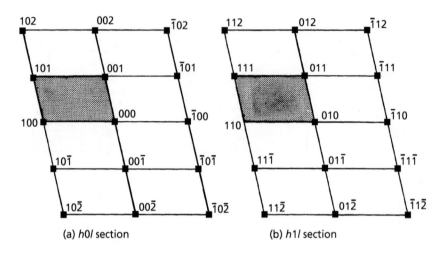

(a) h0l section (b) h1l section

Fig. 6.5. Sections of a monoclinic reciprocal lattice perpendicular to the **b*** vector or y*-axis, (a) h0l section through the origin 000, built up by simply extending the section in Fig. 6.4(c); (b) h1l section (representing planes intersecting the y-axis at one lattice vector **b**) 'one layer up' along the **b*** axis.

Note that **a*** and **c*** are *not* parallel to **a** and **c**, respectively, because the normals to the (100) and (001) planes in the monoclinic lattice are not parallel to **a** and **c**, respectively. Also the angle β* between **a*** and **c*** is the complement of the angle β.

Figure 6.4(c) shows just part of the reciprocal lattice for the few planes we have labelled in Fig. 6.4(a). It can obviously be extended for many more planes and the reciprocal lattice points labelled by adding indices like coordinates, as is shown in Fig. 6.5(a), with the origin labelled 000.

These ideas are readily extended to the third dimension; the **b*** vector is perpendicular to the plane of the paper, normal to the (010) planes which are parallel to the plane of the paper, i.e. in the monoclinic system **b** is parallel to **b*** and the 010 reciprocal lattice point is 'one step above' 000 out of the plane of the paper. Similarly with all the other reciprocal lattice points whose k index is 1; the 011 reciprocal lattice point is 'one step above' 001 and so on. All these points form another layer, or section of the reciprocal lattice as shown in Fig. 6.5(b). Obviously we can build up as many sections of the reciprocal lattice, representing as many planes in the crystal, as we please; the 'ground floor' layer containing the h0l reciprocal lattice points, the section one step along the **b** axis (the 'first floor' layer) containing the h1l reciprocal lattice points and so on.

Now, before expressing these ideas in vector notation, we can reflect on what we have already achieved. First, we are able to represent any set of planes in a crystal by a single reciprocal lattice point: consider how difficult and confusing it would be to have to draw very many different planes in crystal in the same way as in Figs 5.2 or 5.3! Second, the indices (strictly the Laue indices since they have common factors) are represented simply as coordinates of reciprocal lattice points.

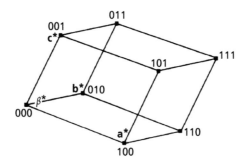

Fig. 6.6. The reciprocal lattice unit cell of a monoclinic P crystal defined by reciprocal lattice vectors \mathbf{a}^*, \mathbf{b}^* and \mathbf{c}^*. β^* is the angle between \mathbf{a}^* and \mathbf{c}^*.

In vector notation, the reciprocal lattice vectors can now be expressed in terms of their components of the reciprocal unit cell vectors \mathbf{a}^*, \mathbf{b}^*, \mathbf{c}^*. For example, for the (102) family of planes (Figs 6.4(b) and (c)) $\mathbf{d}^*_{102} = 1\mathbf{a}^* + 0\mathbf{b}^* + 2\mathbf{c}^*$, or, in general,

$$\mathbf{d}^*_{hkl} = h\mathbf{a}^* + k\mathbf{b}^* + l\mathbf{c}^*.$$

Hence, just as direction symbols $[uvw]$ are the components of a vector \mathbf{r}_{uvw} in direct space (Section 5.2), so also plane indices are the components of a vector \mathbf{d}^*_{hkl} in reciprocal space.

It is worth while rewriting these two equations once again to emphasize their importance and what might be called their symmetry:

$$\mathbf{r}_{uvw} = u\mathbf{a} + v\mathbf{b} + w\mathbf{c}$$

(direction symbols are simply the components of a direct lattice vector)

$$\mathbf{d}^*_{hkl} = h\mathbf{a}^* + k\mathbf{b}^* + l\mathbf{c}^*$$

(Laue indices are simply the components of a reciprocal lattice vector).

The reciprocal lattice unit cell of the monoclinic crystal is drawn in Fig. 6.6 in an orientation where the \mathbf{a}^*, \mathbf{b}^* and \mathbf{c}^* unit cell vectors and the reciprocal lattice points at the corners can be seen clearly. Notice that the idea of 'layers' of reciprocal lattice points applies (of course) to all orientations, for example the 'bottom face' of the cell contains the $hk0$ reciprocal lattice points, the 'top face', one step along \mathbf{c}^* contains the $hk1$ reciprocal lattice points—and so on.

6.4 Reciprocal lattice cells for cubic crystals

In crystals with orthogonal axes (cubic, tetragonal, orthorhombic) the reciprocal lattice vectors \mathbf{a}^*, \mathbf{b}^*, \mathbf{c}^* are parallel to \mathbf{a}, \mathbf{b}, \mathbf{c}, respectively. Hence we have the further identities that $\mathbf{a} \cdot \mathbf{b} = 0$, $\mathbf{a}^* \cdot \mathbf{b}^* = 0$, etc. The reciprocal lattice unit cell of a simple cubic crystal (cubic P lattice) is obviously a cube with reciprocal lattice points only at the corners.

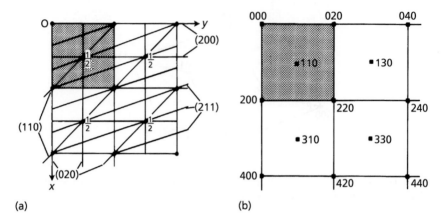

Fig. 6.7. (a) Plan of a cubic I crystal perpendicular to the z-axis and (b) pattern of reciprocal lattice points perpendicular to the z-axis. Note the cubic F arrangement of reciprocal lattice points in this plane.

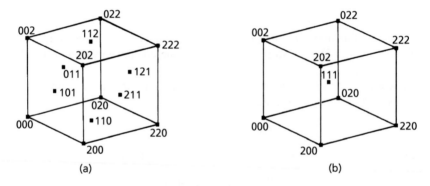

Fig. 6.8. (a) The cubic F reciprocal lattice unit cell of the cubic I (direct) lattice; (b) The cubic I reciprocal lattice cell of the cubic F (direct) lattice.

The reciprocal unit cells of the cubic I and cubic F lattices also have additional lattice points within the unit cells, i.e. they are also not primitive. This is illustrated for the case of the cubic I lattice in Fig. 6.7, which shows four unit cells with the z-axis perpendicular to the plane of the paper and the traces of several $hk0$ planes (parallel to the z-axis) sketched in. Notice (for example) that in the direction of the x-axis the first set of *lattice* planes encountered is not (100) but (200) because of the presence of the additional lattice points in the centres of the cells (see also Fig. 5.4). Hence the reciprocal lattice vector in this direction is equal to \mathbf{d}^*_{200} not \mathbf{d}^*_{100}; similarly, in the direction of the y-axis it is \mathbf{d}^*_{020}. In the [110] direction, however, the first set of lattice planes encountered is (110), which gives a reciprocal lattice point in the centre of the face of the cell and the second-order 220 reciprocal lattice point at the corner. Proceeding in this way we find that this section of the reciprocal lattice is *face-centred*. Repeating this procedure in

the other sections gives a **face-centred cubic reciprocal lattice (cubic *F*)** (Fig. 6.8(a)) for the **body-centred cubic (direct) lattic (cubic *I*)**—the two are reciprocally related. In the same way, we find that the reciprocal lattice of the face-centred cubic lattice is body-centred (Fig. 6.8(b)).

There is another way of looking at these reciprocal relationships. Recall that the cubic lattices may all be referred to rhombohedral axes (primitive cells) with axial angles at 60° (cubic *F*), 90° (cubic *P*), and 109.47° (cubic *I*) (Section 3.2, Fig. 3.2). The reciprocals of these cells are again rhombohedral cells with axial angles 109.47° (i.e. cubic *I*), 90° (i.e. cubic *P*) and 60° (i.e. cubic *F*), respectively.

Finally, just as we can define the environments around lattice points in terms of Voronoi polyhedra, or Wigner–Seitz cells (Section 3.4), so also can we define the environments around reciprocal lattice points in just the same way and in this case the polyhedra are called *Brillouin zones*. Brillouin zones are useful in describing the energy states of electrons in metallic crystals. For a cubic *F* crystal, for which the reciprocal lattice is cubic *I*, the first Brillouin zone is a truncated octahedron and for a cubic *I* crystal, for which the reciprocal lattice is cubic *F*, it is a rhombic dodecahedron (see Fig. 3.7(d) and (e) and Appendix 2).

6.5 Proofs of some geometrical relationships using reciprocal lattice vectors

Some useful geometrical relationships stated without proof in Chapter 5 can be proved very simply using the concept of the reciprocal lattice and an elementary knowledge of vector algebra (see Appendix 5). The results are summarized in Appendix 4.

6.5.1 Relationships between **a**, **b**, **c** *and* **a***, **b***, **c***

Consider, for example, **c*** (Fig. 6.4) and its relationships with **a**, **b** and **c**, which are redrawn for clarity in Fig. 6.9. showing in addition the angle ϕ between **c** and **c***.

Fig. 6.9. Plan of a monoclinic unit cell perpendicular to the *y*-axis emphasizing the geometrical relationships between **a**, **c** and **c*** (see Fig. 6.4 and Section 6.5.1).

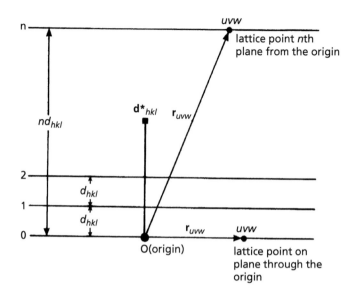

Fig. 6.10. Diagram representing the geometry of the (generalized) zone law. A set of *hkl* planes are drawn 'edge on'. When a lattice point *uvw* lies on a plane through the origin then $hu + kv + lw = 0$; when a lattice point lies on the *n*th plane from the origin then $hu + kv + lw = n$.

6.5.4 d-spacing of lattice planes (hkl)

Recalling (Appendix 5) that the scalar product of a vector multiplied by itself is the modulus squared:

$$\mathbf{d}^*_{hkl} \cdot \mathbf{d}^*_{hkl} = \frac{1}{d^2_{hkl}} = (h\mathbf{a}^* + k\mathbf{b}^* + l\mathbf{c}^*) \cdot (h\mathbf{a}^* + k\mathbf{b}^* + l\mathbf{c}^*).$$

Simple expressions are only obtained for crystals with orthogonal axes (orthorhombic, tetragonal, cubic) where $\mathbf{a}^* \cdot \mathbf{b}^* = 0$, etc., i.e. for orthorhombic crystals;

$$\frac{1}{d^2_{hkl}} = h\mathbf{a}^* \cdot h\mathbf{a}^* + k\mathbf{b}^* \cdot k\mathbf{b}^* + l\mathbf{c}^* \cdot l\mathbf{c}^* = \frac{h^2}{a^2} + \frac{k^2}{b^2} + \frac{l^2}{c^2}$$

$$\text{since } \mathbf{a}^* \cdot \mathbf{a}^* = \frac{1}{a^2}, etc.$$

6.5.5 Angle ρ between plane normals (h₁k₁l₁) and (h₂k₂l₂)

The angle ρ between two vectors \mathbf{a} and \mathbf{b} is given by (Appendix 5):

$$\cos \rho = \frac{\mathbf{a} \cdot \mathbf{b}}{ab}.$$

Hence:

$$\cos \rho = \frac{\mathbf{d}^*_{h_1k_1l_1} \cdot \mathbf{d}^*_{h_2k_2l_2}}{|\mathbf{d}^*_{h_1k_1l_1}| \, |\mathbf{d}^*_{h_2k_2l_2}|}.$$

Again, substituting in terms of components gives a simple expression only for cubic crystals.

6.5.6 Definition of *a**, *b**, *c** in terms of *a*, *b*, *c*

The volume of the unit cell V is given by $\mathbf{a} \cdot (\mathbf{b} \times \mathbf{c})$ (see Appendix 5). Now, as $(\mathbf{b} \times \mathbf{c})$ is a vector parallel to \mathbf{a}^* and of modulus equal to the area of the face of the unit cell defined by \mathbf{b} and \mathbf{c}, then $\mathbf{a}^* = (\mathbf{b} \times \mathbf{c})/V = (\mathbf{b} \times \mathbf{c})/\mathbf{a} \cdot (\mathbf{b} \times \mathbf{c})$, and similarly for \mathbf{b}^* and \mathbf{c}^*. \mathbf{a}, \mathbf{b} and \mathbf{c} can be defined in the same way, i.e. $\mathbf{a} = (\mathbf{b}^* \times \mathbf{c}^*)/V^*$, and similarly for \mathbf{b} and \mathbf{c}, where V^* is the volume of the reciprocal unit cell.

Hence to summarize:

$$\mathbf{a}^* = \frac{\mathbf{b} \times \mathbf{c}}{V}, \quad \mathbf{b}^* = \frac{\mathbf{c} \times \mathbf{a}}{V} \quad \mathbf{c}^* = \frac{\mathbf{a} \times \mathbf{b}}{V},$$

$$\mathbf{a} = \frac{\mathbf{b}^* \times \mathbf{c}^*}{V^*}, \quad \mathbf{b} = \frac{\mathbf{c}^* \times \mathbf{a}^*}{V^*} \quad \mathbf{c} = \frac{\mathbf{a}^* \times \mathbf{b}^*}{V^*}.$$

6.5.7 Zone axis at intersection of planes ($h_1k_1l_1$) and ($h_2k_2l_2$)

The zone axis is defined by \mathbf{r}_{uvw}, the vector product of

$$\mathbf{d}^*_{h_1k_1l_1} \quad \text{and} \quad \mathbf{d}^*_{h_2k_2l_2}.$$

Substituting in terms of their components and remembering the identities $\mathbf{a} = (\mathbf{b}^* \times \mathbf{c}^*)/V^*$, etc., gives

$$u\mathbf{a} + v\mathbf{b} + w\mathbf{c} = \mathbf{a}(k_1l_2 - k_2l_1) + \mathbf{b}(l_1h_2 - l_2h_1) + \mathbf{c}(h_1k_2 - h_2k_1)$$

therefore (see Memogram, Section 5.6.2)

$$u = (k_1l_2 - k_2l_1); \quad v = (l_1h_2 - l_2h_1); \quad w = (h_1k_2 - h_2k_1).$$

6.5.8 A plane containing two directions [$u_1v_1w_1$] and [$u_2v_2w_2$]

The plane is defined by \mathbf{d}^*_{hkl}, the vector product of $\mathbf{r}_{u_1v_1w_1}$ and $\mathbf{r}_{u_2v_2w_2}$. Proceeding as above gives (see Memogram, Section 5.6.3)

$$h = (v_1w_2 - v_2w_1); \quad k = (w_1u_2 - w_2u_1); \quad l = (u_1v_2 - u_2v_1).$$

6.6 Lattice planes and reciprocal lattice planes

Just as we have lattice planes passing through lattice points in the 'direct' lattice, so also we have planes (or layers) passing through reciprocal lattice points in the reciprocal

lattice. There is, as far as I know, no standard terminology to describe these reciprocal lattice planes or layers nor their spacings (unlike the familiar (hkl) and d_{hkl} for lattice planes) so I will call them $(uvw)^*$ reciprocal lattice planes with spacing r^*_{uvw}—the stars to indicate that they are planes and spacings in reciprocal space. In the case of X-ray diffraction, the reciprocal lattice planes correspond to the layer-lines of spots in oscillation photographs (Fig. 9.17) and in the case of electron diffraction they are called Laue zones—the plane or layer of reciprocal lattice points passing through the origin being called the zero order Laue zone; the next plane or layer 'up' from the origin being called the first order Laue zone, and so on (Fig. 11.1).

The spacings r^*_{uvw} of reciprocal lattice planes are defined in a precisely analogous way to that for the spacings of lattice planes, d_{hkl}, so we will write the equations side-by-side to emphasise their 'symmetry':

$$r^*_{uvw} = \frac{1}{|\mathbf{r}_{uvw}|} = \frac{1}{|u\mathbf{a} + v\mathbf{b} + w\mathbf{c}|} \qquad d_{hkl} = \frac{1}{|\mathbf{d}^*_{hkl}|} = \frac{1}{|h\mathbf{a}^* + k\mathbf{b}^* + l\mathbf{c}^*|}.$$

In short, just as lattice planes (hkl) are, by definition, perpendicular to the reciprocal lattice vector \mathbf{d}^*_{hkl} and of reciprocal spacing, so also are reciprocal lattice planes $(uvw)^*$ perpendicular to the direct lattice vector \mathbf{r}_{uvw} and of reciprocal spacing.

These interrelationships are perhaps best understood by means of a particular example. Figure 6.11(a) shows a plan view (perpendicular to the c-axis) of an orthorhombic crystal. The unit cell, lightly shaded, is outlined by lattice vectors \mathbf{a} and \mathbf{b}. The

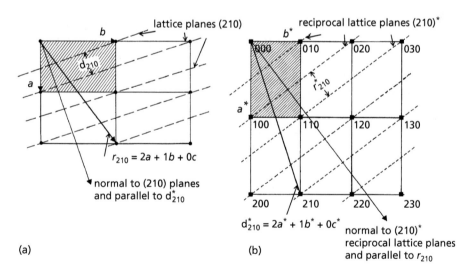

Fig. 6.11. (a) The (direct) lattice of a simple (primitive) orthorhombic crystal with the unit cell shaded, traces of the (210) lattice planes (long dashed lines) and their normal and the lattice vector \mathbf{r}_{210}. (b) The corresponding reciprocal lattice with the unit cell shaded, traces of the (210)* reciprocal lattice planes or layers (short dashed lines) and their normal (parallel to \mathbf{r}_{210}) and the reciprocal lattice vector \mathbf{d}^*_{210} (parallel to the normal to the lattice planes 210).

traces of a particular set of lattice planes, (210), are indicated by the long dashed lines together with their normal and also the lattice vector $\mathbf{r}_{210} = (2\mathbf{a} + 1\mathbf{b} + 0\mathbf{c})$. Note that the normal to the lattice planes is *not* parallel to the direction [210] (see Fig. 5.5).

Figure 6.11(b) shows the corresponding plan view of the reciprocal orthorhombic lattice, the unit cell, lightly shaded, being outlined by reciprocal lattice vectors \mathbf{a}^* and \mathbf{b}^*. (In this case, (orthogonal axes) \mathbf{a}^* is parallel to \mathbf{a} and \mathbf{b}^* is parallel to \mathbf{b}.) The traces of the reciprocal lattice planes (210)* are indicated by the short dashed lines together with their normal and also the reciprocal lattice vector $\mathbf{d}_{210}^* = (2\mathbf{a}^* + 1\mathbf{b}^* + 0\mathbf{c}^*)$. Note, by comparing the diagrams, that \mathbf{d}_{210}^* is (by definition) perpendicular to the lattice planes (210) and that \mathbf{r}_{210} is also (by definition) perpendicular to the reciprocal lattice planes (210)*.

The Weiss zone law or zone equation (Section 6.5.3) also applies to reciprocal lattice planes: for the reciprocal lattice points (hkl) lying in the plane through the origin and perpendicular to the lattice vector \mathbf{r}_{uvw}:

$$hu + kv + lw = 0$$

and for the reciprocal lattice points lying in the nth reciprocal lattice plane from the origin:

$$hu + kv + lw = n.$$

You can easily 'check' these equations with reference to Fig. 6.11(b). For example, reciprocal lattice point (130) lies in the fifth plane in zone [210] and $n = 1.2 + 3.1 + 0.0 = 5$. For crystals with orthogonal axes and primitive unit cells, the spacings d_{hkl} (of lattice planes) and the spacings r_{uvw}^* (of reciprocal lattice planes) may be readily calculated (see Section 6.5.4):

$$d_{hkl} = \frac{1}{|\mathbf{d}_{hkl}^*|} = \frac{1}{\sqrt{(h\mathbf{a}^*)^2 + (k\mathbf{b}^*)^2 + (l\mathbf{c}^*)^2}} = \frac{1}{\sqrt{(h/a)^2 + (k/b)^2 + (l/c)^2}}$$

$$r_{uvw}^* = \frac{1}{|\mathbf{r}_{uvw}|} = \frac{1}{\sqrt{(u\mathbf{a})^2 + (v\mathbf{b})^2 + (w\mathbf{c})^2}} = \frac{1}{\sqrt{(ua)^2 + (vb)^2 + (wc)^2}}.$$

For crystals with centred lattices, analogous conditions apply to the symbols $(uvw)^*$ for reciprocal lattice planes as they do to (hkl) for lattice planes. For example, for the cubic F lattice, for which the reciprocal lattice is cubic I (Section 6.4), $u + v + w =$ an even integer.

Now, having completed this chapter, you should feel as confident and familiar with the concept of reciprocal space as you do with direct space—for as we shall see in subsequent chapters, diffraction patterns reveal, or are imperfect images of, the reciprocal lattices of crystals.

6.7 Summary

In this chapter we have developed the reciprocal lattice as a purely geometrical exercise—as simply a method of representing the orientations and spacings of planes in a crystal and from which the geometrical relationships between planes and directions may be easily worked out. But (as might be expected), the reciprocal lattice has a much greater significance than this. First of all we shall see (in Chapters 8, 9 and 11) how the reciprocal lattice may be used to interpret the geometry of X-ray and electron diffraction patterns. Then we shall see (in Chapters 9 and 10) that reciprocal lattice points may not be 'points' but have finite size and shape which is related to the size, shape and strain in 'real' (as opposed to 'infinitely large perfect') crystals.

Note In the field of electron diffraction reciprocal lattice vectors are often denoted by the symbol \mathbf{g}_{hkl} (no star) rather than \mathbf{d}^*_{hkl}.

Exercises

6.1 Derive the reciprocal lattice section passing through the origin and perpendicular to the [0001] zone for the hexagonal P lattice, and similarly perpendicular to the $<111>$ zone for the cubic I and cubic F lattices.

6.2 Derive expressions for $1/d_{hkl}$ and $\cos \rho$ for orthorhombic crystals.

6.3 Do the relationships derived in Section 6.4 apply also to non-cubic centred Bravais lattices and their reciprocal unit cells?

6.4 Calculate the reciprocal lattice plane spacing r^*_{210} of the (210)* planes in a simple cubic (cubic P) crystal of lattice parameter $a = 3\text{Å}$.

The diffraction of light

7.1 Introduction

As introduced in Chapter 6, the reciprocal lattice is the basis upon which the geometry of X-ray and electron diffraction patterns can be most easily understood and, as we shall see in Chapters 8, 9 and 11, the electron diffraction patterns observed in the electron microscope, or the X-ray diffraction patterns recorded with a precession camera, are simply sections through the reciprocal lattice of a crystal—the pattern of spots on the screen or area detector and the pattern of reciprocal lattice points in the corresponding plane or section through the crystal are identical.

This also applies to the diffraction of light from 'two-dimensional crystals' such as umbrella cloths, net window curtains and the like and in which the warp and weft of the cloth correspond in effect to intersecting lattice planes. The diffraction patterns which we see through umbrellas and curtains are, in effect, the two-dimensional reciprocal lattices of these two-dimensional lattices.

The important point to realize at the outset is that the reciprocal lattice is not merely an elegant geometrical abstraction, or a crystallographer's way of formally representing lattice planes in crystals or two-dimensional nets, but that it is something we can 'see for ourselves'. Indeed, in the case of light it is a familiar part of our everyday visual sensations. Furthermore, as explained in Section 7.2 below, when our eyes are relaxed and focused on infinity we see the patterns of spots and streaks around the lights from distant street lamps when viewed through umbrellas and net curtains, or the circles of haze around them on foggy nights quite clearly. Hence, it might be suggested that diffraction patterns constitute our first visual sensations, before we have learned to near-focus upon the world!

The familiarity and ease of demonstration is the main justification or reason for considering the diffraction of light, but there are three others. First, there is a geometrical analogy between light and electron diffraction: in both cases the wave-lengths (of light and electrons) are small in comparison with the spacings of the diffracting objects (the fibre separations in woven fabrics or the lattice spacings in crystals). For example, the wavelength of green light (\sim0.5 μm) is about 500\times smaller than the fibre spacings in nets and fabrics (\sim0.25 mm); similarly, the wavelength of electrons in a 100kV electron microscope (\sim4.0 pm) is about 100\times smaller than the lattice spacings in crystals (\sim0.4 nm). In both cases therefore (as explained in Sections 7.4 and 11.2) the diffraction angles are also small and electron diffraction patterns can, at least initially, be interpreted as sections through the reciprocal lattice of the crystal.

The geometry of X-ray diffraction patterns is rather more complicated because the wavelengths of X-rays (∼0.2 nm) are roughly comparable with the lattice spacings in crystals. Hence the diffraction angles are large (Sections 8.2, 8.3, 8.4 and 9.5) and X-ray diffraction patterns are in a sense 'distorted' representations of the patterns of reciprocal lattice points from crystals, the nature of the 'distortion' depending upon the particular X-ray technique used.

The second reason for considering the diffraction of light is that it provides a simple basis or analogy for an understanding of how and why the **intensities** of X-ray diffraction beams vary (Section 9.1), **line broadening** and the occurrence of **satellite** reflections (Section 9.6). The analogy is provided by a consideration of the diffraction grating, which is, in effect, a one-dimensional crystal. There are three variables to consider in the diffraction of light from a diffraction grating—the line or slit spacing, the width of each slit and the total number of slits. The slit spacing corresponds to the lattice spacings in crystals and determines the directions of the diffracted beams, or, in short, the geometry of the diffraction patterns. The width of the slits determines, for each diffracted beam direction, the sum total of the interference of all the little Huygens' wavelets which contribute to the total intensity of the light from each slit (Section 7.4), which is analogous to the sum total of the interference between all the diffracted beams from all the atoms in the motif. In short it is the lattice which determines the *geometry* of the pattern and the motif which determines the *intensities* of the X-ray diffracted beams. The analogy, however, must not be pressed too far because it takes no account of the dynamical interactions between diffracted beams, i.e. the interference effects arising from re-reflection (and re-re-reflection, etc.) of the direct and diffracted/reflected beams as they pass through a crystal. This desideratum is particularly important in the case of electron diffraction. Finally, the total number of slits in a diffraction grating determines the numbers and intensities of the satellite or subsidiary diffraction peaks each side of a main diffraction peak—the greater the number of slits, the greater the numbers and the smaller are the intensities of the satellite peaks. In most X-ray and electron diffraction situations the total number of diffracting planes is so large that satellite peaks are unobservable and of no importance, but in the case of X-ray diffraction from thin film multilayers, consisting not of thousands but only of tens or hundreds of layers, the numbers and intensities of the satellite peaks are important and useful.

Third, it is the diffraction of light which sets a limit, 'the diffraction limit,' to the resolving power or limit of resolution of optical instruments, in particular telescopes and microscopes, and is therefore of utmost importance to an understanding of how these instruments work. The diffraction limit however is not unsurmountable and it is an important characteristic of modern microscopical techniques—for example scanning tunnelling, atom force, or near field scanning optical microscopes—that they overcome this limit by virtue of the close approach of a fine probe to a specimen surface.

Finally, to generalize the point made in the first paragraph, the reciprocal relationship between an object and its diffraction pattern is formally expressed by what is known as a Fourier transform, which is a (mathematical) operation which transforms a function containing variables of one type (in our case distances in an object or *displacements*) into a function whose variables are reciprocals of the original type (in our case

1/*displacements*). The reciprocal lattices which we worked out in Chapter 6 using very simple geometry are, in fact, the Fourier transforms of the corresponding (real) lattices.

In this sense diffraction patterns may be described as 'visual representations' or 'images' of the Fourier transforms of objects—irrespective as to whether they are generated using light, X-rays or electrons. In the case of light, diffraction patterns are often described as the 'optical transforms' of the corresponding objects.

To summarize, in Section 7.4 we derive in a very simple way (emphasizing the physical principles involved) the form of the diffraction pattern of a grating: we consider the conditions for *constructive* interference between the slit separation a—which give us the angles of the principal maxima and then the conditions for *destructive* interference, both across a single slit, width d, and also across the whole grating, width $W = Na$ (where N = the number of slits)—which gives us the angles of the minima for a single slit and the minima between the principal and subsidiary maxima respectively. The analysis may be carried out more rigorously using *amplitude— phase diagrams* and this approach is covered separately in Chapter 13 (Section 13.3).

Finally, we show in Section 7.5 that it is diffraction which determines the limit of resolution of optical instruments—the telescope, microscope and of course the eye. Again, this subject is covered in greater depth, by means of Fourier analysis, in Section 13.4.

7.2 Simple observations of the diffraction of light

The diffraction of light is most easily demonstrated using a laser—the hand-held types which are designed to be used as 'pointers' on screens for lecturers are more than adequate and are relatively safe (but you should never look directly at the laser light). Many everyday objects may be used as 'diffraction gratings' either in transmission or reflection—fabrics, nets, stockings (transmission) or graduated metal rulers (reflection) and the resultant diffraction patterns may be projected on to a wall or screen. Such experiments will quickly make apparent the inverse or reciprocal relationships between the spacings of the nets, graduations, etc. and the spacings between the diffracted spots. For example, when laser light is reflected from a graduated metal scale or ruler, in addition to the mirror-reflected beam, diffracted beams will be observed on each side; the spacings of these will change as the angle of the surface of the ruler to the laser beam (and hence the apparent spacings between the graduations) is changed, but, more convincingly, when the beam is shifted from the '$\frac{1}{2}$ mm' to the '1 mm' graduated regions of the ruler, the spacings of the diffraction spots are halved. However, it is probably better to begin with the familiar diffraction patterns which are obtained with a simple point source of light and a piece of fabric with an open weave and which moreover serve to emphasize some important ideas about the angular relationships between the diffracted beams. A point-source of light may in practice be a distant street lamp or, in order to conduct the experiments indoors, a mini torch bulb or a domestic light bulb placed behind a screen with a pinhole (or a fine needle hole) punched into it. Nylon net curtain material is an effective fabric to use—the filaments of nylon which make up the strands are tightly twisted giving sharply defined transparent/opaque boundaries.

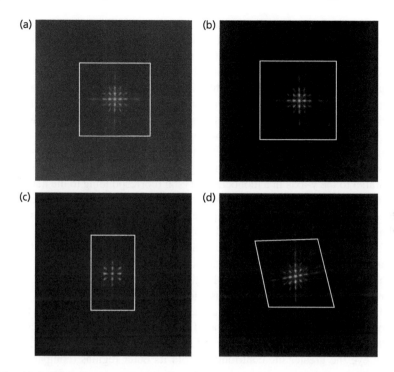

Fig. 7.1. (a) A diffraction pattern from a piece of net curtain with a square weave. The outline of the net is indicated by the white frame. Note that two rows of strong spots and the fainter spots forming a square grid diffraction pattern, (b) The diffraction pattern is identical in scale when the net is brought closer to the observer's eye. (c) Rotating the net about a vertical axis reduces the effective spacing of the vertical lines of the weave, resulting in diffraction spots in the horizontal direction which are further apart. (d) Shearing or distorting the net such that the strands or lines are no longer at 90° to each other results in a diffraction pattern 'sheared' in the opposite sense—the rows of strong diffraction spots remain perpendicular to the lines of the net.

Observe the point source (which should be at least 5 m distant) through the net. The image of the point source will be seen to be repeated to form a grid (or reciprocal lattice) of diffraction spots (Fig. 7.1(a)). In general the spots are of greatest intensity, and are streaked, in directions perpendicular to the lines or strands of the net.

These familiar observations may be supplemented by two more. First the same pattern of spots is seen whether both, or only one, eye is used. Second the size of the pattern— the apparent spacings of the spots—is *independent* of the position of the net. The diffraction pattern appears to be unchanged, irrespective of whether the net is held away from, or close to the eyes (Fig. 7.1(b)).[1] These observations show that the diffraction spots or diffracted beams bear *fixed angular relationships* to the direct beam.

[1] If the light falling on the net is not parallel but is slightly diverging (i.e. the point source is not effectively at infinity), then the apparent spacing of the spots will change slightly as the net is held at different distances from the eyes.

The reciprocal relationship between the net and diffraction pattern may be demonstrated in two ways. First, a net with a finer line spacing will be found to give diffraction spots more widely spaced. If a finer net is not available, the effective spacing of the lines may be decreased by rotating the net such that it is no longer oriented perpendicular to the line of sight to the light (Fig. 7.1(c)). Secondly, the net may be twisted or sheared such that the strands—the weft and the warp—no longer lie in lines at 90° to each other. It will be seen that the rows of strong diffraction spots rotate in such a way that they continue to lie in directions perpendicular to the lines of the net—the grid of diffraction spots is reciprocally related to the sheared lattice (Fig. 7.1(d)).

These reciprocal relationships are analogous to those shown for a set of planes in a zone parallel to the y-axis in a monoclinic crystal (corresponding to the net, Fig. 6.4(a)) and its reciprocal lattice section (corresponding to the diffraction pattern, Fig. 6.4(c)).

The observation that the size of the pattern is independent of the position of the net, and that it is seen to be identical with both eyes, may be explained by reference to Fig. 7.2. The light from a point source radiates out in all directions, but if it is distant then that part incident upon the net is (approximately) parallel.

Figure 7.2 shows a parallel beam of light from a (distant) source with the net held away from the eye (Fig. 7.2(a)) and closer to the eye (Fig. 7.2(b)). One set of diffracted beams from the net is shown. Of all the direct light and diffracted light from the net, only a portion will enter the eye. The portion of the net from which the diffracted light entering the eye originates changes as the net is moved: the further the net is moved away from the eye the greater the contribution from its outer regions (compare Fig. 7.2(a)

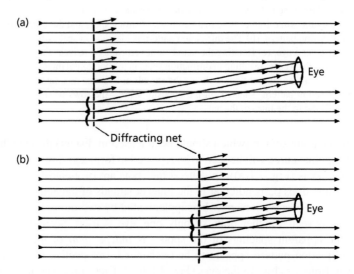

Fig. 7.2. (a) Parallel light (from a distant point source) falls on the net and is diffracted; only one set of diffracted beams is shown. The region of the net contributing to the diffracted light entering the eye of the observer is indicated by brackets, (b) As the net is moved towards the eye, the region of the net which contributes to the diffracted light entering the eye is different. The angular relationship between the direct and diffracted light at the eye remains unchanged.

with Fig. 7.2(b)). However, for incident parallel light, irrespective of the position of the net, the angular relationship at the eye between the direct and diffracted beams, and hence the apparent size of the diffraction pattern, remains unchanged. And what is true for one eye is true for both eyes; you do not need to squint, or keep your head still: the diffraction pattern remains unchanged.

When using lasers to observe diffraction, a narrow, parallel beam of monochromatic light is available, as it were, ready made. Hence, unlike the previous situation, the diffraction pattern can be recorded on a screen because the diffracted beams all originate from the same small area of the net which is illuminated by the laser. (In the previous case, Fig. 7.2, because the diffracted beams originate from a large area of the net, they will correspondingly be 'blurred out' when they fall on a screen.) Only if most of the incident light is 'blocked off such that only a small area of the net is exposed (or illuminated in the case of a laser) will a (faint) diffraction pattern be recorded. These points are best appreciated by 'self-modifying' Fig. 7.2. Remove the 'eye' and extend all the arrows showing the diffracted beams: notice how broad is the width of the total beam. Now 'block off all the beams except those passing through a small (bracketed) area of the net and add a screen, in place of the eye, on the right. The direct and diffracted spots will be about equal to the area of the diffracting net (or the diameter of the laser beam), and their separation will increase the more distant is the screen.

The reciprocal relationship between the diffracting object and its diffraction pattern also extends to the shapes and sizes of the diffracting apertures (the 'holes' in the net) as well as their spacings. However, this is not easy to demonstrate simply without a laser carefully set up on an optical bench, with diffracting apertures of various shapes and sizes and appropriate lenses to focus the diffraction patterns (see Section 7.4 below).

Figure 7.3 shows a sequence of pinholes or apertures and the corresponding diffraction patterns using a laser as a coherent source of light (see Section 7.3). Figures 7.3(a)–(c) show the diffraction patterns (right) from the simplest apertures (left)—single pinholes of various sizes (which may be simply made by punching holes in thin aluminium foil and observing the diffraction pattern on a screen in a darkened room). The diffraction patterns consist of a central disc (of diameter greater than the geometrical shadow of the pinhole) surrounded by much fainter (and sometimes difficult to see) annuli or rings; notice that the diameters of the disc and rings decrease as the diameter of the aperture increases. Figure 7.3(d) shows the diffraction pattern (right) from a rectangular net or lattice (left) of 20 (4 by 5) of the smallest apertures. There are three things to notice about this net and its associated diffraction pattern. First, the reciprocal relationship between the unit cell of the net and that of the diffraction pattern (see also Fig. 7.1(c)); second the intensities of the diffraction spots which vary in the same way as the intensities of the central disc and rings from a single aperture (Fig. 7.3(a)), and third that there are faint spots, or subsidiary maxima, between the main diffraction spots. Finally, Fig. 7.3(e) shows the diffraction pattern (right) from a square net of many small apertures (left). The diffraction spots all lie within the region of the large central disc from a single small aperture and the subsidiary maxima between the main diffraction spots are no longer evident.

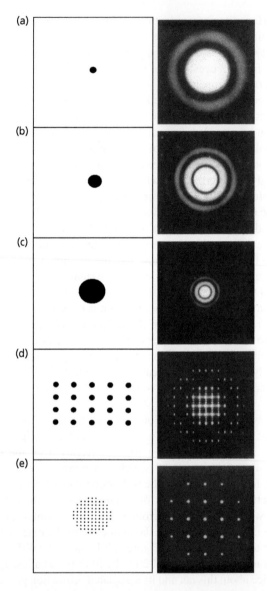

Fig. 7.3. Diffracting apertures (reproduced as black dots on a white background) (left) and their corresponding diffraction patterns (or optical transforms) (right) taken using a laser, (a), (b) and (c) show the diffraction patterns (Airy discs and surrounding haloes or rings) from circular apertures of increasing diameters, (d) shows the diffraction pattern from a 4 × 5 net of small apertures (as in (a)). Note the reciprocal relationship between the net and the diffraction pattern, the 2 × 3 number of subsidiary peaks in each direction and the overall variation in intensity of the diffraction pattern in accordance with that for a single aperture (a), (e) shows that for a net of many apertures the subsidiary maxima are not discernible. (From *Atlas of Optical Transforms* by G. Harburn, C. A. Taylor and T. R. Welberry, Bell and Hyman, 1983, an imprint of HarperCollins.)

7.3 The nature of light: coherence, scattering and interference

In the previous section we surveyed the phenomenon of diffraction: the pattern of beams which occur when light passes through pinholes and nets or is reflected from graduated rulers. We examined the conditions under which diffraction patterns can be observed either by looking at a point source of light through a net, or when using a laser, by projection onto a screen. Finally we observed the inverse or reciprocal relationship between the diffracting object and the diffraction pattern. All these observations need to be explained and given a quantitative basis (Section 7.4), but before doing so we need to know something of the nature of light itself and the notions of coherence, scattering and interference.

The nature of light is expressed through the 'working models' which seek to describe it. Physics provides us with two such models—light as 'particles' (photons or quanta) and light as waves. These working models are not contradictory or opposed; rather they express different aspects of the quantum-mechanical interpretation of light and matter. Visible light occupies a very small part of the whole spectrum of electromagnetic radiation, with a wavelength range from about 400 nm (violet-blue) to about 700 nm (red). In the range below we have ultraviolet (\sim400 nm–1 nm), X-rays (\sim10 nm–10 pm), and the very shortest, γ-rays (\sim10 pm–10 fm). In the range above we have infrared (\sim700 nm–1 mm), microwaves (\sim1 mm–500 mm) and, the very longest, radio waves (\sim500mm–100km).

Light (in common with all other electromagnetic radiation) has both wave-like and particle-like character which are linked by an equation of great simplicity—Planck's equation: $E = hc/\lambda$, where E is the energy of the photon (a particle-like quantity), c is the velocity of light (\sim3 \times 10^8 ms^{-1}), λ is the wavelength (a wave-like quantity) and h is the Planck constant (6.6256 \times 10^{-34} J s). Notice that the smaller is the wavelength, the higher is the energy of the photon—a matter of great physiological and practical importance: we are all continuously bathed in radio waves with no ill effects, but a short exposure to X-rays or γ-rays results in severe damage to our body tissues. Light obviously occupies an important niche in between.

Matter—electrons, protons, particles of any sort—also has wave-like as well as particle-like character, which are similarly linked by de Broglie's equation, $\lambda = h/mv$, where m and v are the mass and velocity of the particle respectively and λ is the wavelength. As before, the 'linking constant' between the particle-like and wave-like character is h, the Planck constant.

The notion that light has both particle-like and wave-like character reaches back to the seventeenth century and the work of Isaac Newton* and Christiaan Huygens*. Newton has been represented as the advocate of the particle or 'corpuscular' theory of light and Huygens the advocate of the wave theory of light. Certainly Huygens' simple and elegant interpretation of reflection and refraction in terms of wavefronts made up of the envelopes of secondary waves or wavelets, which was taken up and extended

* Denotes biographical notes available in Appendix 3.

by Thomas Young* and Augustin Jean Fresnel* early in the nineteenth century, has been far more fruitful than Newton's corpuscular theory. However, it is clear from his own writings that Newton himself was undecided. In his great work *Opticks* there are a number of *Queries* which represent his reflections on his life's work on the properties of light. In Query 13 he asks:

Do not several sorts of Rays make Vibrations of several bignesses, which according to their bignesses excite Sensations of several Colours?

and in Query 17:

considering the lastingness of Motions excited in the bottom of the Eye by Light, are they not of a vibrating nature?

Light originates from the transitions of electrons which occur within excited atoms, excited by virtue of being at a high temperature, as in lamp filaments and the Sun, or by stimulated emission, as in a laser. The physics of the processes in either case does not concern us; the important point is that each atom emits a wavetrain for a period of time (about 10^{-8} s for lamp filaments; or from about 10^{-12} s for pulsed lasers to indefinite times for continuous lasers). In the former cases the wave is therefore not infinite in extent, but begins and ends, like a group of ripples moving outward from a point where a stone has been thrown into a pond. It is called a *wavetrain* or a *photon*, so leading back, in a sense, to the corpuscles of Newton. The length of each wavetrain depends on the velocity of light, 3×10^8 ms^{-1} in air, and the period of emission; for lamp filaments a wavetrain is typically about 3 m long (3×10^8 ms$^{-1} \times 10^{-8}$ s) and for lasers from about 0.3 mm long (3×10^8 ms$^{-1} \times 10^{-12}$ s) to much longer lengths (for 'continuous' lasers).

For lamp filaments there are, in general, no phase relationships between the individual wavetrains; they overlap and follow on each other's heels without any registration between the individual 'peaks' and 'troughs'. Furthermore, the vibration directions of each wavetrain will be more or less random—and the light is said to be **natural** or **unpolarized.**

Light from an extended source (e.g. the Sun or a domestic light bulb) is said to be **incoherent**. Interference will occur spasmodically as it were, between the peaks and troughs and peaks and peaks and troughs and troughs of similarly polarized wavetrains from different regions of the source, but the net effect is zero. **Coherence**, the condition that gives rise to clearly defined diffracted beams is achieved by **point source** in which all the waves emanate from a single point as shown in Fig. 7.4(a). Of course, a point source does not exist, but in practice is closely approximated to by, e.g. a distant lamp or light passing through a small pinhole or slit in a screen or a laser. The boundary condition, i.e. the size of the pinhole or slit above which it can no longer be regarded as the source of coherent light, is specified below.

Figure 7.4(a) shows (idealized) a wavetrain 'on its way' from a point source and its arrival at a screen with a single slit (or pinhole) S. According to Huygens' construction the slit will act, in turn, as the source of secondary waves which, for a narrow slit (of width comparable to the wavelength of the light) will radiate out equally in all

directions as shown. Now consider the single wavetrain falling on a screen with two slits S_1 and S_2 (Fig. 7.4(b))—this is essentially repeating Young's experiment). Each slit is the source of secondary waves which, because they originate from the same incident wave (Fig. 7.4(a)), are 'in step' with each other. As the waves spread out they interfere; where 'peak meets peak' and 'trough meets trough' we have reinforcement or constructive interference and where 'peak meets trough' we have cancellation or destructive interference. The directions along which constructive interference occurs, i.e. the directions of the diffracted beams, are clearly shown and are labelled zero order in

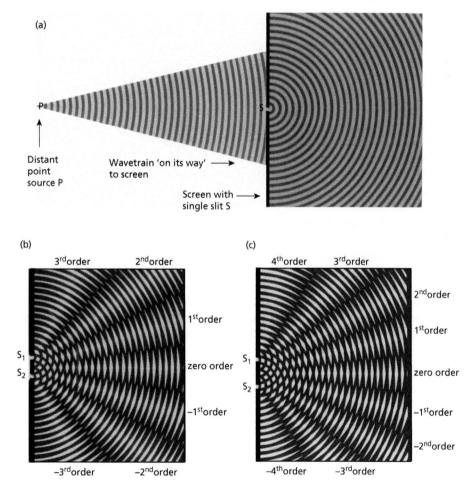

Fig. 7.4. A diagram **(a)** of a distant point source emitting coherent wavetrains and a single slit S in a screen which is the source of secondary waves (Huygens' construction). In **(b)** and **(c)** the wavetrains fall on two slits S_1 and S_2, the waves from each of which interfere to form the diffraction pattern. Note the phase shift of π between alternate diffracted beams and the **smaller** the separation of the slits (Fig. 7.4(b)) the **larger** are the angles of the diffracted beams.

the centre and 1st, 2nd, etc. orders each side. Figure 7.4(c) shows the corresponding pattern for the slits S_1 and S_2 spaced **further apart**: note that the diffracted beams occur at **smaller** angles. The final point to note is that the peaks and troughs in alternate diffracted beams are shifted by $\lambda/2$, i.e. there is a difference in phase of π between alternate beams. This observation is analysed in terms of amplitude-phase diagrams in Section 13.3, but the important point is that here, we see, in this simple representation, the **phases** of the diffracted beams. In Section 7.4 we shall determine the particular angles or directions in which constructive and destructive interference occur. The important point to note now is that the geometrical conditions for interference which apply to just *one* wavetrain apply to *all* the wavetrains, and the resultant pattern of light and dark on the screen is the diffraction pattern. The words 'diffraction pattern' are a very inadequate description of the physical processes which are involved and the words or phrase 'diffraction (or scattering)-followed-by-interference-between-the-diffracted (or scattered)-beams-pattern' would be more accurate, albeit rather long-winded!

Now consider the situation when the slits are illuminated by separate light sources, or by an extended source. *Separate* sets of wavetrains from different parts of the source will arrive independently at the slits and thus the secondary waves emanating from each slit will bear no relationship with each other. To be sure, when 'peak meets peak' constructive interference will occur and where 'peak meets trough' destructive interference will occur—but these effects will be quite random in space and will cancel out: the light in this situation is said to be *incoherent*.

What is the boundary condition between a point source and an extended source of width w? It is given approximately by the condition that the light from opposite edges of the source should arrive at the slits with a phase difference less than $\sim \pi/4$. For a slit-spacing a and slit-source distance r. This condition is approximately given by $wa/r \leq \lambda/8$.

Note: The diffraction patterns in Fig. 7.4 may be simulated by copying Fig. 7.4(a) on to transparency sheets (the clear regions representing peaks) and then overlapping them rather in the same manner as in making Moiré patterns. The inverse relationship between slit spacing and angles of the diffracted beams is then particularly evident.

7.4 Analysis of the geometry of diffraction patterns from gratings and nets

We will now analyse the diffraction phenomena described in Section 7.1 by considering the constructive and destructive interference conditions for all the scattered waves which emerge from nets and apertures of different sizes, shapes and spacings. We will do this step-by-step by first considering diffraction from a grating, which consists of a set of parallel lines engraved or photographed on to a glass or a polished metal plate. Diffraction occurs when light passes through the transparent 'slits' between the lines, or when it is reflected from the metal 'strips'. The constructive and destructive interference conditions in each case are identical, but it is much easier to draw the transmission case because the incident and diffracted beams do not overlap. This situation is analogous to the ray diagrams used in light microscopy; it is simpler to demonstrate image formation in transmitted light microscopy by showing light passing from the condenser

and through the specimen and objective lens to the eyepiece rather than light passing in two directions through the condenser/objective lens as it is reflected from the specimen as in reflected light microscopy.

The conditions for constructive/destructive interference from a set of parallel slits will, fairly obviously (ignoring end effects), be the same all along their length and the diffraction pattern will be a series of light/dark bands (corresponding to constructive and destructive interference conditions, respectively) running parallel to the slits. A net may be regarded as a two-dimensional grating, with a criss-cross pattern of slits, or two gratings superimposed upon each other with their slits running in different directions. Each grating will give rise to its own diffraction pattern of light/dark bands, but the net effect is that constructive interference will only occur when the light bands intersect and reinforce—giving rise to the observed diffraction peaks or spots. Hence, consideration of the, in effect, 'one-dimensional diffraction' case of a grating will lead us to a complete analysis of the 'two-dimensional diffraction' case of a net.

First we will consider diffraction from a grating made up of very narrow or thin slits spaced distance a apart in which each slit is the source of just one Huygens' wavelet across its width; from this we will determine the conditions for constructive interference and therefore the directions of the diffracted beams. By simply extending the analysis to two dimensions we will demonstrate the reciprocal relationship between the diffracting net and the positions of the spots in the diffraction pattern. Then we shall consider diffraction from a wide aperture—a circular hole or wide slit which is the source of not one but many Huygens' wavelets and we shall see that this modifies the intensities, but does not change the positions, of the diffracted spots. Finally we shall consider gratings or nets of limited extent and will show how this leads to the occurrence of subsidiary peaks or spots.

Figure 7.5 shows just part of a diffraction grating—a one-dimensional net— consisting of many narrow slits distance a apart. The incident light is shown as parallel beams from a distant source (i.e. a plane wavefront). This special case is sometimes

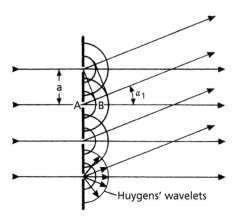

Huygens' wavelets

Fig. 7.5. A set of lines in the net acts as a diffraction grating, each slit acting as the source of a single set of Huygens' wavelets. The diffracted beam is drawn for a path difference of one wavelength.

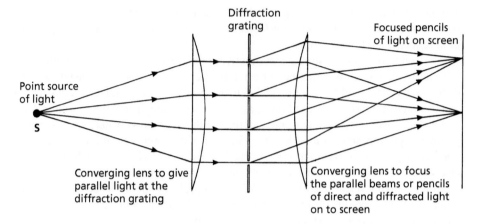

Fig. 7.6. A plane wavefront (parallel light) incident upon a grating may be achieved by inserting a lens (left) with the point source at its focus. Conversely, the parallel 'pencils' of the direct and diffracted light may be sharply focused onto a screen or photographic plate by inserting another lens (right). (Compare with Fig. 7.5.)

called the case of Fraunhofer* diffraction as opposed to the case of Fresnel* diffraction in which the source is not distant and the wavefront is an arc, as shown in Fig. 7.4. Of course they are not different 'types' of diffraction, it is simply that Fraunhofer diffraction is simpler to treat mathematically. However, Fresnel diffraction may in practice be 'converted' to Fraunhofer diffraction by inserting a converging (positive) lens into the light beam with the point source at its focus and from which of course emerges parallel light (Fig. 7.6).

Each narrow slit (Fig. 7.5) acts as a *single* source of Huygens' wavelets spreading out in all directions. Constructive interference occurs in those directions where the path difference (AB) between adjacent slits equals a whole number of wavelengths, i.e.

$$AB = n\lambda = a \sin \alpha_n$$

where a = the line (or lattice) spacing and α_n = the diffracted angle for the nth order diffracted beam. Since $\lambda \approx 0.5 \, \mu m$ and typically $a \approx 0.5$–$2 \, mm$, then, as pointed out in Section 7.1, the diffraction angles are small (because $a \gg \lambda$) and $\sin \alpha_n \approx \alpha_n$. Hence, for the first few visible diffraction orders $AB = n\lambda = a\alpha_n$, from which the angle $\alpha_n = n\lambda/a$: here we have the reciprocal relationship between the diffraction angles, α_n, and lattice spacing, a.

The screen on which the diffracted beams fall (not shown in Fig. 7.5) should be a long distance from the grating compared with its width—and of course the further it is away the greater is the separation between the beams. This may, in practice, be rather inconvenient, but the inconvenience may be overcome by placing a converging (positive) lens after the grating and placing the screen at its focal point such that the

* Denotes biographical notes available in Appendix 3.

'pencils' or parallel light of the direct and diffracted beams are sharply focused there (Fig. 7.6). The presence of such a lens does not modify in any way our understanding of the diffraction and interference phenomena at the grating and is best 'left out' of diagrams because it may appear to complicate the ray paths unnecessarily.

Figure 7.7(a) shows diagrammatically the diffraction pattern from such a grating—a central maximum or peak with the first, second, third order, etc. diffracted beams

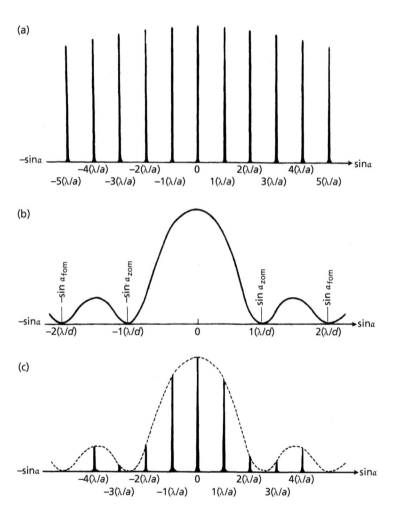

Fig. 7.7. Diagrams of the diffraction patterns from gratings and a single slit as a function of angle α; the intensities of the peaks are represented only qualitatively. (a) The diffracted beams for a narrow-slit grating of spacing a. Constructive interference occurs at $\sin \alpha_n$ angles of $\pm n(\lambda/a)$ where $n =$ the order of the diffracted beams ($n = 0$ represents the direct beam), (b) The diffraction profile for a single slit of width d. Destructive interference occurs at $\sin \alpha$ angles $1(\lambda/d)$ (zero order minimum), $2(\lambda/d)$ (first order minimum), etc. (c) For a wide-slit grating (slit-width d, slit spacing a) the intensities of the diffracted beams Fig. 7.7(a) are modulated by the intensity profile for a single slit Fig. 7.7(b).

each side. Their intensity decreases because as the angle α_n increases the screen is further away from the diffraction grating.

This analysis of diffraction from a grating—a one-dimensional net—may readily be extended to the other set of lines in the net spaced, say, distance b apart, giving a second set of diffracted beams whose diffraction angles are again reciprocally related to b, the observed diffraction spots forming, as stressed before, the two-dimensional reciprocal lattice of the two-dimensional net (Figs 7.1(c) and 7.3(d) and (e)).

Now, the practical requirements for narrow slits which are the source of just one Huygens' wavelet are difficult if not impossible to achieve. Not only must the slit width be a small fraction of the wavelength of light, but it must also be equally thin—otherwise it will be, in effect, not a slit but a little rectangular tunnel—rather like the narrow openings in castle walls! Furthermore, the narrower (and thinner) are the slits, the smaller is the intensity of light.

In practice, therefore, slits are wide, which means that they are the source of many Huygens' wavelets, as shown in Fig. 7.8(a). The problem now is to sum all the contributions of all the wavelets. This may be carried out analytically or graphically by means of amplitude-phase diagrams, which are explained in Chapter 13 (Section 13.4). However, we shall use a simplified approach, which is quite adequate and emphasizes the principles involved.

In the forward (direct beam) direction, all the wavelets are in phase and interfere constructively. At increasing angles to the direct beam the wavelets become progressively

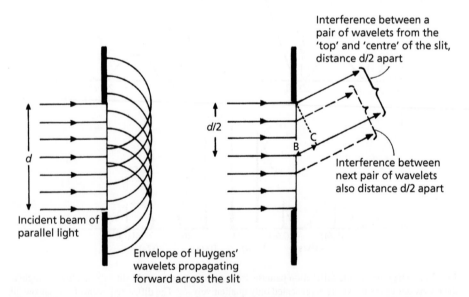

Fig. 7.8. Showing (a) a wide slit as the source of many Huygens' wavelets (shown propagating in the forward direction) and (b) the condition for destructive interference for the zero order peak. For the pair of wavelets separated by distance $d/2$—one at the 'top' edge and one at the centre of the slit (shown bracketed)—destructive interference occurs when the path difference BC $= \lambda/2$. Similarly for the next pair down (indicated by dashed lines and brackets), and so on for pairs of wavelets across the whole slit.

more out of step; they begin to interfere destructively and the intensity falls: we need to work out the angle at which it falls to zero. Now consider a pair of wavelets; one wavelet at the 'top edge' of the slit and the other at the centre of the slit (Fig. 7.8(b)); when the path difference between them is $\lambda/2$ 'trough meets peak' and total destructive interference occurs. Hence from Fig. 7.8(b),

$$BC = \lambda/2 = (d/2)\sin\alpha_{zom}$$

i.e.

$$\lambda = d\sin\alpha_{zom}$$

where d is the slit width and α_{zom} is the angle at which total destructive interference of the direct (zero order) peak occurs, i.e. zero order minimum, α_{zom}).

Now consider the next pair of wavelets again distance $d/2$ apart, one just below that at the top edge and one just below that at the centre, as shown by dashed lines in Fig. 7.8(b). The condition for destructive interference between this pair of wavelets will be the same—and so on for all the pairs of wavelets across the whole slit. Hence the above condition for total destructive interference applies to the *whole* slit and gives the angle at which the intensity of the direct or zero order beam or the 'central maximum' falls to zero. Notice that the equation 'looks' the same as that for constructive interference *between* the slits; the difference is of course that in the former case only two Huygens' wavelets needed to be considered, one from each slit.

As the angle increases further there will be some net constructive interference, giving a first order diffracted beam but of much smaller intensity than the zero order beam. The next condition for total destructive interference may be found by again considering the condition for destructive interference between a pair of wavelets, one wavelet at the 'top edge' of the slit as before and the other a *quarter* of the slit width from the top. Again, when the path difference between them is $\lambda/2$, destructive interference occurs. Continuing as before, i.e. considering pairs of such wavelets across the whole slit width, it is seen that this is again the condition for total destructive interference of all the wavelets across the slit, i.e. for the first order zero or minimum

$$\lambda/2 = (d/4)\sin\alpha_{fom}, \quad i.e. \ 2\lambda = d\sin\alpha_{fom}$$

where α_{fom} is the angle for first order destructive interference (*first* order *minimum* α_{fom}). Note that $\sin\alpha_{fom} = 2(\lambda/d) = 2\sin\alpha_{zom}$, i.e. for small angles where $\sin\alpha \approx \alpha$, the first order minimum occurs at twice the angle of the zero order minimum.[2]

The diffraction pattern from a single slit is therefore a central maximum with much fainter bands of half the width of the central maximum on each side. It is represented diagrammatically in Fig. 7.7(b). The diffraction pattern from a circular hole or aperture

[2] The angle for maximum *constructive* interference for the first order diffracted beam cannot be determined precisely by such a simple approach as we have used. It turns out that the peak is slightly asymmetrical—the peak does not lie at an angle half way between α_{zom} and α_{fom} but is displaced slightly to a lower angle (see Section 13.3).

is, correspondingly, a central disc surrounded by much fainter rings or haloes, as shown in Figs 7.3(a)–(c). The disc is called the Airy disc because its diameter was worked out precisely by Sir George Airy.* His calculation for the angle α_{zom} of the zero order minimum, which takes into account the circular symmetry of a round hole or aperture, only differs from our simple calculation by a factor of 1.22, i.e. $1.22\lambda = d\sin\alpha_{zom}$ A close approximation to the Airy factor can be obtained by comparing the diffraction pattern from a circular aperture diameter d, to a rectangular aperture *of the same area,* one side of which is length d and the other side of length $(\pi/4)\,d$. In comparison with the circular aperture, the *narrower* rectangular aperture gives rise to an Airy disc which is *wider* in the reciprocal ratio, i.e. $4/\pi = 1.27$—a value which is only 4% different from the Airy factor 1.22. The Airy disc is of immense importance in optics since it determines the limit of resolution of telescopes and microscopes, as explained in Section 7.5.

Now we have to consider the diffraction patterns from a grating with wide slits (width d) separated a distance a apart. The result is simply the combination of the narrow-slit diffraction pattern (Fig. 7.7(a)) and the single wide-slit diffraction pattern (Fig. 7.7(b)). The resultant is shown in Fig. 7.7(c). Notice that the angles at which the diffraction peaks occur are unchanged and are determined solely by the slit spacing a, but that their intensities are modulated by the intensity profile of a single slit of width d.

Finally, we have to consider the situation in which the diffraction grating is of limited extent, consisting of a limited number of slits[3]. If W is the width of the grating and N is the number of slits, then $W = Na$. We simply consider the whole grating as a very wide slit, then work out the conditions for destructive interference as before. The sines of the angles are simply

$$1(\lambda/W),\ 2(\lambda/W),\ 3(\lambda/W),\ldots \quad \text{or} \quad \frac{1}{N}(\lambda/a),\ \frac{2}{N}(\lambda/a),\ \frac{3}{N}(\lambda/a)\ \ldots$$

Each of the diffracted peaks or 'principal maxima' will no longer be 'sharp', as indicated in Figs 7.7(a) and (c), but will be broadened and surrounded by fringes or 'subsidiary maxima'. The number of fringes increases with N and their intensities decrease with N, such that they are generally not detectable for large N values.

Figure 7.9 shows the zero, first and second order principal maxima and the subsidiary maxima from a grating with N slits. Note that the (half) angular widths of the principal maxima, and the angular widths of the subsidiary maxima, are simply equal to $1/N$ of the angular separation of the principal maxima. This leads to the simple result that there are $(N-2)$ subsidiary maxima between the principal maxima, as shown in Fig. 7.9. The proof is as follows. The first minimum of the zero order principal maximum (the direct beam) occurs at angle $1/N(\lambda/a)$, the second minimum at $2/N(\lambda/a)$ and so on. Altogether therefore, going from the zero to the first order principal maximum which occurs at angle $1(\lambda/a)$, there will be $(N-1)$ subsidiary minima, the $(N-1)$th minimum defining, in effect, the first minimum of the first order principal maximum. Hence, between the $(N-1)$ minima there will be $(N-2)$ subsidiary maxima.

* Denotes biographical notes available in Appendix 3.
[3] This analysis is also carried out by means of amplitude-phase diagrams (Section 13.4).

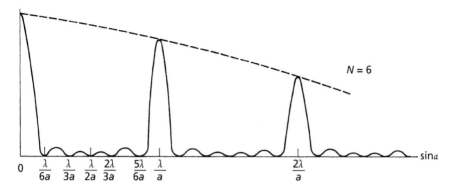

Fig. 7.9. Diagram of the diffraction pattern from a grating (drawn on one side of the direct beam) consisting of N narrow slits of spacing a. Between the principal maxima there are $(N-2)$ subsidiary maxima or fringes. The diagram is drawn for $N = 6$. The dashed line shows the intensity profile for a single narrow slit (modified from *Optics* by E. Hecht and A. Zajac, Addison-Wesley, 1980).

Figure 7.9 is drawn for $N = 6$ and shows the principal maxima at sin α angles $1(\lambda/a)$, $2(\lambda/a)$, the subsidiary minima at sin α angles $1/6(\lambda/a)$, $2/6(\lambda/a)$, $3/6(\lambda/a)$, $4/6(\lambda/a)$ and $5/6(\lambda/a)$ and the $(N-2) = 4$ subsidiary maxima.

All these considerations—the sizes and numbers of diffracting apertures—may be extended to the two-dimensional case of nets, as shown in Figs 7.3(d) and (e). The intensities of the diffraction peaks are modulated by the intensity profiles of the Airy disc and its surrounding fainter rings or haloes and the numbers of subsidiary maxima between the principal maxima are two less than the number of apertures in each direction—and clearly will become insignificant as the number of apertures becomes large. Hence, although we have not determined quantitatively the light intensity distribution in the diffraction patterns from nets, the main features that we noted in Section 7.2 are now explained.

7.5 The resolving power of optical instruments: the telescope, camera, microscope and the eye

One of the most important and useful applications of the study of diffraction is the determination of the resolving power of optical instruments, in particular the telescope and the microscope. By resolving power we mean the ability to distinguish points with small angular separations, such as stars in the telescope, or points which are small distances apart as in the microscope. In both cases the customary term *resolving power* is better expressed by the more precise term *limit of resolution;* either an angular limit of resolution as in the telescope or a distance limit of resolution as in the microscope. In either case the resolving power is a term reciprocally related to the limit of resolution.

There are several other factors which determine the limit of resolution of optical instruments, in particular lens defects, such as spherical and chromatic aberration, and coma. To a large extent these can be eliminated by appropriate lens design or the use of

reflecting elements: mirrors have almost wholly replaced lenses as the principal element in astronomical telescopes and, except for the difficulties in manufacture, could partially replace objective lenses in microscopes. But the limit of resolution is ultimately limited by diffraction—either at the aperture of the telescope or microscope or at closely spaced features in the microscope.

Consider first the limit of resolution of a telescope. Parallel rays from a distant point source—a star—enter the aperture, diameter d, and are focused by the objective lens on to a screen, photographic plate or the focal plane of the eyepiece. As explained in Section 7.4, the image of the distant point source is not a point, but an Airy disc (surrounded by haloes of decreasing intensity), as shown in Figs 7.3(a)–(c) and Fig. 7.10. Now consider light from another star; again its image is an Airy disc, as also shown in Fig. 7.10. The net intensity in the image plane is the sum of the intensities of the two

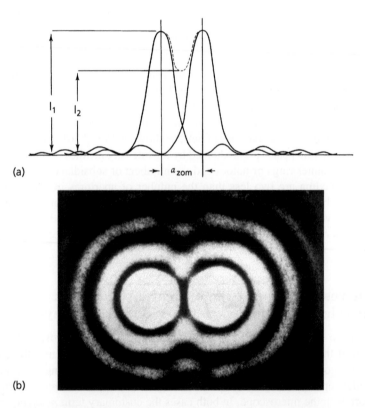

Fig. 7.10. (a) The intensity profile of an Airy disc pattern of two points separated by an angle which just satisfies the Rayleigh criterion for resolution. (Note that the centre of one disc falls on the zero order minimum of the other.) The overall intensity profile is indicated by the dotted line and the minimum, I_2, is about 85% of the maximum, I_1. (b) Photograph of two point sources which are clearly resolved (angular separation of Airy discs $\approx 2\alpha_{zom}$, i.e. about double the Rayleigh limit) (from *An Introduction to the Optical Microscope,* 2nd edn, by S. Bradbury, Oxford University Press/Royal Microscopical Society, 1989).

discs (and their haloes). As the angular separation between the stars decreases their Airy discs overlap and the limit of resolution is the point at which the net intensity does not show a discernible decrease between the central maxima of the Airy discs. Determining this point is a matter of some difficulty, but the criterion proposed by Lord Rayleigh* is most generally adopted. The Rayleigh criterion is that the centre of one Airy disc should coincide with the zero order minimum of the adjacent Airy disc. This criterion is shown in Fig. 7.10(a). The net intensity profile shows a small dip of about 85% of the maxima each side—which is easily discernible by the eye or detectable by a photographic plate. Hence the angular resolution of a telescope is α_{zom}, the zero order minimum angle for destructive interference, i.e. referring to Section 7.4:

$$\text{limit of resolution} = \alpha_{zom} \approx \sin \alpha_{zom} = \frac{1.22\lambda}{d}.$$

Clearly, the greater is the aperture of a telescope, the smaller is the limit of resolution and this explains why astronomical telescopes in particular are constructed with objective mirrors with as large a diameter as possible.

 These considerations are also relevant to an estimation of the limit of resolution of cameras in which the diameter x of the Airy disc depends, additionally, on the focal length of the telescope or camera objective, f, the radius or half-diameter $x/2$ being given approximately by $x/2 = f \sin \alpha_{zom}$ (Fig. 7.11). Substituting from above gives $x = 2.44\lambda(f/d)$. The ratio (f/d) is called the f-number and is the important number engraved on camera lenses and the like. For blue-green light $\lambda \approx 0.5\ \mu\text{m}$ and therefore x is *roughly* equal to f-number expressed as micrometres. 'Stopping down' a camera lens (i.e. increasing f-number) means increasing the size of the Airy disc. How important this is (in comparison with other factors such as the depth of field) depends upon the relation between the size of the Airy disc and the 'grain size' of the photographic film

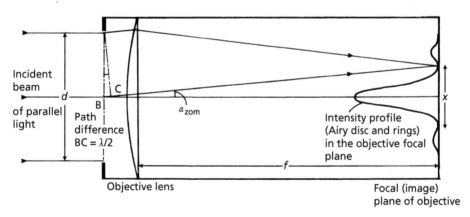

Fig. 7.11. The formation of the Airy disc in the focal plane of a telescope or camera. The half-diameter $x/2$ (shown much exaggerated) is given by $f \sin \alpha_{zom}$. (Compare with Fig. 7.8(b)).

* Denotes biographical notes available in Appendix 3.

or pixel size of the CCD. Clearly, deterioration in image sharpness will only begin to occur when the Airy disc becomes (roughly) equal to the grain or pixel size.

The diffraction-limited resolution of a microscope can be treated in two ways. First, it may be treated in the same way as for a telescope or camera—by considering the separation between the Airy discs of two closely spaced points of light and situated close to the front focal plane of the objective rather than (as in the telescope or camera) a long distance away. In so doing a simplifying assumption is made that the light from the two point sources is incoherent (just as in the case of two stars) and hence interference between them can be neglected (see Section 7.3). The assumption is generally invalid—microscopial specimens do not consist of independent points of light of different intensities across their surfaces. It is more reasonable to assume that the specimen is illuminated with *coherent* light and that interference occurs between the light waves diffracted at the light/dark regions across the specimen surface.[4] This leads us to the second way of treating the diffraction-limited resolution of the microscope which was first developed by Ernst Abbe,* the optical designer for Carl Zeiss.

It turns out from a rigorous analysis that the expressions for the limit of resolution are closely identical irrespective of the assumptions made about the illumination or self-illumination of the specimen. We shall follow both approaches, not least because they serve as useful applications of the diffraction theory we have already learned.

The first approach is represented in Fig. 7.12 which shows two such points of light and their Airy disc images—an axial point (solid lines) and an off-axis point

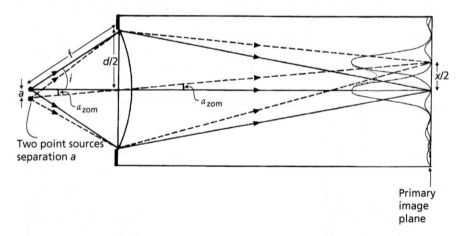

Fig. 7.12. The limit of resolution in a microscope for two points of light separated by distance a and which subtend angle α_{zom} at the objective lens. At this angle the zero order minimum of the Airy disc of one point source corresponds with the central maximum of the Airy disc of the adjacent point source. The sources are situated at a distance from the objective lens close to f, the focal length.

[4] The condition for incoherent illumination is closely approached in the case of fluorescence microscopy. Otherwise, in the usual Köhler illumination arrangement (in which each point in the light source gives rise to a parallel pencil of rays at the specimen as in Fig 7.6), the illumination is coherent.
* Denotes biographical notes available in Appendix 3.

(dashed lines). In a microscope the plane in which the Airy disc images occur is called the primary image plane—the plane which is 'observed' in the eyepiece.

Just as in a telescope, the limit of resolution is given when the angular separation between the two point sources, $\alpha_{zom} = 1.22\lambda/d$ (see Fig. 7.11). We now have to express this limit in terms of a, the separation of the two point sources and i, the semi-angle subtended by the objective lens at the point-sources. Let f be the focal length of the objective lens as indicated in Fig. 7.12, then $\sin i = d/2f$ and $\alpha_{zom} \approx a/f$. Hence $\alpha_{zom} = 1.22\lambda/d = a/f$ from which $a = 1.22\lambda/(f/d)$. Substituting $(f/d) = 1/2 \sin i$ we obtain

$$\text{limit of resolution} = a = \frac{1.22\lambda}{2\sin i} = \frac{0.61\lambda}{\sin i}.$$

In the case of microscope objectives of very small focal length (i.e. 'high power' objectives), the space between the specimen and front element of the lens may be filled with a drop of oil, of refractive index n, which is simply the ratio of the wavelength λ of light in air (or vacuum) to the wavelength λ' in the oil, i.e. by definition $n = \lambda/\lambda'$. Substituting this factor gives:

$$\text{limit of resolution} = a = \frac{0.61\lambda'}{\sin i} = \frac{0.61\lambda}{n\sin i}.$$

The second approach, due to Ernst Abbe, is as follows: in order for the microscope to form an image of the specimen (i.e. in order to resolve the slits or lines of a grating), at least two beams (usually the direct beam and the first order diffracted beam) should enter the objective lens. This situation is shown in Fig. 7.13. The specimen/grating is illuminated with a parallel beam of light (from a distant source or a sub-stage condenser in which the aperture diaphragm is almost fully closed) and the first order diffracted beams (on either side of the direct beam) are just collected by the objective lens.

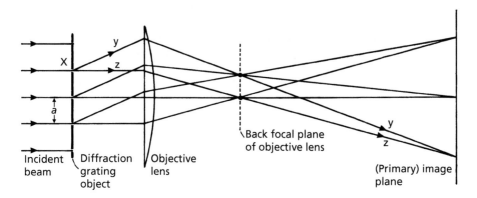

Fig. 7.13. Image formation in the light microscope. The direct and diffracted beams enter the objective lens and are focused to form a diffraction pattern in its (back) focal plane (compare with Fig. 7.6). The rays 'continue on their way' to form an image by the recombination of the diffracted light, e.g. incident ray **X** is separated into z (direct) and y (diffracted) light; these two rays recombine in the image plane.

These parallel beams are then focused by the objective lens to give diffraction spots which are situated, as shown, in its back focal plane (a screen placed here would record this diffraction pattern, just as shown in Fig. 7.6). However, the beams or rays continue on their way and intersect, as shown, to form the (primary) image, (which is then magnified by the eyepiece or photographic projection lens). The important point is that the image is formed by the *recombination* of light which was *separated* at the specimen by diffraction. Consider, for example, the incident beam marked X (Fig. 7.13); this is separated by diffraction into z (direct beam) and y (diffracted beam). These two beams then contribute to the separate (zero and first order) diffraction spots, but then continue on their way to be reunited or recombined and, as Abbe pointed out, it is this recombination that constitutes an image. The greater the number of diffracted beams which are collected by the objective lens and thence recombined, the closer the image approximates to the object, but the limit of resolution is determined by the criterion that at least two beams should enter the objective.

Hence, for the case of normal incidence, in which the direct and two first order beams enter the objective lens (Fig. 7.14(a)), the limit of resolution is simply obtained from the equation derived in Section 7.4, i.e. $1\lambda = a \sin \alpha_1 = a \sin i$, i.e. the limit of resolution $a = \lambda / \sin i$, where i is the semi-angle of the objective lens subtended at the specimen which is set equal to α_1 the angle between the direct and first order diffracted beams.

This equation may be modified in two ways. First, Fig. 7.13 shows an *overfulfilled* case in which, because the incident beam is normal to the specimen and axial with the microscope, *two* first-order diffracted beams, as well as the direct or zero order beam, enter the objective. This situation is sketched for emphasis in Fig. 7.14(a). However, if the incident beam is inclined to the axis of the microscope as shown in Fig. 7.14(b), then the first order beam on one side of the direct beam is lost but a first order beam on the other side of the direct beam of greater diffraction angle can enter the objective lens. The extreme or limiting case arises when the incident beam is inclined at angle i to the microscope axis (Fig. 7.14(c)), in which case the direct or zero order beam just passes inside one edge of the objective lens and one first order diffracted beam, now at twice the angle, $2i$ to the direct beam, just passes inside the opposite edge. The Abbe criterion is still satisfied because two beams (zero order and first order) enter the objective lens and are recombined to form an image in the same way as shown in Fig. 7.13.

We must now work out the path difference for this inclined case. The geometry is shown in detail in Fig. 7.15, which should be compared with that for normal incident light shown in Fig. 7.5. The path difference between the direct and first order diffracted beam is now $AB + BC = 1\lambda$ for constructive interference, i.e. $a \sin i + a \sin i = 2a \sin i$. This equation, it should be noted, is identical to the Bragg equation (see Section 8.3) since both describe the same geometrical conditions: in both cases the incident and diffracted beams are at the same angle to the diffraction grating or lattice planes (i.e. $i = \theta$) and the grating spacing a corresponds to the interplanar spacing d.

Rearranging the equation gives:

$$\text{limit of resolution} = a = \frac{\lambda}{2 \sin i} = \frac{0.5\lambda}{\sin i},$$

which differs from our previous equation by a factor of 0.5.

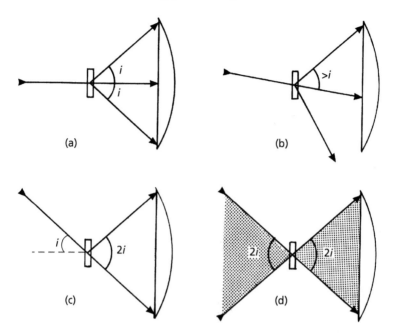

Fig. 7.14. The Abbe criterion for the limit of resolution of an objective lens of (semi-angular) aperture i. (a) normal incidence, parallel light (condenser diaphragm closed down); the direct and two first-order beams at angle i enter the objective and contribute to the image; (b) as (a) but a slightly inclined incident parallel beam, the direct beam and just one first-order beam enter the objective lens but at an angle greater than i; (c) as (b) but the incident beam is inclined at the maximum angle i; it passes inside one edge of the objective lens and one diffracted beam at angle $2i$ passes inside the opposite edge; (d) the practical arrangement; the condenser diaphragm is opened to admit a convergent cone of light on to the specimen (in practice rather less than $2i$) to admit direct and diffracted light to the objective over a whole range of angles.

Second, the space between the specimen and the front element of the objective lens may be filled with oil of refractive index n. Proceeding as before:

$$\text{limit of resolution} = a = \frac{0.5\lambda'}{\sin i} = \frac{0.5\lambda}{n \sin i}.$$

In practice we do not in general illuminate the specimen with a single parallel inclined beam as shown in Fig. 7.14(c), but we open the aperture diaphragm of the condenser lens to give a highly convergent beam on the specimen of semi-angle i (in practice slightly less) as shown in Fig. 7.14(d). This is analogous, in terms of illumination, to the case of the point sources of light radiating out in all directions as described above and the equations for the limit of resolution which we have derived—one for incoherent illumination (diffraction at the objective aperture) and the other for coherent illumination (diffraction at the specimen)—are, except for the Airy factor 1.22, precisely the same. As mentioned above, a much more rigorous analysis also predicts closely identical solutions.

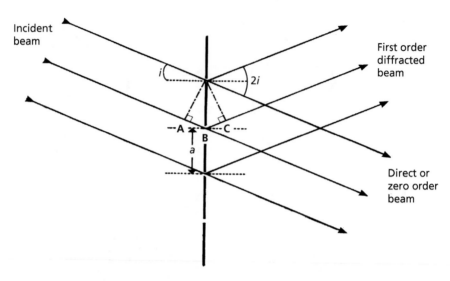

Fig. 7.15. The geometry for constructive interference when the incident and diffracted beams make equal angle i to the diffraction grating. For constructive interference (first order) the path difference = $AB + BC = a \sin i + a \sin i = 2a \sin i = 1\lambda$. This condition is equivalent to Bragg's law where a is the interplanar spacing d and i is θ, the Bragg angle.

The important quantity in the equations $n \sin i$ was called by Abbe the **numerical aperture, NA** and fulfils, in effect, the same role with respect to the microscope as does the aperture d of the objective lens (or mirror) with respect to the telescope. To summarize:

$$\text{Angular–limit of resolution of a telescope} = \alpha_{\text{zom}} = \frac{1.22\lambda}{d}$$

$$\text{Distance–limit of resolution of a microscope} = a = \frac{0.61\lambda}{\text{NA}} \text{ or } \frac{0.5\lambda}{\text{NA}}.$$

The concept of numerical aperture $\text{NA} = n \sin i$ is of enormous importance in optics. It is closely related to the f-number, the ratio of focal length/aperture, (f/d) of a lens. Referring again to Fig. 7.12, since $\sin i = d/2f$ then $(f/d) = 1/2 \sin i$ since $\sin i = \text{NA}/n$,

$$f\text{-number}, (f/d) = \frac{n}{2\text{NA}}.$$

The Abbe criterion applies equally to light microscopes and electron microscopes. The enormous decrease in the limit of resolution obtainable in the transmission electron microscope arises because of the much smaller values of λ for electrons (typically 4 pm for 100 kV instruments compared with light (typically 500 nm)). This (approximate) 100 000 decrease is however offset because of the high inherent spherical aberrations in electron lenses which limit NA values to the order of 0.01 compared with NA values up to 1.4 for oil immersion light objectives.

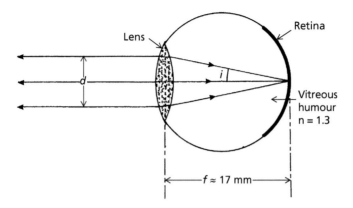

Fig. 7.16. The components of the eye (a) considered as a telescope or camera (light path to the right), the retina being the image plane. The angular-limit of resolution (for $d = 3\,mm$) = 0.77 min of arc or 0.07 mm at the near distance of 250 mm. (b) The eye considered as a water-immersion microscope objective (light path to the left), the retina being the object plane. The numerical aperture NA = 0.115 and the distance-limit of resolution = 2.9 μm (see text for calculations). These values correspond (a) for visual acuity of ≈0.1 mm at 250 mm and (b) to the spacing of the central retinal cone cells.

Finally, we may apply these ideas to those most important and precious optical instruments—our eyes.

We may treat the eye either as a telescope or camera to determine the angular-limit of resolution between two distant points, or as a microscope, with the direction of the light in effect reversed, to determine the distance-limit of resolution at the retina, the space between the objective (eye) lens and the retina (corresponding to the microscope object or specimen-plane) being filled with liquid (the vitreous humour) of refractive index n (see Fig. 7.16).

If we take $f = 17\,mm$, $n = 1.3$, d (the diameter of the pupil) ≈3 mm (which varies of course roughly in the range $1 \sim 4\,mm$ depending on the intensity of the light) and λ (for green light) ≈0.55 μm, then the f-number of the eye is $(17/3)\,mm$ = 5.67 and the numerical aperture NA is $1.3/2(5.67) = 0.115$.

Substituting these values of λ and d into the expression for the angular-limit of resolution α_{zom} gives 0.22×10^{-3} rads or 0.77 min of arc which is close to that for normal vision (~1 min of arc or 0.07 mm at the near distance of 250 mm).

Similarly, substitution of these values of λ and d into the expression for the distance-limit of resolution a at the retina gives 2.9 μm—which is approximately equal to the spacing of our visual receptors (the cone cells) in the central part (the *fovea centralis*) of our retina. In short, our eyes have evolved, as we should expect, with a perfect balance between the sizes of their physiological components.

A note on light and radio telescopes

It is of interest to compare the performance of light and radio telescopes with regard to the observation of extragalactic objects. The equation for the angular-limit of resolution,

$\alpha_{zom} = \dfrac{1.22\lambda}{d}$, applies (approximately) equally to radio telescopes where d is now the diameter of the paraboidal receiving dish and λ is in the range 5 cm upwards. For $\lambda = 1\,\text{m}$ and $d = 100\,\text{m}$ we obtain $\alpha_{zom} = 0.012\,\text{rads} = 42\,\text{min}$ of arc—a resolution inferior to that of the human eye. However, it should be noted that substantially decreased limits of resolution are obtained by interferometric techniques, i.e. by linking the signals from widely spaced radio telescopes. Resolutions down to a thousandth of an arc second or so are obtainable using the technique of *very long baseline interferometry* (VLBI). This is superior to anything that can be presently obtained using optical telescopes. Radio telescopes also have the advantage of using larger area collecting dishes than is currently possible with optical telescopes. They therefore enable very faint radio sources to be detected.

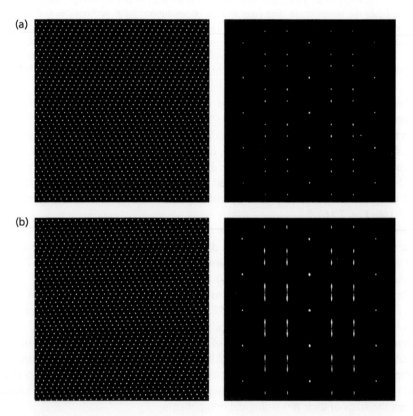

Fig. 7.17. (a) A net with a single twin, and (b) a net with several twins and their corresponding diffraction patterns (from *Atlas of Optical Transforms* by G. Harburn, C. A. Taylor and T. R. Welberry, Bell & Hyman, 1983, an imprint of HarperCollins).

Exercises

7.1 Figure 7.17(a) shows a net with a single twin and the corresponding diffraction pattern (optical transform) and Fig. 7.17(b) shows a net with several twins and the corresponding diffraction pattern (see Figs. 1.18 and 1.21). Using a tracing paper overlay on the diffraction pattern in Fig. 7.17(a), outline the reciprocal lattice unit cells of the two twin orientations and note those reciprocal lattice points which are common to both twin orientations.

 The diffraction pattern in Fig. 7.17(b) is similar to that in Fig. 7.17(a) except that some of the spots are streaked in a direction perpendicular to the twin plane. Explain this streaking qualitatively after you have read Section 9.3.

7.2 With reference to Fig. 7.7(c), describe the diffraction pattern which you would obtain with a diffraction grating in which the slit width d is equal to half the slit spacing a.

8

X-ray diffraction: the contributions of Max von Laue, W. H. and W. L. Bragg and P. P. Ewald

8.1 Introduction

The experimental technique which has been of the greatest importance in revealing the structure of crystals is undoubtedly X-ray diffraction.

The story of the discovery of X-ray diffraction in crystals by Laue,* Friedrich* and Knipping* in Munich in 1912 and the development of the technique by W. H. Bragg* and W. L. Bragg* in Leeds and Cambridge in the years preceding the First World War is well known. But why did the Braggs make such rapid advances in the analysis of X-ray diffraction photographs in comparison with Laue and his co-workers? An important factor in the answer seems to be that Laue envisaged crystals in terms of a three-dimensional network of rows of atoms and based his analysis on the notion that the crystal behaved, in effect, as a three-dimensional diffraction grating. This approach is not wrong, but it is in practice rather clumsy or protracted. On the other hand, the Braggs (and here the credit must go to W. L. Bragg, the son) envisaged crystals in terms of layers or planes of atoms which behaved in effect as reflecting planes (for which the angle of incidence equals the angle of reflection), strong 'reflected' beams being produced when the path differences between reflections from successive planes in a family is equal to whole number of wavelengths. This approach is not correct in a physical sense—planes of atoms do not reflect X-rays as such—but it is correct in a geometrical sense and provides us with the beautifully simple expression for the analysis of crystal structure:

$$n\lambda = 2d_{hkl}\sin\theta,$$

where λ is the wavelength, n is the order of reflection, d_{hkl} is the lattice plane spacing and θ is the angle of incidence/reflection to the planes.

What led W. L. Bragg to this novel perception of the diffraction? Simply his observation of the elliptical shapes of the diffraction spots, which he noticed were also characteristic of the reflections from mirrors of a pencil-beam of light (see the Laue photograph on p. xiv). Only connect!

* Denotes biographical notes available in Appendix 3.

The Basics of Crystallography and Diffraction. Fourth Edition. Christopher Hammond.
© Christopher Hammond 2015. Published in 2015 by Oxford University Press.

Finally, we come to the contribution of P. P. Ewald,* a physicist who never achieved the recognition that was his due. The story is briefly recorded in his autobiographical sketch in *50 Years of X-ray Diffraction*. Ewald was a 'doctorand'—a research student working in the Institute of Theoretical Physics in the University of Munich under Professor A. Sommerfeld. The subject of his thesis was 'To find the optical properties of an anisotropic arrangement of isotropic oscillators'. In January 1912, while he was in the final stages of writing up his thesis, he visited Max von Laue, a staff member of the Institute, to discuss some of the conclusions of his work. Ewald records that Laue listened to him in a slightly distracted way and insisted first on knowing what was the distance between the oscillators in Ewald's model; perhaps 1/500 or 1/1000 of the wavelength of light, Ewald suggested. Then Laue asked 'what would happen if you assumed very much shorter waves to travel through the crystal?' Ewald turned to Paragraph 6, Formula 7, of his thesis manuscript, saying 'this formula shows the results of the superposition of all wavelets issuing from the resonators. It has been derived without any neglection or approximation and is therefore valid also for short wave-lengths.' Ewald copied the formula down for Laue shortly before taking his leave, saying that he, Laue, was welcome to discuss it. Laue's question, of course, arose from his intuitive insight that if X-rays were waves and not particles, with wavelengths very much smaller than light, then they might be diffracted by such an array of regularly spaced oscillators.

The next Ewald heard of Laue's interest was through a report which Sommerfeld gave in June 1912 on the successful Laue–Friedrich–Knipping experiments. He realized that the formula which he had copied down for Laue, and which Laue had made no use of, provided the obvious way of interpreting the geometry of the diffraction patterns—by means of a construction which he called the reciprocal lattice and a sphere determined by the mode of incidence of the X-rays on the crystal (the Ewald or reflecting sphere). Ewald's interpretation of the geometry of X-ray diffraction was not published until 1913, by which time rapid progress in crystal structure analysis had already been made by W. H. and W. L. Bragg in Leeds and Cambridge.

8.2 Laue's analysis of X-ray diffraction: the three Laue equations

Laue's analysis of the geometry of X-ray diffraction patterns has been referred to in Section 8.1. What follows is a much simplified treatment which does not take into consideration Laue's interpretation of the origin of the diffracted waves from irradiated crystals.

Consider a simple crystal in which the motif is one atom and the atoms are simply to be regarded as scattering centres situated at lattice points. The more general situation in which the motif consists of more than one atom and in which the different scattering amplitudes of the atoms and the path differences between the atoms have to be taken into account is discussed in Section 9.2. The crystal may be considered to be built up

* Denotes biographical notes available in Appendix 3.

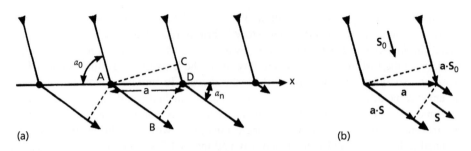

Fig. 8.1. (a) Diffraction from a lattice row along the x-axis. The incident and diffracted beams are at angles α_0 and α_n to the row respectively. The path difference between the diffracted beams = (AB - CD). (b) The incident and diffracted beam directions and the path difference between the diffracted beams as expressed in vector notation.

of rows of atoms in three dimensions: rows of atoms of spacing a along the x-axis, of spacing b along the y-axis and of spacing c along the z-axis. Consider first of all the condition for constructive interference for the waves scattered from the row of atoms along the x-axis—which may simply be reduced to a consideration of the path differences between waves scattered from adjacent atoms in the row (Fig. 8.1(a)).

For constructive interference the path difference (AB − CD) must be a whole number of wavelengths, i.e.

$$(\text{AB} - \text{CD}) = a(\cos\alpha_n - \cos\alpha_0) = n_x\lambda$$

where α_n, α_0 are the angles between the diffracted and incident beams to the x-axis, respectively, and n_x is an integer (the order of diffraction).

This equation, known as the first Laue equation, may be expressed more elegantly in vector notation. Let \mathbf{s}, \mathbf{s}_0 be unit vectors along the directions of the diffracted and incident beams, respectively, and let \mathbf{a} be the translation vector from one lattice point (or atom position) to the next (Fig. 8.1(b)). The path difference $a(\cos\alpha_n - \cos\alpha_0)$ may be represented by the scalar product $\mathbf{a} \cdot \mathbf{s} - \mathbf{a} \cdot \mathbf{s}_0 = \mathbf{a} \cdot (\mathbf{s} - \mathbf{s}_0)$. Hence the first Laue equation may be written

$$a\,(\cos\alpha_n - \cos\alpha_0) = \mathbf{a} \cdot (\mathbf{s} - \mathbf{s}_0) = n_x\lambda.$$

Now Fig. 8.1 is misleading in that it only shows the diffracted beam at angle α_n *below* the atom row—but the same path difference obtains if the diffracted beam lies in the plane of the paper at angle α_n *above* the atom row—or indeed *out of the plane of the paper* at angle α_n to the atom row. Hence all the diffracted beams with the same path difference occur at the same angle to the atom row, i.e. the diffracted beams of the same order all lie on the surface of a cone—called a Laue cone—centred on the atom row with semi-apex angle α_n. This situation is illustrated in Fig. 8.2 which shows just three Laue cones with semi-apex angle α_0 (zero order, $n_x = 0$), semi-apex angle α_x (first order, $n_x = 1$) and semi-apex angle α_2 (second order, $n_x = 2$). Clearly there will be a whole set of such cones with semi-apex angles α_n varying between $0°$ and $180°$.

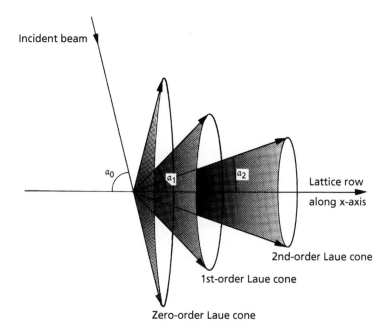

Incident beam

a_0

a_1

a_2

Lattice row
along x-axis

2nd-order Laue cone

1st-order Laue cone

Zero-order Laue cone

Fig. 8.2. Three Laue cones representing the directions of the diffracted beams from a lattice row along the x-axis with $0\lambda(n_x = 0)$, $1\lambda(n_x = 1)$ and $2\lambda(n_x = 2)$ path differences. The corresponding Laue cones for $n_x = -1$, $n_x = -2$ etc. lie to the left of the zero order Laue cone.

The analysis is now repeated for the atom row along the y-axis, giving the second Laue equation:

$$b\,(\cos\beta_n - \cos\beta_0) = \mathbf{b}\cdot(\mathbf{s} - \mathbf{s}_0) = n_y\lambda,$$

and for the atom row along the z-axis giving the third Laue equation:

$$c\,(\cos\gamma_n - \cos\gamma_0) = \mathbf{c}\cdot(\mathbf{s} - \mathbf{s}_0) = n_z\lambda,$$

where the angles β_n, β_0, γ_n, γ_0 and the integers n_y and n_z are defined in the same way as for α_n, α_0 and n_x.

Now, for constructive interference to occur simultaneously from all three atom rows, all three Laue equations must be satisfied simultaneously. This is equivalent to the geometrical condition that diffracted beams only occur in those directions along which three Laue cones, centred along the x-, y- and z-axes, intersect. Each diffracted beam may be identified by three integers n_x, n_y and n_z which, as pointed out above, represent the order of diffraction from each of the atom rows. We shall find in Section 8.3 that these integers are simply h, k and l Laue indices (see Section 5.5) of the reflecting planes in the crystal.

8.3 Bragg's analysis of X-ray diffraction: Bragg's law

Laue's analysis is in effect an extension of the idea of a diffraction grating to three dimensions. It suffers from the severe practical disadvantage that in order to calculate the directions of the diffracted beams, a total of six angles α_n, α_0, β_n, β_0 and γ_n, γ_0, three lattice spacings a, b and c, and three integers n_x, n_y and n_z need to be determined. As discussed in Section 5.5, W. L. Bragg envisaged diffraction in terms of reflections from crystal planes giving rise to the simple relationship (Bragg's law, derived below):

$$n\lambda = 2d_{hkl}\sin\theta.$$

It can be seen immediately, by comparing the Laue equations with Bragg's law, that the number of variables needed to calculate the directions of the diffracted beams are much reduced.

Bragg's law may be derived with reference to Fig. 8.3(a) which shows (as for the derivation of the Laue equations) a simple crystal with one atom at each lattice point. The path difference between the waves scattered by atoms from adjacent (hkl) lattice planes of spacings d_{hkl} is given by

$$(AB + BC) = (d_{hkl}\sin\theta + d_{hkl}\sin\theta) = 2d_{hkl}\sin\theta.$$

Hence for constructive interference:

$$n\lambda = 2d_{hkl}\sin\theta,$$

where n is an integer (the order of reflection or diffraction).

As explained in Section 5.5, n is normally incorporated into the lattice plane symbol, i.e.

$$\lambda = 2\left(\frac{d_{hkl}}{n}\right)\sin\theta = 2d_{nh\,nk\,nl}\sin\theta$$

where $nh\,nk\,nl$ are the Laue indices for the reflecting planes of spacing d_{hkl}/n. In other words, to repeat the important point made in Section 5.5, n is not written separately

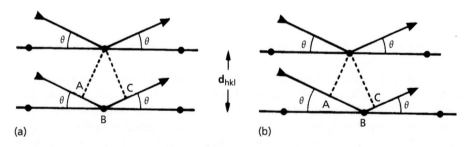

Fig. 8.3. (a) Bragg's law for the case of a rectangular grid, i.e. $AB = BC = d_{hkl}\sin\theta$; the path difference $(AB + BC) = 2d_{hkl}\sin\theta$. (b) Bragg's law for the general case in which $AB \neq BC$. Again, the path difference $(AB + BC) = 2d_{hkl}\sin\theta$.

but is represented as the common factor in the Laue indices. For example a *third* order reflection from the (111) lattice planes (Miller indices—in parentheses) is represented as a *first* order reflection from the 333 planes (Laue indices—no parentheses), the 333 planes having l/3rd the spacing of the (111) planes.

Figure 8.3(a) represents a particularly simple geometrical situation in which the lattice is shown as a rectangular grid and the atoms are symmetrically disposed with respect to the incident and diffracted beam, i.e. AB = BC. With reference to the diffraction of light this corresponds to the case in which the incident and diffracted beams make equal angles to the diffraction grating—i.e. the situation shown in Fig. 7.15 where the angle i corresponds to θ and the slit spacing a corresponds to d_{hkl}. Figure 8.3(b) shows a more general situation in which the lattice is not rectangular and the distance AB does not equal BC. However, the sum (AB+BC) is unchanged and is again equal to $2d_{hkl}\sin\theta$ (see Exercise 8.3). The important point is that Bragg's law applies irrespective of the positions of the atoms in the planes; it is solely the spacing between the planes which needs to be considered. It follows as a corollary that the path difference between the waves scattered by the atoms in the same plane is zero—i.e. all the waves scattered from the same plane interfere constructively. This is *only* the case (to emphasize the significance of Bragg's law once more) when the angle of incidence to the planes equals the angle of reflection. Finally, note that (unlike the Laue equations), Bragg's law is wholly represented in two dimensions: the incident and diffracted beams and the normal to the reflecting planes (Fig. 8.3) all lie in a plane—i.e. the plane of the paper.

Bragg's law may also be expressed in vector notation. Again, let \mathbf{s}, \mathbf{s}_0 be unit vectors along the directions of the diffracted and incident beams, then (with reference to Fig. 8.4) the vector $(\mathbf{s} - \mathbf{s}_0)$ is parallel to \mathbf{d}^*_{hkl}, the reciprocal lattice vector of the reflecting planes. Comparing the moduli of these vectors $|\mathbf{s} - \mathbf{s}_0| = 2\sin\theta$ and $|\mathbf{d}^*_{hkl}| = 1/d_{hkl}$, it is seen from Bragg's law that their ratio is simply λ. Hence Bragg's law may be written:

$$\frac{(\mathbf{s} - \mathbf{s}_0)}{\lambda} = \mathbf{d}^*_{hkl} = h\mathbf{a}^* + k\mathbf{b}^* + l\mathbf{c}^*.$$

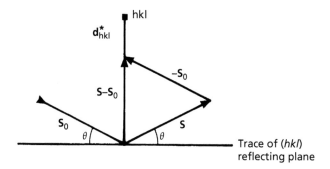

Fig. 8.4. Bragg's law expressed in vector notation. Vectors $(\mathbf{s} - \mathbf{s}_0)$ and \mathbf{d}^*_{hkl} are parallel and the ratio of the moduli is λ. Hence Bragg's law is expressed as $(\mathbf{s} - \mathbf{s}_0)/\lambda = \mathbf{d}^*_{hkl}$.

Hence constructive interference occurs, or Bragg's law is satisfied, when the vector $(\mathbf{s} - \mathbf{s}_0)/\lambda$ coincides with the reciprocal lattice vector \mathbf{d}^*_{hkl} of the reflecting planes.

The vector form of Bragg's law may be combined with each of the three Laue equations: i.e. for the first Laue equation:

$$\mathbf{a} \cdot (\mathbf{s} - \mathbf{s}_0) = n_x\lambda = \mathbf{a} \cdot \mathbf{d}^*_{hkl} \cdot \lambda = \mathbf{a} \cdot (h\mathbf{a}^* + k\mathbf{b}^* + l\mathbf{c}^*)\lambda.$$

Hence $n_x = h$ (since $\mathbf{a} \cdot \mathbf{a}^* = 1$, $\mathbf{a} \cdot \mathbf{b}^* = 0$, etc), and similarly $n_y = k$ and $n_z = l$ for the other Laue equations. The integers n_x, n_y and n_z of the Laue equations are simply the Laue indices h, k, l of the reflecting planes.

Bragg's law, like Newton's laws, and all such uncomplicated expressions in physics, is deceptively simple. Its applicability and relevance to problems in X-ray and electron diffraction only unfold themselves gradually (to teachers and students alike!). Newton was once asked how he made his great discoveries: he replied 'by always thinking unto them'. The student of crystallography could do no better with respect to Bragg's law![1]

8.4 Ewald's synthesis: the reflecting sphere construction

Ewald's synthesis is a geometrical formulation or expression of Bragg's law which involves the reciprocal lattice and a 'sphere of reflection'. It is best illustrated and understood by way of an example. Consider a crystal with the (hkl) reflecting planes (Laue indices) at the correct Bragg angle (Fig. 8.5). The reciprocal lattice vector \mathbf{d}^*_{hkl} is also shown. Now draw a sphere (the reflecting or Ewald sphere) of radius $1/\lambda$ (where λ is the X-ray wavelength) with the crystal at the centre. Since Bragg's law is satisfied it may be shown that the vector **OB** (from the point where the direct beam exits from the sphere to the point where the diffracted beam exits from the sphere) is identical to \mathbf{d}^*_{hkl}: i.e. from the triangle AOC, $|\mathbf{OC}| = (1/\lambda)\sin\theta = (1/2)|\mathbf{d}^*_{hkl}| = 1/2d_{hkl}$, i.e. $\lambda = 2d_{hkl}\sin\theta$. Hence, if the origin of the reciprocal lattice is shifted from the centre of the sphere (A) to the point where the direct beam exits from the sphere (O), then $\mathbf{OB} = \mathbf{d}^*_{hkl}$ and Bragg's law is equivalent to the statement that the reciprocal lattice point for the reflecting planes (hkl) should intersect the sphere; the diffracted beam direction being given by the vector **AB**—i.e. the line from the centre of the sphere to the point where the reciprocal lattice point \mathbf{d}^*_{hkl} intersects the sphere. Conversely, if the reciprocal lattice point does *not* intersect the sphere then Bragg's law is not satisfied and no diffracted beams occur. Finally, note the equivalence of the Ewald reflecting sphere

[1] The notion that the spots in a Laue diffraction pattern arise from reflections of the primary beam is also due to Torahiko Terada (1878–1935), Professor at the Imperial University in Tokyo, who carried out Laue experiments on a wide range of crystals. By using a broad (5~10 mm diameter) primary beam he was able to observe the shapes and movements of the spots (corresponding to rotations of the crystal) on a fluorescent screen. The work was presented in a paper to the Tokyo Physiomathematical Society on 3rd May 1913, at which time Bragg's papers would not have reached Japan. Terada was also a distinguished man of letters and top essayist of the Meiji period in Japan.

The contribution of Georg Wulff (1863–1925) in Russia is discussed in the biographical notes in Appendix 3.

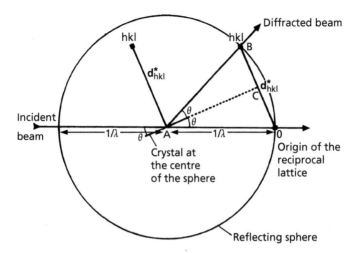

Fig. 8.5. The Ewald reflecting sphere construction for a set of planes at the correct Bragg angle. A sphere (the Ewald or reflecting sphere) is drawn, or radius $1/\lambda$ with the crystal at the centre. The vector **OB** is identical to d^*_{hkl}. The origin of the reciprocal lattice is fixed at O and the reciprocal lattice point *hkl* intersects the sphere at the exit point of the diffracted beam.

construction to the vector form of Bragg's law $\dfrac{(\mathbf{s} - \mathbf{s_o})}{\lambda} = (\mathbf{k} - \mathbf{k_o}) = d^*_{hkl}$ (Section 8.3). The vector AO $= \mathbf{k_o}$, the vector AB $= \mathbf{k}$ and again the origin of d^*_{hkl} is shifted from A to O (Fig. 8.5).[2]

Figure 8.5 shows the construction of just one reflecting plane and one reciprocal lattice point. It is a simple matter to extend it to all the reciprocal lattice points in a crystal. Figure 8.6(a) shows a section of the reciprocal lattice of a monoclinic crystal perpendicular to the $\mathbf{b^*}$ reciprocal lattice vector (i.e. the *y*-axis—see Fig. 6.5(a)). All the reciprocal lattice points in this section have indices of the form $h0l$. An incident X-ray beam is directed along the $\mathbf{a^*}$ reciprocal lattice vector (i.e. along a direction in the crystal perpendicular to the *y*-and *z*-axes—see Fig. 6.4(a)). The centre of the reflecting sphere is at a distance $1/\lambda$ from the origin of the reciprocal lattice along the line of the incident beam: note again that the origin of the reciprocal lattice is *not* at the centre of the sphere but is at the point where the direct beam exits from the sphere.

In the section shown, Fig. 8.6(a), the reflecting sphere intersects the 201 reciprocal lattice point: hence the only plane for which Bragg's law is satisfied is the (201) plane and the direction of the 201 reflected beam is as indicated.

Figure 8.6(a) only shows one layer or section of the reciprocal lattice (through the origin) and a (diametral) section of the reflecting sphere. But the reciprocal lattice

[2] An alternative vector notation, widely used in transmission electron microscopy (Chapter 11), is to write $\mathbf{k_o} = \mathbf{s_o}/\lambda$ and $\mathbf{k} = \mathbf{s}/\lambda$. Hence Bragg's law in vector notation is $\mathbf{k} - \mathbf{k_o} = d^*_{hkl} = h\mathbf{a^*} + k\mathbf{b^*} + l\mathbf{c^*}$. The advantage of this notation is that the moduli of $\mathbf{k_o}$ and \mathbf{k} are equal to the radius of the Ewald reflecting sphere (see Section 8.4). Further, the symbol \mathbf{g}_{hkl} (no star) is widely used instead of $\mathbf{d^*_{hkl}}$.

sections above and below the plane of the page need to be taken into account and also the smaller non-diametral sections of the reflecting sphere which intersects them. Fig. 8.6(b) shows the 'next layer above' or $h1l$ section of the reciprocal lattice (see Fig. 6.5(b)) and the smaller, non-diametral section of the reflecting sphere which it intersects. Notice that this section of the sphere does *not* pass through 010 (the reciprocal lattice point immediately 'above' the origin 000). The sphere intersects the $21\bar{1}$ reciprocal lattice point, hence the $(21\bar{1})$ plane also satisfies Bragg's law and the direction of the $21\bar{1}$ reflected beam is from the centre of the sphere (which is *not* the point in

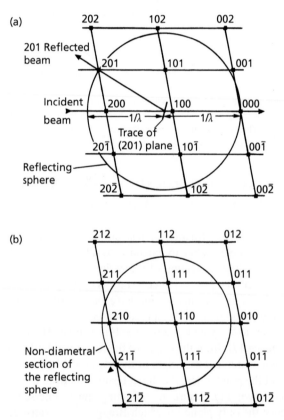

Fig. 8.6. The Ewald reflecting sphere construction for a monoclinic crystal in which the incident X-ray beam is directed along the \mathbf{a}^* reciprocal lattice vector, (a) Shows the $h0l$ reciprocal lattice section (through the origin and perpendicular to the \mathbf{b}^* reciprocal lattice vector or y-axis—see Fig. 6.5(a) and a diametral section of the reflecting sphere radius $1/\lambda$. The 201 reciprocal lattice point intersects the sphere and the direction of the 201 reflected beam is indicated, (b) Shows the $h1l$ reciprocal lattice section (i.e. the layer 'above' the $h0l$ section—see Fig. 6.5(b)) and the smaller, non-diametral section of the reflecting sphere. The $21\bar{1}$ reciprocal lattice point intersects the sphere, the direction of the $21\bar{1}$ reflected beam is 'upwards' from the centre of the sphere (in the $h0l$ section below) through the $21\bar{1}$ reciprocal lattice point as indicated by the arrow-head.

the centre of this smaller circle but the point in the centre of the sphere in the reciprocal lattice section below—Fig. 8.6(a)) and the $21\bar{1}$ reciprocal lattice point. The reflected beam direction therefore cannot be drawn in Fig. 8.6(b) because it is directed upwards, out of the plane of the page.

Clearly, the construction can be extended to other reciprocal lattice sections, i.e. through the $h2l$, $h3l$, etc. sections (above); the $h\bar{1}l, h\bar{2}l, h\bar{3}l$ etc. sections (below) and so on. The further the reciprocal lattice section is from the origin, the smaller is the section of the reflecting sphere which it intersects. In the example above, the sphere only intersects the $h0l$ section (Fig. 8.6(a)), the $h1l$ section (Fig. 8.6(b)) and the $h\bar{1}l$ section (below), but no more. Hence only these sections, and the reciprocal lattice points within them, need to be considered.

In Fig. 8.6 the relative sizes of the reciprocal lattice and the sphere happen to be such that just one reciprocal lattice point in each section intersects the sphere. Clearly if the diameter of the sphere were made a little larger (i.e. the X-ray wavelength was made a little smaller) then no reciprocal lattice points would intersect the sphere and no planes in the crystal would be at the correct Bragg angle for reflection; if the diameter of the sphere were continuously made larger (or smaller) than that shown in Fig. 8.6 then other planes would reflect as their reciprocal lattice points successively intersected the sphere. This is the basis of Laue's original X-ray experiment; 'white' X-radiation was used which contains a range of wavelengths, which correspond to a range or 'nest' of spheres of different diameters. Any plane whose reciprocal lattice point falls within this range will therefore satisfy Bragg's law for one particular wavelength.

The situation is illustrated in Fig. 8.7, again for the $h0l$ reciprocal lattice section of the monoclinic crystal shown in Fig. 8.6(a). The shaded region indicates a 'nest' of spheres with diameters in the wavelength range from largest wavelength (smallest sphere diameter) to smallest wavelength (largest sphere diameter). All the planes whose reciprocal lattice points lie within this region satisfy Bragg's law for the particular sphere on which they lie.

The Laue technique is unique in that it utilizes white X-radiation. All the others utilize monochromatic or near-monochromatic (K_α) X-radiation. In order to obtain diffraction, therefore, the crystal and the sphere (of a fixed diameter) must be moved relative to one another; whenever a reciprocal lattice point touches the sphere then 'out shoots' a diffracted (or reflected) beam from the centre of the sphere in a direction through the reciprocal lattice point. As described in Chapter 9 there are several ways in which these relative movements may be achieved in practice and several ways in which the diffracted beams may be recorded. The crystal may be oscillated (oscillation method), precessed (precession method), the film may be arranged cylindrically round the crystal or flat, it may be stationary or it may be moved in some way as in the Weissenberg method (in which the cylindrical film movement is linked to the oscillation of the crystal) or, as in the precession method, precessed with the crystal. The geometry of these X-ray diffraction methods may *appear* to be complicated, but the basis of them all—the Ewald reflecting sphere construction—is the same.

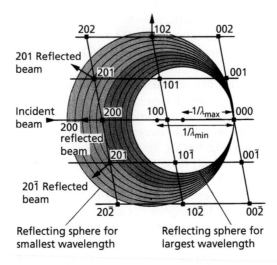

Fig. 8.7. The Ewald reflecting sphere construction for the *h0l* reciprocal lattice section of the mono-clinic crystal shown in Fig. 8.6(a) for the 'white' X-radiation, i.e. for a range of wavelengths from the smallest (largest sphere) to the largest (smallest sphere) giving rise to a 'nest' of spheres (shaded re-gion) all passing through the origin. The 102, 201, 200 and 20$\bar{1}$ reciprocal lattice points lying within this region satisfy Bragg's law, each for the particular wavelength or sphere which they intersect. The directions of the reflected beams from these planes are indicated by the arrows.

A note on light diffraction and Bragg's law

The *d*-spacings, in both organic and inorganic crystals, are smaller than the spectral range of the wavelengths of light (\sim0.4–0.7 μm) by a factor of about 10^3. Hence, the path lengths between scattered light waves are also smaller by this factor and the con-ditions for constructive/destructive interference do not occur. However, some naturally occurring minerals, such as (precious) opal have complex microstructures which in opal consist of an amorphous silica matrix in which there are distributions of close-packed, or nearly closed packed, arrays of silica spheres of diameter \sim0.3 μm (close to the limit of resolution of the light microscope). The light-scattering of these close-packed arrays can be described in terms of Bragg's law; the spacings between the close-packed lay-ers (\sim0.24 μm) corresponding to the Bragg *d*-spacings. Since diffraction is occurring within the mineral, refractive index n_p (\approx 1.45 for opal), Bragg's law is expressed as $n\lambda = 2n_p d\sin\theta$. For the above spectral range of white light, only the first and second orders of reflection occur, each with a dispersion of $\sin\theta$ values corresponding to the whole spectral range giving rise to the characteristic 'iridescent' colours.

Bragg scattering of light is made use of in **photonic crystals** (precious opal is a photonic crystal). Consider an idealized one-dimensional crystal with scattering centres spaced *d* along the length. Light of wavelength 2*d* travelling parallel to the line of scat-terers will be wholly reflected ($\theta = 90°$) and will not be transmitted through the crystal (a photonic band-gap). From this simple exercise it can be seen that photonic crystals can be designed as optical 'gates'.

Exercises

8.1 Compare Figs 7.5 and 8.3; both show conditions for constructive interference, one for light at a diffraction grating with line spacing a (Fig. 7.5) and one for X-rays reflected from planes of spacing d_{hkl} (Fig. 8.3). Show that the equations describing the conditions for constructive interference in each case ($n\lambda = a \sin\alpha_n$ and $n\lambda = 2d_{hkl} \sin\theta$) are equivalent.

8.2 Iron (bcc, $a = 0.2866$, nm (2.866 Å)) is irradiated with CrKα X-radiation ($\lambda = 0.2291$ nm (2.291 Å)). Find the indices $\{hkl\}$ and d-spacings of the planes which give rise to X-ray reflections.

 (*Note:* In body-centred lattices, reflection from planes for which $(h + k + l)$ does *not* equal an even integer are forbidden (see Appendix 6).)

 (*Hint*: Prepare a table listing the indices and d-spacings of the allowed reflecting planes in order of decreasing d-spacings and determine the θ angles for reflection using Bragg's law.)

8.3 It is stated without proof with respect to Bragg's law that when the atoms are not symmetrically disposed to the incident and reflected beams (Fig. 8.3(b)), the path difference $(AB + BC) = 2d_{hkl} \sin\theta$. Prove, using very simple geometry, that this is indeed the case.

9

The diffraction of X-rays

9.1 Introduction

In Chapter 8, the Laue equations and Bragg's law were derived on the basis that single atoms, of unspecified scattering power, were situated at each lattice point. Now we need to consider the physics of the scattering process. Since it is almost exclusively the electrons in atoms which contribute to the scattering of X-rays we have to sum the contributions to the scattered amplitude of all the electrons in all the atoms in the crystal, a problem which may be approached step-by-step. First the scattering amplitude of a single electron and the variation in scattering amplitude with angle is determined.[1] Then the scattering amplitude of an atom is determined by summing the contributions from all Z electrons (where Z = the atomic number of the atom)—the summation taking into account the path or phase differences for all the Z scattered waves. The result of this analysis is expressed by a simple number, f, the **atomic scattering factor**, which is the ratio of the scattering amplitude of the atom divided by that of a single (classical) electron, i.e.

$$\text{atomic scattering factor } f = \frac{\text{amplitude scattered by atom}}{\text{amplitude scattered by a single electron}}.$$

At zero scattering angle, all the scattered waves are in phase and the scattered amplitude is the simple sum of the contribution from all Z electrons, i.e. $f = Z$. As the scattering angle increases, f falls below Z because of the increasingly destructive interference effects for all the Z scattered waves. Atomic scattering factors f are plotted as a function of angle (usually expressed as $\sin \theta/\lambda$). Figure 9.1 shows such a plot for the oxygen anion O^{2-}, the neon atom Ne, and the silicon cation Si^{4+}—all of which contain 10 electrons. When $\sin \theta/\lambda = 0, f = 10$ but with increasing angle f falls below 10. The extent to which it does depends upon the relative sizes of the atoms or ions; the silicon cation is small, hence the phase differences are small and the destructive interference between the scattered waves is least—and conversely for the large oxygen anion.

[1] The elastic, or coherent scattering, of X-rays by a single (free) electron (no change in wavelength)—which is of importance in X-ray diffraction—was first worked out by J.J. Thomson in 1903. We need not make explicit use of the Thomson equation except to note that it includes a factor '$e^2/4\pi\varepsilon_0 c^2 m$', which has the dimensions of length, numerically equal to 2.82×10^{-15}m; referred to in classical electromagnetic theory as the radius of the electron. Inelastic or incoherent scattering, in which the scattered X-rays have longer wavelengths than the incident X-rays, is known as Compton scattering. These topics are discussed in detail in, e.g. *X-ray Crystallography* by M.M. Woolfson, Cambridge University Press (2nd edn. 1997).

The Basics of Crystallography and Diffraction. Fourth Edition. Christopher Hammond.
© Christopher Hammond 2015. Published in 2015 by Oxford University Press.

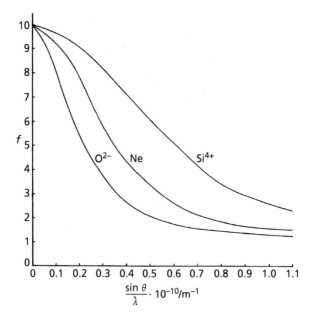

Fig. 9.1. The variation in atomic scattering factor f with scattering angle (expressed as $\sin \theta/\lambda$) for atoms and ions with ten electrons. Note that the decrease in f is greatest for the (large) O^{2-} anion and least for the (small) Si^{4+} cation.

The scattering amplitude of a unit cell is determined by summing the scattering amplitudes, f, from all the atoms in the unit cell (equivalent, in the case of a primitive unit cell, to all the atoms in the motif). Again, the summation must take into account the path or phase differences between all the scattered waves and is again expressed by a dimensionless number, F_{hkl}, the **structure factor**, i.e.

$$\text{structure factor } F_{hkl} = \frac{\text{amplitude scattered by the atoms in the unit cell}}{\text{amplitude scattered by a single electron}}.$$

F_{hkl} must not only express the amplitude of scattering from a reflecting plane with Laue indices hkl but must also express the phase angle of the scattered wave, an important concept which is explained in Section 9.2 below. F_{hkl} is therefore not a simple number, like f, but is represented as a vector or mathematically as a complex number (see Appendix 5).

Crystal structure determination is a two-part process: (a) the determination of the size and shape of the unit cell (i.e. the lattice parameters) from the geometry of the diffraction pattern and (b) the determination of the lattice type and distribution of the atoms in the structure from the (relative) intensities of the diffraction spots. Part (a) is in principle a straightforward process; part (b) is not, because films and counters record intensities which are proportional to the *squares* of the amplitudes. The square of a complex number, F_{hkl} is always real (i.e. a simple number) and hence the information about

the phase angles of the diffracted beams is lost. A major problem in crystal structure determination is, in effect, the recovery of this phase information and requires for its solution as much insight and intuition as mathematical and crystallographic knowledge. A graphic account of the problems involved in the epoch-making determination of the structure of DNA is given by one of those most closely involved, James D. Watson, in his book *The Double Helix* (see Further Reading).

The distinction between the direction-problem and intensity-problem of diffracted beams from crystals has its corollary in the diffraction of light from diffraction gratings, which may be expressed as follows. A (primitive) crystal with one atom per lattice point (Fig. 9.2(a)) may be regarded as being analogous to a 'narrow slit' diffraction grating in which each slit can be regarded as the source of a single Huygens' wavelet which propagates uniformly in all directions (Fig. 7.5). Only the interference effects of light emanating from different slits (equivalent to single atoms at lattice points) need to be taken into account and these determine the directions of the diffracted beams (Section 7.4). The intensities of the diffracted beams are proportional to the squares of the scattered amplitudes of the Huygens' wavelets (for light) or the squares of the atomic scattering factors of the single atoms (for X-rays).

A (primitive) crystal with a motif consisting of more than one atom (Fig. 9.2(b)) may be regarded as being analogous to a 'wide slit' diffraction grating in which the interference effects from all the atoms in the motif may be regarded as being analogous to the interference effects between all the Huygens' wavelets distributed *across* each slit. Of course, the problem is rather more complicated because the atoms in the motif do not all lie in a plane or surface as do the Huygens' wavelets, but the principle—of summing the contributions with respect to phase differences—is the same.

As shown in Section 7.4 and Fig. 7.7, the diffraction pattern from a wide-slit diffraction grating may be expressed as that from a narrow-slit grating in which the intensities of the diffracted beams are modulated by the intensity distribution predicted to occur from a single wide slit. Similarly, in the case of X-ray diffraction from crystals it is the structure factor, F_{hkl}, which expresses the interference effects from all the atoms in the unit cell and which modulates, in effect, the intensities of the diffracted beams.

The final step is to sum the contributions from all the unit cells in the crystal. This is a difficult problem because we have to take into account the fact that the incident X-ray beam is attenuated as it is successively scattered by the atoms in the crystal—such that atoms 'deeper down' in the crystal encounter smaller amounts of incident radiation—a process known as primary extinction. Furthermore, the reflected beams also propagate through the crystal at the Bragg angle θ (see Fig. 9.2(c)) and are hence 're-reflected' in a direction parallel to that of the incident beam. These re-reflected beams then interfere destructively[2] with the incident or direct beam, attenuating it still further—a process known as secondary extinction. This is covered in a comprehensive analysis, known as the dynamical theory of X-ray diffraction because it takes into account the dynamical

[2] When considering the interference between the direct and re-reflected beams, the 180° or π phase difference of the re-reflected beams needs to be taken into account. This complication does not arise when we are considering interference between the reflected or diffracted beams alone.

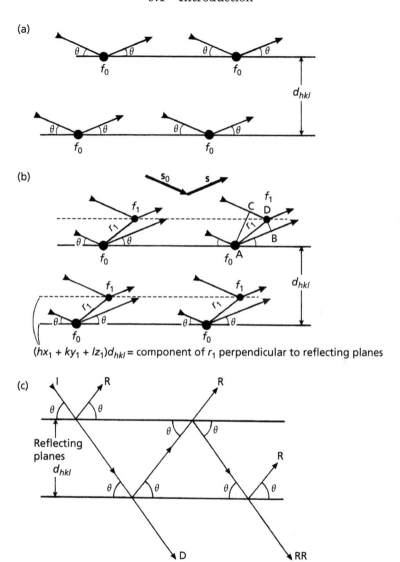

Fig. 9.2. (a) Part of a crystal lattice with atoms with atomic scattering factor f_0, situated at each lattice point, and a particular set of (hkl) planes through the lattice points. Incident/reflected beams at the Bragg angle θ to these planes are indicated by the arrows. (b) As for (a), but with another atom with atomic scattering factor f_1 defined by position vector \mathbf{r}_1. The path difference (shown for simplicity for one motif) is given by (AB − CD), \mathbf{s} and \mathbf{s}_0 are unit vectors along the reflected and incident beam directions and the component of r_1 perpendicular to the (hkl) planes is also indicated. (c) The process of reflection and re-reflection for a single incident beam I; as it passes through the crystal at the Bragg angle θ it is (partially) reflected by each successive plane. The reflected beams R are then re-reflected from the 'undersides' of the planes giving rise to beams RR with interfere *destructively* with the direct beam D. Clearly, the process is repeated for all the beams throughout the crystal.

interactions between the direct, reflected and re-reflected, etc., beams. It is much simpler to consider the case in which the size of the crystal is sufficiently small such that the attenuation of the direct beam is negligible and the intensities of the diffracted beams are small in comparison with the direct beam.[3] This case is fairly readily achieved in X-ray diffraction and enables the observed relative intensities of the diffracted spots to be assumed to be equal to the relative intensities of the squares of the F_{hkl} values.[4] In electron diffraction, however, dynamical effects are always important and the above assumption cannot be made, but they are an important source of specimen contrast in the electron microscope.

Finally, we have to consider the connection between the diffraction pattern and the type of unit cell (whether it is centred or not) and the symmetry elements present. It turns out that the presence of the centring lattice points and translational symmetry elements (glide planes and screw axes—see Section 4.5) results in 'zero intensity' or *systematically absent* reflections from certain planes with Laue indices *hkl*. This topic, and that of *double diffraction,* which is of particular importance in the case of electron diffraction, are discussed in Appendix 6.

9.2 The intensities of X-ray diffracted beams: the structure factor equation and its applications

We shall begin by considering the simplest case of a primitive crystal with just one atom at each lattice point. Figure 9.2(a) shows part of such a crystal with atoms with atomic scattering factor f_0 at each lattice point (the subscript 0 indicating that each atom is at an origin). Consider an X-ray beam incident at the correct Bragg angle θ in a particular set of lattice planes (*hkl*) as indicated. For all the atoms lying in one plane the path differences between the reflected beams are zero, and for the atoms lying in successive planes spaced d_{hkl}, $2d_{hkl}$ etc. apart the path differences are λ, 2λ, etc. (Fig. 8.3). In all cases constructive interference occurs and the total scattered amplitude (relative to that of a single electron—see Section 9.1) is simply the sum of the atomic scattering factors. Hence, since we have a primitive crystal with one lattice point and therefore one atom per unit cell, the scattered amplitude from one cell, F_{hkl}, is simply equal to the atomic scattering factor, f_0.

Now consider a crystal with a motif consisting of two atoms, one at the origin with atomic scattering factor f_0, as before and another with atomic scattering factor f_1 at a distance from the origin defined by vector \mathbf{r}_1 (Fig. 9.2(b)); \mathbf{r}_1 is called a **position vector** because it specifies the position of an atom within the unit cell. It may be expressed in terms of its components or fractional atomic coordinates along the unit cell vectors

[3] The crystal size in this context is better expressed by the notion of coherence length—the dimensions over which the scattering amplitudes from the unit cells can be summed. Crystal imperfections—dislocations, stacking faults, subgrain boundaries—within imperfect single crystals effectively limit or determine the coherence length as well as grain boundaries in perfect crystals (see Section 9.3.3).

[4] A number of physical and geometrical factors also need to be taken into account—temperature factor, Lorentz-polarization factor, multiplicity factor, absorption factor, etc. These are described in standard textbooks such as *Elements of X-ray Diffraction* by B. D. Cullity and S. R. Stock, (2001).

a, b, c in the same way as for a lattice vector \mathbf{r}_{uvw}, i.e. $\mathbf{r}_1 = x_1\mathbf{a} + y_1\mathbf{b} + z_1\mathbf{c}$; the important difference being that the components $x_1y_1z_1$ are *fractions* of the cell edge lengths, whereas the components uvw of a lattice vector \mathbf{r}_{uvw} are *integers*.

The path difference (P.D.) between the waves scattered by these two atoms is AB – CD (Fig. 9.2(b)), which, expressed in vector notation, is:

$$\text{P.D.} = \text{AB} - \text{CD} = \mathbf{r}_1 \cdot \mathbf{s} - \mathbf{r}_1 \cdot \mathbf{s}_0 = \mathbf{r}_1 \cdot (\mathbf{s} - \mathbf{s}_0)$$

where \mathbf{s}, \mathbf{s}_0 are unit vectors along the direction of the reflected and incident beams, respectively. Two substitutions can be made in this equation. First, \mathbf{r}_1 may be expressed in terms of its components and second (since Bragg's law is satisfied) the vector $(\mathbf{s} - \mathbf{s}_0)$ may be expressed in terms of λ and \mathbf{d}^*_{hkl} (see Section 8.3), i.e.

$$(\mathbf{s} - \mathbf{s}_0) = \lambda\mathbf{d}^*_{hkl} = \lambda(h\mathbf{a}^* + k\mathbf{b}^* + l\mathbf{c}^*).$$

Hence

$$\text{P.D.} = \lambda(x_1\mathbf{a} + y_1\mathbf{b} + z_1\mathbf{c}) \cdot (h\mathbf{a}^* + k\mathbf{b}^* + l\mathbf{c}^*)$$

multiplying out, and remembering the identities $\mathbf{a} \cdot \mathbf{a}^* = 1$, etc., $\mathbf{a} \cdot \mathbf{b}^* = 0$, etc.:

$$\text{P.D.} = \lambda(hx_1 + ky_1 + lz_1).$$

This equation is another manifestation of the Weiss zone law (Section 6.5.3); in this case the *number* within the brackets $(hx_1+ky_1+lz_1)$ represents the component of \mathbf{r}_1 perpendicular to the lattice planes as a fraction of the interplanar spacing, d_{hkl} (Fig. 9.2(b)). It is clearly the important number with respect to constructive/destructive interference conditions; when for example $(hx_1+ky_1+lz_1) = 0$ (the two atoms lying within the (hkl) planes), complete constructive interference occurs and when $(hx_1 + ky_1 + hz_1) = 0.5$ (the second atom lying halfway between the (hkl) planes), destructive interference occurs, which is complete when the atomic scattering factors of the two atoms in the motif are equal.

In general, in adding the contributions of the two atoms, we need to use a vector-phase diagram in which the lengths or moduli of the vectors are proportional to the atomic scattering factors of the atoms, and the phase angle between them, ϕ_1, is equal to $2\pi/\lambda$ (P.D.), i.e. in the above case

$$\phi_1 = 2\pi(hx_1 + ky_1 + lz_1).$$

This is shown in Fig. 9.3(a). The resultant is the structure factor, F_{hkl}.

The analysis may be simply extended to any number of atoms in the motif with position vectors $\mathbf{r}_1, \mathbf{r}_2, \mathbf{r}_3$ etc. and phase angles (referred to the origin) ϕ_1, ϕ_2, ϕ_3 etc. The resultant, F_{hkl} is found by adding all the vectors, representing the atomic scattering factors of all the atoms, as shown in Fig. 9.3(b). Note that the phase angles, ϕ, are all measured with respect to the origin (horizontal line in Fig. 9.3); they are *not* the angles *between* the vectors. Note also that, although for simplicity we began with an

(a) (b)

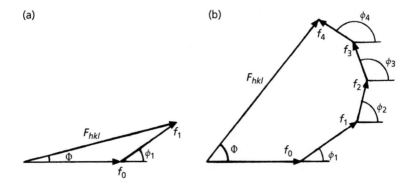

Fig. 9.3. Vector-phase diagrams for obtaining F_{hkl}. The atomic scattering factors f_1, f_2, \ldots are represented as vectors with phase angles ϕ_1, ϕ_2, \ldots with respect to a wave scattered from the origin: (a) result for two atoms and (b) result for several atoms.

atom at the origin, there need not be one there. The length or *modulus* of the vector F_{hkl} represents the resultant amplitude of the scattered or reflected beam and the angle Φ which it makes with the horizontal line is the resultant phase angle.

Adding vectors graphically in this way is obviously not very convenient; in practice vector-phase diagrams, such as Fig. 9.3, are substituted by Argand diagrams in which F_{hkl} is represented as a complex number (see Appendix 5), i.e.

$$F_{hkl} = \sum_{n=1}^{n=N} f_n \exp 2\pi \mathrm{i}(hx_n + ky_n + lz_n)$$

where f_n, is the atomic scattering factor and $2\pi(hx_n + ky_n + lz_n)$ is the phase angle ϕ_n of the nth atom in the motif with fractional coordinates $(x_n \; y_n \; z_n)$.

Many students are deterred at first sight of equations such as this. It is important to realize that it *merely* represents an analytical way of adding vectors 'top to tail', the convenience and ease of which is soon appreciated by way of a few examples.

Example 1: CsCl structure (Fig. 1.12). The $(x_n \; y_n \; z_n)$ values are (000) for Cl, atomic scattering factor f_{Cl} and $\left(\frac{1}{2}\frac{1}{2}\frac{1}{2}\right)$ for Cs, atomic scattering factor f_{Cs}. Substituting these two terms in the equation:

$$F_{hkl} = f_{Cl} \exp 2\pi \mathrm{i}(h0 + k0 + l0) + f_{Cs} \exp 2\pi \mathrm{i} \left(h\tfrac{1}{2} + k\tfrac{1}{2} + l\tfrac{1}{2}\right)$$
$$= f_{Cl} + f_{Cs} \exp \pi \mathrm{i}(h + k + l).$$

Two situations may be identified: when $(h + k + l)$ = even integer, $\exp \pi \mathrm{i}$ (even integer) = 1, hence $F_{hkl} = f_{Cl} + f_{Cs}$ and when $(h + k + l)$ = odd integer, $\exp \pi \mathrm{i}$ (odd integer) = -1, hence $F_{hkl} = f_{Cl} - f_{Cs}$. These two situations may be simply represented

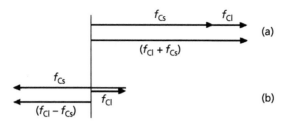

Fig. 9.4. Argand diagrams for F_{hkl} for CsCl (a) $(h + k + l)$ = even integer, $F_{hkl} = f_{Cl} + f_{Cs}$; (b) $(h + k + l)$ = odd integer, $F_{hkl} = f_{Cl} - f_{Cs}$.

on the Argand diagram as shown in Fig. 9.4. Note that in both cases F_{hkl} is a real number; the imaginary component is zero. This arises because CsCl has a centre of symmetry at the origin, as explained below.

Example 2: bcc metal structure. The atomic coordinates are (000), $(\frac{1}{2}\ \frac{1}{2}\ \frac{1}{2})$—the same as for CsCl—but the atomic scattering factors are equal. Proceeding as before we find that when $(h + k + l)$ = odd integer, F_{hkl} is zero (see Appendix 6).

Example 3: A crystal with a centre of symmetry at the origin. This is an important case because the structure factor for all reflections is real (as in Examples 1 and 2).

For every atom with fractional coordinates (xyz) and phase angle $+\phi$ there will be an identical one on the *opposite* side of the origin with fractional coordinates $(\bar{x}\bar{y}\bar{z})$ and phase angle $-\phi$. For these two atoms:

$$F_{hkl} = f \exp 2\pi i(hx + ky + lz) + f \exp 2\pi i(h\bar{x} + k\bar{y} + l\bar{z})$$
$$= f \exp 2\pi i(hx + ky + lz) + f \exp -2\pi i(hx + ky + lz).$$

The second term is the complex conjugate of the first, hence the sine terms cancel and

$$F_{hkl} = 2f \cos 2\pi (hx + ky + lz)$$

as shown graphically in Fig. 9.5.

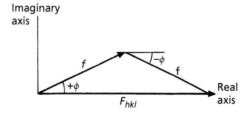

Fig. 9.5. The Argand diagram for a centrosymmetric crystal. The phase angle $+\phi$ for the atom at (xyz) is equal and opposite to the phase angle $-\phi$ for the atom at $(\bar{x}\bar{y}\bar{z})$, hence F_{hkl} is real.

Fig. 9.6. hcp metal structure, centres of A and B layer atoms as indicated. The origin of the primitive hexagonal cell may be chosen at an atom position (solid lines) or at centre of symmetry (dashed lines).

Example 4: hcp metal structure. Here we have a choice of unit cells (Fig. 5.8). It is best to refer to the primitive hexagonal cell, Fig. 5.8(a), which contains two identical atoms, atomic scattering factor f. We also have a choice of origin. We can choose a cell with one atom at the origin (000) (an A-layer atom) and the other with fractional atomic coordinates ($\tfrac{1}{3}$ $\tfrac{2}{3}$ $\tfrac{1}{2}$) (a B-layer atom), Fig. 9.6. In this case the origin is *not* at a centre of symmetry. Alternatively, we can place the origin at a centre of symmetry which are at positions equidistant between an A-layer and a *next nearest* B-layer atom (not, it should be noted between adjacent A and B-layer atoms). One such choice of origin is also shown in Fig. 9.6 and the fractional atomic coordinates in the cell become ($\tfrac{1}{3}$ $\tfrac{2}{3}$ $\tfrac{3}{4}$) (A-layer atom) and ($\tfrac{2}{3}$ $\tfrac{1}{3}$ $\tfrac{1}{4}$) (B-layer atom). These correspond to the fractional atomic coordinates denoted by Wyckoff letter d in space group $P6_3/mmc$ (Fig. 4.14).

It is a useful exercise to apply the structure factor equation to both of these choices of origin. First we choose the origin at an A-layer atom. Substituting coordinates (000) and ($\tfrac{1}{3}$ $\tfrac{2}{3}$ $\tfrac{1}{2}$) in the equation:

$$F_{hkl} = f \exp 2\pi i(h0 + k0 + l0) + f \exp 2\pi i \left(h\tfrac{1}{3} + k\tfrac{2}{3} + l\tfrac{1}{2} \right)$$
$$= f \left(1 + \exp 2\pi i \left(h\tfrac{1}{3} + k\tfrac{2}{3} + l\tfrac{1}{2} \right) \right).$$

Now let us apply this to some particular (hkl) planes, e.g. (002) ≡ (0002); (100) ≡ (10$\bar{1}$0) and (101) ≡ (10$\bar{1}$1):

$$F_{002} = f(1 + \exp 2\pi i) = 2f$$
$$F_{100} = f \left(1 + \exp \tfrac{2}{3}\pi\, i \right) = f \left(1 + \cos \tfrac{2}{3}\pi + i \sin \tfrac{2}{3}\pi \right) = f(0.5 + i0.866)$$
$$F_{101} = f \left(1 + \exp 2\pi i \left(\tfrac{1}{3} + \tfrac{1}{2} \right) \right) = f \left(1 + \cos \tfrac{5}{3}\pi + i \sin \tfrac{5}{3}\pi \right) = f(1.5 - i0.866).$$

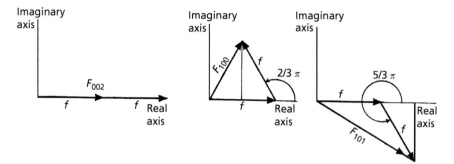

Fig. 9.7. Argand diagrams for an hcp metal for (left to right) F_{002}, F_{100} and F_{101} showing also how these structure factors are obtained graphically by vector addition. The origin is not at the centre of symmetry.

These results are shown graphically in Fig. 9.7. Note that F_{100} and F_{101} are complex numbers.

The intensities I_{hkl} of X-ray beams are proportional to their amplitudes squared, or F_{hkl} multiplied by its complex conjugate F^*_{hkl} (see Appendix 5). For the hcp metal example above:

$$I_{002} = 2f \cdot 2f = 4f^2$$
$$I_{100} = f(0.5 + i0.866)f(0.5 - i0.866) = f^2$$
$$I_{101} = f(1.5 - i0.866)f(1.5 + i0.866) = 3f^2.$$

Again, it should be stressed that I_{hkl} is a real number, the phase information expressed in F_{hkl} is lost.

For the origin at a centre of symmetry we use the simplified equation (Example 3) and the coordinates of either an A-layer or a B-layer atom. For ($\frac{1}{3}$ $\frac{2}{3}$ $\frac{3}{4}$):

$$F_{hkl} = 2f \cos 2\pi \left(h^1/3 + k^2/3 + l^3/4\right).$$

Again, for the particular (hkl) planes (002), (100) and (101) we have:

$$F_{002} = 2f \cos 3\pi = -2f; \qquad I_{002} = 4f^2$$
$$F_{100} = 2f \cos {}^{2\pi}/3 = -f; \qquad I_{100} = f^2$$
$$F_{101} = 2f \cos \pi\,{}^{13}/6 = \sqrt{3}f; \quad I_{101} = 3f^2.$$

Note that the structure factors are real (but not all of the same sign) and (of course) the intensities are the same as those obtained in the previous case.

Example 5: Non-centrosymmetric crystal: Friedel's Law and anomalous scattering. It follows from Example 3 that the diffraction pattern from a centrosymmetric crystal is also centrosymmetric. However, except for the case of anomalous scattering discussed

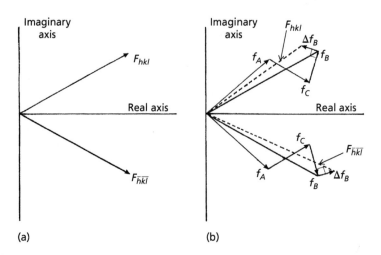

Fig. 9.8. Argand diagram for a non-centrosymmetric crystal: (a) no anomalous scattering; (b) anomalous scattering for atom B (see Fig 9.9).

below, even if a crystal does not possess a centre of symmetry, the diffraction pattern will still be centrosymmetric. This is known as Friedel's law which may be proved with reference to the Argand diagram in Fig. 9.8(a). We have to show that the intensity of the reflection from the hkl planes, i.e. I_{hkl} is equal to that from the $\bar{h}\bar{k}\bar{l}$ planes, i.e. $I_{\bar{h}\bar{k}\bar{l}}$.

As can be seen, $F_{\bar{h}\bar{k}\bar{l}}$ is the complex conjugate of F_{hkl}; i.e. $F_{\bar{h}\bar{k}\bar{l}} = F^*_{hkl}$ and similarly $F_{hkl} = F^*_{\bar{h}\bar{k}\bar{l}}$.

Hence $I_{hkl} = F_{hkl} \cdot F^*_{hkl} = F^*_{\bar{h}\bar{k}\bar{l}} \cdot F_{hkl} = I_{\bar{h}\bar{k}\bar{l}}$.

The presence of a centre of symmetry in the diffraction pattern means that non-centrosymmetric crystals cannot be distinguished from those with a centre of symmetry. There are eleven centrosymmetric point groups (Table 3.1, page 93) and hence eleven symmetries which diffraction patterns can possess. These are called the eleven Laue groups and are identified by the corresponding point group symbol of the centrosymmetric point group. They are listed in Table 9.1.

Friedel's law holds so long as the phases of the scattered waves are those expected for scattering at the atomic centres, i.e. precisely at the positions specified by the fractional atomic coordinates $(x_n\ y_n\ z_n)$. However, if the energy of the incident X-rays is just sufficient to displace the innermost K-electrons from their orbitals, i.e. if the wavelength is just below the K-absorption edge wavelength of an atom, then the phase of the scattered wave differs from that expected from scattering at the atomic centre. It is as if the atom were displaced and the sense of the displacement depends on whether the reflection is from the 'top' or 'underside' of the reflecting planes. This is the condition for anomalous scattering. Figure 9.9 (due to W.L. Bragg) shows the geometry involved for just three atoms, A, B and C, one of which, B, scatters anomalously.

Table 9.1 The eleven Laue point groups or crystal classes.

Crystal system	Laue point group and centrosymmetric point group	Non-centrosymmetric point groups belonging to the Laue point group		
Cubic	$m\bar{3}m$	432	$\bar{4}3m$	
(two Laue point groups)	$m\bar{3}$	23		
Tetragonal	$4/mmm$	422	$4mm$	$\bar{4}2m$
(two Laue point groups)	$4/m$	4	$\bar{4}$	
Orthorhombic	mmm	222	$mm2$	
Trigonal	$\bar{3}m$	32	$3m$	
(two Laue point groups)	$\bar{3}$	3		
Hexagonal	$6/mmm$	622	$6mm$	$\bar{6}m2$
(two Laue point groups)	$6/m$	6	$\bar{6}$	
Monoclinic	$2/m$	2	m	
Triclinic	$\bar{1}$	1		

Fig. 9.9. Representation of anomalous scattering of atom labelled B: (a) hkl reflection, B displaced to B′; and (b) $\bar{h}\bar{k}\bar{l}$ reflection, B displaced to B″.

For the hkl reflection, Fig. 9.9(a), B scatters as if it were situated at B′ and for the $\bar{h}\bar{k}\bar{l}$ reflection, Fig. 9.9(b), B scatters as if it were at B″. The resultant amplitudes, obtained by combining the effects of all the atoms, will clearly be different.

The effect may also be represented in the Argand diagram; the alteration of phase of an anomalously scattering atom is equivalent to combining the normal f of the atom with a vector Δf at right angles to it as shown in Fig. 9.8(b), where the resultant structure factors F_{hkl} and $F_{\bar{h}\bar{k}\bar{l}}$ which are now different, are shown by the dashed lines.

Anomalous scattering (or absorption) can be 'put to use' to distinguish centrosymmetric and non-centrosymmetric point groups. In particular it can be used to distinguish the right and left-handed (dextro and laevo) crystals in the enantiomorphic point groups—e.g. tartaric acid (Fig. 4.7, page 109). In so doing a salt is synthesized which contains an anomalously scattering atom for the X-ray wavelength used. The first such experiments were carried out by J. M. Bijvoet* using crystals of sodium rubidium

* Denotes biographical notes available in Appendix 3.

tartrate. It is very fortunate (even chances!) that the dextro- and laevo-configurations thus determined agree with those of chemical convention—otherwise all the dexro- and laevo-conventions would have had to be interchanged.

To summarize, these simple examples show how the amplitude, F_{hkl}, and hence the intensity, I_{hkl}, of the reflected X-ray beam from a set of hkl planes can be calculated from the simple 'structure factor' equation on p. 216; all we need to know are the positions of the atoms in the unit cell (the x_n y_n z_n values) and their atomic scattering factors, f_n. The great importance of the equation is that it can be applied, as it were, 'the other way round': by measuring the intensities of the reflections from several sets of planes (the more the better), the positions of the atoms in the unit cell can be determined. This is the basis of crystal structure determination, which has developed and expanded since the pioneering work of the Braggs, so that, at the time of writing, some 400 000 different crystal structures are known. Many of these are very complex, for example protein crystals in which the motif may consist of several thousand atoms. Again, it should be emphasized that the procedures are invariably not straightforward because the phase information in going from the F_{hkl} to the (measured) I_{hkl} values is lost, e.g. as in Example 4, Fig. 9.7. This is called the **phase problem** in crystal structure determination, which may be understood with reference to Fig. 9.3(b). All F_{hkl} vectors with the same modulus or amplitude will give the same observed intensity I_{hkl}; the value of the phase angle Φ, which is an essential piece of information in the vector-phase diagram, is lost. In short, we do not know in which direction the vector F_{hkl} 'points', e.g. as in Fig. 9.7. In some cases (as in Example 4), the problem is simply solved if we are able to arrange the origin to coincide with a centre of symmetry in the crystal in which case, as shown in Examples 3 and 5 the phase angle ϕ is zero or 180° and the structure factor F_{hkl} is a real number with no imaginary component. However, in the many cases where the crystal does not possess a centre of symmetry, we must resort to more subtle procedures, the details of which are beyond the scope of this book. One method is to arrange a heavy atom (possibly substituted in the crystal structure for a light atom) to be at the origin. Then, in terms of our vector-phase diagram (Fig. 9.3(b)), f_0 is so large that it dominates the contributions of all the other atoms such that the phase angles Φ for all the F_{hkl} values are small and therefore can more easily be guessed at. In all cases the structure factor equation is expressed as it were in a 'converse' form (or transform of that on p. 216 in which atomic positions (expressed as electron (X-ray scattering) density) are expressed in terms of the F_{hkl} values of the reflections.

The notion of *electron density* provides a much more realistic representation of atomic structure. Atoms, which are detected by X-rays from the scattering of their con-stituent electrons, have a finite size and the atomic coordinates essentially represent those positions where the amount of scattering (the electron density) is the highest. In our two-dimensional plan views (Section 1.8) we may therefore represent the atoms as hills—a contour map of electron density; the 'higher the hill' the greater the atomic scat-tering factor of the atom. These ideas, which involve the application of Fourier analysis, are introduced in Chapter 13, but it is a subject of great complexity which is covered in more detail in those books on crystal structure determination which are listed in Further Reading.

9.3 The broadening of diffracted beams: reciprocal lattice points and nodes

In Chapter 8 we treated diffraction in a purely geometrical way, incident and reflected beams being represented by single lines implying perfectly narrow, parallel beams and reflections only at the Bragg angles. Of course, in practice, such 'ideal' conditions do not occur; X-ray beams have finite breadth and are not perfectly parallel to an extent depending upon the particular experimental set-up. Such *instrumental factors* give rise to broadening of the reflected X-ray beams: the reflections peak at the Bragg angles and decrease to zero on either side. However, broadening is not solely due to such instrumental factors but much more importantly also arises from the crystallite size, perfection and state of strain in the specimen itself. The measurement of such broadening (having accounted for the contribution of the instrumental factor) can then provide information on such specimen conditions.

We now consider the effects of crystal size on the broadening and peak intensity of the reflected beams, which lead to the Scherrer equation (Section 9.3.1) and the notion of integrated intensity (Section 9.3.2). In Section 9.3.3 we consider the imperfection (or mosaic structure) of real crystals. The effect of lattice strain on broadening is covered in Section 10.3.4.

9.3.1 The Scherrer equation: reciprocal lattice points and nodes

In Section 7.4 we found that when the number N of lines in a grating, or the number of apertures in a net, were limited, the principal diffraction maxima were broadened and surrounded by much fainter subsidiary maxima. These phenomena are shown in Fig. 7.3(d) and diagrammatically in Fig. 7.9. Precisely the same considerations apply to X-ray (and electron) diffraction from 'real' crystals in which the number of reflecting planes is limited: and the broadening and occurrence of the subsidiary maxima can be derived by similar arguments. This, in turn, leads us to modify our concept of reciprocal lattice *points*, which are not geometrical points, but which have finite size and shape, reciprocally related, as we shall see, to the size and shape of the crystal. There is, as far as I know, no common term to express the fact that reciprocal lattice points do have a finite size and shape, except in special cases such as the reciprocal lattice streaks or 'rel-rods' which occur in the case of thin plate-like crystals. Reciprocal lattice **nodes** seems to be the closest approximation to a common term.

The broadening of the reflected beams from a crystal of finite extent is derived as follows. Consider a crystal of thickness or dimension t perpendicular to the reflecting planes, d_{hkl}, of interest. If there are m planes then $md_{hkl} = t$. Consider an incident beam bathing the whole crystal and incident at the exact Bragg angle for first order reflection (Fig. 9.10 (a)). For the first two planes labelled 0 and 1, the path difference $\lambda = 2d_{hkl} \sin \theta$; for planes 0 and 2 the path difference is $2\lambda = 4d_{hkl} \sin \theta$ and so on—constructive interference between all the planes occurs right through the crystal. Now consider the interference conditions for an incident and reflected beam deviated a small angle $\delta\theta$ from the exact Bragg angle (Fig. 9.10(b)). For planes 0 and 1 the path

difference will be very close to λ as before and there will be constructive interference. However, for planes 0 and 2, 0 and 3, etc. the 'extra' path difference will deviate increasingly from 2λ, 3λ etc.—and when it is an additional half-wavelength *destructive* interference between the pair of planes will result.

The condition for destructive interference for the whole crystal is obtained by notionally 'pairing' reflections in the same way as we did for the destructive interference of Huygens' wavelets across a wide slit (Fig. 7.8). Consider the constructive and destructive interference condition between planes 0 and $(m/2)$ (halfway down through the crystal). At the exact Bragg angle θ (Fig. 9.10 (a)), the condition for constructive interference is $(m/2)\lambda = (m/2)2d_{hkl}\sin\theta$. The condition for *destructive* interference at angle $(\theta + \delta\theta)$ is given by $(m/2)\lambda + \lambda/2 = (m/2)2d_{hkl}\sin(\theta + \delta\theta)$. Now this is also the condition for destructive interference between the next pair of planes 1 and $(m/2) + 1$—and so on through the crystal. This equation gives us, in short, the condition for destructive interference for the *whole* crystal and the angular range $\delta\theta$ (each side of the exact Bragg angle) of the reflected beam.

Expanding the sine term and making the approximations $\cos\delta\theta = 1$ and $\sin\delta\theta \approx \delta\theta$ gives:

$$(m/2)\lambda + \lambda/2 = (m/2)2d_{hkl}\sin\theta + (m/2)2d_{hkl}\cos\theta\delta\theta.$$

Cancelling the terms $(m/2)\lambda$ and $(m/2)2d_{hkl}\sin\theta$ and substituting $md_{hkl} = t$ gives

$$2\delta\theta = \frac{\lambda}{t\cos\theta} = \frac{2d\sin\theta}{t\cos\theta} = \frac{2\tan\theta}{m}.$$

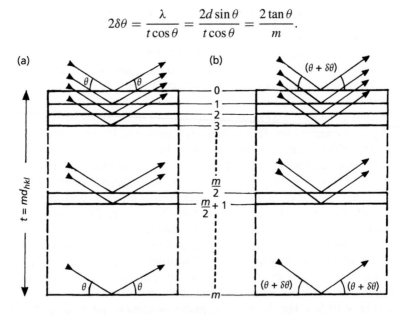

Fig. 9.10. Bragg reflection from a crystal of thickness t (measured perpendicular to the particular set of reflecting planes shown). The whole crystal is bathed in an X-ray beam (a) at the exact Bragg angle θ and (b) at a small deviation from the exact Bragg angle, i.e. angle $(\theta + \delta\theta)$. The arrows represent the incident and reflected beams from successive planes 0,1,2,3 ... $(m/2)$ (half-way down) and ... m (the lowest plane).

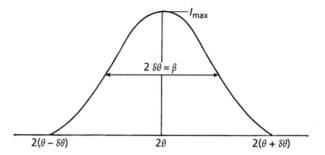

Fig. 9.11. A schematic diagram of a broadened Bragg peak arising from a crystal of finite thickness. The breadth at half the maximum peak intensity, β, is approximately equal to $2\delta\theta$. Note that the angular '2θ' scale is measured in relation to the direct and reflected beams.

This is the basis of the Scherrer* equation which relates the broadening of an X-ray beam to the crystal size t or number of reflecting planes m. The broadening is usually expressed as β, the breadth of the beam at half the maximum peak intensity and in which the angles are measured relative to the direct beam. As indicated in Fig. 9.11, β is approximately equal to $2\delta\theta$, hence

$$\beta = \frac{\lambda}{t\cos\theta} = \frac{2\tan\theta}{m}.$$

The broadening can be represented in the Ewald reflecting sphere construction in terms of the extension of the reciprocal lattice point to a node of finite size. Figure 9.12 shows

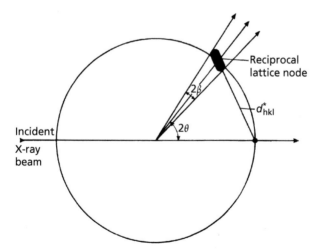

Fig. 9.12. The Ewald reflecting sphere construction for a broadened reflected beam, β, which corresponds to an extension $1/t$ of the reciprocal lattice node.

* Denotes biographical notes available in Appendix 3.

a diffracted beam (angle 2θ to the direct beam) which is broadened over an angular range $2\beta \approx 4\delta\theta$. This is expressed by extending the reciprocal lattice point into a node of finite length which, as the crystal rotates, intersects the reflecting sphere over this angular range. Let $\delta(d^*_{hkl})$ represent the extension of the reciprocal lattice point about its mean position. Now since

$$|\mathbf{d}^*_{hkl}| = d^*_{hkl} = \frac{2\sin\theta}{\lambda};$$

$$\delta(d^*_{hkl}) = \delta\left(\frac{2\sin\theta}{\lambda}\right) = \frac{2\cos\theta}{\lambda}\delta\theta.$$

Substituting for $\delta\theta$ from above gives:

$$\delta(d^*_{hkl}) = \frac{2\cos\theta}{\lambda} \cdot \frac{\lambda}{2t\cos\theta} = \frac{1}{t}$$

i.e. the extension of the reciprocal lattice node is simply the reciprocal of the crystal dimension perpendicular to the reflecting planes. This applies to all the other directions in a crystal with the result that the *shape* of the reciprocal lattice node is reciprocally related to the *shape* of the crystal. For example, in the case of thin, plate-like crystals (e.g. twins or stacking faults), the reciprocal lattice node is a rod or 'streak' perpendicular to the plane of the plate.

The extension of the reciprocal lattice node $\delta\left(d^*_{hkl}\right)$ may also be expressed as a ratio or proportion of d^*_{hkl}, i.e. since

$$\delta\left(d^*_{hkl}\right) = \frac{1}{t} = \frac{1}{md_{hkl}} = \frac{\left(d^*_{hkl}\right)}{m},$$

then

$$\frac{\delta\left(d^*_{hkl}\right)}{d^*_{hkl}} = \frac{1}{m},$$

i.e. the ratio is simply equal to the reciprocal of the number of reflecting planes, m.

The above is a simplified treatment, both of the Scherrer equation and the extension of reciprocal lattice points into nodes. The Scherrer equation is normally applied to the broadening from polycrystalline (powder) specimens and includes a correction factor K (not significantly different from unity) to account for particle shapes.[5] Hence the Scherrer equation is normally written:

$$\beta = \frac{K\lambda}{t\cos\theta} = \frac{K\lambda\sec\theta}{t}.$$

Finally, it should be noted that the reciprocal lattice nodes are also surrounded by subsidiary nodes (or satellites, maxima, or fringes) just as in the case of light diffraction

[5] Note that the observed breadth at half the maximum peak intensity, β_{OBS}, includes additional sources of broadening arising from the experimental set up and instrumentation, β_{INST}, which must be 'subtracted' from β_{OBS} in order that β can be determined (see Section 10.3.3).

from gratings with a finite number of lines (Sections 7.4 and 13.4). In most situations these subsidiary nodes are very weak because the number of diffracting planes contributing to the beam is large. However, in the case of X-ray diffraction from specimens consisting of a limited number of diffracting planes, superlattice repeat distances or multilayers, the subsidiary nodes (or satellites, maxima or fringes) are observable and are important in the characterization of the specimen, as described in Section 9.6.

9.3.2 Integrated intensity and its importance

In Section 9.3.1, we have seen the effect of crystal thickness on broadening. What is its effect on the peak intensity, I_{max} (Fig. 9.11)? To answer this question we will carry out a 'thought experiment'. Let us suppose that in our crystal, of thickness t, there are a total of n unit cells each of which contributes a scattering amplitude $|F_{hkl}|$. The total scattered amplitude is therefore $n|F_{hkl}|$ and the total scattered intensity, I_{max}, is proportional to $n^2|F_{hkl}|^2$. Now let us separate the crystal into two halves, each of thickness $t/2$. The X-ray peaks from each crystal are twice the breadth of that for the single crystal (half the thickness, twice the breadth). The total scattered amplitude from each crystal is $n/2\,|F_{hkl}|$ and hence the peak intensity is proportional to $n^2/4\,|F_{hkl}|^2$. Added together, the two crystals give a peak intensity, $I_{max} = n^2/2\,|F_{hkl}|^2$, i.e. *half* that of the single crystal but a peak breadth *twice* that of the single crystal. What is constant and independent of the 'state of division'—or crystallite size in the specimen—is the *area* under the diffraction peak. This quantity, usually measured in arbitrary units, is called the *integrated intensity* of the reflection or just the *integrated reflection*. It is not a measurement of intensity, but rather a measurement of the total energy of the reflection. Furthermore, except for situations in which dynamical interactions need to be taken into account (i.e. large, perfect crystals in which the reflected beams are comparable in intensity with the direct beam—see Section 9.1), the integrated reflection is a measure of $|F_{hkl}|^2$ (and of course crystal volume).

9.3.3 Crystal size and perfection: mosaic structure and coherence length

Except for rather special cases there is rarely a continuity of structure throughout the whole volume of a 'single' crystal, but rather is separated into 'blocks' of slightly varying misorientation (Fig. 9.13(a))—a situation recognized in the early days of X-ray diffraction from the discrepancy between the intensities predicted for 'perfect' crystals and those actually observed. Ewald termed this a 'mosaic' structure—but of course the nature of the mosaic blocks and the boundaries were unknown. It is now evident, from transmission electron microscopy, that crystals contain dislocations which may be distributed uniformly throughout the structure, or arrange themselves (as shown in the simple case in Fig. 9.13(b)), into sub-grain boundaries. However, it remains the case that the relationship between sub-grain size, as measured by electron microscopy, and coherence length or mosaic size, as measured by X-ray broadening, is by no means clear.

(a) (b)

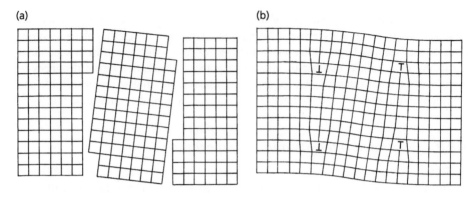

Fig. 9.13. Representation of a single crystal: (a) divided into three mosaic blocks; and (b) the boundaries as 'walls' of edge dislocations.

In an X-ray experiment, as a crystal is 'swept' through its diffracting condition, the individual crystallites of the mosaic structure reflect at slightly different angles and the total envelope of the diffraction profile, or integrated reflection, is the sum of all the separate reflections.

An expression for the integrated reflection, ρ_{hkl}, may be derived as follows. Let I_o be the **intensity** of the incident beam and let the **power** of the reflected beam at a particular angle θ be $\left(\frac{dE_{hkl}}{dt}\right)_\theta$. This power is clearly proportional to I_o and their ratio we define as the **reflecting power**, $P(\theta) = \frac{\left(dE_{hkl}/dt\right)_\theta}{I_o}$, a quantity which has the dimensions of area.

$P(\theta)$ varies across the whole angular diffracting range (i.e. as in Fig. 9.11) and the **integrated reflection**, ρ_{hkl} is found by integrating $P(\theta)$ over the whole diffraction profile, i.e.

$$\rho_{hkl} = \int P(\theta)\, d\theta = \frac{1}{I_o} \int \left(\frac{dE_{hkl}}{dt}\right)_\theta d\theta = \frac{1}{I_o} \int dE_{hkl} \left(\frac{d\theta}{dt}\right)$$

Now $\int dE_{hkl} = E_{hkl}$ and $\left(\frac{d\theta}{dt}\right) = \omega$, the angular velocity of rotation. Hence $\rho_{hkl} = E_{hkl}\frac{\omega}{I_o}$.

E_{hkl}, the total energy of the reflected beam, is (to repeat) dependent on the total volume of the crystallites, the physical factors in the Thomson equation, $|F_{hkl}|^2$, and the Lorentz, polarization and temperature factors (see footnote 4, in Section 9.1, p. 214).

9.4 Fixed θ, varying λ X-ray techniques: the Laue method

X-ray single crystal techniques may be classified into two groups depending upon the way in which Bragg's law is satisfied experimentally. There are two variables in Bragg's law—θ and λ—and a series of fixed values, d_{hkl}. In order to satisfy Bragg's law for any of the d-values, either λ must be varied with θ fixed, or θ must be varied with λ fixed. The former case has only one significant representative—the (original) Laue method and its variants, whereas there are many methods based upon the latter case.

The geometry of the Laue method, in terms of the reflecting sphere construction, has already been explained in Section 8.4. Now we need to consider the practical applications of the technique. The important point to emphasize is that each set of reflecting planes with Laue indices hkl (see Section 8.4) gives rise to just one reflected beam. Of all the white X-radiation falling upon it, a lattice plane with Miller indices (hkl) reflects only that wavelength (or 'colour') for which Bragg's law is satisfied, a reflecting plane of half the spacing with Laue indices $2h\ 2k\ 2l$ reflects a wavelength of half this value, and so on. In other words the reflections from planes such as, for example, (111) (Miller indices for lattice planes) 222, 333, 444, etc. (Laue indices for the parallel reflecting planes) are all superimposed.

The spots in the Laue method reveal **only** the angles between different $\{hkl\}$ planes (the measured angles being twice those between the planes themselves) and from which, and also from the symmetry relations between the spots which may reveal or confirm the crystal structure, they may be indexed and the orientation of the crystal, with respect to the incident beam, determined. The method does **not** allow us to determine d_{hkl}-spacings since the wavelength(s) giving rise to each spot are unknown. In short, crystals with identical structures but different lattice parameters give identical Laue patterns.

The method falls into two distinct groups: the (original) transmission method and the back-reflection method. In both cases reflections from planes in the same zone lie on the surface of a circular cone centred about the zone axis. The intersection of the cone with the film (image plate) is a conic section. For small reflection angles (transmission), i.e. the planes in the zone making small angles to the incident beam, the conic section is an ellipse which passes through the origin (i.e. the direction of the incident beam). As the angle of the cone increases the ellipse 'spreads out' as a parabola, then a hyperbola, and when the axis of the cone is at 90° to the incident beam the spots fall in straight lines passing through the origin. All these features are (partially) revealed in the Laue photograph of zinc blende (see page xvi of this book).

Although it was the first, the transmission Laue method became little used (in comparison with 'fixed λ, varying θ' methods described in the following sections). However, with the advent of 'wiggled', varying wavelength synchrotron radiation (see Section 9.8.2) it has had something of a renaissance particularly for protein crystals since data can be acquired in a fraction of a second, i.e. before any degradation of the crystal can take place. Figure 9.14(a) shows an example in which the ellipses and parabolae of spots are clearly evident.

In comparison, the back-reflection technique has been widely used in metallurgy/ materials science to determine the orientations and orientation relationships between crystals, e.g. the orientations of crystals in turbine blades or the orientation relationships between parent and product crystals in phase transformations. Again, the spots from planes in the same zone lie in hyperbolae except that (unlike the transmission case) these do **not** intersect the centre of the film/image plate except for the special case where the zone axis is 90° to the incident beam. A simplified picture of the set-up showing just one reflected beam in the plane of the paper is shown in Fig. 9.14(b). In analysing the film it is necessary to determine the projection of the normal (or reciprocal lattice vector) of each of the reflecting planes on to the film from each reflection S and

The diffraction of X-rays

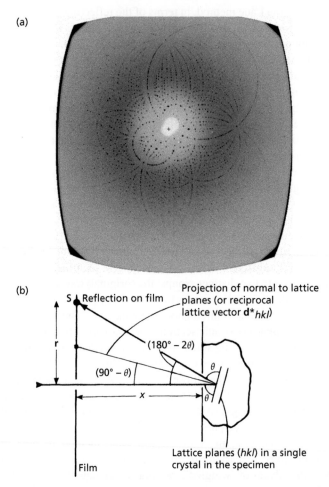

Fig. 9.14. (a) A synchrotron X-ray transmission Laue photograph (wavelength range 0.34–1.6Å), exposure time 1 ms of the protein Hydroxymethylbilane Synthase showing zones of reflections (ellipses and parabolae), all of which pass through the origin (direction of the incident beam). Prominent reflections, common to several zones, occur at the intersections of the ellipses/parabolae. (Photograph courtesy of Professor John Helliwell (University of Manchester) and Dr Alfons Hädener (University of Basle). From Y.P. Nieh *et al. J. Synchrotron Rad.* (1999), **6** 995–1006.) (b) The geometry of the back-reflection Laue method for a particular set of *hkl* planes in a single crystal. The wavelength of the reflected beam is that for which Bragg's law is satisfied for the particular fixed θ and d_{hkl} value, (c) A back-reflection Laue photograph of a single crystal of aluminium oriented with a ⟨100⟩ direction nearly parallel to the incident X-ray beam showing the reflections *S* from many (*hkl*) planes. They lie on a series of intersecting hyperbolae (close to straight lines in this photograph), each hyperbola corresponding to reflections from planes in a single zone, [*uvw*]. Note the four-fold symmetry of the intersecting zones indicating the ⟨100⟩ crystal orientation. (Photograph by courtesy of Prof. G. W. Lorimer.)

(c)

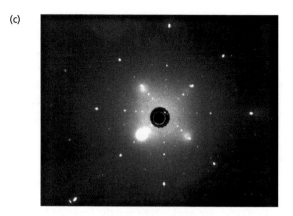

Fig. 9.14. (*continued*)

then to plot these on a stereographic projection. By measuring the angles between the normals, and then comparing them with lists of angles such as given for cubic crystals in Section A4.4 (Appendix 4), it is possible to identify the reflections. In practice such manual procedures[6] have largely been replaced by computer programs which determine the orientation of the crystal using as input data the positions of spots on the film, film-specimen distance and (assumed) crystal structure.

An example of a back-reflection Laue photograph (hyperbolae are nearly straight lines) is shown in Fig. 9.14(c).

Figure 9.14(b) should be compared with Fig. 8.7. Consider for example the 201 reciprocal lattice point which is situated in the 'nest' of spheres representing the wavelength range of the incident 'white' X-radiation. Bragg's law is satisfied for the particular wavelength represented by the sphere which passes through the 201 reciprocal lattice point and the direction of the reflected beam (indicated by the arrow) is from the centre of this sphere (which can be found by construction) and the 201 reciprocal lattice point. This beam makes angle $(180° - 2\theta)$ with the incident beam and the reciprocal lattice vector \mathbf{d}^*_{201} makes angle $(90° - \theta)$ with the incident beam.

9.5 Fixed λ, varying θ X-ray techniques: oscillation, rotation and precession methods

In Section 8.4, Fig. 8.6, we showed the Ewald reflecting sphere construction for the case where the incident X-ray beam was incident along the \mathbf{a}^* reciprocal lattice vector (from the left). Figure 8.6(a) shows the $h0l$ section of the reciprocal lattice through the

[6] The manual procedures are fully described in *Elements of X-ray Diffraction* by B.D. Cullity and S.R. Stock (2001). In brief, a **Leonhardt chart** is used for transmission and a **Greninger chart** for back-reflection. These, calibrated for a given specimen-film distance, show arcs representing ellipses, parabolae, and hyperbolae for different inclinations of the zone axes and are used in conjunction with a stereographic projection.

centre of the sphere and the origin of the reciprocal lattice and Fig. 8.6(b) shows the
$h1l$ section of the reciprocal lattice and the smaller (non-diametral) section of the Ewald
sphere. For the particular wavelength and particular incident beam direction drawn in
Fig. 8.6, only two planes—201 and 21$\bar{1}$ satisfy Bragg's law. Now, instead of varying λ,
as in the Laue case (Fig. 8.7), let us change the direction of the incident beam such that
it is no longer incident along the \mathbf{a}^* reciprocal lattice vector direction. This variation in
the Bragg angle is accomplished in practice by moving the crystal.

9.5.1 The oscillation method

Consider the crystal, Fig. 8.6, oscillated say $\pm10°$ about an axis parallel to the \mathbf{b}^* re-
ciprocal lattice vector (or y-axis), i.e. perpendicular to the plane of the paper. As the
crystal is (slowly) oscillated, the angles between the incident beam and hkl planes vary
and whenever Bragg's law is satisfied for a particular plane 'out shoots' momentarily a
reflected beam. The oscillation of the crystal can be represented by an oscillation of the
reciprocal lattice about the origin: imagine a pencil fixed to the crystal perpendicular
to the page with its point at the origin of Fig. 8.6(a). As you oscillate the pencil the
reciprocal lattice oscillates about the origin but the Ewald sphere remains fixed, centred
along the direction of the incident beam from the left: whenever a reciprocal lattice
point intersects the sphere 'out shoots' a reflected beam.

The relative movement of the sphere and reciprocal lattice is most easily represented,
not by drawing the whole reciprocal lattice in its different positions, but by drawing the
Ewald sphere at the extreme limits of oscillation say $10°$ in one direction and $10°$ in
the other. These limits are shown in Fig. 9.15 for (a) the $h0l$ and (b) the $h1l$ reciprocal
lattice sections. The shaded regions, called 'lunes' because of their shape, represent
the regions of reciprocal space through which the surface of the sphere passes as it is
oscillated. Only those reciprocal lattice points which lie within these shaded regions
give rise to reflected beams.

There are several ways of recording the reflected beams. The simplest is to arrange
a cylindrically shaped film coaxially around the crystal (with a hole to allow the exit
of the incident beam) such that reflections at all angles can be recorded. Now, all the
reflected beams from the $h0l$ reciprocal lattice section (Fig. 9.15(a)) lie in the plane
of the paper and thus lie in a ring around the film. The reflected beams from the $h1l$
reciprocal lattice section lie in a cone whose centre is the centre of the Ewald sphere
and these beams also intersect the film in a ring 'higher up' than the ring from the $h0l$
reciprocal lattice section. Figure 9.16(a) shows this geometry for the reflected beams
from the $h0l$, $h1l$, $h\bar{1}l$, $h2l$, etc. sections of the reciprocal lattice.

When the film is 'unwrapped' and laid flat, the rings of spots appear as lines, called
layer lines (Fig. 9.16(b)), the 'zero layer line' spots from the $h0l$ reciprocal lattice sec-
tion through the centre, the 'first layer line' spots from the $h1l$ reciprocal lattice section
next above—and so on for all the reciprocal lattice sections perpendicular to the axis
of oscillation. The larger unit cells characteristic of biological structures and hence the
much smaller reciprocal lattices as compared with the size of the Ewald sphere (for
X-ray wavelengths ≈ 1.5Å), give rise to many more, and more closely spaced, Bragg

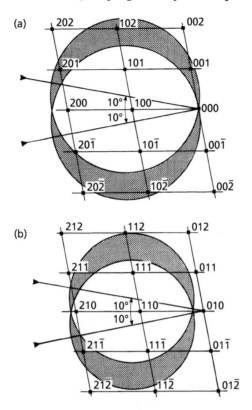

Fig. 9.15. The Ewald reflecting sphere construction for a monoclinic crystal in which the incident X-ray beam is oscillated ±10° from the direction of the **a*** reciprocal lattice vector about the [010] direction (y-axis, perpendicular to the plane of the paper). The directions of the X-ray beams and the corresponding surface of the reflecting sphere are shown at the limits of oscillation. Any reciprocal lattice points lying in the regions of reciprocal space through which the reflecting sphere passes (the shaded regions, called 'lunes') give rise to reflections, (a) The *h0l* reciprocal lattice section through the origin and a diametral section of the reflecting sphere and (b) the *h1l* reciprocal lattice section (i.e. the layer 'above' the *h0l* section) and the smaller, non-diametral section of the reflecting sphere (compare with Fig. 8.6). In these two sections the planes which reflect are 201, 102, 001, $20\bar{1}$, $10\bar{2}$ (zero layer line) and 112, $21\bar{1}$ (first layer line above).

reflections. Hence, in rotation photographs many spots 'overlap' in the more close-spaced layer lines. To avoid this, the oscillation technique is used but with very much smaller angles of oscillation—e.g. 1° rather than the 10° shown in Fig. 9.15. Since that only a small region of reciprocal space is sampled, many such photographs are needed for a complete data set. This is a situation in which the Weissenberg method (mentioned in Section 8.4 on p. 207) which gives rise to a greater separation of the Bragg reflections, finds particular application.

In practice it is necessary to set up a crystal such that some prominent zone axis is along the oscillation axis, such as, in our case, the [010] axis. This implies that the

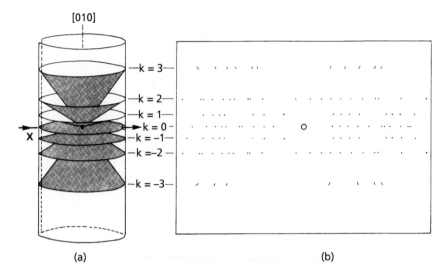

Fig. 9.16. (a) The oscillation photograph arrangement with the crystal at the centre of the cylindrical film. The oscillation axis [010] is co-axial with the film and the reflections from each reciprocal lattice section ($h0l$), ($h1l$), ($h\bar{1}l$), ($h2l$) etc. lie on cones which intersect the film in circles. **X** indicates the incident beam direction and in (b), showing the cylindrical film 'unwrapped', **O** indicates the exit beam direction through a hole at the centre of the film.

reciprocal lattice sections or layers perpendicular to the oscillation axis are well defined and give rise to clearly defined layer lines on the film. If, for example, we set a cubic crystal with the z-axis or [001] direction along the axis of oscillation, then the reciprocal lattice sections giving the layer lines would be $hk0$ (zero layer line), $hk1$ (first layer line above) and so on (see Figs 6.7 or 6.8). If on the other hand the crystal was set up in no particular orientation, i.e. no particular reciprocal lattice section perpendicular to the axis of rotation, the layer lines would, correspondingly, hardly be evident.

9.5.2 The rotation method

As the angle of oscillation increases, the Ewald reflecting sphere sweeps to and fro through a greater volume of reciprocal space, the lunes (Fig. 9.15) become larger and overlap and more reflections are recorded. The extreme case is to oscillate the crystal $\pm180°$—but this is the same as rotating it 360°, in which case the Ewald reflecting sphere makes a complete circuit round the origin of the reciprocal lattice. As the Ewald sphere sweeps through the reciprocal lattice, a plane will reflect twice: once when it crosses 'from outside to inside' the reflecting sphere and once when it crosses out again, the reflections being recorded on opposite sides of the direct beam. Rotation photographs are generally used to determine the orientation of small single crystals. An example is given in Fig. 9.17.

Fig. 9.17. A single crystal *c*-axis rotation photograph of α-quartz showing zero, *hk*0, first order, *hk*1 and *hk*$\bar{1}$ and second order *hk*2 and *hk*$\bar{2}$ layer lines. (Photograph by courtesy of the General Electric Company.)

9.5.3 The precession method

This is an ingenious and useful technique invented by M. J. Buerger * by means of which (unlike oscillation and rotation photographs), undistorted sections or layers of reciprocal lattice points may be recorded on a flat film. It is probably easiest to appreciate the geometrical basis of the method by using, as an example, a simple cubic crystal.

Figure 9.18(a) shows an (*h*0*l*) section of a simple cubic crystal (drawn with the *y*-axis or **b*** reciprocal lattice unit cell vector perpendicular to the page). The X-ray beam is incident along the *x*-axis or a* reciprocal lattice unit cell vector and the Ewald reflecting sphere for a particular wavelength is centred about the incident beam direction. The flat film is set perpendicular to the X-ray beam. We are now going to consider the conditions by which we can obtain reflections from the reciprocal lattice points lying in the plane 0*kl*—i.e. the plane through the origin and perpendicular to the *x*-axis or **a*** reciprocal lattice vector. Figure 9.18(a) shows only one row of these reciprocal lattice points—001, 002, 003, 00$\bar{1}$, 00$\bar{2}$, 00$\bar{3}$, etc.; the others are above and below the plane of the diagram. Now, ignoring for the time being the reciprocal lattice points not lying in this plane 0*kl*, let us consider the movements necessary to bring 001, 002, 003, etc. reciprocal lattice points into reflecting positions. Clearly, we need to tilt or rotate the crystal anticlockwise such that first 001, then 002, then 003, etc. successively intersect the sphere. As we do so we tilt or rotate the film anticlockwise by the same angle (the film and crystal rotation mechanisms being coupled together). The 001, 002, 003 reflections then strike the film successively at equally spaced distances from the origin 000. The situation in which 003 is in the reflecting position at a tilt or rotation angle φ is drawn in Fig. 9.18(b). Clearly, if we rotate the crystal and film clockwise we would record the 00$\bar{1}$, 00$\bar{2}$, 00$\bar{3}$, etc. reflections on the opposite side of the film from the origin. The result (ignoring reflections from reciprocal lattice points not lying on this row) would be a row of equally spaced spots on the film corresponding to the equally spaced reciprocal lattice points in

* Denotes biographical notes available in Appendix 3.

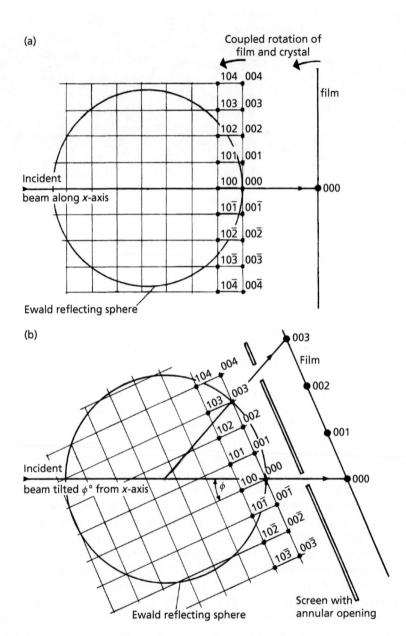

(a)

Coupled rotation of
film and crystal

film

104 004
103 003
102 002
101 001
100 000
10$\bar{1}$ 00$\bar{1}$
10$\bar{2}$ 00$\bar{2}$
10$\bar{3}$ 00$\bar{3}$
10$\bar{4}$ 00$\bar{4}$

Incident
beam along x-axis

000

Ewald reflecting sphere

(b)

003

Film

002

001

000

104 004
103 003
102 002
101 001
100 000
10$\bar{1}$ 00$\bar{1}$
10$\bar{2}$ 00$\bar{2}$
10$\bar{3}$ 00$\bar{3}$

Incident
beam tilted $\phi°$ from x-axis

ϕ

Ewald reflecting sphere

Screen with
annular opening

Fig. 9.18. The geometry of the precession method, (a) The incident beam normal to the 0*kl* section of the reciprocal lattice (indicated by the row of reciprocal lattice points 001, 002, 003, etc.). In order to bring these points into reflecting positions the crystal and the film are rotated anti-clockwise as indicated. The situation in which the 003 reciprocal lattice point is in a reflecting position is shown in (b). (c) A plan view of the film at this angle φ with the circle representing the intersection of the 0*kl* plane and the reflecting sphere and the projection of the annulus on to the screen. The only reflections to reach the film are those falling within the annulus (shaded region), As the crystal + screen + film are precessed about the angle φ the annulus effectively sweeps through the 0*kl* section of the reciprocal lattice as indicated by the arrow and all reciprocal lattice points within the large circle give rise successively to reflections.

the row through the origin. These are shown on a plan view of the film (Fig. 9.18(c)), as the vertical row of spots through the centre.

Now we have to consider the reciprocal lattice points in the 0*kl* section of the reciprocal lattice which intersect the Ewald reflecting sphere above and below the plane of Fig. 9.18(b). The sphere intersects the reciprocal lattice section 0*kl* in a circle, Fig. 9.18(c); any reciprocal lattice point lying in this circle gives rise to a reflected beam. Now we precess the crystal and the film about the incident beam direction such that the film is always parallel to the 0*kl* reciprocal lattice section (the precession movement means that the *x*-axis of the crystal rotates about the incident beam direction at the fixed angle φ). As we do so the circle moves through the reciprocal lattice section in an arc centred at the origin 000 of the reciprocal lattice, i.e. about the direction of the direct X-ray beam. Hence, in a complete revolution, all the reciprocal lattice points lying within the large circle of Fig. 9.18(c) reflect—and the pattern of spots on the film

corresponds to the pattern of reciprocal lattice points in the $0kl$ section of the reciprocal lattice, as shown in Fig. 9.18(c).

Finally, we have to eliminate the complicating effects of the reflections from reciprocal lattice points not lying in the $0kl$ plane through the origin. This is achieved by placing a screen with an annular opening between the crystal and the film, the size of the annulus being chosen to allow reflections to pass to the screen only from reciprocal lattice points lying in the circle (Fig. 9.18(c)). The screen is indicated in cross-section in Fig. 9.18(b) for the case in which the furthest reciprocal lattice point from the origin which can be recorded is 003. The screen is also linked to the crystal–film precession movements.

The precession method can be modified to record reciprocal lattice sections which do not pass through the origin (non-zero-level photographs), details of which can be found in the books on X-ray diffraction techniques listed in Further Reading. Its main application is in crystal structure determination by measurement of the intensities of the X-ray reflections. Figure 9.19 shows an example of a zero-level photograph (of a reciprocal lattice section passing through the origin) which shows very convincingly the pattern of reciprocal lattice points.

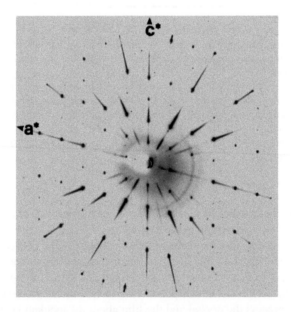

Fig. 9.19. A (zero level) precession photograph of tremolite, $Ca_2Mg_5Si_8O_{22}(OH)_2$ (monoclinic $C2/m$, $a = 9.84$ Å, $b = 18.05$ Å, $c = 5.28$ Å, $\beta = 104.7°$), showing reflections from the $h0l$ reciprocal lattice section (incident beam along the y-axis). The *orientations* of the reciprocal lattice vectors \mathbf{a}^* and \mathbf{c}^* are indicated. Compare this photograph with the drawing of the $(h0l)$ reciprocal lattice section of a monoclinic crystal (Fig. 6.4(c)). The streaking arises from the presence of a spectrum of X-ray wavelengths in the incompletely filtered MoKα ($\lambda = 0.71$ Å) X-radiation. (Photograph by courtesy of Dr. J. E. Chisholm.)

9.6 X-ray diffraction from single crystal thin films and multilayers

Thin films and multilayer specimens are invariably studied using the X-ray diffractometer, the geometrical basis of which is described in detail in Section 10.2.

The normal arrangement, as used in X-ray powder diffractometry, is the symmetrical (Bragg–Brentano) one in which only the d-spacings of those planes parallel to the specimen surface are recorded. This is achieved by arranging the X-ray source and X-ray detector and their collimating slits such that the incident and reflected beams make equal angles to the specimen surface. The arrangement is shown diagrammatically in Fig. 9.20(a). As the angle θ is varied (either by keeping the specimen fixed and rotating the source and detector in opposite senses as indicated in Figs 10.3(a) and (b), or by keeping the source fixed and rotating the detector at twice the angular velocity of the specimen), reflections occur whenever Bragg's law is satisfied. Clearly, for a single crystal, with only one set of planes parallel to the surface, there will only be one Bragg reflection. For multilayer specimens, which may consist of a sequence of thin crystal films (say, of copper and cobalt) mounted on a single crystal substrate, there will be

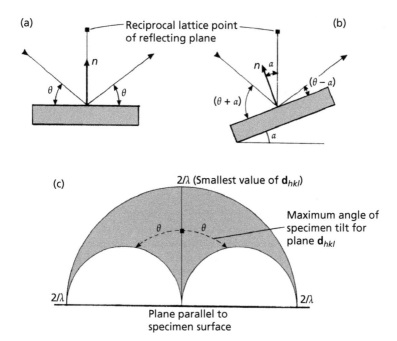

Fig. 9.20. The X-ray diffractometer arrangements for a single crystal. (a) The symmetrical setting: the reciprocal lattice vectors of the reflecting planes are all parallel to the specimen surface normal n. (b) The asymmetrical setting, the reciprocal lattice vectors of the reflecting planes are inclined at angle α to n. (c) The region of reciprocal space (shaded) within which lie the reciprocal lattice points of possible reflecting planes.

three Bragg reflections—one from the substrate and one each from the thin films (of different crystal structure).

However, there are more diffraction phenomena to be recorded. First, since the layers are generally thin—of the order of 1–10 nm—the (high angle) Bragg peaks are substantially broadened and this may be used to estimate the thickness of the layers by means of the Scherrer equation (Section 9.3). Second, the repeat distance, or superlattice wavelength Λ of the layers can be determined from the angles of 'satellite' reflections which occur (given a multilayer specimen with a long range structural coherence) on either side of the (high angle) Bragg peaks. The value of Λ may be determined from the equation

$$\Lambda = \frac{\lambda}{2(\sin\theta_2 - \sin\theta_1)}$$

where λ = the X-ray wavelength and $\sin\theta_1$, $\sin\theta_2$ are the Bragg angles of adjacent satellite peaks. This equation may be rearranged to give

$$\frac{1}{\Lambda} = \frac{1}{d_2} - \frac{1}{d_1}$$

where d_1, d_2 are the notional d-spacings of the satellite peaks. An example of a cobalt-gold multilayer specimen is given in Fig. 9.21(a).

The repeat distance Λ of the layers also gives rise to low-angle Bragg reflections (of the order $2\theta = 2 - 5°$), between which occur a further set of satellite peaks or fringes called Kiessig fringes which may be used to determine the total multilayer film thickness $N\Lambda$ using an analogous equation to that above, viz.

$$\frac{1}{N\Lambda} = \frac{1}{d_{k2}} - \frac{1}{d_{k1}}$$

where d_{k1}, d_{k2} are the notional d-spacings of any pair of adjacent Kiessig fringes. An example of a cobalt–copper multilayer specimen is given in Fig. 9.21(b).

Fig. 9.21. (a) A high angle X-ray diffraction trace (CuKα_1 radiation) from a cobalt–gold multilayer specimen showing the satellite peaks or fringes Ⓢ each side of the Au 111 Bragg reflection. The repeat distance Λ of the layers is determined by measuring the (notional) d-spacings d_1 and d_2 of adjacent fringes and using the equation on page 231. The notional d-spacings (Å) of the fringes and the Au 111 peak are indicated (from which a value of $\Lambda \approx 64$ Å is obtained). The group of higher angle reflections are from Co 111, Co 0002, the substrate GaAs 220 and the 'buffer' layer Ge 220. (b) A low angle X-ray diffraction trace (CuKα_1 radiation) from a cobalt–copper multilayer specimen showing the low angle Bragg reflection at 4.525° and the Kiessig fringes each side. The Bragg peak gives the multilayer or superlattice repeat distance Λ and the number $N - 2$ of Kiessig fringes between the Bragg peak and zero angle gives $N\Lambda$, the total thickness. In this example $\Lambda = 19.512$ Å and $(N - 2) = 22$ (only those fringes from about $2\theta \approx 1.5°$ are shown). Hence the total thickness $N\Lambda = 468.3$ Å. $N\Lambda$ may also be determined by measuring the (notional) d-spacings, d_{K1} and d_{K2} of adjacent Kiessig fringes and using the equation on page 231. The notional d-spacings (Å) of fringes 11 to 22 are indicated, from which values of $N\Lambda$, in fair agreement with that given above (subject to experimental scatter), are obtained.

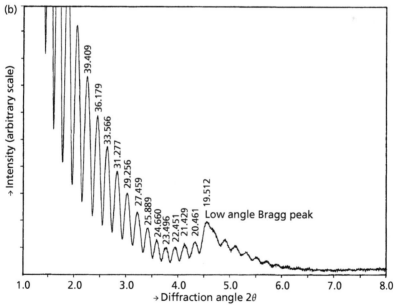

The occurrence of the low-angle Bragg peaks and Kiessig fringes may be understood by analogy with the diffraction pattern from a limited number N of slits of width a (Section 7.4); Fig. 7.9 shows a particular case for $N = 6$. The principal maxima correspond to the low-angle Bragg peaks in which the slit spacing a corresponds to the repeat distance of the layers, or superlattice wavelength Λ. The subsidiary maxima, of which there are $(N - 2)$ between principal maxima, correspond to the Kiessig fringes; by counting the number of Kiessig fringes (plus two) between the low-angle Bragg peaks the number of multilayers or superlattice wavelengths N can be determined and by measuring the angles between adjacent fringes the total thickness $N\Lambda$ of the multilayers can be determined using the equation given above. Its validity can be checked by applying it, say, to the subsidiary maxima in Fig. 7.9; for the zero and first order minima the $\sin \alpha \approx \alpha \approx 2\theta$ values are $1(\lambda/6a)$ and $2(\lambda/6a)$. Substituting these values in the above equation gives $N\Lambda = 6a$, the total width of the grating.

In practice in X-ray diffraction the Kiessig fringes are most clearly defined between the direct beam (zero order peak) and first low-angle Bragg peak (Fig. 9.21(b)).

The other arrangement used in X-ray diffractometry is the asymmetrical one in which, for any particular Bragg angle (source and detector fixed), the specimen can be rotated such that the incident and diffracted beams make different angles to the specimen surface (the angle between them remaining fixed at 2θ). The arrangement is shown diagrammatically in Figs 9.20(b) and 10.3(c). It has the advantage in that reflections from planes which are not parallel to the surface can be recorded (and is also a useful means of 'fine tuning' the angle of the specimen to maximize the intensity of reflections from planes which may not be exactly parallel to the specimen surface). Clearly, the maximum angle of tilt either way is θ, otherwise the surface of the specimen will block or 'cut off either the incident or the diffracted beam.

The Ewald reflecting sphere construction, which shows the extent to which the symmetrical and asymmetrical techniques sample a volume of reciprocal space, is shown in Fig. 9.20(c). The construction is essentially two-dimensional, since only those planes whose normals lie in the plane of the diagram (co-planar with the incident and reflected beams) can be recorded. In the symmetrical case (Fig. 9.20(a)), the reciprocal lattice points of the reflecting planes lie along a line perpendicular to the specimen surface whose maximum extent corresponds with the reciprocal lattice vector of the plane of smallest d_{hkl}-spacing that can be measured, i.e. that for which $\theta = 90°$; hence $|\mathbf{d}^*_{hkl}| = 2/\lambda$.

In the asymmetrical case (Fig. 9.20(b)) the crystal can be tilted or rotated clockwise or anticlockwise such that the reciprocal lattice vector of the reflecting planes is also tilted or rotated with respect to the specimen surface. The angular range, $\pm\theta$ (limited by specimen cut-off), increases as $|\mathbf{d}^*_{hkl}| = 2 \sin \theta/\lambda$ increases, the maximum being $\pm 90°$ when $\sin \theta = 1$. This is shown as the shaded region in Fig. 9.20(c). Clearly, the choice of the wavelength of the X-radiation may be important, particularly in the asymmetrical case, since it determines which crystal planes in the specimen may, or may not, be recorded.

9.7 X-ray (and neutron) diffraction from ordered crystals

In metal alloys, solid solutions may be of two kinds—*interstitial solid solutions* such as, for example, carbon in iron (see p. 18) and *substitutional solid solutions*, such as, for example, α-brass where zinc atoms substitute for copper atoms in the ccp structure. In some cases complete solid solubility is obtained across the whole compositional range from one pure metal to another—such as, for example, copper and nickel, both of which have the ccp structure; but in most alloys the solid solubility is limited and a series of different solid solutions and intermetallic compounds occur across the whole composition range. Copper-zinc alloys for example exhibit a range of such structures or phases: α (ccp, stable up to about 35 at% Zn); β (which is a bcc solid solution stable over a narrow composition range close to equal atomic proportions of copper and zinc and which is also described as the intermetallic compound CuZn); γ (a complex cubic structure); ε (a complex hexagonal structure); and finally η (an hep solid solution of copper in zinc). The study of such alloy phases belongs to physical metallurgy; our interest is solely concerned with the arrangements of the atoms.

In some phases, e.g. α-brass, the copper and zinc atoms are randomly distributed amongst the lattice sites and the structure is said to be *disordered*. In others there is a tendency, which increases with decreasing temperature, for the atoms to occupy specific sites. Two situations may be distinguished: *clustering,* where atoms tend to be surrounded by atoms of their own type (such as, for example, solid solutions of zinc in aluminium); and the converse *short range ordering* where atoms tend to be surrounded by atoms of opposite type. β brass provides a simple example of short range ordering; at high temperatures the structure is disordered but as the temperature decreases (to about 460°C), short range ordered regions develop, the copper (or zinc) atoms tending to occur either at the corners or the centres of the bcc unit cell. Below this temperature the short-range ordered regions extend throughout the crystal and impinge. We now have *domains* of *long-range order.* Within each domain the copper and zinc atoms are at the corners or centres of the unit cell and the crystal structure is the same as that for CsCl (Fig. 1.12(b)). The boundaries (called antiphase domain boundaries) are where the ordering 'changes over'—in one domain it is, say, the copper atoms which occupy the centres of the cell and in the other it is the zinc atoms. It is, of course, possible for a single domain to extend across the whole crystal, in which case of course the distinction between corners and centres in the bcc unit cell disappears.

Long-range ordering also occurs in many ccp alloys, of which the copper–gold system provides 'type' examples. For example, at high temperatures the alloy Cu_3Au is disordered but at low temperatures the gold atoms occupy the corners, and the copper atoms the face-centres of the unit cell. In the alloy CuAu the atoms order so as to occupy alternate (002) planes.

The phenomenon of long-range ordering is often described as *superlattice formation,* but this is a bad name because it suggests the existence of lattices other than the 14 Bravais lattices. Ordering rather consists of a *change* of lattice type. For β-brass (Fig. 9.22(a)) it is a change from Cubic I to Cubic P (the motif, as in CsCl, being Cu + Zn), and for Cu_3 Au (Fig. 9.22(b)) it is a change from Cubic F to Cubic P (the

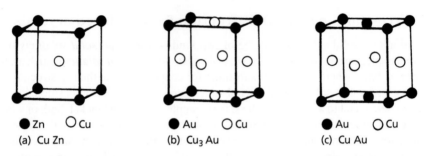

Fig. 9.22. Unit cells of ordered structures, atom positions as indicated, (a) CuZn (bcc-like structure, cubic P lattice), (b) Cu_3Au (fcc-like structure, cubic P lattice), (c) CuAu—the ordering of Cu and Au atoms on alternate 002 planes in the ccp structure results in a reduction of symmetry from a cubic to a tetragonal P lattice.

motif being 3Cu + Au). In the case of CuAu (Fig. 9.22(c)), the change is from Cubic F to Tetragonal P as a consequence of the size difference between the copper and gold atoms, which results in a small reduction in the interlayer spacing perpendicular to the 002 planes giving a c/a ratio less than unity.

Long range ordering and antiphase domain boundaries are also characteristic of ferromagnetic (and antiferromagnetic) materials where in each domain it is the magnetic dipoles or moments which are ordered or orientated in a particular direction.

Ordering was one of the earliest phenomena to be detected by X-ray diffraction techniques and provides further simple examples of the application of the structure factor equation (Section 9.2).

Example 6: Cu_3Au. For the disordered case the atomic scattering factor for each atom site, f_{Av}, is taken as the weighted average of f_{Cu}, and f_{Au}, i.e. $f_{Av} = \frac{1}{4}(f_{Au} + 3f_{Cu}.)$ and the structure factor F_{hkl} is that for an fcc crystal (see Appendix 6), i.e.

$$F_{hkl} = 4 \cdot \frac{1}{4}(f_{Au} + 3f_{Cu}) = (f_{Au} + 3f_{Cu}) \quad \text{for } h, k, l \text{ all odd or all even}$$
$$F_{hkl} = 0 \quad \text{for } h, k, l \text{ mixed.}$$

For the fully ordered case (Fig. 9.22(b)) the atomic positions $u_n v_n w_n$ for the gold atoms are (000) and for the copper atoms $\left(\frac{1}{2}\frac{1}{2}0\right), \left(\frac{1}{2}0\frac{1}{2}\right) \left(0\frac{1}{2}\frac{1}{2}\right)$. Hence:

$$F_{hkl} = f_{Au} \exp 2\pi i(h0 + k0 + l0) + f_{Cu} \exp 2\pi i \left(h\frac{1}{2} + k\frac{1}{2} + l0\right)$$
$$+ f_{Cu} \exp 2\pi i \left(h\frac{1}{2} + k0 + l\frac{1}{2}\right) + f_{Cu} \exp 2\pi i \left(h0 + k\frac{1}{2} + l\frac{1}{2}\right)$$
$$= f_{Au} + f_{Cu}((\exp \pi i(h + k) + \exp \pi i(h + l) + \exp \pi i(k + l).)$$

Hence

$$F_{hkl} = (f_{Au} + 3f_{Cu}) \quad \text{for } h, k, l \text{ all odd or all even}$$
$$F_{hkl} = (f_{Au} - f_{Cu}) \quad \text{for } h, k, l \text{ mixed.}$$

Hence, in the ordered case reflections occur when h, k, l are mixed and it is the occurrence of these additional or 'superimposed' reflections in the diffraction pattern that gave rise to the term 'superlattice'. Their intensities, in proportion to the 'base' reflections for which h, k, l are all odd or all even, can be estimated by approximating $f \approx Z$; i.e. $f_{Au} = 79$ and $f_{Cu} = 29$. Hence:

$$I_{hkl} = F_{hkl} \cdot F^*_{hkl} = (79 + 3.29)^2 = (166)^2 \quad h, k, l \text{ all odd or all even}$$

$$I_{hkl} = F_{hkl} \cdot F^*_{hkl} = (79 - 29)^2 = (50)^2 \quad h, k, l \text{ mixed.}$$

Hence the intensities of the 'superlattice' reflections are $50^2/166^2 \times 100\% = 9\%$ of the intensities of the 'base' reflections and are easily detectable.

Partial ordering is expressed in terms of the long range order parameter S which is defined in terms of p, the fraction of sites occupied by (in the fully ordered case) the number of 'right' atoms and r, the fraction of such sites, whence $S = (p - r)/(1 - r)$. We may apply this equation either to the Au or the Cu atoms. For the Au atoms $r = 0.25$. Suppose, for example that $\frac{2}{3}$ of these sites are occupied by the 'right' Au atoms (and $\frac{1}{3}$ by the 'wrong' Cu atoms). Then $p = \frac{2}{3}$ and $S = 0.56$.

For this partially ordered case, for h, k, l mixed indices $F_{hkl} = S(f_{Au} - f_{Cu})$ and the intensity of the reflections is proportional to S^2.

Example 7: *CuZn (β-brass)*. For the disordered case $f_{Av} = \frac{1}{2}(f_{Cu} + f_{Zn})$ and, as for a bcc crystal, reflections only occur when $(h + k + l) =$ even integer. For the fully ordered case we follow the same procedure as for CsCl (Example 1), i.e. $F_{hkl} = f_{Cu} + f_{Zn}$ for $(h + k + l) =$ even integer and $F_{hkl} = f_{Cu} - f_{Zn}$ for $(h + k + l) =$ odd integer. Approximating $f_{Cu} = 29$ and $f_{Zn} = 30$, the intensities of the 'superlattice' reflections are $-1^2/59^2 \times 100\% = 0.03\%$ of the intensities of the 'base' reflections and are much too small to be readily detected by X-ray diffraction techniques.

This situation is illustrative of a general problem in X-ray diffraction: since atomic scattering factors are proportional to Z then the positions in the unit cell of atoms of similar atomic number are not easy to determine, nor are the positions of atoms of low atomic number which scatter X-rays very weakly. It is a situation in which **neutron diffraction** finds a 'niche market'. Unlike X-rays, neutrons in most cases are scattered by the nuclei rather than the electrons in atoms and the scattering amplitudes (expressed as scattering lengths) vary in an irregular way with atomic number. The relatively large scattering amplitudes of, for example, hydrogen and oxygen atoms in comparison with heavy metal atoms enables these atoms to be located within the unit cell and also allows the small distortions which occur below the Curie temperature in ferroelectric perovskite structures (see Sections 11.1.1 and 4.4) to be determined.

Figure 9.23 shows, in a most visually convincing way, the differences between the scattering amplitudes, and therefore the absorptions, of neutrons by hydrogen, oxygen and heavy metal atoms. It is a neutron radiograph of a rose-stem within a thick-walled lead container; the lead is almost invisible to the neutrons, but the absorption in the rose and stem reveals the delicacy of the flower.

Fig. 9.23. A neutron radiograph of a rose-stem within a thick-walled lead cylinder. (Photograph by courtesy of Dr Hans Priesmeyer, Christian-Albrechts University, Kiel.)

9.8 Practical considerations: X-ray sources and recording techniques

The discovery of X-rays by W. C. Röntgen* at the University of Würtzburg in 1895 is a wonderful example of serendipity in science[7]. Röntgen was interested in the nature of cathode rays generated when a high voltage from an induction coil was discharged across a discharge tube and which gave rise to faint luminescence in gases and fluorescence in some crystals placed close to the tube. He enclosed the tube in a light-tight cardboard box (for what experiments we do not know). On discharge of the induction coil he noticed that, on a table a considerable distance away, a barium platinocyanide crystal (which happened to be there) gave a flash of fluorescence. This clearly was not due to the cathode rays, which would have been absorbed in the glass walls of the tube and in the air, but originated from rays (or particles) emanating from the point where the cathode rays struck the walls of the tube. In a feverish period of work, Röntgen

* Denotes biographical notes available in Appendix 3.

[7] The word comes from the fairy-tale *The Three Princes of Serendip* 'who were always making discoveries by accidents and sagacity, of things they were not in quest of'. (From the *Shorter Oxford English Dictionary.*)

established that the rays travelled in straight lines, were not refracted or diffracted (by optical diffraction gratings), were more strongly absorbed the denser the material through which they passed and could be recorded on photographic plates. He took the very first medical radiograph—a beautiful image of the bones in his wife's hand and clearly showing her wedding ring. Röntgen coined the term 'X-rays'—a term used ever since, even though we now know that they are short-wavelength ($\sim 0.1 \sim 10$ Å) electromagnetic waves.

Modern X-ray tubes are descendants of Röntgen's discharge tube except that the cathode rays (electrons) are provided by a heated filament and the target, or anode, is a heavy metal, typically copper, iron, molybdenum or tungsten. By far the greatest proportion of the energy of the incident electrons is converted into heat (phonons) and hence the anode must be water-cooled or, in the case of micro-focus tubes in which the beam is concentrated into a small area, by also rotating the anode to prevent incipient melting. Heating is a 'nuisance' in X-ray tubes and largely limits the intensity of the beam which can be obtained—but it is also the basis of electron beam melting and welding techniques.

9.8.1 The generation of X-rays in X-ray tubes

As the electrons pass into the anode they suffer collisions with the atoms and are eventually brought to rest. At each collision the loss of energy ΔE gives rise to an X-ray photon of energy hc/λ. The energy losses ΔE have a wide range of values, giving rise to a wide range of λ values. This is the origin of the 'continuous', 'white' or Bremsstrahlung (which is German for braking) radiation.

Figure 9.24 shows a typical X-ray spectrum. Notice that there is a 'cut-off' at a short wavelength called the short-wavelength limit (swl). This arises from electrons which lose all their energy in one single collision, i.e. $\Delta E = eV = hc/\lambda_{swl}$, where V is the voltage of the X-ray tube and e is the charge on the electron. The intensity of the continuous radiation peaks at a wavelength of about $^4/_3 \lambda_{swl}$. Superimposed on this continuous radiation are a series of sharp peaks. These arise from electron transitions between energy levels in the atom. The innermost (K-shell) has one energy level, the L-shell has three and the M-shell has five such energy levels. If (and only if) the incident electrons have sufficient energy can they 'knock out' an electron from its shell. The greatest energy (designated W_K, the work function), is required to 'knock out' an electron from the K-shell. In this situation the incident electrons are strongly absorbed and the corresponding wavelength is called the K absorption edge, designated λ_K. Hence $\lambda_K = hc/W_K$. This ionized state of the atom is short lived and the atom returns to its ground state as a result of the electrons 'tumbling down' from outer to inner shells, each transition being accompanied by the emission of an X-ray photon of energy, and therefore wavelength, characteristic of the difference between the energy levels. The spectra are designated the K_α-series (for transitions from the L to the K-shell), the K_β-series (for transitions from the M to the K-shell), the L_α-series (for transitions from the M to the L-shell), and so on. The most important, in relation to X-ray diffraction, is the K_α-series. Although there are three energy levels in the L-shell, one transition (between the s-orbitals in the K and

L shells) is forbidden by what are known as **Selection Rules**. Hence, there are just two $L \to K$ transitions giving two closely separated $K\alpha$ wavelengths, $K\alpha_1$ and $K\alpha_2$. These are not equally strong, $K\alpha_1$ is about twice as intense as $K\alpha_2$ and comprise what is termed the $K\alpha$-doublet (Fig. 9.24). For copper radiation, $\lambda_{K\alpha_1} = 1.541\text{Å}$, $\lambda_{K\alpha_2} = 1.544\text{Å}$, and the weighted average when these components are unresolved is $\lambda_{K\bar{\alpha}} = 1.542\text{Å}$.

Except for Laue-type experiments, use is generally made only of the $K_{\bar{\alpha}}$ or K_{α_1} radiation and this is achieved by the use of a crystal monochromator set to reflect only this particular wavelength (and its sub-multiples, i.e. $\lambda_{K_{\alpha_1}} = 2d \sin\theta = 2(\lambda_{K_{\alpha_1}}/2)$, etc.) Older 'filter' methods which are unable to discriminate between K_{α_1} and K_{α_2} wavelengths, make use of a thin sheet of a metal whose K absorption edge lies in between the $K\alpha$ and $K\beta$ wavelengths of the X-ray spectrum (Fig. 9.24). For example, for copper radiation ($\lambda_{k\beta} = 1.392\text{Å}$) a nickel filter, $\lambda_K = 1.488\text{Å}$ is used.

The characteristic X-ray spectra were discovered by W H Bragg but it was the work of H G J Moseley in 1913–14 that established the relation between their wavelengths and atomic number Z. Moseley's Law, which for the K_α series may be expressed $1/\lambda_{K_\alpha} = {}^3/_4 R(Z - \sigma)$ where σ is a constant and R is the Rydberg Constant for the optical spectral series was a key element in the development of the Bohr Theory of the atom.

9.8.2 Synchrotron X-ray generation

Very high intensity X-ray beams, \sim100–10,000 times more intense than the K_{α_1} radiation from X-ray tubes, are generated in a synchrotron, a type of particle accelerator. Electrons, travelling at velocities close to the speed of light, are confined to travel in near-circular paths in a 'storage ring' by the action of magnets placed at intervals around

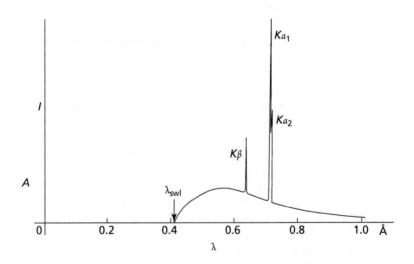

Fig. 9.24. An X-ray spectrum showing the continuous (white) radiation which peaks at about $^4/_3\lambda_{swl}$ and the sharp $K_{\alpha 1}$ and $K_{\alpha 2}$ peaks at wavelengths characteristic of the anode element (molybdenum). (From *Contemporary Crystallography* by Martin J. Buerger, McGraw-Hill, 1970.)

the ring. The 'synchrotron radiation', which arises as a result of the continuous inward radial acceleration of the electrons, is outputted tangentially from the ring and covers a wavelength range from infrared to very short X-ray wavelengths. The radiation then passes to a crystal monochromator, set to reflect the particular wavelength required. Apart from the higher intensity, the advantage of synchrotron radiation is that, unlike X-ray tubes where one is restricted to the particular K_{α_1} wavelength of the anode element, the monochromator can be 'tuned' to reflect X-rays either well away from the absorption edge of (e.g.) a heavy element in the specimen to minimize absorption effects, or, contrariwise, close to an absorption edge to maximize the effects of anomalous absorption (see Example 5). Furthermore, by the insertion of a device into the beam known as a **wiggler**, the synchrotron output can be rapidly 'wiggled' over a wavelength range (typically 0.5~2.0 Å), thus simulating 'white' X-radiation. The radiations also differ in their states of polarization; that from an X-ray tube is almost wholly unpolarized whereas that from a synchrotron is wholly polarized in the plane of the storage ring.

The advent of synchrotron radiation has not only allowed very much smaller crystals to be examined (in the micrometre, rather than tens or hundreds of micrometre range) but has also allowed the examination of very short-lived crystal structures occurring in chemical reactions.

9.8.3 X-ray recording techniques

The recording of X-ray reflections has been revolutionized by the advent of image-plate systems, CCD (charge-coupled-device) and position sensitive area detectors and the associated developments in computer software. They represent, in a sense, a return to the older film methods in that they record reflections in the whole of reciprocal space (or that portion intercepted by the detector) simultaneously rather than sequentially as was the case in Bragg's early X-ray spectrometer or the complex 4-circle goniometers which until recently were dominant in single crystal X-ray work. Image-plate systems (which were developed from medical diagnostic radiography) are particularly useful for recording the diffraction patterns from weakly-scattering biological crystals. They are based upon a process known as photostimulated luminescence: when exposed to X-rays some electrons in the phosphor-coated plate are excited and trapped into higher energy states which on exposure to visible light lose energy and emit ultraviolet light. Similarly, in X-ray powder diffractometry (Section 10.2), scintillation and proportional counters are being superseded by position sensitive detectors although the future of the diffractometer, as an item of hardware, seems secure.

It is of course very easy for a beginner to the subject of X-ray diffraction to be dazzled by the technology: the basic principles remain unchanged irrespective of developments in recording techniques.

Exercises

9.1 In the Laue experiment, a bcc crystal, lattice parameter $a = 0.4$ nm (4 Å) is irradiated in the [100] (or \mathbf{a}^*) direction with an X-ray beam which contains a continuous spectrum of

wavelengths in the range between 0.167 nm (1.67 Å) and 0.25 nm (2.5 Å). Use the reflecting sphere construction to determine the indices of the planes in the crystal for which Bragg's law is satisfied and draw the direction of the reflected beam for *one* plane in the [001] zone. (*Hint:* Make a scale drawing of the section of the reciprocal lattice normal to the [001] (or **c***) direction and which passes through the origin 000 (i.e. the section which contains the *hk*0 reciprocal lattice points as shown in Fig. 6.7(b)). A convenient scale to use between the reciprocal lattice dimensions and A4 size drawing paper is 1 nm^{-1} =0.5 cm (1 Å$^{-1}$ = 5.0 cm). Draw a line indicating the [100] direction of the incident beam and draw in the two reflecting spheres representing the limits of the wavelength range. Remember that the origin of the reciprocal lattice is located at the point where the beam *exits* from the spheres—hence the centres of the spheres are obviously not coincident. Shade in the region of the reciprocal lattice between the two spheres; planes whose reciprocal lattice points lie in this region satisfy Bragg's law. For one reciprocal lattice point in this region, find, by construction, the sphere which it intersects. The direction of the reflected beam is from the centre of this sphere through the reciprocal lattice point, and the radius of the sphere gives the particular wavelength reflected. Draw sections of the reciprocal lattice normal to the [001] (or **c***) direction and which pass through the *hk*1, *hk*2, *hk*$\bar{1}$, *hk*$\bar{2}$, etc. reciprocal lattice points (see Fig. 6.8(a)). In these sections of the reciprocal lattice 'above' and 'below' that through the origin, the sections of the sphere are reducing in size—simple trigonometry will show by how much. Again, planes whose reciprocal lattice points lie in the region between the spheres satisfy Bragg's law.)

9.2 In an oscillating crystal experiment the bcc crystal described in Exercise 9.1 is irradiated in the [100] (or **a***) direction with a monochromatic X-ray beam of wavelength 0.167 nm (1.67 Å). The crystal is then oscillated $\pm 10°$ about the [001] (or **c***) direction. Find the indices *hk*0 of the planes in the [001] zone which give rise to reflections during the oscillation of the crystal.

(*Hint*: Make a scale drawing of the section of the reciprocal lattice through the origin 000 and normal to the [001] (or **c***) direction. Draw a line indicating the [100] direction of the incident beam and a single sphere corresponding to the single X-ray wavelength (see Exercise 9.1). Oscillating the crystal (at the centre of the sphere) is equivalent to oscillating the reciprocal lattice (at the origin). The simplest way to represent the relative changes in orientation between the crystal and the X-ray beam is to 'oscillate' the beam. The directions of the beam at the oscillation limits are $\pm 10°$ from the [100] direction in the plane of the reciprocal lattice section. Draw in the reflecting sphere at these limits and shade in the lunes or the regions of reciprocal space through which the surface of the sphere passes. Planes whose reciprocal lattice points lie in these regions reflect the X-ray beam during oscillation.)

9.3 The kinetic energy of neutrons emerging in thermal equilibrium from a reactor is given by 3/2 *kT* where k = Boltzmann's constant, 1.38×10^{-23} JK^{-1} and T is the Kelvin temperature. Given that the (rest) mass M_n of a neutron = 1.67×10^{-27} kg and Planck's constant $h = 6.63 \times 10^{-34}$ Js, estimate the wavelength of neutrons in thermal equilibrium at 100 °C.

9.4 Determine the F_{hkl} values for reflections for the ZnS (zinc blende or sphalerite structure, Fig. 1.14(c). Show that they fall into three groups: $(h + k + l) = 4n$; $(h + k + l) = 4n + 2$; $(h + k + l) = 2n + 1$.
Hint: ZnS has a cubic F lattice, each lattice point being associated with a motif consisting of one Zn atom at (000) and one S at ($^1/_4$ $^1/_4$ $^1/_4$) . F_{hkl} is determined (i) by writing down the reflection condition for cubic F crystal (see Appendix A6) and (ii) by substituting f_{Zn} at (000) and f_S at ($^1/_4$ $^1/_4$ $^1/_4$).

9.5 In diamond the atom positions are identical to those in ZnS. Hence determine the conditions
 for reflection in the diamond cubic lattice.
 Hint: We may simply proceed as before. However, since all the atoms are now of the same
 type, diamond has the centrosymmetric point group $m\bar{3}m$, the centres of symmetry lying
 equidistant between nearest neighbour atoms. We may therefore choose an origin at ($^1\!/_8$
 $^1\!/_8$ $^1\!/_8$)—called 'origin choice 2' in space group $Fd\bar{3}m$, No. 227—and use the simplified
 structure factor equation in Example 3, page 217.
9.6 Determine the F_{hkl} values for reflections for the NaCl structure (Fig. 1.14(a)). Show that
 they fall into two groups: h, k, l all even and h, k, l all odd.
 Note: Proceed as in Exercise 9.4.
9.7 In Fig. 9.17, measure the $hk1$ and $hk2$ layer line spacings and hence determine the c-axis
 repeat distance in α-quartz.
 Note: The photograph was taken using X-rays of wavelength $\lambda = 1.541$Å and a camera of
 30 mm radius. To account for the reduced scale of the photograph in printing, multiply your
 layer-line spacing measurements by a factor 1.33.
9.8 In CaF$_2$ the F atoms are situated in the centres of the tetrahedral sites such that in the
 {100} planes they lie exactly half-way between the Ca atoms (see Fig. 1.14(b)). In FeS$_2$ the
 S atoms are shifted to one of the 'empty' corners of the tetrahedral sites (see discussion in
 Section 4.7 on p. 121–122 and Fig. 4.16). The resulting sequence of the {100} atomic layers,
 for all the atoms in the unit cell, is shown in Fig. 9.25 where x is the fractional distance of
 the Fe-Fe atom layer spacing.

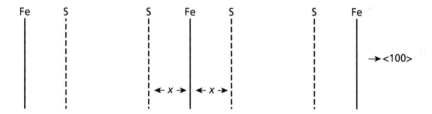

Fig. 9.25. The atomic layers in the {100} planes in FeS$_2$.

 In 1913, W.L. Bragg measured the intensities of the reflections from the {100} planes
and showed that 'first order reflection is large, the second and third too small to detect
and the fourth and fifth are easily measurable'. Assuming that the squares of the atomic
scattering factors for the 1st, 2nd, etc. order reflections decrease in the ratios 1.0, 0.2, 0.07,
0.03, and 0.01 (Bragg's estimates), determine the ratios of the intensities of the reflections
for values of $x = {}^1\!/_4, {}^1\!/_5, {}^1\!/_6$, of the Fe-Fe spacing, and show that a value of $^1\!/_5$ is consistent
with the observed intensities.
9.9 Given the following wavelengths for copper radiation: $\lambda_{K\alpha} = 1.542$Å, $\lambda_{K\beta} = 1.392$Å, and
 λ_K (absorption edge) = 1.381Å, estimate the wavelengths of the L and M absorption edges,
 and $\lambda_{L\alpha}$ radiation.

10

X-ray diffraction of polycrystalline materials

10.1 Introduction

The preparation or synthesis of single crystals which are sufficiently large—of the or-
der of a tenth of a millimetre or so—to be studied using the X-ray diffraction methods
described in Chapter 9 is often a matter of great experimental difficulty. This is particu-
larly the case for proteins and other complex organic crystals, the preparation of which
requires considerable ingenuity and skill. However, in many situations the preparation
of large single crystals is neither possible nor desirable. In materials science and pet-
rology, for example, the crystal structures of interest are frequently those of metastable
phases which occur on a very fine scale as a result of precipitation (or exsolution) from
metal, ceramic or mineral matrices. As such phases grow, either as a result of natural or
artificial ageing processes, their crystal structures invariably change as they evolve into
more stable phases. These changes are best studied using the electron diffraction tech-
niques outlined in Chapter 11, since electron beams can be focused down to diameters
of the order of 1–10 nm, compared with beam diameters of the order of 0.1–1.0 mm
for X-rays. Electron diffraction has the further advantage that crystallographic relation-
ships between the phases and the matrices in which they occur can be investigated. Its
disadvantage lies in the fact that the accuracy with which d_{hkl}-spacings can be measured
is low compared to that for X-ray diffraction.

Polycrystalline or 'powder' X-ray diffraction techniques (Section 10.2) were devel-
oped by Debye* and Scherrer* and independently by Hull* in the period 1914–1919.
They may be classified as 'fixed λ, varying θ' techniques (see Section 9.5) in which
the 'varying θ' is achieved by having a sufficiently large number of more-or-less ran-
domly orientated crystals in the specimen such that *some* of the *hkl* planes in *some* of
them will be orientated, by chance, at the appropriate Bragg angles for reflection. All
the planes of a given d_{hkl}-spacing reflect at the same 2θ angle to the direct beam and
all these reflected beams lie on a cone of semi-angle 2θ about the direct beam. The
various 'powder' X-ray diffraction techniques may be classified as to the ways in which
the cones of diffracted beams are intercepted and recorded. In situations in which the
crystals are randomly orientated, the diffracted intensity in the cones will be uniform
and hence only part of the cones need to be recorded. This is the case with what might
be called the 'classical' powder camera and diffractometer techniques (Section 10.3).

* Denotes biographical notes available in Appendix 3.

The Basics of Crystallography and Diffraction. Fourth Edition. Christopher Hammond.
© Christopher Hammond 2015. Published in 2015 by Oxford University Press.

However, in situations in which the crystals are not randomly orientated the diffracted intensity in the cones will not be uniform and in order to determine the extent of the 'non-randomness' the whole, or a large part, of the cones needs to be recorded. In metallurgy and materials science the 'non-randomness' of crystals is known as *texture* or *preferred orientation* whereas in earth science it is known as *fabric* or *petrofabric*. (The word texture in earth science has quite a different meaning and refers rather generally to grain shapes and grain size distributions—a different nomenclature which is a possible source of confusion.)

The analysis of preferred orientation (texture or fabric) is important since it almost invariably arises as a consequence of the processes of crystallization and re-crystallization, sintering, extrusion and hot deformation which occur either in the short time-scale in materials processing or in the long time-scale in the earth's crust and mantle. The simplest experimental techniques are the 'fibre' X-ray techniques in which the crystals are orientated with a particular crystallographic direction (or directions) along the fibre axis. In the case of rolled/recrystallized metal sheets or strata in the earth's crust the analysis is more complicated—in addition to preferred orientation along some reference (e.g. rolling or shear) direction there may also be preferred orientation of the crystals with respect to the plane of the metal sheet or rock strata. Such textures may be represented by stereographic projections referred to as pole figures, fabric or petrofabric diagrams (Section 10.4) or, more quantitatively, by what are known as orientation distribution functions.

Fibre techniques play an essential experimental part in the determination of the structures of long-chain organic molecules in which the aim is to draw out a fibre from a solution which contains a sufficient number of molecules, all perfectly aligned along the fibre axis, and which give rise to reflections of measurable intensity. A simple description of the structure of DNA, which was discovered by such a technique, is given in Section 10.5.

Finally, X-ray (and neutron) powder diffraction techniques are being used increasingly in crystal structure determination, once the sole preserve of single crystal techniques. This has arisen from the necessity of determining the structures of the many materials of technological importance—e.g. clay minerals, ferrihydrites, etc.—which are finely crystalline and/or poorly ordered and which cannot be grown into crystals of large enough size (\sim0.2 mm) to be investigated by single crystal techniques. However, powder techniques are only able to deal with situations in which the structure (depending on its complexity) is known approximately and are thus better described as *structure refinement* techniques. The principal of these—known as the Rietveld method—is outlined in Section 10.6.

10.2 The geometrical basis of polycrystalline (powder) X-ray diffraction techniques

We begin, perhaps rather surprisingly, with a theorem which we learned off-by-heart (or should have done) in our early geometry lessons at school—never imagining that it would ever come in useful! The theorem comes from *The Elements of Euclid,* the

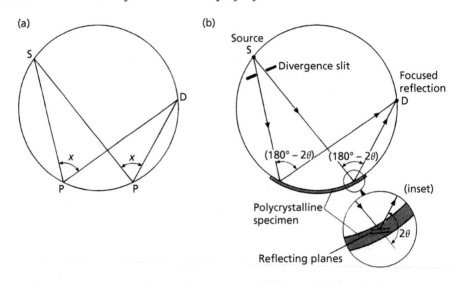

Fig. 10.1. (a) The geometrical basis of X-ray powder diffraction techniques; for two given points S and D on the circle, the angle x at the circumference is constant, irrespective of the position of P. (b) Application to an X-ray focusing camera. S is a source of monochromatic X-rays, the angular spread of which is collimated by a divergence slit to strike a thin layer of a powder specimen on the circumference at P...P. Angle $x = (180° - 2\theta)$, hence any crystal planes in the right orientation for Bragg's law to be satisfied reflect to give a focused beam (as shown in the inset for a particular d_{hkl}-spacing). Note that the reflecting planes are not in general tangential to the circumference of the camera, although in practice they are closely so.

mathematics textbook which has remained in print ever since it was first written in *c*. 300 B.C. The theorem, which is proved in Proposition 21, Book III of *The Elements,* states that 'the angles in the same segment of a circle are equal to one another'. It follows the other equally well known theorem (Proposition 20) that 'the angle at the centre of a circle is double that of the angle at the circumference on the same base, that is, on the same arc'.

The former theorem is illustrated (without Euclid's proof) in Fig. 10.1(a) and may be simply restated by saying that for any two points S and D on the circumference of a circle, the angle shown as x is constant irrespective of the position of point P. Now let us make use of this geometry in devising an X-ray camera (Fig. 10.1(b)). Let S be a point or slit source of monochromatic X-rays (the slit being perpendicular to the plane of the paper) and let part of the circumference P... P be 'coated' with a thin film of a polycrystalline specimen.

The X-rays diverge from the source S and in order that they should only strike the specimen we limit or collimate them with a (divergence) slit as shown. Now the angle x is, in terms of Bragg's law, the angle $(180° - 2\theta)$—and whenever this is the correct Bragg angle for reflection for a particular set of *hkl* planes, reflections occur from all those crystals in the specimen which are in the right orientation and the reflected beams, as proved by Euclid, all converge to point D. In short D is a point where all the reflected beams for a particular d_{hkl} spacing are focused. Different positions of D on the

circumference correspond to different x, hence different $(180° - 2\theta)$ values. Hence we will have a series of focused diffraction lines, one for each d_{hkl}-spacing in the specimen, around the circumference of the circle.

The Bragg reflecting geometry is shown in detail in the little 'inset' diagram below Fig. 10.1(b) which shows the orientation of the reflecting planes in one part of the specimen and the angle 2θ between the direct and reflected beams. Notice that the reflecting planes are not (except for some particular part of the specimen) tangential to the circumference.

A film placed around the rest of the circumference of the circle will record all the reflected beams and this arrangement is the basis of the Seeman–Bohlin camera and the variants of it devised by Guinier and Hägg. They are called **focusing X-ray methods** because they exploit the focusing geometry of Euclid's circle which is thus called the **focusing circle.**

The focusing geometry can also be used to provide a monochromatic source of X-rays (Fig. 10.2). The source S is now the line focus of an X-ray tube from which a spectrum of X-ray wavelengths (the 'white' radiation with superimposed characteristic wavelengths, $K\alpha_1$, $K\alpha_2$, $K\beta$ etc.) is emitted. A single plate-shaped crystal is cut such that a set of strongly reflecting planes is parallel to the surface. The crystal is curved or bent to a radius of curvature *twice* that of the focusing circle and then (ideally) is shaped or 'hollowed out' such that its inner surface lies along that of the focusing circle. The position of the source S, and hence the angle $(180° - 2\theta)$ is, together with the d_{hkl} spacing of the crystal planes, chosen to reflect, say, only the strong $K\alpha_1$ component of the whole X-ray spectrum which is focused to a point or line D, symmetrical

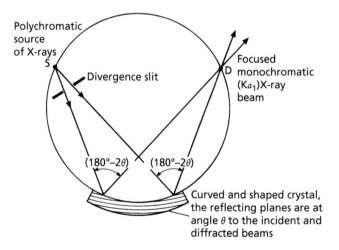

Fig. 10.2. The principle of the X-ray monochromator. The crystal planes are curved to a radius twice that of the focusing circle such that they make the same angle (θ) to the divergent incident beam. The surface of the crystal is also shaped so as to lie precisely on the focusing circle. The source and focus of the reflected beams are symmetrical (i.e. equal distances) with respect to the crystal and θ and the d_{hkl}-spacing of the crystal planes are chosen such that only the strong $K\alpha_1$ X-ray wavelength component is reflected.

with the source S. D becomes, in effect, the line source of a beam of monochromatic X-rays, which can be used 'upstream' of the Seeman–Bohlin camera as the source S of monochromatic X-rays.

The symmetrical arrangement is also made use of in the X-ray diffractometer, which has become by far the most important classical X-ray powder diffraction technique. The diffractometer has become, in effect, the X-ray data acquisition end of a computer in which the data can be stored, analysed, plotted, compared with standard powder data files, etc. The advantages are enormous; the disadvantage, if it can be called such, is that students may simply regard the diffractometer as a 'black box' to generate data, without understanding the principles on which the instrument works, nor the parameters underlying the data analysis procedures.

The symmetrical or Bragg–Brentano focusing geometry of the diffractometer is shown in Figs 10.3(a) and (b). As described in Section 9.6, the source of monochromatic

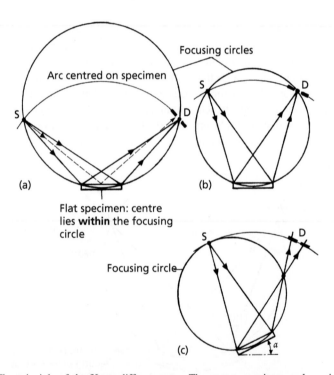

Fig. 10.3. The principle of the X-ray diffractometer. The source–specimen and specimen–detector slit distances are fixed and equal, (a) The situation for a small θ angle and (b) for a large θ angle, the source and detector moving round the arc of a circle centred on the specimen. Notice the change in the diameter of the focusing circle and the fact that, since the specimen is flat, complete focusing conditions are not achieved (dashed line in (a)), (c) An asymmetrical specimen arrangement for the same 2θ angle as in (b), the position and diameter of the focusing circle have changed such that the detector slit is no longer coincident with it. The reflecting planes are now those which make (approximately) angle α to the specimen surface (see Fig. 9.20).

X-rays, S (from a monochromator upstream of the diffractometer) is at an equal distance and equal angle to the specimen surface as D, the 'receiving slit' of the detector (a proportional counter). The 2θ angle is continuously varied by the source and detector slit tracking round the arc of a circle centred on (and therefore at a fixed distance from) the specimen. Figure 10.3(a) shows the situation at a low θ angle and Fig. 10.3(b) shows the situation at a high θ angle. In some instruments the specimen is fixed and the source and detector slit rotate in opposite senses; in others the source is fixed and the specimen and detector slit rotate in the same sense, the detector slit at twice the angular velocity of the specimen—but the result geometrically is the same.

However, the important point to note is that the polycrystalline specimen as used in a diffractometer is flat, and not curved to fit the circumference of the focusing circle. Focusing therefore is not perfect—the reflected beams from across the whole surface of the specimen do not all converge to the same point: those from the centre converge to a point a little above those from the edges, as shown in Fig. 10.3(a). The diffractometer is hence called a *semi-focusing X-ray method*. In practice the deviation from full focusing geometry is only important (in the symmetrical arrangement) at low θ angles and in which a large width of specimen contributes to the reflected beam. It is not however laziness or experimental difficulty which prevents specimens being made to fit the circumference of the focusing circle, but the fact that, unlike the situation for the Seeman–Bohlin camera, the radius of the focusing circle changes with angle as is shown by a comparison with Figs 10.3(a) and (b).

As pointed out above, in the symmetrical arrangement, in which the specimen surface makes equal angles to the incident and reflected beams, the only crystal planes which contribute to the reflections are those which lie (approximately) parallel to the specimen surface, particularly when the divergence angle of the incident beam, and therefore the irradiated surface of the specimen, is small. In situations in which the d_{hkl} spacings of planes which lie at large angles to the specimen surface need to be measured (e.g. in determining the variation of lattice strains from planes in different orientations), the specimen is rotated as shown in Fig. 10.3(c); the angle of rotation α is of course (approximately) equal to the angle of the reflecting planes to the specimen surface (Fig. 9.20). In doing so, however, the symmetrical Bragg–Brentano focusing geometry is lost; the focusing circle changes both in diameter and position, as shown in Fig. 10.3(c), and the detector slit is no longer at the approximate line of focus, but in this case beyond it where the beam diverges and broadens. This deviation from the focusing geometry may be serious because the observed 'broadening' at the detector slit could be misinterpreted as arising from a variation in reflection angle as a result of a variation in lattice strain. The problem may be partially overcome in two ways. First, the detector slit may be moved along a sliding-arm arrangement so as to coincide more precisely with the line of focus. However, this may itself introduce errors in the angular measurements because of the difficulty of achieving a precise radial alignment of the slider. Second, the angular divergence of the incident beam may be restricted to very low values—say 0.25°–0.5°—such that the broadening of the beam at the detector slit is very small, and this is now the preferred option.

10.2.1 *Intensity measurement in the X-ray diffractometer*

As the incident beam penetrates the specimen it is progressively absorbed such that, beyond a certain distance, the weak diffracted beams (which are further absorbed on their way out) make a negligible contribution to the overall intensity of the diffracted beam. This distance, t, is normally set as the reciprocal of the linear absorption coefficient, μ, of the material[1]. It varies from about 5–20 μm for strongly absorbing materials (e.g. metals) to 0.5–2 mm for weakly absorbing materials (e.g. loosely compacted oxide powders). The area of the specimen that is irradiated is fixed (1) by a series of thin metal plates, called Soller slits, which limit the axial divergence of the beam (i.e. along the axis of the diffractometer) and (2) the width of the divergence slit (e.g. as in Fig. 10.2). Clearly, the greater the θ angle the smaller is the length of specimen irradiated. Figure 10.4 shows the geometry involved. The incident beam, width w, (shown as a parallel beam for simplicity) covers a length $l = {}^w\!/_{\sin\theta}$. Conversely the depth of the specimen d, which contributes to the diffracted beam, increases with increasing θ angle, i.e. $d = t\sin\theta$. Hence ld and the volume irradiated, is constant.

This situation is satisfactory for highly absorbing specimens of thickness greater than d and where d is so small that the diffracted intensity can be considered to arise from a single plane (the reference plane) along the axis of the diffractometer. However, for weakly-absorbing specimens much of the diffracted intensity arises from regions of the specimen **below** the reference plane that leads both to peak shift and peak broadening.

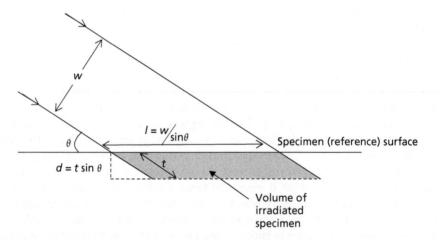

Fig. 10.4. An incident beam of width w irradiates a length $l = {}^w\!/_{\sin\theta}$ of the specimen surface. Diffracted beams arise from a region up to a distance $d = t\sin\theta$ below the specimen surface that is situated along the axis of the diffractometer (reference surface).

[1] The relevant equation is $I = I_o \exp -\mu t$ where I_o is the incident intensity and I is the intensity at thickness t. For $t = {}^1\!/_\mu$, $I = I_o \exp^{-1} = {}^{I_o}\!/_{2.72}$, i.e. the incident beam intensity $\approx {}^1\!/_3 I_o$ and thus the diffracted beams have about ${}^1\!/_9$ th of the intensity of the surface-diffracted beams.

In order to avoid this, and also to maintain constant-volume conditions, (1) a very thin layer (fixed thickness) of the specimen is laid across a non-diffracting substrate (or a single crystal substrate from which any diffraction peaks are readily recognizable[2]) and (2) the width of the divergence slit is linked to the θ angle such that it increases as θ increases and the length l of the specimen that is irradiated is constant. Such continually-adjustable divergence slits (and corresponding anti-scatter slits between the specimen and receiving slit) are now standard items of equipment.

Finally, it may be noted that in order to minimize the time needed for experiments, and for low resolution work, the incident beam monochromator can be dispensed with to be replaced by a metal filter that has a K absorption edge wavelength in between the K_α and K_β wavelengths of the X-ray tube (Section 9.8.1) and that preferentially absorbs the K_β (and white) X-radiation.

An example of an X-ray diffractometer 'chart' for quartz (SiO_2) is given in Fig. 10.5. The d-spacings and relative intensities of the reflections may be used to identify the material as described in Section 10.3 below.

Fig. 10.5. An example of an X-ray diffractometer chart of intensity of the reflected beams (ordinate) vs 2θ angle (abscissa). The intensity scale is non-linear (square root) to emphasize the weaker reflections. The specimen is polycrystalline SiO_2 (quartz), $CuK\alpha_1$, radiation ($\lambda = 1.541$ Å). The d-spacings (Å) of the reflections are indicated (compare with Figs 10.7(a) and 10.8). (Courtesy of Mr. D. G. Wright.).

[2] As a young research student, I found that very thin flakes of a titanium alloy could most easily be examined by pressing them on to a substrate of 'plasticine'—a modelling clay. A spectacular number of diffracted beams were observed that I indexed according to hitherto unknown titanium alloy structures. Happily, these 'results' were never published.

10.2.2 Back reflection and Debye–Scherrer powder techniques

There are two other 'classical' X-ray powder techniques of interest—the back reflection method in which the whole or part of the diffraction cones are recorded on flat film (an experimental arrangement identical to that for the Laue back reflection method) and the Debye–Scherrer technique in which a thin rod-shaped specimen or a powder in a capillary tube is set at the centre of a cylindrical camera.

The geometry of the back reflection powder method is shown in Fig. 10.6. The effective position of the source S is determined by the design of the collimator tube which passes through a hole in the centre of the film. In practice the divergence angle is very small, of the order 0.5°–2°. Notice that, like the diffractometer, it is a semi-focusing method—all the reflected beams from the (flat) specimen surface focus approximately at the circumference of the focusing circle. However, the flat film can be placed only at one distance x from the specimen: if x is the focal distance of the high 2θ angle reflections as shown, it will not be at the focal distance of the lower 2θ angle reflections.

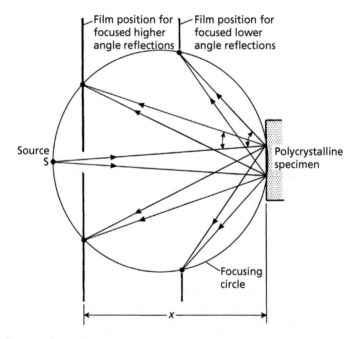

Fig. 10.6. The geometry of the back reflection X-ray powder method; the diverging X-ray beam from an (effective) source S passes through a hole at the centre of the film. The reflected beams for two reflections are shown. The film is shown placed at a specimen–film distance x so as to intercept the higher angle reflections such that they are sharply focused. In order to intercept the lower angle reflections such that they are sharply focused it would need to be moved closer to the specimen as indicated. In practice a 'compromise' value for x is chosen which is acceptable since the divergence angle, and the de-focusing of the incident beam, is small.

In practice a 'compromise' value of x is chosen such that the diffraction rings of particular interest are most sharply focused.

Although the Debye–Scherrer (cylindrical camera) technique is now near-obsolete, it has played such a major role in X-ray diffraction studies, and provides such an immediately visual record of a powder pattern, that it merits some discussion. Moreover, the semi-focusing geometry also applies to its extension to the X-ray diffractometer. The geometry and specimen-film arrangement are shown in Fig. 10.7(a). The cylindrical specimen (a powder in a capillary tube or a cemented powder 'rod') is in the centre of the camera. Two holes are punched in the film, one to match the 'entrance', the other the 'exit' directions of the X-ray beam.

This 'asymmetrical' arrangement or 'setting' has the advantage that a complete range of 2θ angles—from 0° to 180°—may be recorded on one side of the film without a 'cut-off occasioned by the presence of a film end. In this way accurate d_{hkl}-spacings may be determined and any effects of film shrinkage accounted for.

The position of the effective source S is determined by the design of the collimator. The focusing circle for a particular reflection is shown; however, this only represents the physical situation very approximately since the reflected beams arise to different extents from the whole volume of the specimen and the position of the source S and the focusing effect is by no means as precise as that drawn in Fig. 10.7(a).

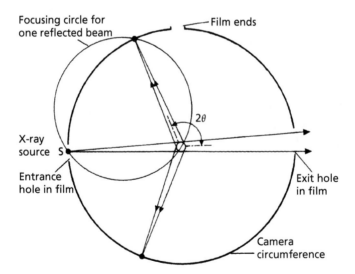

Fig. 10.7(a). The geometry of the Debye–Scherrer camera; the reflected beams from a cylindrical specimen *tend* towards a focus on the film around the circumference of the camera and the focusing circle for a particular reflected beam is shown. In practice deviations from this idealized situation occur because the reflected beams originate with varying intensities from the whole volume of the specimen and the effective source S may not coincide with the circumference of the camera. The geometry is identical to that for a cylindrical specimen in the diffractometer.

An example of a Debye–Scherrer film (asymmetrical setting) is shown in Fig. 10.7(b). The $2\theta = 0°$ and $2\theta = 180°$ positions can be worked out exactly by determining the centres of the 'exit' and 'entrance' rings, the distance between them on the film corresponding to 180° exactly. Hence the camera diameter does not need to be known, although in practice cameras are made with diameters 57.53 mm or 114.83 mm such that (and accounting for film thickness) the circumference is 180 mm or 360 mm. Hence for quick work the angles can simply read in terms of millimetres on the film. A cylindrical specimen may be used instead of a 'flat-plate' specimen in the diffractometer. The detector takes the place of the film but otherwise the semi-focusing geometry shown in Fig. 10.7(a) is the same. Moreover, since the specimen is completely bathed in the X-ray beam the diffracting volume is constant.

10.3 Some applications of X-ray diffraction techniques in polycrystalline materials

10.3.1 Accurate lattice parameter measurements

The accurate measurement of d_{hkl}-spacings is now largely carried out with the X-ray diffractometer using monochromatized ($K\alpha_1$) radiation, narrow divergence and receiving slit systems, internal calibration systems and peak profile measurement procedures. However, as shown below, it is the high 2θ angle reflections which are most sensitive to small variations in d_{hkl}-spacings and for this reason the Debye–Scherrer powder camera and back reflection flat film techniques in which 2θ angles up to nearly 180° can be recorded, are still used. By contrast, the largest 2θ angles measurable with a diffractometer are about 150° because the near approach and possible contact between the source and receiving slit assemblies.

Given the crystal system and the *hkl* values of the reflections, the lattice parameter(s) can be determined using the appropriate equation (Appendix A4.1). This procedure is clearly simplest for the cubic system in which the d_{hkl}-values depend only upon the one variable, *a*. In other systems several d_{hkl}-spacings need to be measured and the values inserted into simultaneous equations from which values of the lattice parameters are extracted.

Fig. 10.7(b). An example of a Debye–Scherrer X-ray film (CuKα X-radiation) for SiO$_2$ (quartz). S is the 'entrance' hole and O the 'exit' hole. The 'spotty' appearance of the rings arises because of the relatively coarse crystallite size in the specimen. At high 2θ angles (near the entrance hole) the Kα_1/Kα_2 doublets are resolved but at low 2θ angles (near the exit hole) they are unresolved. The streaks around the exit hole arise from the presence of residual 'white' X-radiation in the incident beam.

The sensitivity of change of the 2θ angle for a reflection with a change in d_{hkl}-spacing is a measure of the resolving power of the diffraction technique. The resolving power, or rather the limit of resolution, $\delta d/d$, can be obtained by differentiating Bragg's law with respect to d and θ (λ fixed), i.e.

$$\lambda = 2d \sin \theta$$

differentiating:

$$0 = 2d \cos \theta \delta\theta + 2 \sin \theta \delta d$$

hence

$$\delta d/d = -\cot \theta \delta\theta.$$

This equation may be 'read' in two ways. For a given value of $2\delta\theta$, which may be expressed in terms of the minimum resolved distance between two reflections on the film, chart, etc., for the smallest limit of resolution we want $\cot \theta$ values to be as small as possible; or, for a given value of $\delta d/d$, we want the angular separation of the reflections $2\delta\theta$ to be as large, and hence again $\cot \theta$ values to be as small as possible.

A glance at cotangent tables shows that $\cot \theta$ values are high at low θ angles and rapidly decrease towards zero as θ approaches $90°$. Hence d_{hkl}-spacing measurements are most accurate using high-angle X-ray diffraction methods and least accurate with low-angle electron diffraction methods.

10.3.2 Identification of unknown phases

For powder specimens in which the crystals are randomly oriented, the set of d-spacings and their relative intensities serves as a 'fingerprint' (or 'genetic strand') from which the phase can be identified by comparison with the 'fingerprints' of phases in the *X-ray Powder Diffraction File*—a data bank of over 300,000 inorganic and 450,000 organic and organometallic phases. The File is administered by the International Centre for Diffraction Data (ICDD)—formerly the Joint Committee for Powder Diffraction Standards (JCPDS)—and new and revised sets of data are published annually in book form (replacing the original 5 inch by 3 inch cards) or as a computer database. In addition, there are published Indexes with the phases arranged alphabetically by chemical name, search manuals and abstracts from the File (including Indexes and Search Manuals) or Frequently Encountered Phases (FEP).

The entry for each phase in the File or computer database (which differ slightly in their layout) includes: file number, chemical formula and name, experimental conditions, physical and optical data (where known), references and the set of d-spacings (given in Å), their Laue indices and their relative intensities arranged in order of decreasing d-spacing. Finally, the ICDD editors assign a 'quality mark' to indicate the reliability of the data (e.g. a '*' indicates that the chemistry, structure and X-ray data of the phase is well characterized); or whether the X-ray data has been derived by

46-1045 ★

SiO₂					

SiO$_2$

Silicon Oxide

Quartz, syn

Rad. CuKα$_1$ λ 1.540598 **Filter** Ge Mono.**d-sp** Diff.
Cut off **Int.** Diffractometer I/I$_{cor.}$ 3.41
Ref. Kern, A., Eysel, W., Mineralogisch-Petrograph.Inst., Univ.
Heidelberg, Germany, ICDD Grant-in-Aid, (1993)

Sys. Hexagonal **S.G.** P3$_2$21 (154)
a 4.91344(4) **b** c 5.40524(8) **A** **C** 1.1001
α β γ **Z** 3 **mp**
Ref. Ibid.

D$_x$ 2.65 **D$_m$** 2.66 **SS/FOM** F$_{30}$ = 539(.002,31)
εα nωβ 1.544 εγ 1.553 **Sign** + **2V**
Ref. Swanson, Fuyat, Natl. Bur. Stand. (U.S.), Circ. 539,324(1954)

Color White
Integrated intensities. Pattern taken at 23(1) C.Low tempertature quartz. 2θ determination based on profile fit method. O$_2$Si type. Quartz group. Silicon used as internal standard. PSC: hP9. TO replace 33-1161. Structure reference: Z.Kristallogr., **198** 177(1992)

See following card.

dÅ	Int	hkl	dÅ	Int	hkl
4.2550	16	100	1.1530	1	311
3.3435	100	101	1.1407	<1	204
2.4569	9	110	1.1145	<1	303
2.2815	8	102	1.0816	2	312
2.2361	4	111	1.0638	<1	400
2.1277	6	200	1.0477	1	105
1.9799	4	201	1.0438	<1	401
1.8180	13	112	1.0346	1	214
1.8017	<1	003	1.0149	1	233
1.6717	4	202	0.9896	<1	115
1.6592	2	103	0.9872	<1	313
1.6083	<1	210	0.9783	<1	304
1.5415	9	211	0.9762	<1	320
1.4529	2	113	0.9608	<1	321
1.4184	<1	300	0.9285	<1	410
1.3821	6	212	0.9182	<1	322
1.3750	7	203	0.9161	2	403
1.3719	5	301	0.9152	2	411
1.2879	2	104	0.9089	<1	224
1.2559	3	302	0.9009	<1	006
1.2283	1	220	0.8972	<1	215
1.1998	2	213	0.8889	1	314
1.1978	<1	221	0.8814	<1	106
1.1840	2	114	0.8782	<1	412
1.1802	2	310	0.8598	<1	305

Fig. 10.8. The Powder Diffraction File Card No. 46-1045, for quartz. The left-hand side gives crystallographic optical and physical information and the source of the data. The right-hand side lists the *d*-spacings, in descending order, with the strongest line assigned an intensity of 100. Formerly (up to set 24) the cards included along the top left edge the *d*-spacings of the three strongest lines and the largest *d*-spacing recorded in the pattern. (International Centre for Diffraction Data.)

calculation, rather than observed (a 'c' mark). Figure 10.8 shows an example of a Powder Diffraction File card, that for quartz, set number 46, card number 1045.

A comparison of an unknown phase with so many others is potentially a daunting task! It is greatly facilitated by a search procedure based on that first devised by J. D. Hanawalt in 1936 and now refined and speeded up using computer search procedures.

The phases are grouped together (into what is known as Hanawalt groups) according to the *d*-spacing of their strongest reflection. There are forty Hanawalt groups covering the whole range of *d*-spacings. For example, the strongest line of quartz (relative intensity 100) is that at 3.34 Å (Figs 10.4 and 10.8); quartz therefore lies in the Hanawalt group with *d*-spacings in the range from 3.39 to 3.32 Å (Fig. 10.9—first column). Within each Hanawalt group the patterns are listed in the second column in order of decreasing *d*-spacings of the second strongest line which for quartz is that at 4.26 Å; on reading down the list of patterns in the second column of the Hanawalt group a potential 'match' can be made with the second strongest line and then (from the number of possibilities available) final confirmation made with the *d*-spacings and intensities of the third, fourth etc. strongest lines in the subsequent columns. Figure 10.9 is from the Hanawalt Inorganic Search Manual showing some entries in the Hanawalt group 3.39–3.32 Å which includes that for quartz, File No. 46-1045.

In practice the procedure is by no means as straightforward as this. First, many minerals and metallic alloys give very nearly the same diffraction pattern because of the existence of solid solution ranges, which result in very small shifts in lattice parameter.

3.39 – 3.32 (±.02)

QM	Strongest Reflections								PSC	Chemical Formula	Mineral Name;Common Name	PDF#	I/I_c
	3.40_x	4.29_9	4.09_9	6.63_7	4.82_6	3.31_6	2.67_6	2.33_6	o*144	ClF_2OBF_4		26-235	
	3.39_x	4.29_9	5.88_8	3.12_8	2.98_8	2.30_8	1.96_6	1.78_6	hP15	$Li_{2.33}Fe_{0.67}S_2$		36-1089	
	3.39_x	4.29_4	2.47_1	2.31_1	2.23_1	2.14_1	1.99_1	1.83_1	h**	BPO_4		11-237	
O	3.36_9	4.29_x	2.85_7	5.60_5	4.02_5	2.58_5	1.98_5	1.90_5		$V_{0.87}Cr_{0.13}O_{2.17}$		35-332	
i	3.34_8	4.29_x	10.5_6	2.64_6	2.71_5	5.27_4	5.25_4	3.79_3	oI32	$Na_2Ta_3O_6F_5$		25-871	
	3.32_x	4.29_6	2.99_6	2.87_6	2.52_6	4.70_5	2.38_5	2.30_5		$Cs_2Al(NO_3)_5$		28-274	
O	3.31_x	4.29_x	3.68_x	2.91_x	2.34_x	2.06_x	1.99_x	1.93_8		$Pu(OH)CO_3•xH_2O$		33-982	
i	3.37_x	4.28_3	1.84_2	1.55_1	2.47_1	2.31_1	1.39_1	2.14_1	hP18	$AlPO_4$	Berlinite, syn	10-423	
	3.35_8	4.28_x	3.71_6	2.85_6	2.17_6	4.07_4	3.03_4	2.73_4		$K_2N_2V_6O_{21}$		20-1379	
i	3.34_x	4.28_3	4.27_3	3.38_3	2.77_3	2.33_2	2.79_2	2.32_2	mP28	$CsLiSeO_4$		29-411	2.60
i	3.33_x	4.28_x	3.24_x	8.58_8	2.91_8	2.37_8	1.87_8	4.95_6	hP20.70	$Cs_{0.35}V_3O_7$		30-385	
	3.36_x	4.27_3	1.82_2	2.28_2	1.54_2	2.46_1	2.13_1	1.37_1	hP9	SiS_2		47-1376	
i	3.34_x	4.27_4	3.19_2	2.70_2	7.28_2	4.91_2	1.82_2	3.13_1	mP100	$CaAl_2Si_2O_8•4H_2O$	Gismondine	20-452	
i	3.30_x	4.27_7	3.71_7	3.45_7	3.92_5	3.34_5	3.27_5	2.96_5	aP26	$KFeSi_3O_8$		16-153	
	3.38_x	4.26_x	4.16_x	1.68_x	6.70_8	2.97_8	2.86_8	2.56_8		$Cs_2VCl_5•4H_2O$		20-301	
i	3.37_x	4.26_x	4.18_8	2.98_8	8.56_6	6.76_6	3.49_6	3.09_6	mC40	$Cs_2VCl_5•4H_2O$		34-322	
O	3.36_x	4.26_x	2.86_x	5.90_8	5.55_8	3.26_8	2.94_8			$KAlH(PO_3)_2$		38-55	
O	3.36_x	4.26_x	2.98_x	5.95_8	5.55_8	4.02_6	3.26_6	2.87_6		$KCrHP_3O_{10}$		41-442	
	3.34_x	4.26_x	4.33_5	3.40_5	3.38_4	2.05_3	4.89_2	2.02_2	oP36	$KNaSiF_6$		43-1313	
	3.34_x	4.26_9	3.66_9	2.61_9	2.48_5	1.80_5	3.97_8	3.05_8	hP168	$3PbCO_3•2Pb(OH)_5•H_5O$		9-356	
i	3.34_x	4.26_2	2.13_6	7.40_5	3.49_5	2.58_5	2.24_5	2.21_5	hP70	$Ca_3Ge(SO_4)_2(OH)_6•3H_2O$	Schaurteite	19-225	
	3.34_x	4.26_2	2.13_8	7.40_6	2.57_6	2.03_6	3.49_4	2.24_4	hP70	$Ca_3Mn^{+4}(SO_4)_2(OH)_6•3H_2O$	Despujolsite	20-226	
O	3.33_x	4.26_4	3.27_4	4.31_3	2.89_3	2.75_3	3.88_2	3.19_2		$K_2Ca(VO_3)_4$		37-180	
	3.30_x	4.26_4	4.27_3	3.66_7	2.89_6	2.30_6	2.63_5	2.32_5		$Sm_2O(CO_3)_2•xH_2O$		28-994	
i	3.30_x	4.26_2	2.89_x	4.45_5	3.14_6	2.56_4	3.89_2	2.66_2	mP16	CsO_3	cesium ozonide	28-994	
*	3.34_x	4.26_2	1.82_1	2.46_1	1.54_1	2.28_1	1.38_1	2.13_1	hP9	SiO_2	Quartz, syn	46-1045	3.41
	3.31_9	4.25_x	4.09_9	3.72_7	2.98_7	2.20_7	1.93_7	5.47_6	o*66	$ClPtOF_8$		26-415	
	3.30_9	4.25_x	4.09_9	3.67_9	7.42_6	5.47_6	2.96_6	1.91_6	o*66	$(ClF_2O)AsF_6$		26-117	
	3.40_x	4.24_8	5.00_4	5.72_3	2.65_3	2.85_2	2.37_2	0.00_1	tI24	$KAuF_4$		27-395	
*	3.37_x	4.24_3	2.32_2	2.12_2	2.45_1	1.84_1	1.83_1	1.54_1	hP18	$GaPO_4$		8-497	
O	3.36_x	4.24_x	2.86_x	5.90_6	5.55_8	3.26_8	2.94_6	6.60_6		$KAlH(PO_3)_2$		38-55	
*	3.33_x	4.24_x	3.39_4	2.64_3	3.25_3	2.60_3	4.38_3	3.44_3	mP44	$SrUMo_4O_{16}$		38-1360	
i	3.41_8	4.23_x	2.87_x	2.49_x	2.06_8	1.99_8	1.95_8	1.93_6	mC12	$α\text{-}WP_2$		35-1467	
	3.39_9	4.23_x	5.06_9	3.45_9	3.18_9	3.05_9	5.25_7	3.76_7	mC*	$RbAl_2(H_2P_3O_{10})(P_4O_{12})$		37-261	
O	3.39_9	4.23_x	2.99_9	2.89_8	2.77_7	5.93_6	3.24_5	4.75_4		$RbGaHP_3O_{10}$		34-1347	
	3.38_x	4.23_9	3.83_4	2.50_3	3.10_2	1.71_2	3.31_2	2.21_2	mP62	$K_2Mn(SeO_4)_2•6H_2O$		53-1151	
O	3.37_x	4.23_x	3.46_x	3.00_x	2.13_x	6.00_8	2.69_8	2.45_6		$CsSnFBrI$		30-369	
O	3.36_x	4.23_x	2.37_x	2.29_x	2.15_x	1.90_x	9.50_8	3.95_x		$Al_8La_2S_{15}$		41-734	
O	3.36_x	4.23_5	1.64_5	2.72_4	2.44_3	2.22_3	1.93_3	3.14_2		$Mo_3O_8•xH_2O$	Ilsemannite	21-574	
	3.35_8	4.23_x	3.12_x	3.43_8	3.84_6	3.75_6	3.39_6	3.37_6		$β\text{-}Cs_2Cu(SO_4)_2$		28-288	
i	3.32_x	4.23_x	1.88_7	3.45_7	2.76_3	2.25_2	2.12_2	2.64_2	oP36	$β\text{-}NaSrGaF_7$		44-1352	
O	3.40_x	4.22_8	3.20_7	3.14_7	5.81_6	3.62_5	3.03_4	2.64_4		$Na_3VO_2(SO_4)_2$		41-175	
	3.41_8	4.21_x	3.11_x	2.75_8	7.58_7	5.60_7	4.46_7	3.31_7		$Na_3SbF_3SO_4$		28-1038	
O	3.41_9	4.21_x	2.99_9	3.24_8	2.91_8	2.77_7	5.95_6	4.77_6		$RbCrHP_3O_{10}$		41-443	
O	3.40_9	4.21_x	3.30_8	3.22_8	6.97_5	4.76_5	3.11_5	2.58_5	a*15.54	$UO_3•0.393H_2O$		41-443	
O	3.39_x	4.21_9	3.92_9	2.78_8	1.89_6	3.78_5	2.96_5	2.08_5		$Pb_2Sb_2S_5$		36-94	
O	3.34_x	4.21_4	3.15_4	2.11_4	2.71_3	2.05_2	4.73_2	2.56_1		$Cs_3Zr_{1.5}(PO_4)_3$		52-1181	
	3.33_8	4.21_x	3.82_7	2.96_7	2.47_7	2.34_7	2.01_7	1.89_7		$CsXeO_2F_5$		35-1267	
O	3.30_8	4.21_x	3.40_8	3.22_8	6.97_5	4.76_5	3.11_5	2.58_5	a*15.54	$UO_3•0.393H_2O$		43-346	
O	3.41_9	4.20_x	3.65_7	2.77_7	2.71_7	2.96_5	2.11_5	4.00_4		$Tl_2Si_2S_5$		38-914	

Fig. 10.9. Part of the Hanawalt Search Manual for Inorganic Phases, showing some entries in the Hanawalt Group 3.39–3.32 Å. The intensities of the reflections are indicated by subscripts $x = 100$, $9 = 90$, $8 = 80$, and so on. Notice that some patterns are entered in this Hanawalt Group according to their *second*-strongest or *third*-strongest lines; e.g. $K_2N_2V_6O_{21}$ (20-1379) or $RbGaHP_3O_{10}$ (34-1347) respectively. This is because they have intensities over 75% of that of the strongest line and may experimentally be identified as the strongest lines. (International Centre for Diffraction Data.)

Second, the crystals in the powder specimen may not be randomly orientated—the specimen may be textured or may show preferred orientation, which implies that the relative intensities of the reflections may be altered as compared with those in the File. Third, the unknown specimen may consist of more than one phase—and if they are in roughly equal proportions the strongest lines may be 'mixed up'.

These problems are partly (but not wholly) circumvented by a series of 'entry rules', the details of which are given in the Search Manual. First, those patterns in which the d-spacing of the strongest reflection lies close to one of the boundary values of the Hanawalt group are entered in both groups—the one above and the one below. Second, those patterns in which the intensity of the second strongest reflection is greater than 75% of the intensity of the strongest (and in which there is therefore a strong possibility of the mis-identification of the strongest reflection) are entered twice; the second strongest reflection determining the Hanawalt group and the position in the group listed (as before) in accordance with the (decreasing) d-spacing of the second strongest

reflection. Similar rules apply to the third and fourth strongest reflections such that a pattern may be listed several times. Examples of such permutations are shown in Fig. 10.9.

10.3.3 Measurement of crystal (grain) size

This is based on the determination of β, the line broadening in the Scherrer equation (Section 9.3) and an estimate of K, the 'shape' correction factor (which does not differ significantly from unity). However, this is not a straightforward business because the *observed* broadening, β_{obs}, also includes 'instrumental' factors (see Section 9.3) such as detector slit width, area of specimen irradiated, possible presence of a $K\alpha_2$ component in the X-ray beam, etc. The contributions from these factors have to be 'subtracted' from the observed peak breadth (or, strictly speaking, they have to be deconvoluted from the peak). The usual procedure is to measure the instrumental broadening, β_{inst}, in a large-grained material in which grain size broadening is assumed to be negligible, and to determine β either by a simple subtraction, due to Scherrer, i.e. $\beta = \beta_{obs} - \beta_{inst}$, or by a subtraction of squares, due to Warren, i.e. $\beta^2 = \beta_{obs}^2 - \beta_{inst}^2$. In practice, attempts to measure 'absolute' grain sizes by X-ray diffraction are fraught with difficulty because of the presence of imperfections (particularly sub-grain boundaries), lattice strains, etc. which contribute to the broadening of the X-ray peaks.

10.3.4 Measurement of internal elastic strains

The elastic strains to be measured can be distinguished into those occurring on a 'macro' scale and those occurring on a 'micro' scale. Macroscale refers to the situation where the whole material is subject to some directional residual tension or compression; the resultant strains, which are manifested in terms of increases (tension) or decreases (compression) in the d_{hkl}-spacings are therefore also directional and the measured values vary with the orientation of the reflecting planes—those parallel to the stress axis for example suffering no change.

In order to measure such strains (and to determine the stress axis) in the diffractometer, the specimen needs to be rotated from the symmetrical (Bragg–Brentano) geometry as described above. In practice, measurements of peak shift are made for various settings of the specimen and from which the direction and magnitude of the stress can be extrapolated.

Microscale refers to the situation in which the directions (and magnitudes) of the internal strains vary from crystal to crystal. The subsequent elastic strains give rise, not to a shift, but to a broadening of the diffraction peaks from which an estimate of the micro-stress can be made. The relevant equation is that derived for the limit of resolution (Section 10.3.1), in which e, the elastic strain, is substituted for $\delta d/d$, i.e.

$$\epsilon = -\cot\theta\,\delta\theta.$$

Hence the broadening, β, expressed as the half-height peak width $2\delta\theta$, is:

$$\beta = -\frac{2\epsilon}{\cot\theta} = -2\epsilon\tan\theta.$$

Notice that this equation is very similar to the Scherrer equation for broadening arising from crystal size; the important difference being that in this case β varies as $1/\cot\theta$ or $\tan\theta$ and in the Scherrer equation β varies as $1/\cos\theta$ or $\sec\theta$. As in the case of grain size broadening (Section 10.3.3), the instrumental broadening, β_{inst}, measured for a stress-free material, must be subtracted from β_{obs}, the observed broadening for the micro-strained material. Having measured the strain ϵ, the stress may be determined by substituting the appropriate value of the Young modulus, E.

Finally, the elastic strain ε may be expressed in terms of the extension of reciprocal lattice points into 'nodes' just as we did for a crystal of finite thickness t (Section 9.3.1). Since

$$\varepsilon = \frac{\delta\,(d_{hkl})}{(d_{hkl})} = \frac{\delta\left(1/d^*_{hkl}\right)}{1/\left(d^*_{hkl}\right)} = -\frac{\delta\left(d^*_{hkl}\right)}{\left(d^*_{hkl}\right)}$$

$\left(\text{remembering the differential } d\left(\frac{1}{x}\right) = -\frac{dx}{x^2}\right)$ then the extension is

$$\delta\left(d^*_{hkl}\right) = -\varepsilon d^*_{hkl} = -\frac{\varepsilon}{d_{hkl}}$$

i.e. unlike the extension arising from crystal size in which every reciprocal lattice point is extended the same amount, $1/t$, in this case the reciprocal lattice points are more extended the further we go from the origin in reciprocal space.

These two cases may be distinguished in practice by plotting a graph with $\beta\cos\theta$ as ordinate and with $\sin\theta$ as abscissa. In the case of elastic strain the equation $\beta = -2\epsilon\tan\theta$ may be rearranged as $\beta\cos\theta = -2\epsilon\sin\theta$. Hence, a plot of $\beta\cos\theta$ versus $\sin\theta$ gives a straight line passing through the origin of slope -2ε. In the case of crystal size the Scherrer equation $\beta = K\lambda/t\cos\theta$ (Section 9.3.1, p. 226) may be rearranged as $\beta\cos\theta = K\lambda/t$, which plots as a horizontal line that intercepts the ordinate at $K\lambda/t$. When broadening arises both from crystal size and elastic strain, the plot is a straight of slope -2ε that intercepts the ordinate at $K\lambda/t$.

10.4 Preferred orientation (texture, fabric) and its measurement

There are essentially two methods of recording preferred orientation: film methods which make use of flat-plate film cameras, and counter methods which make use of a special design of diffractometer called a *texture goniometer*. Film methods have the advantage that the cones of reflected beams from planes of different d_{hkl} spacings can be recorded 'all at once' so that you can see at a glance, from the varying intensities around the diffraction rings, whether the crystals are in random orientation or not. In practice the film may be set 'behind' the specimen as in Fig. 10.5 such that the cones of high-angle reflections are recorded (the back reflection X-ray method; or the film may be set 'in front of the specimen such that the cones of low-angle reflections are recorded) the transmission X-ray method. In the latter case the specimen must

be thin enough for the reflected beams to pass through with sufficient intensity to be recorded—which sets a practical limitation in the case of 'bulk' specimens which cannot be cut into thin sections.

However, film methods have the disadvantage that, except for the case of fibres, the data cannot easily be represented in a quantitative way—the varying intensities around the diffraction rings depend also on the orientation of the specimen with respect to the incident beam which means that several exposures need to be made and the results collated and the intensities analysed in terms of the 'blackening' of the films—a tedious business. In the case of fibres only one orientation and hence only one photograph is needed—the fibre axis normally being set vertical and perpendicular to the direction of the (horizontal) incident beam. (the same set-up as used for the photograph of DNA, p. xvii).

In the texture goniometer the detector is set at a fixed 2θ angle to record the diffracted intensity from just one set of d_{hkl} planes. All the movement takes place at the specimen which can be rotated about two axes as described below such that the detector in effect scans a wide range of orientations of the selected d_{hkl} planes. The detector may then be re-set at another 2θ angle to record another set of d_{hkl} planes and the process repeated. It is therefore, in contrast to film methods, a sequential procedure but has the enormous advantage that the output from the detector can be represented in the form of pole figures or processed numerically to give what is known as an orientation distribution function (ODF) which provides a complete description of the texture or fabric of the material.

10.4.1 Fibre textures

Figure 10.10 shows the simple experimental set-up for the transmission X-ray method to record the low 2θ angle (highest d_{hkl}) reflections. The fibre, or reference direction is set vertical and perpendicular to the incident X-ray beam (the same set-up as used for the photograph of DNA, p. xvii). Figure 10.11 is an example of the application of

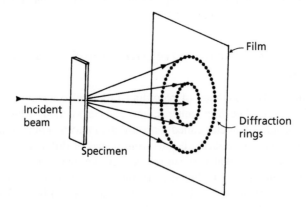

Fig. 10.10. Transmission X-ray powder method. The diffraction rings consist of reflections from many crystallites as indicated by the dots. Each ring corresponds to a cone of reflected beams of semi-angle 2θ.

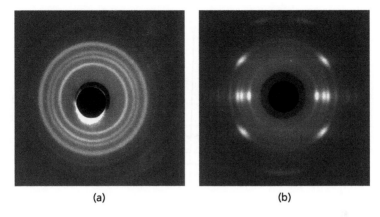

(a) (b)

Fig. 10.11. Transmission X-ray photographs of polypropene $n(CH_2–CHCH_3)$: (a) as-crystallized from the melt (crystallites in random orientation); (b) after 500% plastic deformation in tension (crystallites orientated with the polymer chains along the (vertical) tensile axis). (Photographs by courtesy of Dr. J. Way.)

the technique in the study of polymers. Figure 10.11(a) shows the diffraction pattern of a thin section of an injection-moulded specimen of polypropene $n(CH_2–CHCH_3)$ which partially crystallizes on cooling from the melt; the diffraction rings are of uniform intensity which shows that the crystallites are in random orientation. Figure 10.11(b) is of the same material which has been plastically deformed in tension to a strain of about 500%. The diffraction rings are no longer continuous but consist of peaks of intensity which show that the crystallites now lie in a particular orientation with respect to the tensile axis. The diffraction pattern approaches that of a rotation or oscillation photograph of a single crystal (Fig. 9.17), the peaks of intensity lying in layer-lines perpendicular to the axis of rotation/oscillation. Given the structure and unit cell of polypropene and the $\{hkl\}$ reflections which occur, we can index the reflections and hence determine which crystallographic direction is orientated along the tensile axis. Intuitively we may guess that the crystallites are orientated such that the polymer chains lie in the direction of the tensile axis and the demonstration that this is indeed the case is given in Exercise 10.4.

10.4.2 Sheet textures

The texture goniometer consists essentially of a cradle in which a flat-surface specimen (as in a diffractometer) can be rotated about two axes as indicated in Fig. 10.12(a), one of which, O ... O is normal to the specimen surface and the other B ... B lies in the specimen surface and in the plane containing the incident and diffracted beams (i.e. perpendicular to the direction A ... A). In the 'starting position' the specimen is set in the symmetrical or Bragg-Brentano focusing arrangement as shown in Fig. 9.20(a) and all the reflected intensity received by the detector arises from the crystal planes lying parallel to the specimen surface. The specimen is then tilted a small angle β (say

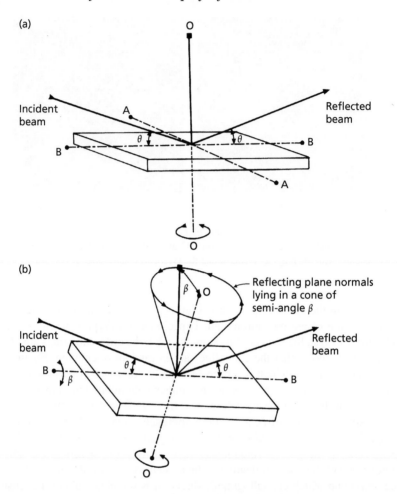

Fig. 10.12. The geometry of the texture goniometer, the specimen is rotated about the axis O... O normal to the specimen surface and tilted about the axis B ... B lying in the specimen surface and in the plane containing the incident and diffracted beams, (a) indicates 'starting position' (reflecting plane normals perpendicular to specimen surface) and (b) at angle of tilt β (reflecting plane normals lying in a cone at angle β to specimen surface).

5°) about the axis B... B. The reflected intensity received by the detector now arises from the crystal planes also tilted 5° to one side of the specimen surface. Note that this angle of tilt is *not* the same as the angle α in Fig. 9.20(b) which is in effect a tilt about the other direction A... A in Fig. 10.12(a). The specimen is then rotated about the normal O... O; in a complete revolution of 360° the detector records the reflected intensity from the planes tilted in all directions at 5° to the specimen surface. Now the specimen is tilted about the axis B ... B say another 5° and again rotated 360° about its normal; the recorded intensity now arises from all the planes tilted 10° from the specimen surface—and so on.

Figure 10.12(b) shows the geometry involved for an angle of tilt β; the diffracted intensity arises from the planes whose normals lie in a cone normal to the specimen surface of semi-apex angle β. The 'vertical' edge of the cone (in the plane of the incident and diffracted beams) corresponds to the normals of the planes which are reflecting at any point during the 360° rotation.

Tilting about the axis B ... B maintains the symmetrical Bragg–Brentano focusing geometry (unlike, as in Fig. 9.20(b), tilting about the other axis A ... A) but clearly as β increases the incident and diffracted beams lie at an increasingly small 'glancing' angle to the specimen surface such that the practical limit of tilt is about 80°. Hence the planes which lie perpendicular to the specimen surface (normals lying in the specimen surface) cannot be recorded. This problem is overcome either by using a transmission method or by sectioning the specimen in another orientation.

Fig. 10.13. (a) Graphitized carbon tape showing the orientations of the graphite layers at a fracture surface (left) at the edge of the tape and (right) in the centre (photograph by courtesy of Dr S. Lu and Professor B. Rand). (b) The corresponding pole figure, the plane of projection is the plane of the tape and the **fibre** or tape direction is as indicated. The pole figure extends to \sim 80° and the limit of the spiral is shown by the thin line. The intensity-contours are indicated in the legend.

In practice the mechanics of the texture goniometer are such that the specimen is tilted and rotated at the same time—say a 5° tilt for 360° rotation, such that the recorded intensity arises from planes lying in a spiral of 5° pitch. The data is then plotted in a stereographic projection—a pole figure (Section 12.5.3) and presented in terms of contours of intensity data. Finally, in order that the X-ray beam may sample as large an area as possible, the specimen is oscillated in its cradle, or shunted to and fro, by a mechanical device rather like the connecting-rod to the wheels in a steam engine. It is the one feature of the texture goniometer which students always notice first!

Figure 10.13(a) is a scanning electron micrograph of a graphitized carbon tape showing clearly the graphite layers lying parallel to the surface of the tape except at the edges where they 'curl over'—but always lying parallel to the long direction of the tape. Figure 10.13(b) is the corresponding pole figure, the 2θ angle being set to record the 0002 graphite planes. They peak in the centre (the planes parallel to the surface of the tape) and also towards the transverse directions (the planes curled over at the edges). This is a nice example because the relationship between the pole figure and the micrograph can be visualized very easily. In most other materials the micrographs do not reveal the orientations of the reflecting planes at all and their orientations are only revealed in the pole figure.

10.5 X-ray diffraction of DNA: simulation by light diffraction

The 1952 X-ray photograph of deoxyribonucleic acid (DNA) is of such importance in the history of science that I have placed it, along with the 1912 X-ray photograph of zinc blende, at the very beginning of this book (p. xv). The caption indicates how some of the features of this pattern relate to the structure of DNA. We are, of course, unable to attempt a complete analysis of the positions and intensities of the diffraction spots from a structure of such complexity, but we can make some progress towards that goal by making use of the simple ideas about diffraction that we covered earlier in this book together with a simplified (two-dimensional) model of DNA itself.[3]

Figure 10.14 shows a projection of a simplified DNA model showing schematically the two helical backbones of the phosphate groups (indicated by dots along the sine waves) and the equidistant planar bases (indicated by horizontal bars), $p = 3.4$ nm is the helical repeat distance; the two helices are separated by 3/8 p and the distance between the bases is 1/10 p. The diameter of the helices is 0.6 $p \approx 2.0$ nm.

We will now represent this structure in the simplest possible way—as a row of horizontal lines of length 0.6 p and separation p (Fig. 10.15(a)). The corresponding

[3] The analysis which follows is largely based on the work of Prof. Amand Lucas and his colleagues of the University of Namur, Belgium, to whom I am very grateful. Their analysis entitled 'Revealing the backbone structure of B-DNA from laser optical simulations of its X-ray diffraction diagram' is published in the *Journal of Chemical Education* **76**, 378–83, 1999. It is also available from the authors as a videotape (together with a diffraction mask) entitled DNA—*from light to life*, and is also included in the Vega Science Film Series (Videotape VRS-7 'How X-rays cracked the structure of DNA'), available from the Vega Science Trust, University of Sussex, UK, BN1 3QJ and via the vega website; www.vega.org.uk.

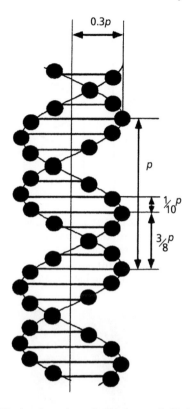

Fig. 10.14. A model of DNA showing schematically the two helical backbones of the phosphate groups (indicated by dots) and the equidistant planar bases (indicated by horizontal bars), $p = 3.4$ nm is the helical repeat distance, the two helices are separated by 3/8 p and the distance between the bases is 1/10 p. The diameter of the helices is $0.6\,p = 2.0$ nm (from A. A. Lucas, Ph Lambin, R. Mairesse and M. Mathot. *J. Chem. Educ.***76**, 378, 1999).

diffraction pattern (or optical transform) is shown in Fig. 10.16(a). It is produced in the same way as the patterns shown in Fig. 7.3 except that, in practice, the diffracting mask is made up not of one, but of many sets of such lines, irregularly spaced, so as to give sufficient diffracted light intensity. It is also made to such a scale that the distance p between the lines and the wavelength of laser light is in the same proportion to the actual helix repeat distance (3.4 nm) and typical X-ray wavelengths (\sim0.15 nm).

The vertical separation between the laterally-streaked spots is given by the simple diffraction grating equation (Section 7.4): $\sin a_n = n\lambda/p$ and since $\lambda \ll p$, $\sin \alpha_n \approx \alpha_n$ and the spots are nearly equally spaced.

Now we will improve our model by inclining the lines at (approximately) the same angle as the 'arms' or the 'zigs' and 'zags' of the helices (Fig. 10.15(b)). The diffraction pattern from *one* of these again shows laterally-streaked spots with the *same* vertical separation but with maximum intensities in a direction perpendicular to the inclined lines (Fig. 10.16(b)). Clearly, for the arms of the helices inclined in the opposite sense

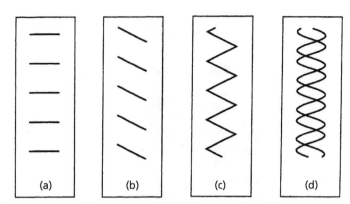

Fig. 10.15. Representational models of the DNA structure. (a) Lines spaced the helical repeat distance p and length $0.6\,p$; (b) lines inclined at (approximately) the same angle as one of the arms of the helices; (c) a zig-zag pattern made by combining the two arms of the helices and roughly representing the single sine wave, in projection, of one of the helices; (d) two helices (represented as sine waves rather than zig-zags) separated by a distance $3/8\,p$. (From A. A. Lucas *et al.* loc. cit.)

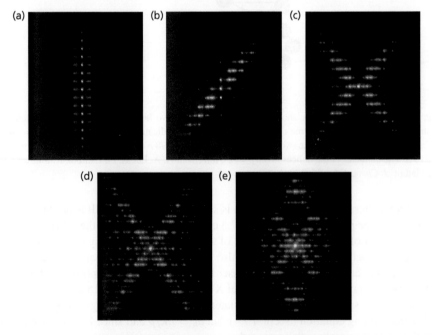

Fig. 10.16. The optical transforms or diffraction patterns corresponding to the models shown in Fig 10.15(a)–(d) and Fig 10.14: (a) from horizontal lines spacing p and length $0.6\,p$; (b) from lines inclined as in one of the arms of the helices, the intensity maxima lie in a direction normal to the lines; (c) from two sets of lines (zig-zags) representing the sine-wave projection of a single helix; (d) from two sine waves separated by a distance $3/8\,p$; and (e) from the double-helix model as represented in Fig. 10.14. (Photographs by courtesy of Prof. A. A. Lucas.)

the diffraction pattern will be mirror-related to that in Fig. 10.16(b) in a vertical line. Now we combine the two sets of lines to give a zig-zag pattern (Fig. 10.15(c)) which roughly represents the projection of a single helix; the resulting pattern (Fig. 10.16(c)) now shows (as expected) the two characteristic rows or arms of spots radiating out from the centre and the angle between them, which is the angle between the arms of the helices and can, together with the repeat distance p, be used to estimate the width of the helices, i.e. the diameter of the DNA structure. We may refine our model by representing the helix more properly as a sine wave but the resulting pattern is only changed in detail.[4]

Next, we combine two zig-zags or two sine waves separated by the distance $3/8\,p$ (Fig. 10.15(d)). The diffraction pattern is similar, but with the crucial difference that there are gaps, or systematic absences, in the sequence of lines—the 4th (12th and 20th) lines are missing (Fig. 10.16(d)). We can very easily understand this by drawing a diffraction grating of slit separation p, working out the diffraction angles as in Section 7.4 and then considering the contributions to the diffracted intensities of a second set of slits separated by distance $3/8\,p$. It is *precisely* the same procedure that we used to determine the diffracted amplitudes from a crystal with just two atoms per unit cell (Section 9.2 and Fig. 9.3(a)).

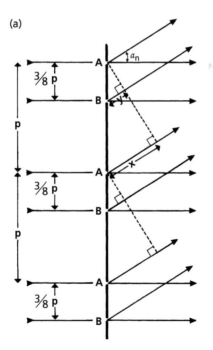

Fig. 10.17. (*continued*)

<hr />

[4] Examples of diffraction patterns of sine waves of various forms are given in the *Atlas of Optical Transforms* (see Chapter 7).

(b)

| n | Path diff, $\frac{3}{8}n\lambda$ | Phase angle | Amplitude, F | Intensity, $|F|^2$ |
|---|---|---|---|---|
| 0 | 0 | 0 | | $4f^2$ |
| 1 | $\frac{3}{8}\lambda$ | 135° | | $0.586f^2$ |
| 2 | $\frac{3}{4}\lambda$ | 270° | | $2f^2$ |
| 3 | $1\frac{1}{8}\lambda$ | 405° (45°) | | $3.414f^2$ |
| 4 | $1\frac{1}{2}\lambda$ | 540° (180°) | | 0 |
| 5 | $1\frac{7}{8}\lambda$ | 675° (315°) | | $3.414f^2$ |
| 12 | $3\frac{1}{2}\lambda$ | 1620° (180°) | | 0 |
| 20 | $7\frac{1}{2}\lambda$ | 2700° (180°) | | 0 |

Fig. 10.17. (a) A diffraction grating of repeat distance p made up of two sets of slits A... A and B... B... separated by distance $3/8\,p$. Maxima occur when the path difference $x = n\lambda = p\sin\alpha_n$. The amplitudes or intensities of the diffracted beams are modulated by interference between the waves diffracted at each pair of slits A and B. (b) The vector-phase diagram for values of $n = 0, 1, 2, 3, 4$, indicating zero intensity for $n = 4$ (and 12, 20, etc.).

Figure 10.17(a) shows the geometry involved. Let f be the scattered amplitude from each slit and consider first the interference between the diffracted beams from the slits A---A---. Maxima occur when the path difference $x = n\lambda = p\sin\alpha_n \approx p\alpha_n$ and the amplitude per unit length p (or per unit cell) is f. Now consider slits B---B---. We only need consider the contribution of one such slit (i.e. in one cell), since what applies to one applies to all. The path difference between each A and B slit, $y = 3/8x = 3/8n\lambda$ and we can now construct the amplitude- or vector-phase diagrams for $n = 0, 1, 2, 3,$ 4... The results, and corresponding intensities in terms of f^2 are shown in Fig. 10.17(b). The cyclic variations in intensities with zeros at $n = 4, 12, 20$, etc. are clear.

Figure 10.16(e) shows the diffraction pattern of the double helix with the phosphate groups and planar bases shown as in Fig. 10.14. Its similarity to well-defined X-ray diffraction patterns of DNA is remarkable, especially so since it was arrived at by such simple procedures.

However, there is a further problem to be solved. It was known (prior to 1952) that the sugar-phosphate groups of nucleic acids were *polar*; the sequence of phosphate groups along one direction is reversed in the opposite direction—in an analogous way as the sequence of Zn and S atoms along a ⟨111⟩ (polar) direction in zincblende (Section 4.7). How, therefore, are the two helices (Fig. 10.14) related; do the DNA backbones run in the same, or in opposite directions? Francis Crick showed that they in fact ran in *opposite* directions. The story is as follows[5]. As mentioned above, within living cells the DNA is surrounded by the high water content of the cell as in a B-DNA fibre. However, when the water content of the fibre is reduced, the strands shrink slightly, but more importantly, they pack together in nearly, but not quite, an hexagonal array (A-DNA). In order to achieve the most efficient packing, the helical 'grooves' of one strand locate into the 'projections' of the next—in just the same way that a bundle of ordinary screws (or threaded rods) pack together; the 'tops' of the threads of one screw fitting in to the 'bottoms' of the threads of the next. This results (in the case of A-DNA) in a C-face centred monoclinic, rather than an hexagonal unit cell. Gosling and Franklin determined the space group to be C2 (a 'permissible' space group—see Table 4.1): the only symmetry elements which the unit cell possesses are the diad (and screw diad) axes running perpendicular to the strands. Crick realized that, since the cell possessed diad symmetry, the DNA molecules must also possess diad symmetry and that the only way this could be achieved was for the polar backbones to run in opposite directions. This is readily seen in Fig. 10.18, the black and white triangles, related by diad symmetry, representing the opposite, or counter-oriented molecular chains. This deduction, arrived at by such simple crystallographic reasoning, was of crucial importance in guiding Watson and Crick in their model building of the structure of DNA.

10.6 The Rietveld method for structure refinement

Powder diffraction data has long been used to identify and characterize crystalline materials, particularly through the use and development of the *X-ray Powder Diffraction File* (Section 10.3.2). However, until the advent of the Rietveld method it was little used for the determination of all but the simplest of crystal structures largely on account of overlapping peaks and the consequent difficulty of accurately measuring integrated intensities, this being the first step in determining the structure factors of the reflections which can be then used, as in single-crystal data, to solve or refine structures. This *peak profile fitting* procedure clearly only works for relatively simple structures which yield

[5] Again, I am indebted to Professor Amand Lucas on whose paper, A-DNA and B-DNA: Comparing their Historical X-ray Fibre Diffraction Images, *Journal of Chemical Education* **85**, 737 (2008) I have largely based this brief account. Professor Lucas and his colleagues have also prepared (available on request) two slides; one consisting of nine optical masks for B-DNA and the other twelve optical masks consisting of the same nine masks for B-DNA plus two additional masks for A-DNA and one mask to disprove the co-oriented model of double-stranded DNA. The use of these masks, in demonstrating the relations between the structures and diffraction patterns of DNA, is explained in the paper.

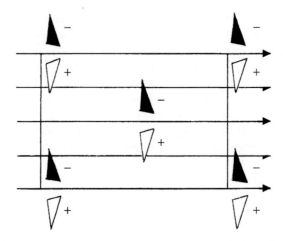

Fig. 10.18. A portion of a pattern based on space group $C2$. An (open) triangle at height $+$ is turned over by the operation of a horizontal diad axis (double-headed arrows along the y-axis) to one at height—(black triangle). The pair of triangles is repeated by the C-face-centring lattice point. Screw diad axes (single-headed arrows) arise between the diad axes.

diffraction patterns with minimal peak overlap. One may say that the first such peak profile-fitting procedure was carried out (albeit with single crystals) by W. L. Bragg in 1913 who established, from relative peak intensities, the location of the S atoms in FeS_2 (Section 4.7, Fig. 4.16).

The Rietveld method, in contrast, does not measure, nor attempt to measure, integrated intensities as such but records the whole powder diffraction pattern: the measurement of intensity at each 2θ step constitutes a data point and the whole set of such data points is compared with those calculated. The calculation takes into account not only structural parameters (unit cell size, atom positions and occupancy factors) but also scale factors, background intensity coefficients and profile parameters describing peak shapes and widths. These parameters are then varied in a least-squares procedure until the calculated pattern best matches the observed pattern. The quantity R to be minimized is given by:

$$R = \sum_i w_i (Y_{io} - Y_{ic})^2$$

where Y_{io} is the observed intensity at step i, Y_{ic} is the calculated intensity and w_i is a weighting assigned to each step—the stronger the observed intensity (i.e. the smaller the counting error), the larger is w_i.

Such is the process of Rietveld refinement, but in order for it to be successful a large number of conditions, both theoretical and experimental, need to be met. First the 'input' structure (unit cell dimensions, atom positions, occupancy of atomic sites,

etc.) needs to be a close approximation of that which is ultimately to be determined by the refinement—how close depends of course on the complexity of the structure. Then, as mentioned above, account must be taken of all the other factors which affect the observed intensities: fluorescence from the specimen, detector noise, thermal diffuse and incoherent scattering, scattering from the sample holder, beam collimator and slits and preferred orientation (Section 10.4). The Rietveld method is able to accommodate moderate amounts of preferred orientation, which may otherwise be difficult or impossible to eliminate (see point 3 below), by the inclusion of refining parameters which model the effect. In so doing preferred orientation may both be quantified and also removed as a confounding influence on the other variables.

The experimental conditions are no less demanding. They may be summarized as follows:

(1) Accurate alignment of diffractometer (to ensure accurate 2θ values).

(2) Correct positioning of specimen (both in the diffractometer and Debye–Scherrer arrangement).

(3) Correct choice of Soller slits (which limit lateral beam divergence in the diffractometer) to avoid low-angle peak 'tails' in low-2θ reflections.

(4) Elimination (as far as possible) of preferred orientation effects; even the process of pressing a powder into a holder, or introducing it into a capillary tube, can give rise to preferred orientation.

(5) Constant volume condition. In the Debye–Scherrer arrangement this is achieved automatically (except in the case of strongly-absorbing specimens). In the diffractometer it is achieved if the incident beam remains within the area of the specimen (a potential problem at low 2θ angles) and if the beam does not penetrate to the underlying holder (i.e. is regarded as 'infinitely thick'). The constant volume condition arises because the area of specimen irradiated varies as $1/\sin\theta$ and the depth of operation varies as $\sin\theta$, hence the volume is independent of θ. Some diffractometers incorporate divergence slits which open out as 2θ increases—resulting in a gain in intensity, and hence reduced counting errors, at high 2θ angles but also an increase in irradiated volume which must be accounted for.

(6) A powder particle size of \sim1–5 μm; for sizes smaller than this line broadening needs to be taken into account and for sizes much larger than all crystallite orientations may not contribute equally to the intensity, a problem akin to preferred orientation.

All these considerations indicate the complexity of the computer programs required in a Rietveld refinement.

The technique was invented, and the first programs written, by Hugo M. Rietveld to refine structures recorded by neutron, rather than X-ray, diffraction. He records that when he first announced the technique at the Seventh Congress of the IUCr in

Moscow in 1966, 'the response was slight, or rather non existent'.[6] Since then it has become a major tool in structure determination, particularly since increasing computer power has led to its application in X-ray diffraction (where line profiles are more complex than those in neutron diffraction). Apart from being able to solve the structures of intrinsically-fine crystalline materials it has found particular application in, e.g., perovskites and superconductors in which the various structural forms arise from displacive transformations (see Section 1.11). It is also able to deal with multiphase specimens and to detect the presence of amorphous components.

The final word must come from Rietveld himself '... the method is not to be treated as a black box. One must be continually aware of the limitations, not only of this method, but in general of all least squares methods.'[6]

Exercises

10.1 Show, by simple geometrical reasoning, that in the X-ray monochromator (Fig. 10.2) the curved reflecting crystal planes should have a radius twice that of the focusing circle.

10.2 Show, by simple geometrical reasoning, that in the back-reflection X-ray powder photograph (Fig. 10.5), the reflections are most sharply focused when $R^2 = xy$ where $R =$ the radius of the ring, $x =$ the specimen–film distance and $y =$ the distance of the (effective) source S behind the film.

10.3 Determine the d-spacings of the strongest low-angle reflections in the Debye–Scherrer film of quartz (Fig. 10.7(b)) and check that they correspond (within the limits of experimental accuracy) with those indicated in Fig. 10.4 or listed in Fig. 10.8.

(*Hint:* the precise locations at which $2\theta = 0$ and $180°$ are *not* found from the centres of the holes in the film (which might be slightly displaced) but by determining the mid-points of the low-angle and high-angle reflections, respectively. The distance between these mid-points corresponds to an angle of $180°$ and the linear scale can be related to the angular scale and the 2θ values. Hence the d-spacings of the reflections can be found using Bragg's law. (The film is taken using filtered CuKα radiation which includes both the Kα₁ and Kα₂ components. At low angles these are unresolved, $\lambda CuK\bar{\alpha} = 1.542\,\text{Å}$, but at high angles they are resolved, $\lambda CuK\alpha_1 = 1.541\,\text{Å}$, $\lambda CuK\alpha_2 = 1.544\,\text{Å}$.)

10.4 Crystalline polypropene may be assigned a monoclinic unit cell with lattice parameters $a = 6.65\,\text{Å}$, $b = 20.96\,\text{Å}$, $c = 6.50\,\text{Å}$ and $\beta = 99.3°$ (see Section 4.8.2). The polymer chains are orientated along the c-axis. The table shows the d spacings, relative intensities and corresponding $hk0$, hkl and $hk2$ indices (data from Natta, G., Corradini, P. and Cesari, M. (1956), *Rend. Acad. Naz. Lincei* **21**, 365).

Given that the X-ray photographs in Fig. 10.11 were taken with CuKα radiation ($\lambda = 1.542\,\text{Å}$) and that the specimen-film distance (corrected for printing = 36 mm), index the reflections and show that in Fig. 10.11(b) the polymer chains are orientated along the tensile axis.

[6] H.M. Rietveld (1993) *The early days, a retrospective view*, in The Rietveld Method, pp. 39–42, ed. R.A. Young. IUCr/Oxford Science Publications, Oxford University Press, Oxford.

d/A	(hk0)	Intensity†	d/A	(hk1)	Intensity†	d/A	(hk2)	Intensity†
6.21	110	s	4.14	111	m	3.06	022	mf
5.20	040	vs	4.06	131	s	3.05	112	mf
4.74	130	s	4.03	041	s	2.50	132	f
3.50	150	m				2.11	202	mf
3.46	060	m						
3.25	200	mf						
3.11	220	mf						

† Intensities: s = strong, v = very, m = medium, f = faint.

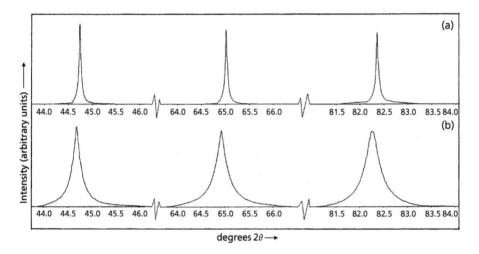

Fig. 10.19. Diffractometer charts (CuKα₁ radiation, $\lambda = 1.541$ Å) showing the first three reflections for a steel specimen (α-Fe, bcc) in (a) the annealed condition and (b) after heavy surface abrasion giving broadened and shifted X-ray diffraction peaks which arise from residual stresses on the micro- and macroscale and a reduction in crystallite size. (Data by courtesy of Mr D. G. Wright.)

10.5 With reference to Fig. 10.19, determine β (expressed in radians) for each of the peaks from the half-height peak widths of the annealed (β_{Inst}) and abraded (β_{OBS}) specimens. On the assumption that the broadening arises solely from microstrain, determine the microstrain and hence residual stress given the Young modulus for iron $E = 207$ GPa (see Section 10.3.4).

10.6 With reference to Fig. 10.20, electron microscopy indicates that the grain size of the milled powder (a) is in the range 10–50 nm and that of the consolidated powder (b) is 5 μm and above. On the basis that the increased broadening in (a) arises solely as a result of the fine crystallite size, show that the X-ray data is consistent with the microscopical data.

Fig. 10.20. Diffractometer charts (CuKα_1 radiation, $\lambda = 1.541$ Å) showing the first three reflections for a titanium alloy (Ti-6% Al-4% V) powder specimen in (a) the mechanically milled condition and (b) following consolidation and heat treatment which gives rise to an increase in grain size and precipitation of the β phase (bcc) in the α phase (hexagonal) alloy matrix. (Data by courtesy of Mr D. G. Wright and Dr A. Brown.)

11

Electron diffraction
and its applications

11.1 Introduction

In Section 7.3 we discussed the wave- and particle-like characters of light which are linked by Planck's equation $E = hc/\lambda$; and also the wave-like character of moving particles which is linked to their mass and velocity by de Broglie's equation $\lambda = h/mv$. The common or 'linking factor' in both equations being Planck's constant $h = 6.63 \times 10^{-34}$ Js. In the case of neutrons (Section 9.7 and Exercise 9.3), it is the thermal conditions in the reactor which give rise to the neutron velocities, and hence mv values which, when substituted in de Broglie's equation, give wavelengths very similar to those for X-rays (Section 9.7). In the case of electrons in the transmission electron microscope the velocity and hence momentum mv of the electrons (m being relativistic-ally corrected) depends on the accelerating voltage V and is determined by equating the potential energy eV (lost) where e is the charge on the electron with the kinetic energy $\frac{1}{2}mv^2$ (gained). The resulting λ values are very small (see Exercise 11.1), for example in a microscope operating at 300 kV, $\lambda = 1.97$ pm (0.0197 Å) and this of course is the basis of the high resolving power (or small limit of resolution) of the electron microscope in comparison with the light microscope as described in Section 7.5. Finally, it should be emphasized that de Broglie's equation applies to all moving particles, not just neutrons and electrons—including pieces of chalk thrown at inattentive students in lectures! However, as a simple calculation shows, the associated wavelengths are very small.

Whereas X-rays are scattered by the electrons in atoms, the incident electrons are primarily scattered by the protons in the nuclei, the scattering amplitude being proportional to Z. The electrons do, however, play a role insofar as they 'shield' the nuclei from the incident electrons by an amount proportional to f, the X-ray atomic scattering factor. Hence, the electron atomic scattering amplitude, f_e is proportional to $(Z - f)$. It is normally expressed as a scattering length (rather than a number like f) and the electron scattering intensities are thus expressed as areas or atomic scattering cross-sections. We may summarize this by saying that electrons are scattered by, or interact with, the variations in **electrostatic potential** in the specimen, the electrostatic potential arising from the net charges on the atoms. In contrast, X-rays interact only with variations in **electron density** and neutrons only with the **nuclei**.

The important consequence is that on average $f_e \propto \sqrt[3]{Z}$ (rather than $f \propto Z$ for X-rays) so that light elements scatter more strongly, in comparison to heavy elements, than X-rays. More important still the *amount* of scattering is much greater than for X-rays by a factor of 10^3 or 10^4 such that electron beams are attenuated much more rapidly.

The Basics of Crystallography and Diffraction. Fourth Edition. Christopher Hammond.
© Christopher Hammond 2015. Published in 2015 by Oxford University Press.

Hence in electron diffraction the dynamical interactions discussed in Section 9.1 are much more important than in X-ray diffraction and must always be taken into account in any attempt to relate observed electron beam intensities to atomic positions. Hence, in solving crystal structures from electron diffraction data, only the thinnest specimens can be used in which the intensity-disturbing dynamical interactions are negligible.

The scattering described above is the dominant scattering process in the 'thin foil' (~1 nm–100 nm) crystalline specimens used in transmission electron microscopy (TEM). It is described as an *elastic* scattering process because the scattered electrons lose no energy and therefore suffer no change in wavelength. It is the physical basis of the 'spot' electron diffraction patterns described in Sections 11.2 and 11.3.

There are in addition a number of *inelastic* scattering processes in which the incident electrons lose energy and therefore suffer a change in wavelength. They are of increasing importance in 'thicker thin-foil' specimens in TEM and the 'bulk' specimens used in scanning electron microscopy (SEM). We have already described some such inelastic scattering processes in Section 9.8; the large energy losses suffered by electrons as they enter the anode of an X-ray tube give rise to the continuous and characteristic X-radiation (Fig. 9.24). There are in addition several inelastic scattering processes which involve much smaller energy losses—thermal diffuse or phonon scattering, plasmon scattering, single valence electron scattering—the physics of which are described in the books on electron microscopy listed in Further Reading. The important point for us is that, since the energy losses are small, the wavelength changes are also small and such inelastically scattered electrons can then be elastically scattered, giving rise to Kikuchi patterns in the TEM and **electron backscattered d**iffraction (EBSD) patterns in the SEM. The geometry of these diffraction patterns and their importance in the precise determination of crystal orientation are described in Section 11.5.

Finally, we should note, as with light microscopy, the importance of diffraction in relation to image formation and resolution. This is a very large topic which is introduced very briefly in Section 11.6.

11.2 The Ewald reflecting sphere construction for electron diffraction

Since electron wavelengths are *very much* smaller than the lattice parameters of crystals, which are normally in the region of 0.3 nm upwards, the diameter of the reflecting sphere is *very large* in comparison with the size of the unit cell of the reciprocal lattice. Referring to the reciprocal lattice sections in Figs 8.6 and 9.15(a) and (b) for example, for the case of electron diffraction, the centre of the reflecting sphere would be situated along the direction of the incident beam at a point far to the left of the origin of the reciprocal lattice, well beyond the left-hand edge of the paper.

The situation for a simple cubic crystal (as for Fig. 9.18(a)) is illustrated in Fig. 11.1, which shows again that part of the reciprocal lattice section perpendicular to the \mathbf{b}^* reciprocal lattice vector or y-axis (out of the plane of the paper) which lies close to the surface of the reflecting sphere, only part of which near the origin can be shown since its

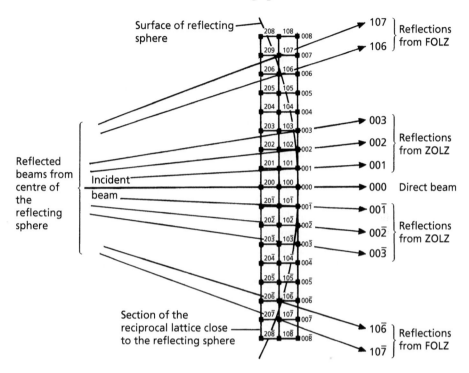

Fig. 11.1. The Ewald reflecting sphere construction for a simple cubic crystal in which the electron beam is incident along the **a*** reciprocal lattice vector (*x*-axis). The **b*** reciprocal lattice vector (*y*-axis) is perpendicular to the plane of the page. A section of the sphere (centre beyond the left-hand edge of the page) and the reflected beams passing at small angles from the centre of the sphere through the reciprocal lattice points in ZOLZ (giving spots close to the direct beam) and FOLZ (giving spots further away from the direct beam) are indicated.

radius is so large. The incident electron beam is directed along the **a*** reciprocal lattice vector or *x*-axis and the centre of the reflecting sphere is far to the left of the diagram.

Close to the origin, 000, the curvature of the sphere is so small that it approximates to a plane perpendicular to the direction of the electron beam and hence passes (in this section) very close to the 001, 002, 003 and 00$\bar{1}$, 00$\bar{2}$, 00$\bar{3}$, reciprocal lattice points.

Now in order for Bragg's law to be satisfied exactly, a reciprocal lattice point must intersect the Ewald reflecting sphere. However, since reciprocal lattice points are broadened into nodes (Section 9.3) they may be extended sufficiently to intersect the sphere and hence give rise to diffracted beams. This situation also arises in X-ray diffraction but to a lesser extent than in electron diffraction (a) because the reciprocal lattice points of the thin foil crystal sections used in electron microscopy are extended perpendicular to the foil surface and are therefore approximately parallel to the electron beam and (b) because thin foils are often slightly bent or 'buckled' such that the orientation of the electron beam varies slightly over the surface of the foil—i.e. it is never incident precisely along one crystallographic direction as indicated in Fig. 11.1.

Hence (referring again to Fig. 11.1), the reciprocal lattice points 001, 002, 003, 00$\bar{1}$, 00$\bar{2}$, 00$\bar{3}$ which lie in a reciprocal lattice row perpendicular to the electron beam direction all give rise to reflected beams, which are directed from the centre of the sphere (to the far left of the page) as indicated. Similar considerations apply to the reciprocal lattice row in the 'next layer above the page', i.e. the row through 010 (above 000), 011, 012, 013, etc. and also the next layer below, i.e. through 0$\bar{1}$0 (below 000), 0$\bar{1}$1, 0$\bar{1}$2, 0$\bar{1}$3, etc. All of these reciprocal lattice points lie in the (0*kl*) plane or section of the reciprocal lattice which is perpendicular to the electron beam direction. In summary, therefore, the reciprocal lattice points close to the origin, in the (0*kl*) reciprocal lattice section perpendicular to the electron beam direction, give rise to reflected beams.

Since the diffraction angles are small (Fig. 11.1), the pattern of spots on the screen perpendicular to the beam corresponds to the pattern of reciprocal lattice points. Figure 11.2 shows this pattern of spots from the (0*kl*) reciprocal lattice section grouped around the direct beam. For simplicity and clarity, only those reflected beams of type 00*l* (in the section perpendicular to the *y*-axis, Fig. 11.1) are indexed.

These diffraction spots are said to belong to the zero order Laue zone (ZOLZ for short) since they all arise from reciprocal lattice points in a section through the origin or, equivalently, they all arise from planes in a zone—the zone axis being parallel (or nearly parallel) to the electron beam direction.

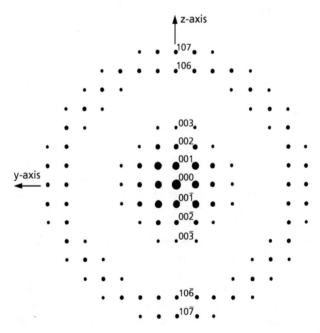

Fig. 11.2. The diffraction pattern on a screen perpendicular to the electron beam direction. The 0*kl* ZOLZ spots are grouped around the direct beam (centre spot) and the 1*kl* FOLZ spots are grouped roughly in a ring outside. Only the *h0l* spots in the section perpendicular to the *y*-axis (Fig. 11.1) are indexed (compare with Fig. 9.18(c)).

The pattern of electron diffraction spots (for a particular incident beam direction) in the zero order Laue zone is identical to that in the zero level X-ray precession photograph (when the crystal is precessed by the appropriate angle about the same incident beam direction). This is seen by comparing Fig. 9.18(c) with Fig. 11.2. The far simpler set-up (a stationary crystal) in electron diffraction 'works' because of the 'flatter' (large diameter) Ewald reflecting sphere, the extension of the reciprocal lattice nodes and the slight buckling of the thin foil specimens—all of which serve to provide the requisite diffraction conditions without the necessity of precessing the specimen over a very much smaller angular range.

Referring again to Fig. 11.1, as we proceed along the surface of the reflecting sphere away from the origin, we begin to intersect the next row of reciprocal lattice points which does not pass through the origin. The reciprocal lattice points in this row which lie close to the sphere, i.e. 106, 107, 10$\bar{6}$ and 10$\bar{7}$ (and in the corresponding rows above and below the plane of the diagram, all of which lie in the (1kl) section of the reciprocal lattice) will also give rise to diffraction spots as indicated. These spots, which lie outside those from the zero order Laue zone away from the centre of the pattern (Fig. 11.2) are said to belong to the first order Laue zone (FOLZ for short). Continuing outwards towards the edge of Fig. 11.1 we come to reciprocal lattice points in the second order Laue zone (SOLZ for short). These are just the first two in a series of higher order Laue zones (HOLZ for short).

A note on the precession electron diffraction (PED) and rotation electron diffraction (RED) techniques

These techniques, which are described in *Electron Crystallography* by X. Zhou *et al.* (2011) (see Further Reading) provide a very nice link to the long-established X-ray precession and oscillation techniques, respectively.

In PED, invented by Roger Vincent and Paul Midgley in 1984, the incident beam is precessed over a small angular range, typically 1~3°. In doing so higher-order reflections in the ZOLZ are sequentially excited as they pass through the Ewald sphere—the extended diffraction pattern 'moves around in a circle' and is recorded at each step. Since a greater number of reflections are collected in PED, the resolution is much enhanced and since fewer reflections are excited at any one moment, dynamical effects are much reduced.

Reflections in the HOLZ (particularly the FOLZ) Laue zones may also be recorded. In cases where the spacing of the reciprocal lattice layers perpendicular to electron beam diffraction is quite small and the FOLZ reflections lie in a ring not far from the inner ZOLZ reflections (as in Fig. 11.2), overlapping of reflections may occur that complicates the interpretation of the patterns.

RED is the 'electron microscopy analogue' of the X-ray oscillation method (Section 9.5.1 and Fig. 9.15). It was developed by Daliang Zhang, Peter Oleynikov, Sven Hovmöller, and Xiadong Zou in 2010. The crystal is kept fixed, the electron beam is oscillated over a small angular range, typically ±3°, and the sequence of diffraction patterns recorded. Then the crystal itself is tilted in the goniometer specimen-holder and the sequence repeated. From these results a three-dimensional reconstruction of the reciprocal lattice of the crystal can be made.

11.3 The analysis of electron diffraction patterns

The analysis of electron diffraction patterns is essentially a process of identifying the reciprocal lattice sections which give rise to them. Several computer programs are now available (see Appendix 1) which generate reciprocal lattice sections, given as input data the lattice parameters, type of unit cell (and in some cases the space group and atomic positions) and the electron beam (zone axis) direction. Several examples showing the ZOLZ only are given in Fig. 11.3. Care must be exercised in using such data because a number of different electron diffraction patterns from different types of high-symmetry crystals have identical symmetry and only differ in scale. For example the $\langle 111 \rangle_{fcc}$ and $[0001]_{hcp}$ patterns shown in Figs 11.3(b) and (d) both consist of hexagonal patterns of points.

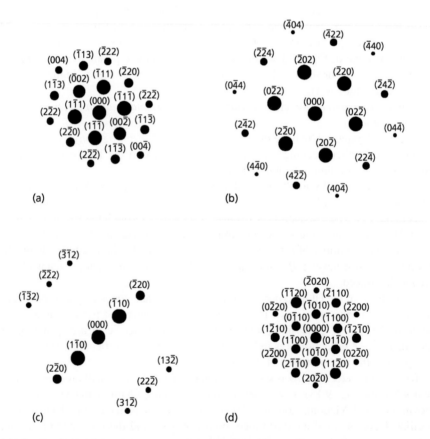

Fig. 11.3. Examples of the computer-generated diffraction patterns, drawn to the same scale ($\lambda L =$ 16.1 Å mm). (a) Aluminium (fcc), zone axis $= \langle 110 \rangle$; (b) aluminium (fcc), zone axis $= \langle 111 \rangle$; (c) iron (bcc), zone axis $= \langle 112 \rangle$; (d) titanium (hcp), zone axis $= [0001]$.

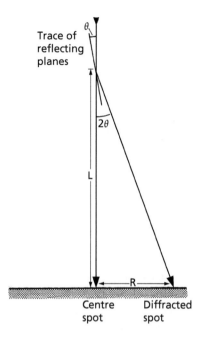

Fig. 11.4. Idealized representation of the direct and diffracted ray paths in the electron microscope between the specimen and the screen. L is the camera length and R is the distance between the centre spot (direct beam) and the diffracted spot.

The first step in the analysis of electron diffraction patterns is measurement of the d-spacings. This is an easy matter. Figure 11.4 shows a set of reflecting planes at the Bragg angle θ (much exaggerated) to the electron beam. The diffracted beam, at angle $\theta + \theta = 2\theta$, to the direct beam, falls on the screen at a distance R from the centre spot. Figure 11.4 does not represent the *actual* ray paths in the electron microscope, which are of course determined and controlled by the lens settings: hence L, the camera length, is not the *actual* distance between the specimen and screen, but is a 'projected' distance which varies with the lens settings.

From Fig. 11.4, $\tan 2\theta = R/L$. Since θ is small, $\tan 2\theta \approx 2\theta$ (2θ of course being measured in radians). Similarly for Bragg's law, $\sin \theta \approx \theta$ hence $\lambda \approx 2d_{hkl}\theta$. Eliminating θ from these equations we have:

$$d_{hkl} = \lambda L/R$$

where λL is known as the camera constant and which, like L, varies with lens settings in the microscope. The units of λL are best expressed in terms of the product of R, the distance measured on the screen (e.g. mm) and d_{hkl} (e.g. nm, Å or pm). Although the ångström (Å) is not an SI (Système International) unit, it is coherent with the SI system ($1 \text{ Å} = 10^{-10}$ m) and still remains in widespread use in crystallography (a) because it is roughly equal to atomic diameters and (b) because X-ray d_{hkl}-spacing data in the *Powder Diffraction File* (see Section 10.3) is recorded in ångströms.

The accuracy with which the d_{hkl}-spacings can be measured is very limited because of the size or 'diffuseness' of the diffraction spots. Moreover the camera constant, λL, will only be known to a high degree of accuracy if a 'standard' specimen is used which gives a diffraction pattern with spots of accurately known d_{hkl}-spacings. Otherwise the small variations in lens settings from specimen to specimen and from day-to-day will limit the accuracy of λL values at best to three significant figures. Hence, as pointed out above, patterns which have identical symmetry but which only differ slightly in scale with respect to the d_{hkl}-spacings of the diffraction spots are not readily distinguished.

The spots may then be indexed by reference to tables of data such as are contained in the *Powder Diffraction File*, which show the (Laue) indices, *hkl*, corresponding to the measured d_{hkl}-spacings. In doing so, a further step is required: the indexing must be such that the addition rule (Section 6.5.2) is satisfied. A given set or family of planes {*hkl*} (which are invariably listed in data files simply as *hkl*—no brackets) will consist of a number of variants of identical d_{hkl}-spacings. For example, in the cubic system (see Section 5.4), there are six variants of the 'plane of the form' {100}, all of which have identical d_{hkl}-spacings. The procedure consists of *choosing* the appropriate variants such that the addition rule is satisfied. Consider, for example, Fig. 11.3(a). The four spots of identical d_{hkl}-spacings closest to the centre spot are all reflections from planes of the form {111}. The particular indices have been chosen such that the indices of the remaining spots are determined correctly. For example, $\mathbf{d}^*_{1\bar{1}1} + \mathbf{d}^*_{\bar{1}11} = \mathbf{d}^*_{002}$ and the zone axis, [110], is found by cross-multiplication of any pair of these indices (see Sections 5.6.2 and 6.5.7), the 'choice' only needs to be made once, and having been made, all the other spots in the pattern can be indexed consistently by repeated application of the addition rule; again, with reference to Fig. 11.3(a), $\mathbf{d}^*_{1\bar{1}1} + \mathbf{d}^*_{002} = \mathbf{d}^*_{1\bar{1}3}$ and so on. The 'choice' is of course arbitrary: the number of different choices which can be made, all leading to a self-consistently indexed diffraction pattern, depends on the number of equivalent variants of the zone axis. In a cubic crystal there are twelve variants of the $\langle 110 \rangle$ direction, hence there are twelve ways in which a pattern like that of Fig. 11.3(a) can be indexed.

The availability of computer-generated diffraction patterns has greatly facilitated the indexing and verification of electron diffraction pattern analysis. However, it must be said that many 'useful' diffraction patterns—i.e. those which may be of potential scientific interest, which have been obtained under conditions of experimental difficulty and which may consist of many spots from several crystals of varying degrees of visibility on the film—still require for their solution careful observation and measurement and an awareness of the possible occurrence of additional complicating factors such as double diffraction (see Appendix 6). Those books on electron diffraction techniques in Further reading provide a more detailed and thorough coverage of these topics.

11.4 Applications of electron diffraction

11.4.1 Determining orientation relationships between crystals

One of the greatest advantages of electron diffraction is the facility in the electron microscope of being able to select a particular area of the specimen and to vary the diameter of the beam to obtain diffraction patterns simultaneously from those phases of interest.

Moreover, the reflections from the phases present may be distinguished by the technique of dark field imaging. Each electron diffraction pattern may be indexed according to the procedures described in Section 11.3. The orientation relationship(s) may then be given in terms of parallelisms between zone axes and planes.

Figure 11.5(a) shows an electron diffraction pattern of α-iron $a = 2.866$ Å (bcc— strong reflections) in which are precipitated (or exsolved) particles of Fe_2TiSi, $a = 5.732$ Å (fcc—weak reflections) in a single crystallographic orientation. It is obvious, simply by inspection or *looking*, that there is a clearly defined orientation relationship between the iron matrix and the precipitate and, because of the coincidence of many of the reflections, that the lattice parameters are related in some rational ratio. The d-spacings of the reflections may be measured (Section 11.3) and, given the lattice parameters and the conditions for reflection for the cubic F (fcc) and cubic I (bcc) lattices (Table A6.2), the reflections may be indexed. Figure 11.5(b) shows the patterns indexed according to one (arbitrary) variant of the orientation relationship. Note that the 200 Fe_2TiSi spot is *half* the distance (and therefore *twice* the d-spacing) of the 200 α-Fe spot from the centre; these planes are therefore parallel and the lattice parameter of Fe_2TiSi is twice that of α-Fe. Similarly, the 022 Fe_2TiSi reflection (by vector addition of $\bar{1}11$ and 111) is coincident with 011 α-Fe. The zone axes, obtained by cross-multiplication (Section 11.3), are both $[0\bar{1}1]$.[1] The orientation relationship may be specified by quoting the parallelisms:

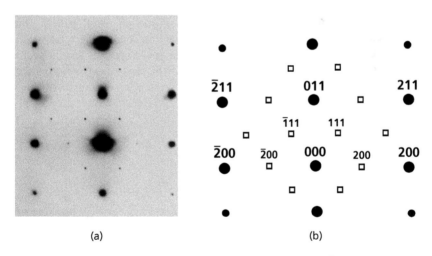

(a) (b)

Fig. 11.5. (a) An electron diffraction pattern of α-iron (bcc, $a = 2.866$ Å: strong reflections) and Fe_2TiSi (fcc, $a = 5.732$ Å: weak reflections) in a single crystallographic orientation. The camera constant $\lambda L = 31.5$ Å mm. (b) The patterns indexed according to one (arbitrary) variant of the 'cube–cube' orientation relationship: $(200)_{Fe_2TiSi} \| (200)_{\alpha-Fe}$; $[0\bar{1}1]_{Fe_2TiSi} \| [0\bar{1}1]_{\alpha-Fe}$. (Photograph by courtesy of Dr. D. H. Jack.)

[1] The *order* of cross-multiplication is important; one should read *anticlockwise*. E.g. for the α-iron diffraction spots write down (211) as the first line in the memogram and then (011) as the second line. This gives (for

$$(200)_{Fe_2TiSi} \| (200)_{\alpha-Fe}$$
$$[0\bar{1}1]_{Fe_2TiSi} \| [0\bar{1}1]_{\alpha-Fe}.$$

This is known as the 'cube-cube' orientation relationship because it is equivalent to the statement that the x, y, z axes of both crystals are mutually parallel.

Other, more complicated, orientation relationships may be determined by the same simple approach, but to go from the parallelisms between the planes and zone axes observed for the particular patterns to the establishment of possible parallelisms between planes not observed in the patterns (i.e. those that are not (nearly) parallel to the electron beam direction), requires a knowledge of the stereographic projection (see Chapter 12).

11.4.2 Identification of polycrystalline materials

The *Powder Diffraction File* may be used to identify polycrystalline specimens from their electron diffraction patterns. However, its use is much more limited owing to the fact that the d-spacings of the diffraction rings can be measured to a much lower level of accuracy and only qualitative estimates can be made of their relative intensities. Hence the use of search procedures, as described in Section l0.3.2, is uncommon. Usually we have a fair idea as to what the specimen might be and make direct comparisons between the experimental data and that of 'possible' patterns in the File. Exercise 11.4 gives an example of this in the case of a single-crystal electron diffraction pattern.

Figure 11.6 gives an example of an electron diffraction 'ring' pattern from a finegrained polycrystalline thin foil specimen of aluminium. It is, in effect, the superposition of many single-crystal patterns. Notice the variations in intensity and the sequence of spacings of the rings. Aluminium can immediately be recognized as having the fcc structure since the rings occur in the characteristic sequence 'two-together', 'one-on-its-own', 'two-together', etc. (see Note 3, Table A6.2). Hence the rings may simply be indexed 'by inspection', i.e.

111,200, 220, 311,222, 400, 331,420.

11.4.3 Identification of quasiperiodic crystals (quasicrystals)

Although the theoretical basis for the existence of quasiperiodic crystals had been established from atom-packing considerations in the 1960s, (see Section 4.9), their discovery (perhaps not the first, but the first to be recognized as such) by Dan Shechtman in 1982 was epoch-making (and also, like Laue's discovery, very simple). In an Al-25wt%Mn (Al_6Mn) alloy, he observed in the electron microscope pentagonal-shaped clusters or nodules that gave rise to diffraction patterns with ten-fold symmetry (Fig. 11.7), i.e. pentagonal symmetry plus a centre of symmetry (Friedel's law). This interpretation was greeted with considerable scepticism by the crystallographic community and delayed

the usual right-handed axial system) the zone axis direction *upwards* (i.e. anti-parallel to the electron beam which is much easier to deal with when plotting the data on a stereographic projection (Chapter 12)).

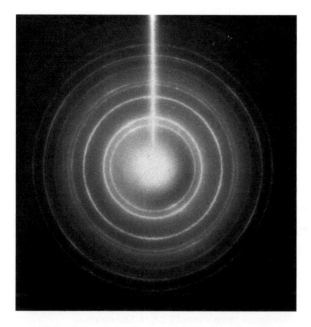

Fig. 11.6. An electron diffraction pattern of polycrystalline aluminium, showing varying intensities of the diffraction rings and the sequence of ring spacings characteristic of the fcc structure.

proper publication of the work until 1984 (see footnote 6, Section 4.9). It was proposed that such patterns could arise from multiply twinned crystals consisting of five twin-related segments radiating out from the central point rather like the slices of a cake (see Fig. A1.4, Appendix 2). The issue is complicated because some alloy systems may exhibit both quasiperiodic crystals or multiply twinned crystals (the microcrystalline state) depending upon heat treatment. For example the $Al_{65}Cu_{20}Fe_{15}$ alloy gives twinned microcrystals when slowly cooled from the melt, which on heating transform to quasiperiodic crystals[2]. However, the fine detail of the spots in electron diffraction patterns—the segments or 'slices of the cake' from multiply twinned crystals are not **perfectly** related by five-fold symmetry: there is some misalignment that results in spots of 'triangular' shape (i.e. made up of a group of little sub-spots, rather than the distinct and clear spots of Fig. 11.7). Furthermore, there is the evidence from dark-field imaging in electron microscopy: in the case of multiply twinned crystals dark-field images arising from different diffraction spots reveal only one segment because of the changes in crystallographic orientation across each twin-face, whereas, as Shechtman showed, each dark-field image revealed the whole quasiperiodic crystal. Finally, there is the evidence from high resolution electron microscopy (HREM); only in microcrystalline twinned crystals are twin boundaries clearly evident.

[2] C. Janot (2012). *Quasicrystals: A Primer.* Second edition. Oxford University Press, Oxford.

Fig. 11.7. Electron diffraction pattern of the first observed icosahedral particle in an Al-25 wt.% Mn alloy, April 1982. (Photograph by courtesy of Prof. D. Shechtman.)

Since Shechtman's discovery (for which he was awarded the Nobel Prize for Chemistry in 2011) a whole new range of quasiperiodic crystals have been identified, prompting the International Union of Crystallography (IUCr) to revise its definition of a crystal.

11.5 Kikuchi and electron backscattered diffraction (EBSD) patterns

11.5.1 Kikuchi* patterns in the TEM

We first consider the trajectories, or paths, of the inelastically scattered electrons within the specimen (Section 11.1). These occur over a range of angles, the scattering being most intense at small angles to the incident beam and decreasing at larger angles, giving in effect a 'pear-shaped' distribution of intensity as shown schematically in the upper part of Fig. 11.8. This distribution of intensity around the incident electron beam (the centre spot) is shown in the lower part of Fig. 11.8. Such a distribution is only observed in practice in 'thicker' specimens (in the range 200 nm upwards) and in which the amount of inelastic scattering is significant. It is almost entirely absent in the thin foil specimens from which the electron diffraction patterns shown in Figs 11.5(a), 11.18, 11.19 and 11.20 were obtained.

Now we describe the elastic scattering, or Bragg reflection, of the weakly inelastically scattered electrons that have suffered negligible changes in wavelength. Consider a specimen in which a set of *hkl* planes is at an angle to the incident beam slightly smaller

* Denotes biographical notes available in Appendix 3.

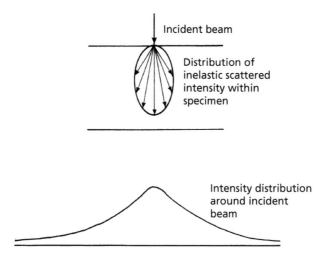

Fig. 11.8. The distribution of inelastic scattered intensity within a specimen and (below) the distribution of intensity around the incident (unscattered) electron beam.

than the Bragg angle (Fig. 11.9(a)). No reflection of the incident beam, or given the extension of the reciprocal lattice point (Section 11.2), only a weak reflection of the incident beam occurs. However, those inelastically scattered beams which are incident to the planes at the Bragg angle are reflected. Figure 11.9(a) shows two such beams, AB and AC, AB being more intense than AC since the inelastic scattering angle is smaller. Beam AB is reflected to D (solid line), giving an increase in intensity; at the same time it is attenuated in the direction to E (dotted line) giving a reduction in intensity. A similar reflection (solid line) and attenuation (dotted line) occurs for AC. However, because beam AC is weaker the overall effect will still be an increase in intensity at D and a decrease in intensity at E. Now these peaks are not spots but are sections through cones of intensity which intersect the screen to give nearly straight lines running parallel to the reflecting planes, i.e. perpendicular to the plane of the paper. These—the 'light' and 'dark' Kikuchi lines—are shown in the plane of the diffraction pattern in Fig. 11.9(b). Also shown is the (weak) *hkl* diffraction spot and the *downwards* projected trace of the planes which lies half-way in between the Kikuchi lines (dashed line). The angle between the Kikuchi lines is 2θ and thus they have the same separation R as the distance between the centre spot and *hkl* diffraction spot.

From Fig. 11.9(a) two angles are of interest; ϕ, the deviation of the planes from the exact Bragg angle, and ρ, the deviation of the planes from exact parallelism with the incident beam. These angles may be determined by measuring the distance a_K (from the dark Kikuchi line to the centre spot) and a_t (from the trace of the planes and the centre spot). Given the camera length L (Fig. 11.4), then $\phi = a_K/L$ and $\rho = a_t/L$.

The geometry may also be represented in terms of the Ewald reflecting sphere construction, Fig. 11.10. **k** and **k₀** are the vectors representing the directions of the

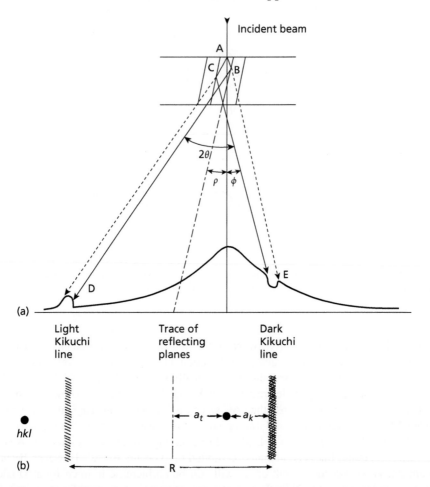

Fig. 11.9. (a) A crystal with a set of planes at an angle to the incident beam slightly smaller than the Bragg angle. AB and AC show two inelastically scattered beams (see Fig. 11.8) which are incident to the planes at the Bragg angle and which are partly reflected (solid lines) and partly attenuated (dotted lines) resulting in light (electron excess) and dark (electron deficient) Kikuchi lines, shown in plan in (b) together with the diffraction and centre spots and the projected trace of the reflecting planes.

diffracted and direct beams and $\mathbf{g}_{hkl} \left(\equiv \mathbf{d}^*_{hkl} \right)$ is the reciprocal lattice vector for the reflecting planes (see footnote, page 198, Section 8.3). The *hkl* reciprocal lattice point lies outside the sphere and the vector \mathbf{s} (closely parallel to \mathbf{k} and \mathbf{k}_0) represents the deviation of \mathbf{g}_{hkl} from the exact Bragg condition, the modulus of which is given by $|\mathbf{s}| = \phi \, |\mathbf{g}_{hkl}|$. Now we see the value of Kikuchi lines in establishing the precise orientation of the reflecting planes: as we (say) rotate the crystal clockwise, the pair of Kikuchi lines move to the left and at the exact Bragg angle ($\phi = 0$), the dark Kikuchi line passes through the centre spot and the bright Kikuchi line passes through the diffraction spot. Similarly,

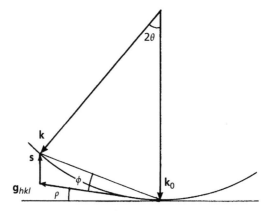

Fig. 11.10. The Ewald reflecting sphere construction for a set of *hkl* planes at an angle to the incident beam (vector \mathbf{k}_o) slightly smaller than the Bragg angle. The reciprocal lattice vector \mathbf{g}_{hkl} lies outside the sphere and the vector **s** represents the deviation from the exact Bragg condition. ϕ is the deviation angle of the planes from the exact Bragg angle and ρ the deviation of the planes from exact parallelism with the incident beam.

when the crystal is rotated such that the planes are exactly parallel to the incident beam, the Kikuchi lines will be at equal distances each side of the centre spot.[3]

In practice we may see many Kikuchi lines—as many pairs as there are diffraction spots (Fig. 11.11). However, just two pairs of lines, and two diffraction spots, are sufficient to determine the orientation of the crystal with respect to the incident electron beam direction. Figure 11.12 shows a pattern with two spots, $h_1k_1l_1$ and $h_2k_2l_2$ and their associated Kikuchi lines; half way in between which are drawn the downwards projected traces of the planes (dashed lines). The downwards direction of the zone axis $[\bar{u}\bar{v}\bar{w}]$ lies at the intersection of the traces as shown; the *upwards* direction $[uvw]$ (i.e. anti-parallel to the electron beam) lies on the opposite side of the centre spot and is found by cross-multiplication of the indices as described in the footnote on p. 291–2. The angle η between $[uvw]$ and the centre spot, 000, is simply found by measuring the distance P, again using the relationship $\eta = P/L$.

The pattern of Kikuchi lines also reflects the symmetry of planes in a crystal and 'maps' may be created showing the patterns of lines for different crystal orientations. Figure 11.13 shows one such map for an fcc crystal centered on [001], the four-fold symmetry of which is clearly evident. Note that the spacings of the pairs of lines correspond to the reciprocal lattice spacings; 200 and 020 passing through the centre being the narrowest, then 220, $2\bar{2}0$, and so on. The value of such maps is that they provide a means of tilting the specimen into any desired orientation: one simply 'moves along' a pair of lines (like railway lines!) to bring the desired zone axis to the centre of the projection.

[3] In this situation the (simplified) explanation of Kikuchi line contrast given above breaks down. A more detailed analysis shows that the lines are still observed.

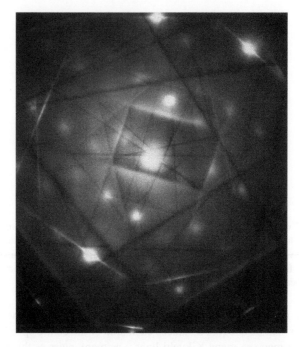

Fig. 11.11. An electron diffraction pattern of an austenitic stainless steel (fcc) showing diffraction spots and Kikuchi lines. (Photograph by courtesy of Dr P.M. Champness and Professor G.W. Lorimer.)

11.5.2 Electron backscattered diffraction (EBSD) patterns in the SEM

In the scanning electron microscope (SEM), EBSD patterns arise from the inelastic-elastic scattering of electrons in the same way as Kikuchi patterns in the TEM—they are in fact 'SEM Kikuchi patterns'. However, the geometrical set up is very different. The incident beam of focused electrons is incident at a small angle, typically 20°, to the specimen surface and only those electrons scattered from depths some 10~20 nm below the surface (depending upon the beam energy and nature of the specimen) emerge from the specimen surface. Hence, EBSD is essentially a surface technique and in order to record clear pairs of Kikuchi lines any surface damage that gives rise to distortions of the reflecting planes must be eliminated by careful surface preparation techniques.

 Unlike the situation in the TEM, the detector (usually a phosphor screen interfaced with a CCD camera) is very close to the specimen, typically 20 mm, and because of space constraints within the specimen chamber (and in order not to obstruct the incident beam), is set, not parallel with the specimen surface, but vertically, i.e. at an angle of ~20° to the specimen surface. The geometry is shown in Fig. 11.14. The inset figure on the left (not to scale) shows a pair of Kikuchi lines arising from a single pair of reflecting planes normal to the specimen surface and from a notional 'point' of inelastic scattering.

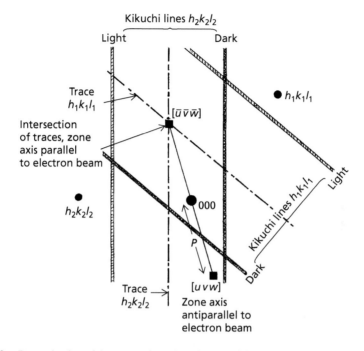

Fig. 11.12. Determination of the exact orientation of a crystal from a spot-Kikuchi line electron diffraction pattern. The *downwards* zone axis, $[\bar{u}\bar{v}\bar{w}]$, lies at the intersection of the projected traces of the reflecting planes $h_1k_1l_1$ and $h_2k_2l_2$. The angle η of the zone axis $[uvw]$ to the incident beam direction is given by $\eta = P/L$.

Overall of course, there are many reflecting planes and many such inelastic scattering 'points' that give rise to the Kikuchi lines. Notice that the trace of these reflecting planes does not intersect the centre of the screen but at the above angle of 20° and the corresponding Kikuchi lines, separation 2θ, lie on either side of this angle. Kikuchi lines that lie on either side of the centre of the screen clearly arise from reflecting planes inclined at (90–20°) = 70° to the specimen surface.

Unlike the situation in the TEM, the Kikuchi lines are detected over a much greater angular range, typically 60~80°, giving a much more complete 'map' of the reflecting planes present in the crystal. Figure 11.15 shows an EBSD pattern of a silicon wafer. Notice that the pairs of Kikuchi lines appear 'much closer together' (a consequence of the small specimen-detector distance) and that the spacings between the lines are (approximately) inversely proportional to the reflecting plane spacings.

As can be seen from Fig. 11.14, the distances of the Kikuchi lines, measured from the centre of the screen, divided by the specimen-screen distance, correspond to the **tangents** of the corresponding angles and since these angles are large the approximation $\tan\theta \approx \sin\theta \approx \theta$ (see Section 11.3, p. 289) can no longer be used. All these 'geometrical-crystallographic' matters are of course 'taken care of' in the computer programs that are invariably used in the crystal identification and orientation analysis of

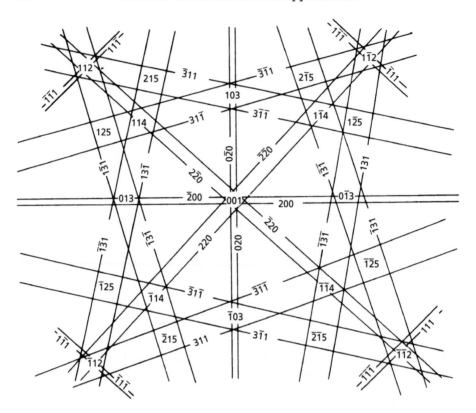

Fig. 11.13. A Kikuchi 'map' for an fcc crystal centered on [001]; note the four-fold symmetry and the relative spacings of the pairs of lines. Tilting the crystal 'along' a pair of Kikuchi lines enables other zone axes to be brought to the centre of the projection (Map by courtesy of Prof. M. H. Loretto.)

EBSD patterns—but the ability to manually measure and index the Kikuchi lines, and to recognize the symmetries between them, must enhance a user's understanding of the EBSD technique.

11.6 Image formation and resolution in the TEM

The Abbe criterion for image formation, discussed in Section 7.5, applies equally to light and electron microscopy. Electron wavelengths are very much smaller (4 pm for 100 kV instruments) than atomic dimensions, but atomic resolution (perhaps the ultimate goal in microscopy) is limited by the severe spherical aberration of electron (electromagnetic) lenses which limits numerical aperture (NA) values to the order of 0.01. Using the equation $a = 0.5\lambda/NA$ gives a limit of resolution of 2Å—a dimension of the same order as atomic dimensions and interatomic spacings. The development of higher kV instruments (smaller λ) and improvements to the corrections (for spherical aberration and astigmatism) in electron lenses have reduced this limit to about 0.5 Å.

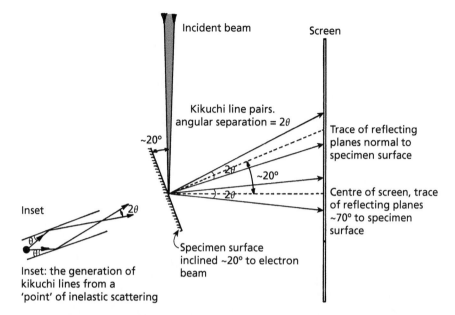

Fig. 11.14. The experimental set-up for EBSD in the SEM. The incident focused electron beam is at an angle of ~20° to the specimen surface and Kikuchi lines arise from the inelastic-elastic scattering of electrons 10~20 nm from below the specimen surface. The detector screen is set vertically such that the trace of reflecting planes normal to the surface intersect the screen at the above angular distance from the centre. The inset figure (left) shows the reflection geometry from a single inelastic scattering 'point' and one pair of reflecting planes.

However, to obtain such Abbe or high resolution images is not a simple matter—neither experimentally nor theoretically. First the specimen must be very thin (of the order 2–20 nm) in order that the inelastic scattering processes described in Section 11.5 are negligible. Second, the specimen must be carefully orientated such that the electron beam is precisely parallel to a prominent zone axis—and of course there are no Kikuchi lines to help guide us! Under such conditions several diffracted beams are excited which then pass through the objective aperture to interfere to form the image. However, there is never an exact one-to-one correspondence between the atomic structure in the specimen and the image, the contrast of which depends upon the amount of defocus of the objective lens and the fact that higher-angle reflections are excluded by the objective aperture and which would otherwise refine the fine detail in the image. Hence, it is standard practice to compare the observed images (obtained at different amounts of defocus) with those obtained by computer simulation. However, despite these problems 'high resolution' TEM images are being routinely obtained in which the peaks of intensity clearly correspond to atom positions. Figure 11.16 shows such a high resolution TEM of germanium which has the diamond cubic structure (Fig. 1.36(a)). The electron beam is aligned precisely along a ⟨110⟩ direction in the crystal and in this orientation the germanium atoms appear to be arranged in pairs—the so-called 'dumbbell' structure.

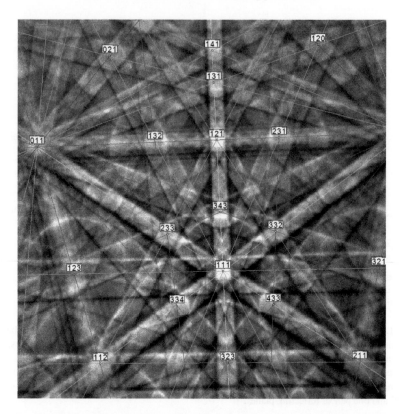

Fig. 11.15. An EBSD pattern of a silicon wafer (diamond cubic structure). Notice the wide angular range of the Kikuchi lines and their different spacings—approximately inversely proportional to the d_{hkl} spacings of the reflecting planes from which they arise. Since the direct beam does not intersect the screen the pattern (unlike Kikuchi patterns in the TEM, Fig. 11.11) is not 'overlain' by an electron diffraction spot pattern. Photograph by courtesy of Dr Ben Britton, Imperial College, London.

So, here we see atoms: how delighted Robert Hooke would have been to actually see his 'bullets' (Fig. 1.1 in this book).

There are, however, other, no less important, modes of image formation available in the electron microscope which occur when only one beam—either the direct beam (bright field mode) or one diffracted beam (dark field mode) passes through an object-ive aperture of small numerical aperture. In the bright field mode the contrast arises from the varying attenuation of the direct beam through different regions of the spe-cimen; the attenuation arising as a result either of inelastic or of elastic scattering (or more generally a combination of the two). The former case gives rise to *mass-thickness contrast* because the amount of inelastic scattering depends on the thickness of the spe-cimen and the variations in atomic number and density of the various atomic species present. The latter case gives rise to *diffraction contrast* which arises from variations in elastic scattering, or Bragg reflection, of the incident beam. The subject is not simple

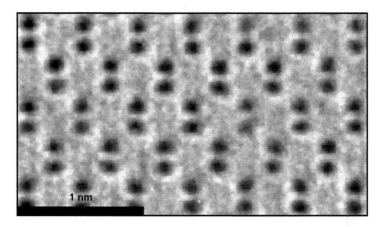

Fig. 11.16. A high resolution TEM 'dumbbell' image of germanium; the electron beam is aligned along a ⟨110⟩ direction in the crystal (diamond-cubic structure). (Photograph by courtesy of TEAM, a collaborative programme between Argonne National Laboratory, CEOS, FEI Company, Frederick Seitz Materials Laboratory, Lawrence Berkeley National Laboratory's National Center for Electron Microscopy, Oak Ridge National Laboratory. TEAM homepage: http://ncem.lbl.gov/TEAM-project/index.html.

because, as noted on p. 212 (Section 9.1) we have to take account of the dynamical interactions between the direct and reflected, re-reflected, etc. beams. The result of these interactions is a 'to-and-fro' or pendulum-type swing of energy between the direct and diffracted beams as they progress down through the specimen, the periodicity of which is called the extinction distance. Hence the tapered edges of specimens, or inclined grain boundaries, show light and dark 'thickness' fringes corresponding to the depths at which (in bright field mode) the direct beam has a maximum or minimum intensity.

Diffraction contrast may be illustrated by two simple examples. Figure 11.17(a) shows a thin metal specimen which is slightly bent or distorted—a very common situation. Hence the planes across the surface are at slightly different angles to the incident beam: when they are close to, or at, the exact Bragg angle, reflection occurs and hence in these regions the direct beam is attenuated. When the direct beam is used to form the image (bright field mode) these regions result in diffuse, dark lines called *Bragg contours*. Correspondingly, when the diffracted beam is used to form the image the Bragg contours appear bright. The contrast from crystal defects arises in a similar way. For example, Fig. 11.17(b) shows an edge dislocation in a specimen; note that around the 'extra' half plane the planes are bent or curved, the distortion decreasing further away from the dislocation. If the curvature is such as to bring the planes to the Bragg angle (as is shown in Fig. 11.17(b)), reflection of the incident beam will occur and thus the dislocation is imaged. Note that we do not image the dislocation as such, we only image, or record, the bending of the planes one side, or the other, of the dislocation.

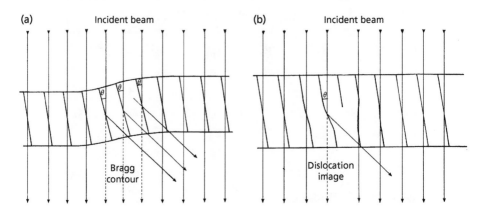

Fig. 11.17. Diffraction contrast image formation. The origin of (a) Bragg contours in a slightly-bent metal thin foil and (b) from the bending of planes around an edge dislocation.

Exercises

11.1 Using de Broglie's equation, derive a simple equation for the (non-relativistically corrected) wavelength λ of electrons in the electron microscope in terms of m, e (electron mass and charge) and V the microscope accelerating voltage. Given the following values of m (rest mass) $= 9.11 \times 10^{-31}$ kg, e (electron charge) $= 1.6 \times 10^{-19}$ C, and h (Planck constant) $= 6.63 \times 10^{-34}$ Js, determine λ for $V = 10$ kV, 100 kV and 1 MV.

11.2 Figure 11.18 is an electron diffraction pattern of a thin crystal of aluminium (fcc, $a = 0.4049$ nm (4.049 Å)) taken using an electron microscope operating at 100 kV, and shows diffraction spots from the ZOLZ (zero order Laue zone) only. Given the electron wavelength $\lambda = 0.0037$ nm (0.037 Å) and the camera length $L = 688$ mm, index the diffraction spots and determine the zone axis of the diffraction pattern. (*Note*: In face-centred lattices, reflections from the planes for which h, k and l are *not* all odd or all even integers are forbidden.)

(*Hint*: Refer to Table A.6.2 listing the indices $\{hkl\}$ and d-spacings of the allowed reflecting planes. Determine the d_{hkl}-spacings of the diffraction spots in the pattern (see Section 11.3) and hence, using the table, find their indices $\{hkl\}$. Using the addition rule in Section 6.5.2, assign a consistent set of indices hkl to the diffraction spots and hence find their common zone axis using the equation in Section 6.5.7.)

11.3 In an electron diffraction experiment the bcc crystal described in Exercises 9.1 and 9.2 is irradiated in the [100] (or \mathbf{a}^*) direction with an electron beam of wavelength 0.002 nm (0.02 Å). Find the reciprocal lattice points which lie close to the sphere in the ZOLZ (zero order Laue zone) and FOLZ (first order Laue zone).

(*Hint*: Using the same reciprocal lattice scale as in Exercises 9.1 and 9.2, it will be found that a very large sheet of paper and very large compasses are needed to draw the complete

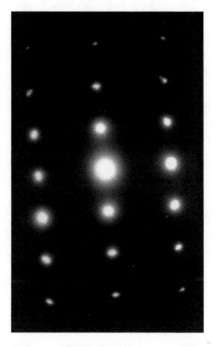

Fig. 11.18. An electron diffraction pattern (ZOLZ) of a fcc crystal. The camera length L (adjusted for the print size) is 688 mm (see Exercise 11.2).

reflecting sphere! However, a sketch of the surface of the sphere near the origin will suffice. In electron diffraction, sections of the reciprocal lattice perpendicular to the electron beam (i.e. normal to the [100] or a^*) direction are most convenient. Draw the section through the origin 000 containing the $0kl$ reciprocal lattice points (i.e. the ZOLZ), and the section containing the $1kl$ reciprocal lattice points (i.e. the FOLZ) (see also Fig. 6.8(a)). Estimate the regions of these reciprocal lattice sections, and hence the reciprocal lattice points, which are close to the surface of the reflecting sphere.)

11.4 Figure 11.19 is an electron diffraction pattern of a particle which has been extracted from a nitrided chromium-containing steel. It is one of three possible compounds, CrN, Fe_3C or $(Cr,Fe)_7C_3$. The d_{hkl}-spacings and indices of the lowest angle reflections are given below.[4] Given $\lambda L = 38.3$ Å mm, identify the compound and index the diffraction pattern.

11.5 Given the lattice parameter of aluminium $a = 4.041$ Å, find the camera constant in Fig. 11.6.

[4] Data for CrN from the ICDD Powder Data File (11–65), for Fe_3C and $(Cr,Fe)_7C_3$ from *Interpretation of Electron Diffraction Patterns*, 2nd edn, 1972 by K. W. Andrews, D. J. Dyson and S. R. Keown, Adam Hilger (1972).

Fig. 11.19. An electron diffraction pattern (ZOLZ) of a particle extracted from a nitrided chromium-containing steel. The camera constant λL (adjusted for print size) is 38.3 Å mm (see Exercise 11.4).

CrN(fcc) $a = 4.14$ Å		Fe$_3$C(orthorhombic) $a = 4.524$ Å $b = 5.088$ Å $c = 6.741$ Å		(Cr,Fe)$_7$C$_3$(hexagonal) $a = 13.982$ Å $c = 4.506$ Å	
d_{hkl}/Å	hkl	d_{hkl}/Å	hkl	d_{hkl}/Å	hkl
2.39	111	6.742	001	12.108	$01\bar{1}0$
2.07	200	5.088	010	6.991	$11\bar{2}0$
1.46	220	4.524	100	6.054	$02\bar{2}0$
1.25	311	4.061	011	4.576	$12\bar{3}0$
1.20	222	3.757	101	4.506	0001
1.04	400	3.381	110	4.223	$01\bar{1}1$
0.950	331	3.371	002	4.036	$03\bar{3}0$
0.926	420	3.022	111	3.878	$11\bar{2}1$
0.845	422	2.810	012	3.615	$02\bar{2}1$
0.800	511	2.703	102	3.495	$22\bar{4}0$
		2.544	020	3.358	$13\bar{4}0$
		2.387	112	3.211	$12\bar{3}1$
		2.380	021	3.027	$04\bar{4}0$
		2.262	200	3.006	$03\bar{3}1$
		2.247	003	2.777	$23\bar{5}0$
		2.218	120	2.767	$22\bar{4}1$
		2.107	121	2.692	$13\bar{4}1$
		2.067	210	2.642	$14\bar{5}1$
		2.031	022	2.512	$04\bar{4}1$
		2.013	103	2.421	$05\bar{5}0$
		1.976	211	2.364	$23\bar{5}1$
		1.872	113	2.330	$23\bar{6}0$
		1.588	130	2.288	$24\bar{6}0$

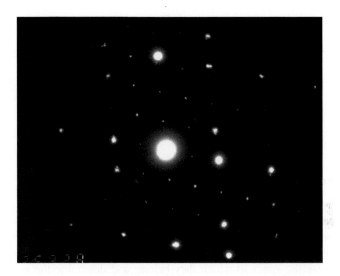

Fig. 11.20. A selected area electron diffraction pattern of a thin foil of a quenched and tempered steel showing reflections from Fe_3C precipitates in a single crystallographic orientation (weak spots) and two twin-related orientations of the ferrite (α-Fe) matrix (strong spots). (Photograph by courtesy of Dr. D. H. Jack.)

11.6 Figure 11.20 is a selected area electron diffraction pattern of a thin foil of a steel (Fe, 1.3% Mn, 0.5% Mo, 0.7% Ni, 0.2% Cr, 0.2% C) which has been quenched from 1300 °C and tempered at 615 °C. The microstructure consists of a matrix of tempered martensite (α-Fe, bcc) and particles of cementite, iron carbide, Fe_3C. The pattern includes α-Fe (strong spots) and Fe_3C (weak spots). Index the patterns and determine the orientation relationship between α-Fe and Fe_3C. The camera constant $\lambda L = 29.4$ Å mm.

 (Hint: A list of d_{hkl}-spacings for α-Fe should be worked out given $a_{\alpha\text{-Fe}} = 2.866$ Å and given the condition for allowed reflections for the Cubic I lattice (Table A6.2). The d_{hkl}-spacings for Fe_3C are given in Exercise 11.4. Notice that the α-Fe pattern also consists of two twin-related orientations (see Fig. 7.17(a) for the analogous case in the diffraction of light.)

11.7 Draw the diamond-cubic pattern of germanium atoms in a $\langle 110 \rangle$ projection (see Fig. 1.36(a)) and show that the pattern corresponds precisely with the high resolution TEM 'dumbbell' image (Fig. 11.16). Given that the lattice parameter a of germanium is 0.566 nm, calculate the separation of the tetrahedrally coordinated atoms and their apparent 'dumbbell' separation in the $\langle 110 \rangle$ projection.

12

The stereographic projection and its uses

12.1 Introduction

In Chapter 6 we showed how both the orientations and the d_{hkl} spacings of planes could be represented in terms of reciprocal lattice vectors and how these vectors could be used to determine the angles between planes and to specify zones and zone axes. We now need a method of representing the planes or faces in a crystal 'all at once' so that we can recognize the zones to which they belong and determine the angles between them without the need for repetitive calculations. The 'crystal drawings' such as Fig. 5.7 are obviously inadequate in this respect; half of the crystal faces are 'hidden from view' and we can only recognize zones with difficulty by looking for the parallel lines of intersections between the (visible) faces. The stereographic projection provides an important method of overcoming these difficulties. It is similar to the reciprocal lattice construction in that in both cases planes or faces are represented by their *normals*.

The stereographic projection is a very ancient geometrical technique; it originated in the second century A.D. in the work of the Alexandrian astronomer Claudius Ptolemy who used it as a means of representing the stars on the heavenly sphere. The original Greek manuscript is lost, but the work comes down to us in a sixteenth-century Latin translation from an Arabic commentary entitled The *Planispherium*. The stereographic projection was first applied to crystallography in the work of F. E. Neumann* and was further developed by W. H. Miller.*

The geometry of the stereographic projection may be described very simply. First, the crystal is imagined to be at the centre of a sphere (the stereographic sphere); the normals to the crystal faces are imagined to radiate out from the centre and to intersect the sphere in an array of points. Each point on the sphere therefore represents a crystal face or plane (and is labelled with the appropriate Miller index) just as a 'point' on the surface of the Earth represents a town or city. The (angular) distance between two points is equal to the angle between the corresponding planes and is determined in the same way that we find the angular distance between, say, Bangkok and New York: we take our private aeroplane and, uninhibited by Traffic Control, fly in a *great circle* between the two. Lines of longitude are simply special cases of great circles which pass through the north and south poles, the angular distances between the poles being of course 180°. Lines of latitude are called *small circles* and represent different angular distances from

* Denotes biographical notes available in Appendix 3.

The Basics of Crystallography and Diffraction. Fourth Edition. Christopher Hammond.
© Christopher Hammond 2015. Published in 2015 by Oxford University Press.

Fig. 12.1. A portion of the surface of the Moon (south at top); above centre is the crater *Tycho* (from which the bright rays radiate) and above this, and seen increasingly foreshortened, is the large crater *Clavius* and smaller craters towards the edge. (Photographed at Mount Wilson Observatory, USA.)

the north and south poles. The largest small circle is the equator which is also a great circle 90° from the north and south poles.

The second step is to find a method of representing this three-dimensional information in two dimensions. One method is to project all the points on the surface of the sphere by parallel lines on to a flat disc called the plane of projection—in the same way as we see in effect all the features on the surface of the Moon, Fig. 12.1. This has the disadvantage that features near the edge of the Moon are seen foreshortened or 'edge on' and that craters, which we know to be circular, are seen as elliptical in shape. The geometry is shown in Fig. 12.2(a), the point of view being taken, in 'Earth' terms, from above the north pole. Five points 15°, 30°, 45°, 60° and 75° from the north pole are shown and it is easily seen how equal angles are represented by *smaller distances* as we move out from the centre. However, in the stereographic projection we *do not* project the points on the surface of the sphere in this way but rather project them to the *south pole* as shown in Fig. 12.2(b). Now the distortion is, as it were, the other way round, equal angles are represented by *larger* distances as we move out from the centre. However, the stereographic projection has one very important geometrical advantage and that is that circles on the surface of the sphere appear as circles, not ellipses, in the plane of projection. The geometry is shown in Fig. 12.3(a). Here a circle in the northern

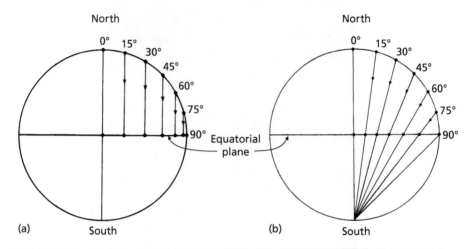

Fig. 12.2. A cross-section of the Earth perpendicular to the equator (the plane of projection) with the north pole at the top and the south pole at the bottom, and points on the surface of the Earth at equal angles 15°, 30°, 45°, 60°, 75° and 90° from the north pole. Viewed from a point a long distance above (as we see the surface of the Moon) the points project on to the equatorial plane as shown in Fig. 12.2(a). However, in the stereographic projection the points are projected to the south pole as in Fig. 12.2(b).

hemisphere (a small circle) is centred about O; all the points on this small circle are at equal angles from O (just as all the points on a line of latitude are at equal angles from the north and south poles). All the points on the circle are projected, as shown, to the south pole giving a cone—not a cone like an ice-cream cone of circular cross-section but one of elliptical cross-section. However, in any such elliptical cone there are two circular cross-sections inclined at equal angles to the axis of the cone, one of which is the circle on the sphere and the other of which lies in the plane of projection (the proof of this is left as Exercise 12.1). Figure 12.3(b) shows a plan view of the equatorial plane, or the stereographic projection of the small circle. Notice that O′, the projection of the centre of the circle is displaced towards the centre of the projection and does not coincide with the point of a drawing-compass. This follows from the fact, pointed out above, that in the projection equal angles are not represented by equal distances.

Figures 12.2 and 12.3 show the construction for points in the northern hemisphere. For points in the southern hemisphere we project instead to the north pole and represent the intersections of the lines with the plane of projection by small unfilled circles instead of dots.

Hence, in summary, and to return to our crystal, the stereographic projection consists of an array of points and circles, called **stereographic poles** or simply **poles** representing plane normals in the northern and southern hemispheres respectively and each labelled with the appropriate Miller index of the crystal face they represent.

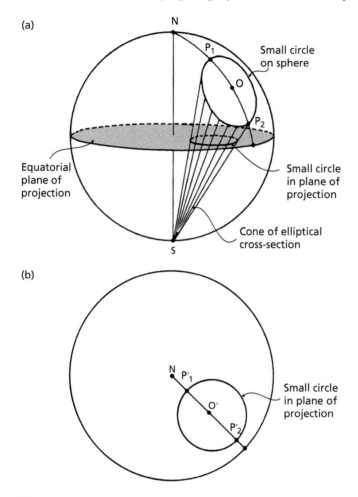

Fig. 12.3. (a) The geometry of the stereographic projection of a small circle (or Moon-crater); all the points on the circle are at equal angles to the centre O. The cone outlined by the lines of projection to the south pole is elliptical in cross-section but its intersection with the equatorial plane of projection is again a circle (see Exercise 12.1). P_1 and P_2 are two points on the small circle on the same line of longitude (great circle through the north pole) which passes through the origin O. (b) The equatorial plane, stereographic projection, or simply stereogram, of the small circle. Notice that the projected points P_1' and P_2' are *not* equidistant from the projected centre of the small circle O'; this follows because equal angles are not represented by equal distances (see Fig. 12.2(b)).

12.2 Construction of the stereographic projection of a cubic crystal

Figure 12.4 shows a cubic crystal at the centre of the stereographic sphere with {100}, {110} and {111} faces and oriented for easy 'Earth' reference with the z-axis in the direction of the north pole and the x and y axes in the equatorial plane (the stereographic

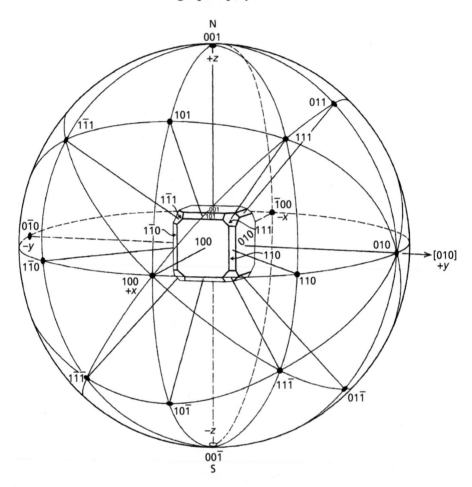

Fig. 12.4. A cubic crystal showing faces of the form {100}, {110} and {111} at the centre of the stereographic sphere with face normals intersecting the surface of the sphere. Zones are shown as great circles (diametral sections) on the sphere, the zone axis being normal to the diametral section. (From *Manual of Mineralogy*, 21st edn, by C. Klein and C. S. Hurlbut Jr; John Wiley, 1993.)

plane of projection). The normals to the crystal faces are shown and their intersections with the sphere labelled with their {*hkl*} indices. The great circles on the sphere represent zones. For example, planes or faces (00$\bar{1}$), (10$\bar{1}$), (100), (101) and (001) all lie in the [010] zone and their normals lie on a great circle—a line of longitude which passes through the north and south poles. Similarly, planes or faces (010), (111), (101), (1$\bar{1}$1) and (0$\bar{1}$0) all lie in the [$\bar{1}$01] zone and their normals all lie on a great circle in this case inclined at 45° to the north and south poles.

Figure 12.5 just shows the poles of planes in the equator and in the northern hemisphere with the crystal and great circles omitted (to avoid the confusion of too many lines). The lines of projection to the south pole are shown and their intersections with

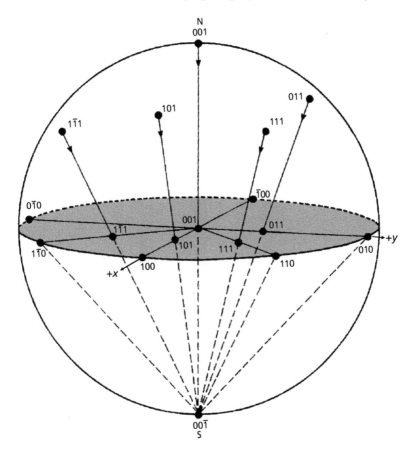

Fig. 12.5. Poles of planes in the northern hemisphere (and in the equatorial plane of projection) and the projection lines to the south pole, the intersections of which in the plane of projection are marked by dots and their corresponding {*hkl*} indices indicate the position of the stereographic poles of the planes.

the stereographic plane of projection marked by dots with their corresponding {*hkl*} indices. Finally, Fig. 12.6 is a 'plan' view of the stereographic projection showing the stereographic poles and again the great circles which project as the arcs of circles passing through them. Now look back to Fig. 12.4. It is very important that you are able to visualize the relationships between the faces in the crystal and their representation as poles in the stereographic projection.

In plotting the exact positions of the poles in Fig. 12.6 it is important to remember that equal angles are not represented by equal distances (except, of course, around the perimeter). For example the (101) plane is 45° between (001) and (100) but projects at a smaller distance to (001) as shown in Fig. 12.2(b). Having plotted the positions of the (101), (011), (0$\bar{1}$1), etc. poles at their proper scale, the great circles through them 'automatically' give the positions of all the other poles or faces in the crystal—both those 'hidden from view' such as ($\bar{1}$11) and also ones not shown in the drawing of the

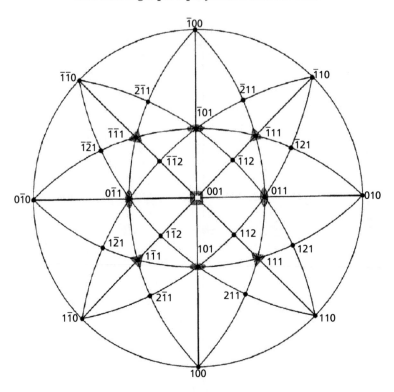

Fig. 12.6. A plan view of the stereographic plane of projection (the stereogram) showing the stereographic poles of the faces of the crystal in Fig. 12.4 and including both those faces 'hidden from view' in Fig. 12.4 and also additional faces (such as, for example, (112)) not drawn in Fig. 12.4. Great circles on the sphere project as the arcs of circles in the plane of projection.

crystal in Fig. 12.4. For example (112) lies at the intersection of the zone through (101) and (011) and also the zone through (111) and (001); i.e. by use of the addition rule (Section 5.6.4) (101) + (011) = (112) and (111) + (001) = (112). Clearly, by drawing more great circles and using the addition rule, planes of higher $\{hkl\}$ values can be located (Exercise 12.2).

Finally, we must now consider the points lying in the southern hemisphere where the projection lines are made to the north pole and their intersections with the plane of projection represented by small unfilled circles. These need not be shown as such in Fig. 12.6 because clearly, for this orientation of the crystal (z-axis or (001) at the centre), a plane such as (10$\bar{1}$) is coincident with (but 'underneath') the plane (101).

12.3 Manipulation of the stereographic projection: use of the Wulff net

We now need a 'stereographic graph paper' in order to locate stereographic poles, zones and zone axes. One such is the *polar net* (Fig. 12.7) which shows great circles like lines

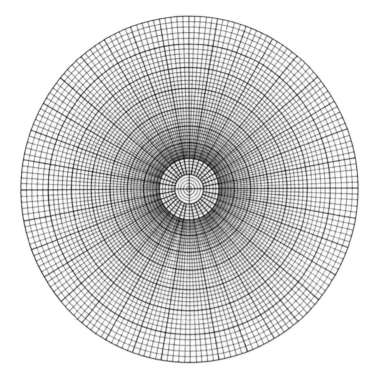

Fig. 12.7. The polar net.

of longitude and small circles like lines of latitude. This net is not very useful. Instead we use the net devised by G. V. Wulff* (Fig. 12.8) which is, in effect, the polar net rotated through 90°; the great circles all pass through diametral points A . . . A in the plane of the projection ('top' and 'bottom' of Fig. 12.8) and the small circles are at increasing angles to these points, i.e. each small circle represents the loci of points at equal angles to a point like 0 in Fig. 12.3 where 0 is now located in the plane of the projection. But the Wulff net is rotatable (about a pin in the centre) and the diametral points A . . . A or axis of the net may be centered along the x-axis, or the y-axis (90° rotation) or indeed any direction in the plane of the projection.

 The use of the Wulff net is best explained by way of some examples. Suppose we wish to find the angle between planes $h_1k_1l_1$ and $h_2k_2l_2$ (Fig. 12.9) (or, in Earth terms, between New York and Bangkok). We rotate the Wulff net until $h_1k_1l_1$ and $h_2k_2l_2$ lie on the *same* great circle as shown (full line) and the angle between them is simply 'read off' from the divisions between the small circles which intersect at 90°. The dashed

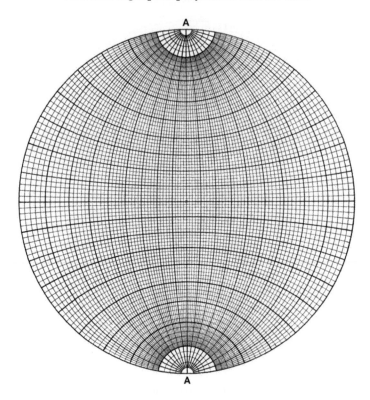

Fig. 12.8. The Wulff net (which may be rotated about the centre of the stereographic projection such that the 'axis' A ... A lies in any direction).

line, symmetrical about the centre, is the continuation of the great circle in the south-ern hemisphere and the angle, say, between $h_2k_2l_2$ and $h_3k_3l_3$ is found by continuing the small-circle division-counting round the edge. The opposite direction of a plane normal, say $h_1k_1l_1$, is found by drawing a line through the centre (arrowed line in Fig. 12.9); the two poles, $h_1k_1l_1$ and $\bar{h}_1\bar{k}_1\bar{l}_1$ being related by a centre of symmetry. Finally, the zone axis is the direction 90° to all points in the zone as indicated by the point ZA.

 Small circles may be used to locate poles in the stereographic projection given their α, β and γ angles to the x, y and z axes (Exercise 12.3). They may also be used to find the positions of the poles when the crystal is rotated away from its 'standard' orientation (z axis along the north pole—Fig. 12.6) to some other orientation. Suppose we wish to rotate the crystal to bring (112) to the centre. The angle of rotation (the angle between (001) and (112)) is 35.3° and the axis of rotation is the $[1\bar{1}0] - [\bar{1}10]$ direction. The Wulff net is orientated such that the small circles are centered along this axis (A ... A, Fig. 12.10) and all the poles rotated 35.3° around their respective small circles, the angle of rotation being calibrated by the intersections of the great circles.

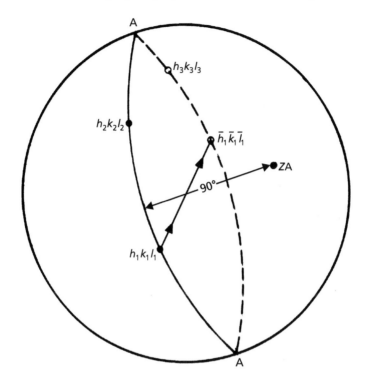

Fig. 12.9. Use of the Wulff net: to find the angle between two planes $h_1k_1l_1$ and $h_2k_2l_2$ in the northern hemisphere, the Wulff net is rotated until the poles lie on the same great circle (solid line, axis along A...A); the angle between the poles is calibrated by the intersections of the small circles. The dashed line represents the continuation of this great circle in the southern hemisphere on which lies the pole $h_3k_3l_3$. The pole $\bar{h}_1\bar{k}_1\bar{l}_1$ is related to $h_1k_1l_1$ by a centre of symmetry indicated by the arrowed line. ZA indicates the zone axis of the zone in which all these poles lie.

12.4 Stereographic projections of non-cubic crystals

Stereographic projections of triclinic, monoclinic and trigonal crystals appear to be rather complicated because the crystal axes in these systems are not orthogonal; we will not consider them further. Stereographic projections of hexagonal crystals in the 'standard' orientation (i.e. the z-axis or (0001) in the centre and the x, y and u axes in the plane of the projection) are, like those of cubic crystals, easily visualized— the hexagonal symmetry of the pattern of poles is immediately evident. Similarly with orthorhombic (and tetragonal) crystals, except that in these cases, as compared to cubic crystals, the {110} planes are no longer at 45° to the {100} planes; the angles must be calculated from the a, b and c lattice parameters. Furthermore, as shown in Figure 5.5, in non-cubic crystals plane normals and directions with the same numerical indices are not, except in special cases, parallel to each other. For example, in hexagonal crystals a plane normal such as $(10\bar{1}0)$ is parallel to the direction $[10\bar{1}0]$ but a plane normal such as $(10\bar{1}1)$ is not parallel to the direction $[10\bar{1}1]$. In orthorhombic

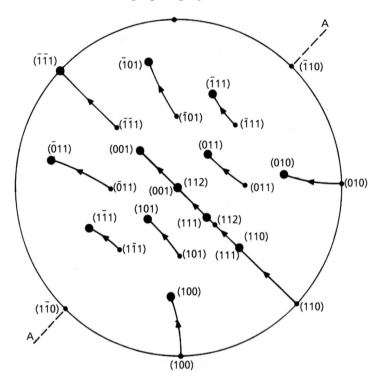

Fig. 12.10. A stereographic projection showing (i) poles of planes in the standard orientation (as in Fig. 12.6) indicated by small dots and indices and (ii) their corresponding positions, indicated by large dots and indices, with (112) rotated to the centre. All the poles rotate 35.3° about small circles centered on the 'axis of rotation' A ... A as indicated by the dashed lines.

crystals a plane normal such as (100) is parallel to the direction [100] but a plane normal such as (110) is not parallel to the direction [110] (see Fig. 5.5). This means that either *two* stereographic projections are needed for non-cubic crystals, one showing plane normals and the other showing directions, or a 'composite' stereographic projection is needed with the positions of plane normals $\{hkl\}$ and directions $[uvw]$ clearly distinguished. Figure 12.11(a) shows a plan view of an orthorhombic crystal (cementite, Fe_3C; $a = 4.524$ Å, $b = 5.088$ Å, $c = 6.741$ Å) perpendicular to the z axis and the angles $\alpha = \tan^{-1} a/b$ and $\rho = \tan^{-1} b/a$ of the (110) plane normal (full line) and the [110] direction (dashed line) to the x-axis respectively. These poles, and those related by symmetry, are plotted on the 'composite' stereographic projection in Fig. 12.11(b). Similarly, the (101), [101], (011) and [011] poles can also be plotted using the Wulff net given the ratios of c/a and c/b. Having located these poles, the positions of the (111), [111] (and also higher-index) poles can be found from the intersections of great circles as before—no further calculations are required.

In general, stereographic projections showing the positions of $\{hkl\}$ plane normals are more useful and where the indices are not bracketed (as in Fig. 12.6) it is understood

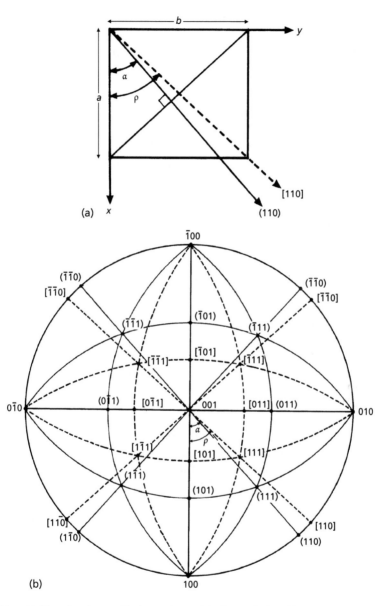

Fig. 12.11. (a) Plan view (perpendicular to the z axis) of an orthorhombic crystal (cementite, Fe_3C) showing the angles α and $\rho = (90° - \alpha)$ of the (110) plane normal (full line) and [110] direction (dashed line) to the x-axis respectively. (b) The 'composite' stereographic projection showing two sets of great circles through the $\{hkl\}$ poles (full lines) and the $[uvw]$ directions (dashed lines).

that they refer to plane normals, not directions. However, in non-cubic crystals a zone axis located as shown in Fig. 12.9 should always be assigned square [*uvw*] brackets because it will not in general coincide with the unbracketed indices which refer to plane normals.

12.5 Applications of the stereographic projection

12.5.1 Representation of point group symmetry

In Chapter 4 we discussed the point group symmetry elements—mirror planes, centres, rotation and inversion axes of symmetry. We can represent the operation of these symmetry elements very easily with the stereographic projection by first placing a single pole (representing a single face) in a *general position* in the projection (i.e. not on an axis or mirror plane) and then seeing how it is repeated by the action of the symmetry elements present. In this way all the 32 point groups can be built up.

Figure 12.12(a) shows a pole reflected in a 'vertical' mirror plane *m* (indicated by a heavy line) and Fig. 12.12(b) shows a pole reflected by a mirror plane in the plane of the projection—one pole being in the 'northern' hemisphere and its reflection (an open circle) in the 'southern' hemisphere below. Fig. 12.12(c) shows the operation of a centre of symmetry and Figs. 12.12(d) and (e) show the operation of a diad axis—a 'vertical' diad as in (d) or a 'horizontal' diad (in the plane of projection) as in (e); the rotation is 180° about a small circle centred along the diad axis giving one pole in the upper hemisphere, and one in the lower hemisphere as shown.

Figure 12.12(f) shows the operation of a 'vertical' tetrad giving a four-fold pattern of poles (the light lines may be regarded as 'guidelines' in making the drawing). When one of these guidelines is made into a mirror plane the resulting pattern of poles is given in Fig. 12.12(g). Notice that the tetrad symmetry not only generates mirror planes at 90° but an additional set at 45° as well (see Section 2.3).

Figure 12.12(h) shows the operation of the inversion tetrad axis of symmetry. The procedure is as follows: starting with the pole marked 1, rotate 90° and invert through the centre (dashed and arrowed lines) to give the next position of the pole marked 2. Then repeat to give poles marked 3 and 4. Notice that the pattern of poles (two 'above' and two 'below') includes the operation of a diad axis, hence the symbol for the inversion tetrad axis includes the diad symbol. Finally, Fig. 12.12(i) shows the pattern of poles when a 'vertical' mirror plane is added, which generates not only a 'horizontal' mirror plane at 90° but also two sets of diads at 45° as indicated. This point group $\bar{4}2m$ represents the symmetry of urea or a tennis ball (Fig. 4.6).

Figure 12.13 shows the pattern of poles for a cubic crystal with minimum symmetry (point group 23) and maximum symmetry (point group $m\bar{3}m$). The principal symmetry elements are the four equally-inclined triads around which the poles rotate like points round the centre of a small circle (Fig. 12.3). In Fig. 12.13(a) the great circles drawn in are all 'guidelines' as in, e.g. Fig. 12.12(f), and around each triad are a set of three poles—two 'above' and two 'below'. Notice that the diad axes are 'automatically' generated along the *x*, *y* and *z* axes. Figure 12.13(b) shows the holosymmetric point group

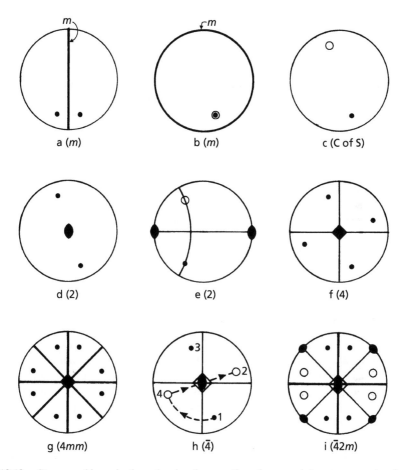

Fig. 12.12. Stereographic projections showing the operation of some point group symmetry elements as indicated. Light lines indicate 'guidelines' for locating the poles, heavy lines indicate mirror planes.

$m\bar{3}m$; all the great circles now correspond to mirror planes, the x, y and z axes correspond to tetrads, the $\langle 110 \rangle$ directions correspond to diads and the triads include a centre of symmetry, hence the symbol $\bar{3}$. These symmetry elements are indicated in the stereographic projection of the cubic crystal shown in Fig. 12.6.

Figures 12.12 and 12.13 show 11 of the 32 point groups. It will help your understanding of symmetry to draw stereographic projections of all the 32 point groups (see Table 3.1) and thereby you will appreciate more fully the relationships between them.

Stereographic poles may also be represented by the same symbolism as used in space groups (Section 4.6); i.e. ○+ and ○– for poles in the 'northern' and 'southern' hemispheres, respectively and ⊙ (comma in a circle) to represent a change of hand.

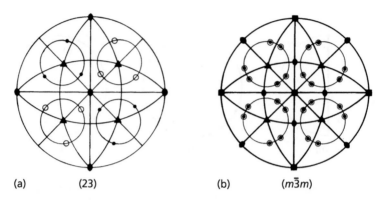

(a) (23) (b) (m̄3m)

Fig. 12.13. Stereographic projections of cubic crystals (a) point group 23 and (b) point group $m\bar{3}m$. The principal symmetry elements are the four equally-inclined triad axes around which the poles are distributed around small circles centred on these axes. In (a) the only additional symmetry elements are the diad axes along the x, y and z-axes, whereas in (b) these become tetrads; There are also six further diads along $\langle 110 \rangle$ directions and all the great circles become mirror planes (heavy lines).

12.5.2 Representation of orientation relationships

The study of solid state transformations in which new crystals nucleate and grow within, or at the boundaries of, parent crystals is of major importance in materials and earth sciences. Our interest as crystallographers is not so much with the mechanisms of the transformations and the conditions which give rise to them but rather with the fact or observation that in almost all cases the planes and directions in the parent and product phases bear definite relationships to each other and that such relationships may provide evidence of the nature of the transformation processes.

That such relationships arise may be seen almost intuitively from the simplest of examples—as we saw in Section 1.5. When iron transforms from its high-temperature ccp form to its low temperature bcc form, a closest-packed {110} plane in the bcc structure (Fig. 1.9(b)) is parallel, or closely parallel, to a close-packed {111} plane in the ccp structure (Fig. 1.4) and a pair of close-packed directions ($\langle 111 \rangle$ in bcc, $\langle 110 \rangle$ in ccp) are also parallel, or closely parallel. Hence, the orientation relationship may be stated in terms of (i) parallelism of planes and (ii) parallelism of directions within these planes, i.e.

$$(111)_{ccp} \parallel (110)_{bcc}$$

$$[1\bar{1}0]_{ccp} \parallel [1\bar{1}1]_{bcc}$$

Similarly, when cementite, Fe_3C, precipitates as fine particles within ferrite (the interstitial solid solution, α, of carbon in bcc iron), then under certain conditions the orientation relationship which occurs is given as:

$$(001)_{Fe_3C} \parallel (112)_{\alpha}$$

$$[010]_{Fe_3C} \parallel [\bar{1}10]_{\alpha}$$

However, such statements of 'parallelisms' do not immediately tell us what *other* planes and directions in the two phases may be parallel (or nearly parallel) to each other, nor do they immediately specify the angles between the two sets of crystal axes—except in the very simplest cases such as the 'cube–cube' orientation relationship between α-iron and Fe_2TiSi (Fig. 11.5) where the diffraction pattern clearly shows that the crystal axes of these two phases are parallel.

The stereographic projection provides a graphical way of analysing the relationships between planes and directions in two phases. For example, if we place the cubic stereogram with the (112) plane in the centre (Fig. 12.10) over the stereogram of the cementite crystal with (001) in the centre (Fig. 12.11(b)) and rotate them until the $[010]_{Fe_3C}$ and $[\bar{1}10]_{\alpha}$ directions (at the edge, i.e. in the plane of the projection) coincide, then we can see, across the whole composite projection, the orientation relationships between the crystal axes and other planes and directions. Furthermore, it may well be (and usually is) the case that the zones we observe in our electron diffraction patterns of the two phases may well be those out of the plane of projection and we may check the occurrence of the orientation relationship by a trial-and-error process of tracing round great circles in the composite projection and seeing whether or not the plane normals of the two phases occur at the observed angles to each other. The diffraction pattern in Fig. 11.5 provides a very simple example. The cube–cube orientation relationship is simply represented by two cubic stereograms in the 'standard' orientation with the x, y and z axes parallel and with the added proviso that for α-Fe (bcc) the planes which reflect are those for which $h + k + l$ is even and for Fe_2TiSi (fcc), h, k, l are all odd or all even. The angles between the two sets of planes may be checked by tracing round the great circles with the common $[0\bar{1}1]$ zone axis.[1]

12.5.3 *Representation of preferred orientation (texture or fabric)*

Pole figures (see Section 10.4), are in effect stereographic 'contour maps', the positions of the 'hills' or 'peaks' in the maps showing the orientations of a particular set of crystal planes and the 'height' and 'steepness' of the hills showing the extent or degree of preferred orientation from a random distribution. In materials science the plane of projection is normally set parallel to the surface of the metal sheet, in Earth science it is normally set horizontal with the points of the compass around the edge.

The geometry of pole figures is best understood by way of a simple example. Let us suppose that in a rolled sheet of a polycrystalline cubic metal the {100} planes of all the crystallites lie closely, but not perfectly, parallel to the sheet surface and that the ⟨100⟩

[1] A more detailed description of the use of stereographic projections in the analysis of orientation relationships, with step-by-step examples, may be found in my article on *Stereographic Techniques*, Module 9.17 in *Procedures in Electron Microscopy* (eds A. W. Robards and A. J. Wilson), John Wiley, 1993.

directions lie closely parallel to the rolling direction (this texture, incidentally, occurs in rolled and recrystallised sheets of iron–silicon alloys used for transformer cores). The pole figure for the {100} planes will appear as shown in Fig. 12.14(a)—the {100} plane normals cluster at the centre (i.e. normal to the sheet surface) and also along the rolling and transverse directions giving contoured 'peaks' in these positions as shown. Now suppose that the crystallites still remain aligned such that the ⟨100⟩ directions lie closely parallel to the rolling direction but that the {100} planes are more randomly orientated with respect to the surface of the sheet. The pole figure will appear as in Fig. 12.14(b)—the contour lines representing the distribution of the {100} plane normals are now spread out towards the transverse directions. This pole figure corresponds closely to that for graphitized carbon tapes (Fig. 10.13(b)) which shows the orientations of the graphite layers.

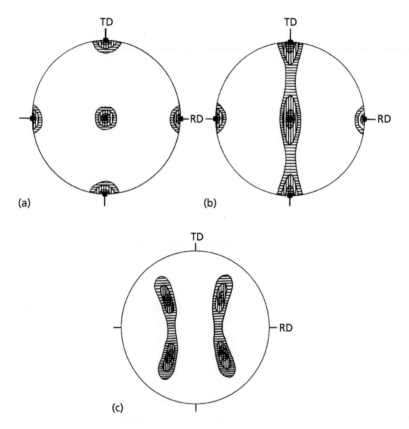

Fig. 12.14. (a) A {100} pole figure in which the {100} planes lie closely parallel to the sheet surface and the ⟨100⟩ directions lie closely to the rolling direction (RD) and transverse direction (TD). The successive levels of shading correspond to the contours of the orientations of plane normals and directions. (b) A similar pole figure to (a) except that the {100} planes are more randomly distributed about the rolling direction—the peaks are extended across the transverse direction. (c) A pole figure of the same texture as (b) but now showing the orientations of the {111} planes.

Pole figures may be prepared for any set of planes. For example the {111} pole figure corresponding to the texture shown in Fig. 12.14(b) would consist of peaks of intensity as shown in Fig. 12.14(c). Pole figures may be prepared in practice by a number of experimental techniques—by the use of the texture goniometer as described in Section 10.4, by recording backscattered electron diffraction patterns in the scanning electron microscope (see Section *11.5.2*) or by recording the orientation of individual grains in thin sections in the light microscope. In general the textures observed are rather more complicated than those of the simple examples described above and may consist of two or more *components*, i.e. two or more sets of planes and directions may be aligned to a greater or lesser extent parallel to the plane of the sheet and along the rolling direction respectively. For example, Fig. 12.15(a) shows the {111} pole figure of heavily cold rolled copper in which the texture may be described as (123)[41$\bar{2}$] and (146)[21$\bar{1}$] where the indices refer to the planes orientated parallel to the rolling plane and the direction symbols refer to the directions orientated along the rolling direction. Finally, Fig. 12.15(b) is a pole figure (or petrofabric diagram[2]) showing the orientations of the *c*-axes of the quartz grains in a metamorphosed psammite rock. They lie, as can be seen, roughly in a great circle, the zone axis of which is closely parallel to the direction of lineation or structural long direction of the rock.

12.5.4 Trace analysis

Crystals, particularly those of interest to mineralogists or metallurgists, may contain internal planar or plate-shaped features such as twins, stacking faults, or (as in partially transformed alloys) martensite plates. The traces of such features refer to their observed lines of intersection with the crystal surface(s). Trace analysis is the stereographic technique by which the interface-planes in the crystal may be identified.

The procedure is perhaps best explained by means of a simple example. Figure 12.16(a) shows a grain in a specimen of α-brass that has been polished and etched on two surfaces at 90°. The line of intersection is a [$\bar{1}$11] direction and the indices of the 'top' and 'side' faces (determined by the Laue back reflection technique) are (112) and ($\bar{1}$10), respectively. The grain shows the traces of a planar feature, AB on the (112) face and CD on the ($\bar{1}$10) face. We now construct a stereographic projection with (112) in the centre, i.e. as in Fig. 12.10. Consider first the trace AB. The pole of the required plane must lie somewhere on a great circle **normal** to the line AB and similarly for the trace CD. These two great circles are shown in Fig. 12.16(b). The required stereographic pole of the feature lies at the intersection of these two great circles. We now measure the angles between this pole and the (100), (010), and (001) poles and find their direction

[2] Petrofabric (or fabric) diagrams may also be represented on *equal area projections* which are, of course, much used in earth science and geography since they show, for example, the sizes of materials in their correct proportions. 'Equal area graph paper' consists, like the Wulff net, of intersecting small and great circles, but these are not arcs of circles and do not appear to intersect at 90°. The geometry of equal-area projection is more complicated than that of the stereographic projection. For example, referring to Fig. 12.2, we need to project the points on the sphere 15°, 30°, etc. from the north pole such that they make *equal* divisions along the equatorial plane, and this cannot be done by projecting lines down to a single point.

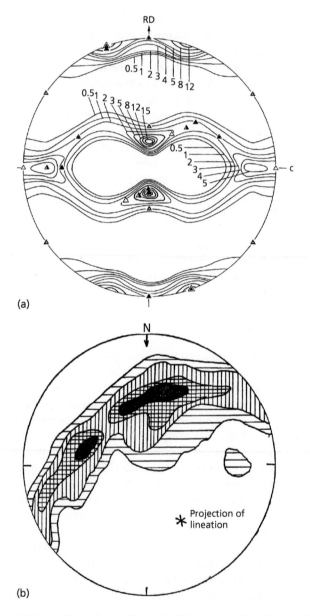

Fig. 12.15. (a) A {111} pole figure for heavily cold rolled copper, rolling direction RD vertical (from H. Hu and S. R. Goodman, *Trans. AIME,* **227**, 627, 1963). (b) A pole figure with contours shaded as in Fig. 12.14, showing the orientations of the *c*-axes (or 0001 planes) of the quartz grains in a metamorphosed psammite rock. These lie in a zone with a zone axis approximately parallel to the lineation of the rock (from *Introduction to Geology* by H. H. Read and Janet Watson, Macmillan, 1968).

cosines, the ratios of which give us the Miller index of the plane in the crystal on which the feature lies (Section 5.5). (In this example, compared with Fig. 12.10 it is clearly $(\bar{1}11)$, i.e. an annealing twin (see Fig. 1.21(d)).)

The trace analysis technique may be applied, as it were, the other way round: we may recognize the interface-planes of twins or stacking faults and from their traces determine the Miller indices of the crystal surface(s).

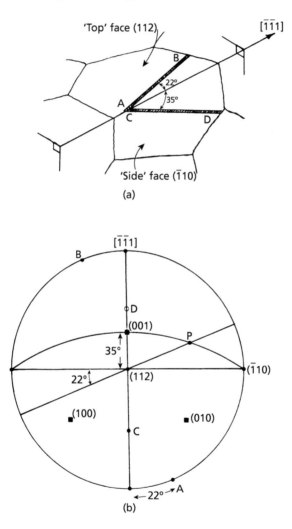

Fig. 12.16. (a) A grain in a specimen of α-brass showing traces of a planar feature AB on the 'top' face (112) and CD on the 'side' face ($\bar{1}$10). The line of intersection of the faces is [$\bar{1}\bar{1}$1]. (b) a stereo-graphic projection with (112) in the centre and the line of intersection vertical. The great circles normal to the directions AB and CD are indicated. The pole of the planar feature P lies at the intersection of these two great circles. The ratios of the direction cosines to the (100), (010), and (001) poles give its the Miller index.

If we only see the trace of a planar feature on a single surface we have only one great circle and the pole cannot be fully determined. Here, we might make a 'reasonable' guess as to where it might be or, if we see the traces intersecting both the top and bottom surfaces of (e.g.) an electron microscope specimen, we may determine the pole from an estimate of the specimen thickness and displacement of the traces.

Exercises

12.1 Prove that the stereographic projection of a small circle on the stereographic sphere is again a circle. (*Hint*: this is an exercise in Euclidean geometry.)

12.2 Use the Wulff net to prepare a projection of a cubic crystal with (001) in the centre as in Fig. 12.6. Find, by drawing in additional great circles, the positions of the (212) and ($\bar{3}\bar{3}2$) planes, measure their angles α, β and γ to the x, y and z axes and hence check that the ratios of their direction cosines $\cos\alpha : \cos\beta : \cos\gamma$ are in the ratios of $h : k : l$ in each case (see Section 5.5).

12.3 Two planes in a cubic crystal make the following angles to the crystal axes

	x-axis	y-axis	z-axis
Plane A	$+36.5°$	$+57.5°$	$+74.5°$
Plane B	$-48.2°$	$-70.5°$	$+48.2°$

Find the positions of these planes on a standard projection from the common intersections of small circles centred about the x, y and z-axes, and determine their Miller indices. Using the Wulff net measure the angles between: Plane A and Plane B; Plane A and ($\bar{1}\bar{1}1$); Plane B and (101); Plane B and ($\bar{1}01$).

12.4 Using sketch stereograms as in Fig. 12.12, show that (a) $\bar{2} \equiv$ a perpendicular mirror plane; (b) $\bar{3} \equiv 3 +$ a centre of symmetry, and (c) $\bar{6} \equiv 3 +$ a perpendicular mirror plane. It is a useful exercise to represent at the same time the poles according to the symbolism used for space groups.

13

Fourier analysis in diffraction and image formation

13.1 Introduction—Fourier series and Fourier transforms

In many areas of physics we encounter physical events, characteristics or properties which repeat at regular intervals and which may be represented by waveforms of various shapes and complexity. Perhaps the most familiar examples are the sound waves produced by musical instruments, where the waveforms represent air pressure variations and the repeat distance or wavelength represents the frequency or pitch of the fundamental note. Figure 13.1 shows three such waveforms; the characteristic 'sound' of each instrument is related to the shape of the waveform in each case. Another example is a simple diffraction grating which has a waveform consisting of flat-topped peaks (corresponding to the light from the slits) with dark regions in between (Fig. 13.12(a) below). Such waveforms are called *periodic functions* and are also of great importance in crystallography because, as we shall see below, crystal structures can be represented in terms of periodic variations of *electron density*, the density peaking at the atom positions and falling to low values in between.

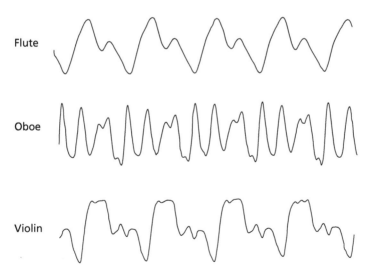

Fig. 13.1. Examples of waveforms of musical instruments (from *Measured Tones* by I. Johnston, Adam Hilger (Institute of Physics), Bristol, 1989).

The Basics of Crystallography and Diffraction. Fourth Edition. Christopher Hammond.
© Christopher Hammond 2015. Published in 2015 by Oxford University Press.

Jean Baptiste Fourier* showed that any periodic waveform or function could be ana-lysed in terms of a series of cosine or sine waveforms (or, more generally, a series of exponential functions). First, a 'constant' term (representing, in sound waves, the un-disturbed air pressure or, in crystals, a uniform 'sea' of electrons); then a term with a wavelength or repeat distance corresponding, in sound, to the fundamental note, or in a crystal to the lattice repeat distance, then a term of half the wavelength (or twice the frequency) corresponding, in sound, to the first harmonic, and so on. Each of these co-sine or sine waves is defined by its amplitude and phase—the phase representing the displacements of the waves with respect to each other.

The process of analysing a complex waveform into a series of such components is called *Fourier analysis*. Conversely, the generation of a complex waveform from such components is called *Fourier synthesis*. Figure 13.2 shows two examples in which the complex waveforms (a) and (b) are synthesized from, or analysed into, three terms: a constant term, a fundamental term and a first harmonic term. Notice that in (a) there is a phase difference between the component waves which gives rise to the synthesized asymmetrical waveform whereas in (b) there is zero (or a 180° phase difference de-pending as to where you take the origin) which gives rise to the symmetrical waveform. In crystallography (a) corresponds to electron density variations in a crystal without a centre of symmetry whereas (b) corresponds to a crystal with a centre of symmetry (or a mirror plane or diad axis normal to the direction in which the electron density is measured).

The amplitude A versus distance x of a periodic function is represented mathematic-ally by a Fourier series:

$$A_x = A_0 + A_1 \cos\left(2\pi 1\left(\frac{x}{a}\right) - \alpha_1\right) + A_2 \cos\left(2\pi 2\left(\frac{x}{a}\right) - \alpha_2\right) + \cdots$$
$$+ A_n \cos\left(2\pi n\left(\frac{x}{a}\right) - \alpha_n\right)$$

i.e.

$$A_x = \sum_{n=0}^{n=N} A_n \cos\left(2\pi n\left(\frac{x}{a}\right) - \alpha_n\right)$$

where a is the repeat distance, the amplitudes A_1, A_2, \ldots, A_n are called the Fourier co-efficients and $\alpha_1, \alpha_2, \ldots, \alpha_n$ are the phase angles. The Fourier series may equally well be written in terms of sines or, more generally, exponential terms, but in cases where the waveform is symmetrical (as in crystals with a centre of symmetry or a simple dif-fraction grating and in which the phase angles are zero or 180°) we shall find the cosine form more convenient: a cosine is a symmetrical function, i.e. $\cos x = \cos(-x)$ whereas a sine is an asymmetrical function, i.e. $\sin x = -\sin(-x)$. Hence the Fourier series may be written:

$$A_x = A_0 + A_1 \cos 2\pi 1\left(\frac{x}{a}\right) + A_2 \cos 2\pi 2\left(\frac{x}{a}\right) + \cdots + A_n \cos 2\pi n\left(\frac{x}{a}\right)$$

* Denotes biographical notes available in Appendix 3.

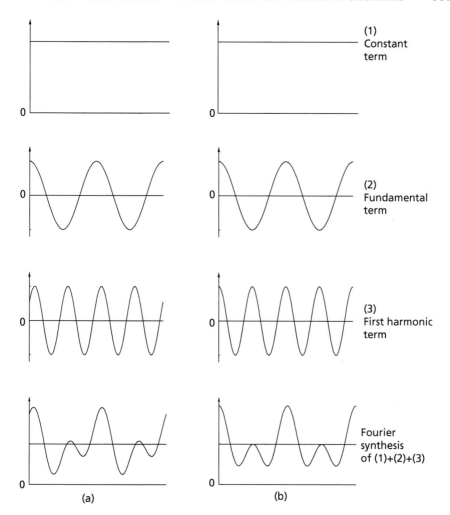

Fig. 13.2. Fourier synthesis of a complex waveform from three components: a constant term (1), a fundamental term (2) and a first harmonic (3). The asymmetrical waveform in (a) arises from the phase difference between the fundamental term and first harmonic. (Courtesy of Mr D.G. Wright.)

i.e.

$$A_x = \sum_{n=0}^{n=N} A_n \cos 2\pi n \left(\frac{x}{a}\right).$$

The first term A_0 represents, in effect, a wave of infinite wavelength (in sound the undisturbed air pressure), the second a wave of amplitude A_1, and wavelength a (in sound the fundamental note), the third a wave of amplitude A_2 and wavelength $a/2$ (in sound the first harmonic), and so on. The Fourier coefficients A_0, A_1, A_2, etc. may also be

represented as components on a frequency scale A_0 (zero frequency), A_1 (frequency $2\pi/a$), A_2 (frequency $4\pi/a$), and so on.

Finally, there is a very important development in Fourier analysis (though not envisaged by Fourier himself) and that is that it can also be applied to *non-periodic* functions such as a single sound pulse or the distribution of light from a single slit or aperture. This may be done by envisaging a non-periodic function as, in effect, a periodic function with a fundamental repeat distance that approaches infinity. In this case the Fourier coefficients, represented as components on a frequency scale, approach closer and closer together (the frequency differences become smaller and smaller) until they merge together. The envelope of all the merged-together Fourier coefficients is called the *Fourier transform* and is treated mathematically as an integral, rather than a summation of the Fourier series for a periodic function. As we shall see, the diffraction pattern from a single slit or aperture (expressed as an amplitude, rather than an intensity distribution) and the Fourier transform are synonymous.

13.2 Fourier analysis in crystallography

In our derivation of the structure factor F_{hkl} (Section 9.2), we considered the atoms to be discrete scattering centres, the atoms being located at specific points in the unit cell denoted, for the nth atom, by the fractional coordinates $(x_n \ y_n \ z_n)$. The atomic scattering factor f_n was derived by considering the interference of all the waves scattered by the Z electrons in the atom and then the structure factor F_{hkl} was obtained by considering the interference of all the scattered waves from all the atoms in the unit cell. Hence we obtained the relation:

$$F_{hkl} = \sum_{n=0}^{n=N} f_n \exp 2\pi i \, (hx_n + ky_n + lz_n).$$

We are now going to determine F_{hkl} starting, as it were, from a different standpoint. Since it is the electrons which scatter X-rays then we are really detecting the distribution, or variation in the 'density' of electrons throughout the unit cell, the electron density being greatest at the atom centres and falling to low values in between. In short we have a three-dimensional periodic variation in electron density, the variation being repeated (as for the lattice) from cell to cell.

It is not easy to represent such electron density variations in three dimensions except for the simplest crystals. It is much easier to represent the *projection* of electron density on to one face of the cell—just as we represented crystal structures in terms of projections or plans (Section 1.8). Figure 13.3 shows a 'classic' example: a projection on the (010) face of the monoclinic mineral diopside, $CaMg(SiO_3)_2$. Figure 13.3(a) shows the atom positions and Fig. 13.3(b) the corresponding electron density 'contour map'—the peaks corresponding to the heavy, overlapping Ca and Mg atoms and the diffuse 'ridges' in between corresponding to the distribution of the lighter O and Si atoms.

We now need to replace the atomic scattering factor f_n in the equation by an equivalent electron-density term. Let ρ_{xyz} be the electron density (number of electrons per

Fig. 13.3. (a) Projection of the atom positions in diopside, $CaMg(SiO_3)_2$, on to the (010) face and (b) the corresponding electron density 'contour map'. Also indicated is a volume element dV, at distance coordinates xyz and edge-lengths dx, dy, dz. (Adapted from W.L. Bragg, *Zeitschrift für Kristallographie* **70**, 475, 1929.)

unit volume) at *distance* coordinates xyz in the unit cell. We use these coordinates in preference to the *fractional* coordinates used in Chapter 9, i.e. x (fractional) $= x$ (distance)/a and similarly for y and z. Now we outline a small volume element with edges dx, dy, dz parallel to the unit cell axes a, b, c as indicated in Fig. 13.3(b). Let dV be the volume of this element and V the volume of the unit cell. The *ratio* dV/V equals the *ratio* $dxdydz/abc$, whence $dV = \dfrac{V}{abc}dxdydz$ (for orthogonal axes of course $dV = dxdydz$). The number of electrons in the volume element is simply $\rho_{xyz} dV = \dfrac{V}{abc}\rho_{xyz} dxdydz$ which we now replace for f_n in the structure factor equation. However, since ρ_{xyz} is a continuous function we must also replace the summation by a triple integral, i.e. along all three axes. Hence the structure factor equation becomes:

$$F_{hkl} = \frac{V}{abc} \int_0^a \int_0^b \int_0^c \rho_{xyz} \exp 2\pi i \left[h\left(\frac{x}{a}\right) + k\left(\frac{y}{b}\right) + l\left(\frac{z}{c}\right) \right] dxdydz.$$

Fig. 13.4. Representation of electron density as a 'slab' of thickness dx parallel to one set of lattice planes of spacing (repeat distance) a.

In order to proceed further with this equation we need to express the electron density ρ_{xyz} in terms of a three-dimensional Fourier series. However, without loss of principle, we shall carry out the analysis in one dimension (corresponding to reflections from one set of crystal planes) and also for a crystal with a centre of symmetry in which the phase angles of the Fourier components are all zero or $180°$ (i.e. the signs of the Fourier coefficients are all $+$ or $-$).

Consider Bragg reflection from one set of lattice planes (100) with a spacing $a\,(= d_{100}$ for orthogonal axes), Fig. 13.4. As we know it is a consequence of Bragg's law (Fig. 8.3) that the interference conditions for the scattered waves are the same irrespective of the positions of the atoms parallel to the planes. Referring to the simple example in Fig. 9.2(b) the atom f_1 may be located anywhere along the dashed line: the phase angle ϕ_1 is solely determined by the component of \mathbf{r}_1 perpendicular to the planes. Hence (Fig. 13.4) we can only specify the electron density, ρ_x, in the x-direction and must therefore replace the volume element with a 'slab' or 'layer' of thickness dx and nominal unit area. The number of electrons in the slab is then simply $\rho_x dx$.

We now express ρ_x in terms of a one-dimensional Fourier series:

$$\rho_x = A_0 + A_1 \cos 2\pi 1 \left(\frac{x}{a}\right) + A_2 \cos 2\pi 2 \left(\frac{x}{a}\right) + \cdots + A_n \cos 2\pi n \left(\frac{x}{a}\right)$$

$$\rho_x = \sum_{n=0}^{n=N} A_n \cos 2\pi n \left(\frac{x}{a}\right).$$

Now we write the structure factor equation for F_{h00} for a crystal with a centre of symmetry and also using distance coordinates instead of fractional coordinates:

$$F_{h00} = \sum_{n=0}^{n=N} f_n \cos 2\pi h \left(\frac{x_n}{a}\right).$$

Substituting f_n by $\rho_x dx$ and replacing, as before, the summation by an integral (ρ_x is a continuous function) we have:

$$F_{h00} = \int_{x=0}^{x=a} \rho_x \cos 2\pi h \left(\frac{x}{a}\right) dx.$$

Then, substituting for ρ_x and interchanging the integral and summation signs we have:

$$F_{h00} = \sum_{n=0}^{n=N} \int_{x=0}^{x=a} A_n \cos 2\pi n \left(\frac{x}{a}\right) \cos 2\pi h \left(\frac{x}{a}\right) dx.$$

Now, remembering the trigonometrical identity:

$$\cos A \cos B = \frac{1}{2} [\cos (A + B) + \cos (A - B)]$$

we have:

$$F_{h00} = \frac{1}{2} \sum_{n=0}^{n=N} A_n \left[\int_{x=0}^{x=a} \cos 2\pi (n + h) \left(\frac{x}{a}\right) dx + \int_{x=0}^{x=a} \cos 2\pi (n - h) \left(\frac{x}{a}\right) dx \right].$$

Now, since n and h are integers the integrals express the total area for one complete cycle of the cosine curve and are thus zero since the areas above and below zero cancel out *except* for the cases where $(n + h)$ and $(n - h)$ are zero, in which case the integrals are $\int_{x=0}^{x=a} \cos 0 \, dx = a$.

Hence, the equation reduces as follows:

$$\text{When } h = n = 0; \quad F_{000} = \frac{1}{2} A_0 [a + a] = A_0 a$$

$$\text{When } n = h; \quad F_{h00} = \frac{1}{2} A_h [0 + a] = A_h \, a/2$$

$$\text{When } n = -h; \quad F_{\bar{h}00} = \frac{1}{2} A_{\bar{h}} [a + 0] = A_{\bar{h}} \, a/2.$$

This is a profound result, first envisaged by W.H. Bragg in 1915. It shows that each Fourier coefficient is equal (apart from the scaling factor a) to the corresponding structure factor. In the general case, where the structure factors are complex numbers, so also are the Fourier coefficients and for reflections hkl the relations are written $F_{hkl} = VA_{hkl}$ where V is the volume of the unit cell.

We may now rewrite the Fourier series equation for ρ_x substituting for $A_0 = (1/a)F_{000}$, $A_h = (2/a)F_{h00}$ and $A_{\bar{h}} = (2/a)F_{\bar{h}00}$ and summing for both positive and negative values of h:

$$\rho_x = \frac{1}{a} \sum_{h=-N}^{h=+N} F_{h00} \cos 2\pi h \left(\frac{x}{a}\right)$$

For three dimensions we have:

$$\rho_{xyz} = \frac{1}{V} \sum_h \sum_k \sum_l F_{hkl} \cos 2\pi \left(h\left(\frac{x}{a}\right) + k\left(\frac{y}{b}\right) + l\left(\frac{z}{c}\right) \right)$$

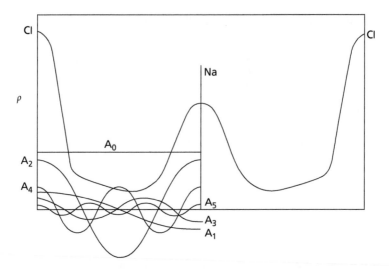

Fig. 13.5. The variation in electron density ρ normal to the (111) planes in NaCl. On the left are shown the Fourier components A_0, \ldots, A_5, each of which correspond to the structure factors $F_{000}, 2F_{111}, \ldots, 2F_{555}$. (Adapted from *X-ray Analysis of Crystals* by J.M. Bijvoet, N.H. Kolkmeyer and C.H. MacGillavry, Butterworths, London, 1951.)

and $\rho_{xyz} = \frac{1}{V} \sum\limits_{h}^{N} \sum\limits_{k} \sum\limits_{l} F_{hkl} \exp{-2\pi i \left(h\left(\frac{x}{a}\right) + k\left(\frac{y}{b}\right) + l\left(\frac{z}{c}\right) \right)}$ for a crystal without a centre of symmetry at the origin. Notice that the minus sign in the exponent arises because the expressions for F_{hkl} and ρ_{xyz} are the Fourier transforms of each other. To summarize, the electron density ρ_{xyz} expressed as a Fourier series with the F_{hkl}s as coefficients is the Fourier transform of the array of structure factors.

Figure 13.5 shows a one-dimensional Fourier synthesis of the electron density perpendicular to the (111) planes of NaCl which consist of alternate layers of Na and Cl atoms. The electron density, which is symmetrical about both Na and Cl atoms, is the sum of five components, A_0, \ldots, A_5 with coefficients corresponding to F_{000}, $2F_{111}, \ldots, 2F_{555}$. For h, k, l all even $F_{hkl} = 4(f_{Cl} + f_{Na})$ and for h, k, l all odd $F_{hkl} = 4(f_{Cl} - f_{Na})$. Hence, using f vs. $\sin\theta/\lambda$ data from the *International Tables* the F_{hkl} values and hence the values of the Fourier coefficients can be determined.

This of course is a simple case because all of the structure factors are real and positive (origin at a Cl atom). Notice that the electron-density curve lies equally above, and below, the A_0 term which represents a uniform 'sea' of electron density.

Hopefully, it is now easy to see how atoms are located at the intersecting maxima of electron density waves from different sets of reflections. Figure 13.6 shows just two waves, $F_{020} \cos 2\pi 2(y/b)$ and $F_{300} \cos 2\pi 3(x/a)$ (where a and b are the cell edge lengths) and (below) their sum and projection as a contour map. Even at this low level of resolution the locations of the atom centres are evident.

A similar procedure was used by W.L. Bragg to locate the atom positions in diopside. By an ingenious photographic process he represented the sinusoidal electron density

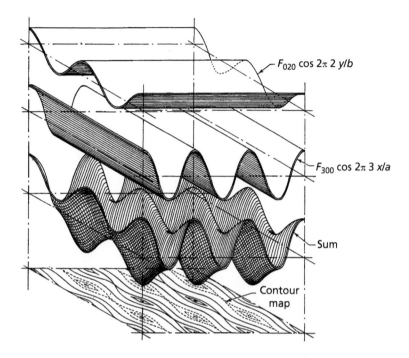

$F_{020} \cos 2\pi\, 2\, y/b$

$F_{300} \cos 2\pi\, 3\, x/a$

Sum

Contour map

Fig. 13.6. A two-dimensional contour map synthesized from just two Fourier components (upper curves) and their sum (lower curve). Even at this low level of resolution the contour map shows discrete maxima (from *X-ray Crystal Structure* by D. McLachlan, McGraw-Hill Book Company, 1957.)

waves as a series of light and dark bands (rather than representing them graphically as in Fig. 13.6. Figure 13.7(a)–(d) shows four such bands projected on the (010) face (as in Fig. 13.3) and Fig. 13.7(e) shows their sum. Again, even at this low level of resolution, the atom positions are clearly evident. (This representational technique, it should be noted, is not 'out of date' but can be readily carried out today by computer simulation.)

13.2.1 X-ray resolution of a crystal structure

The resolution of a crystal structure—i.e. the resolution of closely spaced atoms—is defined in exactly the same way as that for a microscope image (Section 7.5), i.e. $a \approx d = 0.5\lambda/\mathrm{NA} = 0.5\lambda/n \sin i$. Since the refraction of X-rays is very weak, $n \approx 1$, hence this equation is simply a restatement of Bragg's law, $d = \lambda/2 \sin\theta$. Smaller values of d, i.e. larger values of $\sin\theta$, correspond to the inclusion of an increasing number of terms in the Fourier synthesis (Fig. 13.2). Figure 13.8 shows (001) projections of electron density maps for an azidopurine at resolutions (a) 1.5 Å (71 terms) and (b) 0.8 Å (264 terms). For CuK$_{\alpha 1}$ radiation ($\lambda = 1.5405$ Å), these maps correspond to reflections measured up to 2θ values of 62° and 149°. Clearly, in order to achieve a greater resolution, X-rays of smaller wavelength, e.g. MoK$_{\alpha 1}$ ($\lambda = 0.7093$ Å), must be used giving a resolution of ≈ 0.4 Å. This is still greater than the resolution now achieved in high-resolution electron microscopy.

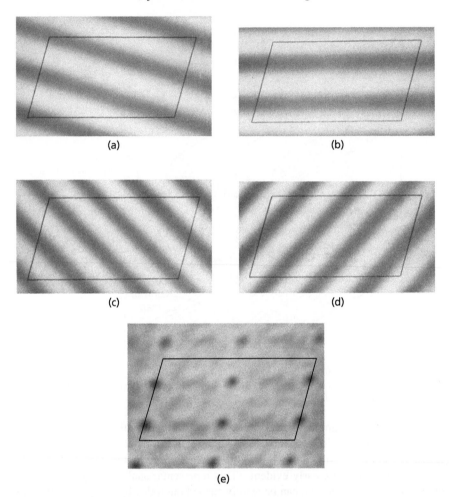

Fig. 13.7. The sinusoidal variations (light and dark) of the Fourier components corresponding to (a) F_{102} (−ive), (b) F_{002} (+ive), (c) F_{302} (−ive), and (d) $F_{30\bar{1}}$ (+ive) in diopside—a centrosymmetric crystal. Hence the phases are either +ive (light at origin) or −ive (dark at origin). (e) shows the sum of these four Fourier components. Compare with the higher-resolution electron density map in Fig. 13.3(b). (From W.L. Bragg, 1929, *Zeitschrift für Kristallographie* **70** p. 475.) *Note* in (a)–(d) the contrast should be *reversed:* W.L. Bragg evidently shows here the black-white *negative* contrast of the projection masks.

13.3 The structural analysis of crystals and molecules

Why water boils at 100° and methane at −161°, why blood is red and grass is green, why diamond is hard and wax is soft, why graphite writes on paper and silk is strong, why glaciers flow and iron gets hard when you hammer it, how muscles contract, how sunlight makes plants grow and how living organisms have been able to evolve into ever more complex forms... The answers to all these problems have come from structural analysis.

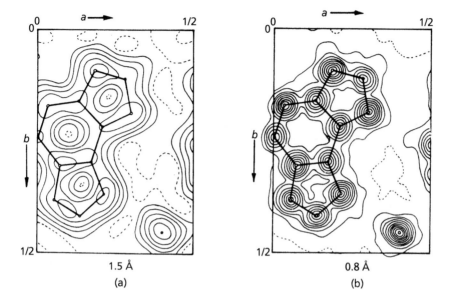

Fig. 13.8. (001) projections of electron density maps for (one quarter) of the unit cell of azidopurine at resolutions (a) 1.5 Å and (b) 0.8 Å. (From *Crystal Structure Analysis: A Primer* by J.P. Glusker and K.N. Trueblood, 3rd edn, 2010, Oxford Science Publications/IUCr. Oxford)

This extract, from a letter (1996) of Max Perutz, to the historian Gerald Holton, encapsulates the supreme importance of structural analysis in our understanding of the nature and operations of the physical world around us, both inorganic and organic. The achievements of structural analysis are, in turn, dependent upon success in overcoming the phase problem: the problem set by nature that only allows us to measure the intensities, $|F_{hkl}|^2$ of diffracted beams.

In this section, we briefly introduce some of the techniques, both experimental and theoretical, by which the phase problem has been overcome. We should note that the subject is expanding rapidly experimentally in that the use of high-energy synchrotron radiation allows the detection and analysis of short-lived structures in chemical reactions (reintroducing a time element) and theoretically in the use of highly sophisticated computer programs that form the basis of the so-called direct and charge flipping methods.

13.3.1 Trial and error methods

These stem from the earliest days of X-ray crystallography. A possible crystal structure is postulated based on the suspected chemical configuration, the possible packing arrangements and interatomic distances between atoms and molecules, and the restrictions on their positions imposed by space group symmetry considerations determined, or partly determined, from the X-ray diffraction evidence. For example, if a centrosymmetric structure contains an odd number of atoms of a particular species then they must

Table 13.1 W.L. Bragg's (1913) observed and calculated intensities for the X-ray reflections from iron pyrites.

Iron pyrites	Observed					Calculated				
Spectra	1st	2nd	3rd	4th	5th	1st	2nd	3rd	4th	5th
(100)	100	0	0	14	4	100	0	0	8	5
(110)	100	50	–	–	–	100	50	11	13	–
(111)	80	100	50	0	22	78	100	35	0	21

be related by a centre of symmetry, one of them situated at the centre of symmetry itself. Similarly, in structures containing mirror planes the atoms must by symmetrically related either side of the mirror plane—and if there is one atom of a particular species it must lie on the mirror plane itself. In short, trial and error methods require the crystallographer to have at his or her disposal an armoury of chemical, physical, and structural knowledge, and also those unquantifiable characteristics—intuition and imagination. And although trial and error methods as such have been superseded almost wholly by the methods described next, a crystallographer still needs 'intuition and imagination'—as is clearly evident in Crick and Watson's discovery of the structure of DNA.

Having postulated a structure, the (moduli of the) calculated structure factors are compared with those observed and the 'correctness' of the structure is estimated by a number called the **Reliability Index**, R, defined as the difference between the sum of the moduli of the observed and calculated structure factors divided by the sum of the moduli of the observed structure factors. R-values for structures that are deemed to be correct lie in the range 0.02–0.06.

The first structure determination by trial and error was that of FeS_2 by W.L. Bragg in 1913 (Section 4.7 and Exercise 9.8). Table 13.1 is extracted from his paper *The Analysis of Crystals by the X-ray Spectrometer* (Proc. Roy. Soc. **A89**, pp. 468–487) and in which, from the $|F|^2$ values, an R-value of 0.08 can be calculated. Brilliant for 1913!

13.3.2 The Patterson function: Patterson or vector maps

The Patterson* function has its origins in the X-ray diffraction analysis of amorphous materials, e.g. liquids and glasses, in which the positions of individual atoms are not measurable. What is measurable, in any radial direction, are the average or mean spacings between them—principally between first and second nearest neighbours. Such information provides evidence for the nature of the atomic packing and the nature of the bonding between the atoms in the liquid or glass.

The Patterson method, introduced in 1935, in conjunction with the heavy atom and isomorphous replacement techniques described in Section 13.3.3 that follows represented a major breakthrough in crystal structure determination and played a key role in (e.g.) Dorothy Crowfoot Hodgkin's* analysis of insulin and vitamin B_{12} and

* Denotes biographical notes available in Appendix 3.

Max Perutz's analysis of haemoglobin: both works of epic proportions in time and effort that put the epics of the Greek myths into the shade! We should also note that these, and other successes in the determination of the structure of biological macromolecules, depend in turn on the experimental skills needed to prepare crystals of sufficient size to give sharp X-ray diffraction peaks: the essential or hoped-for requirement being that all the molecules are oriented in the same way (see Section 4.8). As described by the Scherrer equation (Section 9.3), the smaller the crystal the greater the degree of line broadening (and similarly of course for a diffraction grating with a limited number of lines: see Section 13.3).

First, we describe the construction of a Patterson or vector map and then derive the equation (the Patterson function) upon which it is based. Although (logically) this is the 'wrong way round' it does help us to appreciate immediately the physical significance of the derivation.

Figure 13.9(a) shows a projection of two unit cells of a crystal structure containing just three atoms A, B, and C, and Fig. 13.9(b) shows the corresponding vector map. This is constructed by drawing the six interatomic vectors between the atoms, i.e. A→C, C→A, A→B, etc. All these vectors are referred to an origin or corner of the unit cell

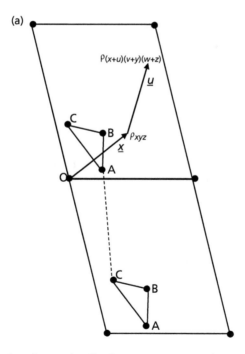

Fig. 13.9. (a) A projection of two unit cells of a non-centrosymmetric structure consisting of three atoms A, B, and C. The vectors **x** and **u**, linked 'tip to tail', are swept throughout the whole volume of the unit cell (varying both **x** and **u**) and the values of **u** that correspond to interatomic vectors are recorded (including the special case where **u** = 0).

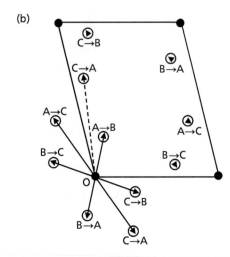

Fig. 13.9. (b) The resulting Patterson or vector map; the interatomic vectors A→C, C→A, A→B are shown by the solid lines referred to one origin. These are repeated (only tips of vectors shown) referred to the three other corners in the projection. Note that any vector from **any** origin to **any** such peak corresponds to an interatomic vector, for example the dashed vector C→A is a vector between atoms in neighbouring unit cells. Each corner or origin of the unit cell corresponds to the **zero** interatomic vectors A→A, B→B, C→C. The map (unlike the structure itself) has a centre of symmetry at the origin and centre of the unit cell.

of the vector map as shown by the full lines. Next, we complete the vector map for this unit cell, by drawing the 'tips' of these vectors referred to the other corners of the unit cell and labelling them accordingly.

There are two things to be noticed about this map. First it is centrosymmetrical: there is a centre of symmetry at the origin and centre of the cell. Second **any** vector from **any** origin of the unit cell corresponds to an interatomic vector in the structure. One such vector, which links atoms C→A in adjacent unit cells, is shown by the dashed line. Third, each origin corresponds to the **zero** interatomic vector between an atom and itself, i.e. A→A, B→B, C→C. To summarize: the map shows the vectors **between** atoms but not the actual **positions** of atoms in the unit cell.

In a Patterson map each vector 'tip', and each origin, corresponds to a peak in electron density. In our three-atom example there are a total of nine such peaks, three at each origin and six within the cell. In general, for a structure consisting of n atoms there are n^2 peaks of which n are at the origin and $(n^2 - n)$ within the cell. Clearly, for a structure of any complexity there will be an enormous number of peaks, many of which may be indistinguishable due to overlapping. Figure 13.10 shows a relatively simple example: Fig. 13.10(a) shows a projection of the structure of azidopurine and Fig. 13.10(b) the corresponding Patterson map. In both figures some of the corresponding vectors are indicated.

Now to derive the Patterson function. Figure 13.9(a) shows two vectors **x** and **u** linked 'tip to tail'. ρ_{xyz} is the electron density at the tip of **x** (or tail of **u**) and $\rho_{(x+u),(y+v),(z+w)}$ is the electron density at the tip of **u**. Now consider the product of these electron densities that we call $p_u = \rho_{xyz} \cdot \rho_{(x+u),(y+v),(z+w)}$. p_u is everywhere zero **except** when **u** connects two regions of high electron density (i.e. when **u** is an interatomic vector **or** (special case) when **u** is zero and the tip/tail of **u** corresponds to a region of high electron density (this special case gives rise to the origin peaks of the Patterson map). Now we 'sweep' these two vectors throughout the unit cell, varying both

Fig. 13.10. (*continued*)

Fig. 13.10. (a) A projection of the structure of azidopurine and (b) the corresponding two-dimensional Patterson map. Some strong peaks are labelled A, B, etc., and the corresponding interatomic vectors from ring to ring shown in (a). (From *Crystal Structure Analysis: A Primer* loc. Sic.)

x and **u** and recording the non-zero values of p_u. However, without loss of principle we shall only do this for the one-dimensional case, i.e. for ρ_x and $\rho_{(x+u)}$ along the a-axis of the unit cell. Writing ρ_x and $\rho_{(x+u)}$ in terms of a Fourier series with F_{hoo} as coefficients (see Section 13.2) we have:

$$\rho_x = \frac{1}{a}\sum_h F_{hoo}\exp{-2\pi ih\left(\frac{x}{a}\right)} \text{ and } \rho_{x+u} = \frac{1}{a}\sum_h F_{h00}\exp{-2\pi ih\left(\frac{x+u}{a}\right)}$$

The Patterson function is the product P_u obtained by integrating x from 0 to a:

$$P_u = \int_0^a \rho_x \cdot \rho_{x+u} dx = \int_0^a \left(\frac{1}{a} \sum_h F_{hoo} \exp -2\pi ih \left(\frac{x}{a} \right) \right)$$

$$\times \left(\frac{1}{a} \sum_{h'} F_{h'oo} \exp -2\pi ih' \frac{(x+u)}{a} \right) dx$$

The two terms in this product have, in general, different indices that we distinguish as h and h'.

We now separate out the variable part of the exponential:

$$P_u = \int_0^a \left(\frac{1}{a} \sum_h F_{hoo} \exp -2\pi ih \left(\frac{x}{a} \right) \right) \left(\frac{1}{a} \sum_{h'} F_{h'oo} \exp -2\pi ih' \left(\frac{x}{a} \right) \exp -2\pi ih' \left(\frac{u}{a} \right) \right) dx$$

$$= \frac{1}{a^2} \sum_h \sum_{h'} F_{hoo} F_{h'oo} \exp -2\pi ih' \left(\frac{u}{a} \right) \int_0^a \exp -2\pi i (h + h') \left(\frac{x}{a} \right) dx$$

When $h = -h'$ the exponential reduces to zero and value of the integral term is a. For all other values it is zero, since the only non-vanishing terms are those for which $h = -h'$. Hence the equation reduces to:

$$P_u = \frac{1}{a^2} \sum_h F_{hoo} F_{\bar{h}oo} \exp -2\pi i(-h) \left(\frac{u}{a} \right) a = \frac{1}{a} \sum_h |F_{hoo}|^2 \exp 2\pi ih \left(\frac{u}{a} \right)$$

(since F_{hoo} and $F_{\bar{h}oo}$ are complex conjugates, i.e. $F_{\bar{h}oo} = F^*_{hoo}$)

Now, in the summation, the sine terms in the exponential:

$$\exp 2\pi ih \left(\frac{u}{a} \right) = \cos 2\pi h \left(\frac{u}{a} \right) + i \sin 2\pi h \left(\frac{u}{a} \right)$$

cancel out because:

$$|F_{hoo}|^2 i \sin 2\pi h \left(\frac{u}{a} \right) = -|F_{\bar{h}oo}|^2 i \sin 2\pi \bar{h} \left(\frac{u}{a} \right).$$

Hence, the Patterson function only contains the cosine terms:

$$P_u = \frac{1}{a} \sum |F_{hoo}|^2 \cos 2\pi h \left(\frac{u}{a} \right)$$

This function has the same value for $+u$ and $-u$ and so it is symmetrical about the origin. In three dimensions the Patterson function becomes:

$$P_{uvw} = \frac{1}{V} \sum_h \sum_k \sum_l |F_{hkl}|^2 \cos 2\pi \left(h \left(\frac{u}{a} \right) + k \left(\frac{v}{b} \right) + l \left(\frac{w}{c} \right) \right)$$

where V is the volume of the cell. This function is usually expressed in terms of **fractional** coordinates (i.e. u (fractional) = $\frac{u}{a}$—see Section 13.2, p. 333), in which case it is written:

$$P_{uvw} = \frac{1}{V} \sum_h \sum_k \sum_l |F_{hkl}|^2 \cos 2\pi (hu + kv + lw)$$

The important point is that the Patterson function is based only upon observed intensities, $|F_{hkl}|^2$; the 'gain' of not needing to know the phases of the reflections is offset by the 'loss' of not knowing the atom positions, only, as described earlier, the vectors between them. Furthermore, since the peaks of the Patterson function are the products of the electron densities of two atoms, say A and B in Fig. 13.9, then the intensity or height of the peak is proportional to $Z_A Z_B$. Hence a carbon–carbon pair will give a non-origin peak of (relative) value of $6 \times 6 = 36$, a carbon–chlorine peak a (relative) value of $6 \times 17 = 102$ and so on. (At the origin of course there are many overlapping peaks.) However, again due to overlapping, such combinations may not be readily identifiable. One or more heavy atoms in a structure, either 'native' to the molecule or introduced experimentally, may give rise to strong, clearly identifiable peaks. This is the basis of the heavy atom and isomorphous replacement techniques.

13.3.3 Interpretation of Patterson maps: heavy atom and isomorphous replacement techniques

The earlier discussion represents 'only the beginning' in the interpretation and analysis of Patterson maps. First, although there are only 24 Patterson space groups (see Table A7.1), detailed analysis of the peak positions can lead to the identification of the space group of the crystal. Second, instead of using $|F_{hkl}|^2$ values in the equation we use the corrected values (known as $|E_{hkl}|^2$ see Section 13.3.4) that enhance the contribution to the pattern of the weak high θ reflections. But most important are the heavy atom and isomorphous replacement techniques that, in conjunction with the Patterson method, were used in the analysis of the first protein structures. Heavy atoms are introduced into the structure in the hope that they (a) will be located at specific sites and (b) will not significantly change the structure. For small molecules the heavy atoms will give strong, clearly recognizable peaks in the map from which their interatomic vectors may be identified, and, given a knowledge of the stereochemistry of the molecule, their positions within the unit cell identified. Furthermore, the presence of heavy atoms helps to 'tie down' the range of values of the phase angles of the reflections. Consider the Argand diagram (Fig. 13.11) for a crystal that contains one heavy atom and a number of light atoms. As can be seen, $f_{(\text{heavy atom})}$ is so large in comparison with $F_{(\text{light atoms})}$ that the resultant phase F is close to, or not far removed from, that of the heavy atom alone. In short, the phase of the heavy atom reflection acts as a **constraint** on the phases of the reflections of the whole crystal structure (Section 13.3.4).

Isomorphous crystals are those with essentially identical cell dimensions and atomic arrangements but with variations the types of one or more of the atoms present. There are innumerable examples in inorganic crystals: in minerals in particular isomorphous

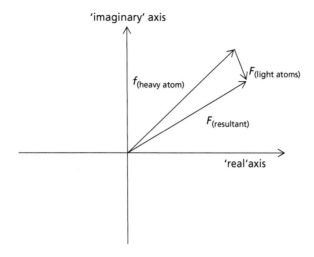

Fig. 13.11. Argand diagram for a structure with one heavy atom, $f_{(heavy\ atom)}$ and a (large) number of light atoms, the sum of the atomic scattering factors of which is shown as $F_{(light\ atoms)}$. The phase of the resultant structure factor, F, is close to, or deviates slightly from, that of the heavy atom alone.

replacement is the rule rather than the exception. For example in the orthopyroxenes space group *Pbca* (inosilicates, Section 1.11.4, p. 40) the composition can range from $MgSiO_3$ (enstatite) to $(Mg, Fe)\ SiO_3$ (hypersthene) to $FeSiO_3$ (orthoferrosilite); the changes in the Mg/Fe ratios being accompanied by small changes in the lattice parameters. Isomorphous crystals give diffraction patterns that differ only in the intensities of the reflections (arising from the different scattering powers of the different atoms) and from which the atom positions can be located.

Heavy atom isomorphous replacement methods may involve single atoms or molecules (single isomorphous replacement, SIR) or many atoms or molecules (multiple isomorphous replacement, MIR). The patterns are analysed by techniques known as difference-Patterson synthesis, the details of which go beyond the scope of this book, except to mention that the heavy-atom group may disturb the structure of the (protein) crystal in its immediate environment such that complete isomorphism is only valid to a limited resolution of the structure.

Finally, following the pioneering work of Bijvoet (Section 9.2), the absolute configuration of enantiomorphous molecules can be determined using synchrotron radiation 'tuned' to wavelengths either close to, or away from, the absorption edges of anomalously scattering atoms. Fortunately, many proteins contain atoms (e.g. Fe, Co) suitable for anomalous scattering experiments. The techniques go under the headings of single wavelength anomalous dispersion (SAD) and multiwavelength anomalous dispersion (MAD)—the latter is now the method of choice for phase determination. The agreement between the data and the right- or left-hand enantiomorphs can be assessed by means of the **Flack parameter**[1] x obtained through the equation:

[1] H.D. Flack (1983) *On enantiomorph-polarity estimation*, Acta. Crystallogr. **A39** pp. 876–881.

$$I_{hkl} = (1 - x)|F_{hkl}|^2 + x\left|F_{\bar{h}\bar{k}\bar{l}}\right|^2$$

When $x \approx 0$, the correct enantiomorph has been obtained. When $x \approx 1$, the signs of all x, y, and z in the structure must be reversed and when $x \approx 0.5$, the crystal may be twinned.

13.3.4 Direct methods

The Patterson/heavy atom/isomorphous replacement methods briefly discussed earlier proceed by a series of deductive steps and the eventual solution of the crystal structure is a result of the skills and insights of individual crystallographers.

However, the word 'direct' is now reserved for those methods in which the structure factor phases and electron density (and atom positions) are derived by mathematical means from a set of X-ray intensities: they may be thought of as representing a 'direct' attack on the phase problem. Their physical basis is that the amplitudes and phases of reflections are **not** independent but are linked (as we see from the equation in Section 13.2 on p. 332) to the electron density. Suppose we assign arbitrary phases to a set of reflections and then combine them with the observed amplitudes to obtain an electron density map. We can tell straight away if the map is nonsense (and our choice of phases is incorrect); (a) if it results in the physically impossible situation of regions of negative electron density and (b) if it does not show peaks that may be identified as atom positions. These considerations represent two **constraints** in the predicted electron density. The mathematical procedures of direct methods are essentially such as to impose further constraints that narrow down the choice of phases, and lead to electron density maps that are physically interpretable and atom positions that are consistent with (e.g.) a knowledge of the stereochemistry of the molecule.

First, instead of using the observed structure amplitudes, $|F_{hkl}|$, which decrease rapidly with angle, **normalized** structure factors $|E_{hkl}|$ are used that represent atoms as point scatterers at rest and that, in turn leads to sharper and more clearly defined peaks in the electron density map. The normalization is carried out by means of a Wilson plot.[2] The decrease in atomic scattering factor with angle arises (1) from the finite size of atoms (Fig. 9.1) and (2) from the thermal vibrations of atoms. The decrease is described by the Debye–Waller equation $f_T = f_o \exp -2B \sin^2 \theta / \lambda^2$ where f_T is the atomic scattering factor at temperature T and f_o is the atomic scattering factor of the atom at rest (close to 0K). B is a temperature-dependent term called the Debye–Waller, or more generally, the **atomic displacement parameter** (ADP). It is related to the mean value of the atomic displacements (see Section 14.4.3).

The most powerful mathematical procedure, which is incorporated into every direct methods program today is the **tangent formula**[3], the basis of which will only be described in outline.

[2] A.J.C. Wilson (1942) *Determination of absolute from relative X-ray intensity data*, Nature **150** pp. 151–152.
[3] J. Karle and H. Hauptman (1956) *A theory of phase determination for four types of non-centrosymmetric space groups*, Acta Crystallogr. **9** pp. 635–651.

In the 1950s it was realized that the phases of specially selected groups of reflections, called triplets, were related. A triplet is a group of three reflections whose *hkl* indices sum to zero (the planes all lie in the same zone). For three such strong reflections there is the probability that their phases also sum to zero. Table 13.2 shows one example of a triple relation and also includes the 'bold italic' symbols for the reflections.

Table 13.2 Triple relations.

h	*k*	*l*	Phase	Symbol
1	$\bar{1}$	2	α	\boldsymbol{h}
3	0	$\bar{1}$	β	\boldsymbol{k}
$\bar{4}$	1	$\bar{1}$	γ	$\boldsymbol{l} = -\boldsymbol{h} - \boldsymbol{k}$
0	0	0	≈ 0	←sum

There are many triple relations that give the phase of \mathbf{h}. If we know the phase of \boldsymbol{k} and of $\boldsymbol{h} - \boldsymbol{k}$ the phase of \boldsymbol{h} can be computed from the tangent formula, viz.:

$$\tan(\alpha_h) \approx \frac{\sum E_k E_{h-k} \sin(\alpha_k + \alpha_{h-k})}{\sum E_k E_{h-k} \cos(\alpha_k + \alpha_{h-k})}$$

The probability that the formula is valid for a given \boldsymbol{h} is:

1. proportional to the magnitudes of the E values;

2. proportional to the number of triplets in the summation (there will be many pairs of reflections \boldsymbol{k} and $\boldsymbol{h} - \boldsymbol{k}$ for each reflection \boldsymbol{h}, i.e. many groups of planes in the same zone with a common \boldsymbol{h});

3. inversely proportional to the square root of the number of atoms in the unit cell. (For inorganic structures with ~10 atoms the probability may be ~95%, but for proteins the probabilities are typically slightly over 50%.)

Direct methods are now the standard technique for solving organic crystal structures by X-ray crystallography and have also been applied to electron diffraction data, although this is less ideal because multiple scattering of electrons makes the relative intensities uncertain and just one single two-dimensional projection is available. However, the situation is much improved by the development of electron diffraction precession and rotation techniques (see Section 11.2).

13.3.5 Charge flipping

Charge flipping is a relatively new (2004) method[4] for determining crystal structures. As with direct methods it depends on the assumption that the electron density (or

[4] G. Oszlanyi and A. Sutó (2004) *Ab initio structure determination by charge flipping. II. Use of weak reflections*, Acta. Crystallogr. **A60** p. 134.

electrostatic potential in electron diffraction) is continuous, and consists of peaks and high electron density (the resolved atoms) surrounded by areas of very low, but never negative, electron density. No assumptions are made about atom types, chemical composition, or even the space group symmetry. The only inputs into the charge-flipping computer algorithms are the observed $|F_{hkl}|$ values of the reflections and the lattice parameters of the unit cell.

The steps in the algorithm may be summarized as follows:

1. A random number generator assigns a phase to each reflection.
2. These phases, and the observed amplitudes $|F_{hkl}|$, are used to compute an electron density map which will fluctuate between positive and negative values.
3. Where the electron density is negative (and below a certain threshold) the sign of the electron density is reversed to positive 'charge flipping'.
4. A reverse Fourier transform of the map is carried out to get new phases that are used as input as at Step (2) and the whole procedure is repeated for hundreds of cycles until a clear pattern of atoms (peaks of high electron density, low in between) emerges and the reliability index R converges to a low value.
5. The **whole** procedure is repeated except with a new set of random phases. If the same pattern of atoms emerges then the structure is solved: the unit cell contents are scanned for symmetry-related features and possible space groups are assigned (rather than the no-symmetry $P1$ space group assumed at the beginning).

13.4 Analysis of the Fraunhofer diffraction pattern from a grating

Figure 13.12(a) shows the 'square top' waveform representing the amplitude of light emerging from a grating consisting of six slits ($N = 6$), width d and spacing a, illuminated by an incident plane wave normal to the grating. Figure 13.12(b) shows the corresponding diffraction pattern. First, we recapitulate the simple analysis given in Section 7.4.

(1) The envelope of the pattern (dashed curve) corresponds to the intensity profile for a single slit with minima at $\sin \alpha$ angles $1^\lambda/_d$, $2^\lambda/_d$, $3^\lambda/_d$, etc.
(2) The principal maxima correspond to the conditions for constructive interference between the slits at $\sin \alpha$ angles $1^\lambda/_a$, $2^\lambda/_a$, $3^\lambda/_a$, etc.
(3) The subsidiary minima correspond to the conditions for destructive interference for the whole grating considered as a single slit of width $6a$, i.e. at $\sin \alpha$ angles $1^\lambda/_{6a}$, $2^\lambda/_{6a}$, $3^\lambda/_{6a}$, etc.
(4) For the given ratio, $a/d = 4$, the first, second, etc. minima of the single slit envelope correspond to the angles of the fourth, eighth, etc. principal maxima (sometimes called 'missing orders').

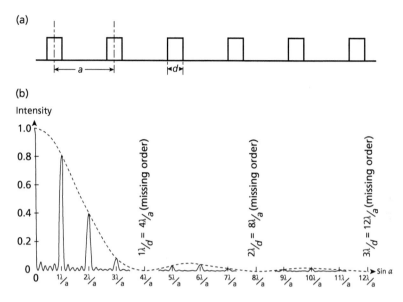

Fig. 13.12. (a) The 'square-top' waveform representing the amplitude of light from a diffraction grating consisting of six slits, width d and spacing a and ratio $a/d = 4$. (b) The corresponding diffraction pattern. Note the fourth, eighth, etc. 'missing orders' and the number $(6-2) = 4$ of the subsidiary maxima between the principal maxima (adapted from *Geometrical and Physical Optics* by R.S. Longhurst, 2nd edn, Longmans, London, 1967.)

It is easy to see the effects of changing the variables d (slit width), a (slit spacing) and N (number of slits). As d increases the single slit profile (dashed line) becomes narrower. As N increases the principal maxima become narrower, i.e. 'sharper' and the subsidiary maxima both increase in number $(N-2)$ and decrease in intensity. As a increases the principal maxima become closer together and the ratio a/d determines the missing orders.

As we have seen in Section 13.2, the amplitudes of the diffracted beams each correspond to a Fourier coefficient of the Fourier analysis of the periodic variation in electron density in a crystal. Precisely the same analysis applies to the amplitudes of the diffracted beams from a grating: they each correspond to a Fourier coefficient of the Fourier analysis of the square-topped waveform. Furthermore, if we take the origin at, say, the centre of a slit, the Fourier coefficients are real (as with the example of NaCl, Fig. 13.5), with signs + (zero phase angle) or − (180° phase angle). Hence, we now have to determine the amplitudes of the diffracted beams as a function of the angle α (rather than simply determining the angles for either destructive or constructive interference). We do this by means of *amplitude–phase diagrams*; first for a single slit and then for a grating of N slits.

*Single slit width **d** (Fig. 13.13)*

When $\alpha = 0$, all the Huygens wavelets are in phase and their sum, a straight line, has amplitude, say A_0 (Fig. 13.14(a)). Now consider the situation at angle α. Taking the centre of the slit as the origin, the path difference, PD, between the wavelet at the centre of the slit and those at the edge is $(d/2)\sin\alpha$ and the phase difference $\phi = 2\pi/\lambda(\text{PD}) = (\pi d/\lambda)\sin\alpha$; $+\phi$ for the wavelet at the lower edge and $-\phi$ for the wavelet at the top edge. The amplitude–phase diagram is now the arc of a circle, length A_0, as shown in Fig. 13.14(b) with the phase differences $+\phi$ and $-\phi$ indicated. The resultant amplitude is given by A, the chord of the circle. From the geometry $A_0 = r2\phi$ and $A = r2\sin\phi$ where r is the radius of the circle. Hence $A = A_0\dfrac{\sin\phi}{\phi}$. This function, $\dfrac{\sin\phi}{\phi}$, occurs so commonly in optics that it is given a special name, sincϕ (pronounced sink ϕ). Its operation can be seen in Fig. 13.14. As ϕ (or α) increases, the arc of the circle 'winds up' until $\phi = \pi$ (or $\sin\alpha = \lambda/d$) at which point $A = 0$ (Fig. 13.14(c)). Then, as ϕ increases further it 'winds on itself' and the resulting amplitude A is reversed in sign (Fig. 13.14(d)), and so on.

The function, sincϕ, plotted vs. $\sin\alpha$ to the right and vs. ϕ to the left is shown in Fig. 13.15 (solid line) together with its square (dashed line), representing the intensity distribution from a single slit.

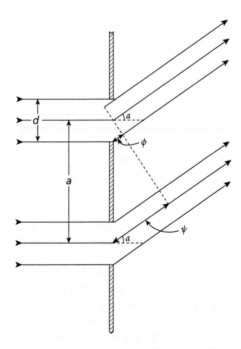

Fig. 13.13. A diffraction grating indicating the path difference $(d/2)\sin\alpha$ (phase difference ϕ) between a Huygens wavelet at the centre and either edge of a slit and the path difference $a\sin\alpha$ (phase difference ψ) between the slits.

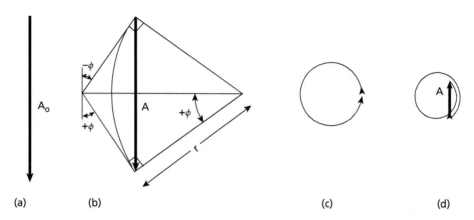

(a) (b) (c) (d)

Fig. 13.14. The amplitude–phase diagram for a single slit: (a) $\alpha = 0$; all the Huygens wavelets are in phase; (b) for phase difference $\pm\phi$ between the wavelet at the centre and those at the edges of the slit; (c) for phase difference $\phi = \pi$, the amplitude A is zero; (d) for $\phi > \pi$, the amplitude A is reversed in sign.

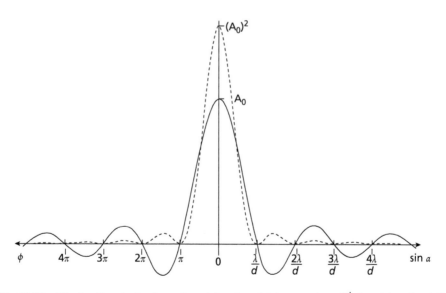

Fig. 13.15. The function $A_0 \sin c\phi$ vs. $\sin \alpha$ (plotted to right) or vs. $\phi = \frac{\pi d}{\lambda} \sin \alpha$ (plotted to left) representing the amplitude distribution from a single slit (or aperture) (solid line) and its square representing the intensity (dashed line). In this example $A_0 = 1.5$. (Courtesy of Mr D.G. Wright.)

N slits, width d, spacing a (Fig. 13.13)

The path difference, PD, between corresponding points between adjacent slits is $a \sin \alpha$ and the phase difference $\psi = 2\pi/\lambda(\text{PD}) = \dfrac{2\pi a}{\lambda} \sin \alpha$. Again, when $\alpha = 0$, all the slits scatter in phase and the amplitude $A_T = A_0$. At angle α we again construct an amplitude–phase diagram—but not as before as the arc of a circle of constant length, but as a series of chords, each of amplitude A and differing in phase by ψ (Fig. 13.16). For each chord $A/2 = R \sin \psi/2$ where R is the radius. For N chords half the total angle is $N\psi/2$ and $A_T/2 = R \sin N\psi/2$. Substituting for R and A we have:

$$A_T = A_0 \frac{\sin \phi}{\phi} \cdot \frac{\sin N\psi/2}{\sin \psi/2} \qquad (\phi = (\pi d/\lambda)\sin \alpha, \ \psi = (2\pi a/\lambda)\sin \alpha \text{ and } \psi/\phi = 2a/d).$$

The first term in this expression is called the *single-aperture term* and the second is called the *grating term*. In computing the amplitude, A_T, and intensity, $(A_T)^2$, distributions the equation may be simplified by using only two variables, ϕ and N, and a given ratio of slit spacing/slit width, a/d; i.e. $\sin N\psi/2 = \sin N\phi\, a/d$ and $\sin \psi/2 = \sin \phi\, a/d$, hence

$$A_T = A_0 \frac{\sin \phi}{\phi} \cdot \frac{\sin N\phi\, a/d}{\sin \phi\, a/d}.$$

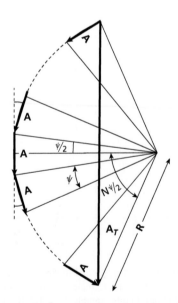

Fig. 13.16. The amplitude–phase diagram for a grating of N slits and for phase difference ψ between the slits, total amplitude $= A_T$. Note that (unlike the single slit) the series of chords (the arc of the circle-dashed line) is not of constant length.

For $N = 6$ and $a/d = 4$ this function is shown in Fig. 13.17(a) and its square, representing the intensity distribution for the diffraction grating, is shown in Fig. 13.17(b) below. Notice that the phases of the amplitudes of the diffraction maxima alternate between $+$ and $-$ (Fig. 13.17(a)) and as shown in Fig. 7.4 for two slits but that this phase information is lost in the intensity distribution (Fig. 13.17(b), which corresponds of course to the experimental intensity profile of Fig. 13.12(b)).

Figure 13.17(c) shows the effect of increasing the number of slits, N, from 6 to 20. The principal maxima are now sharper and the subsidiary maxima are now greater in number but much reduced in intensity. Finally, Fig. 13.17(d) shows the effect of increasing a/d, from 4 to 10 and for $N = 4$. The principal maxima approach closer together and, as pointed out above, merge into one continuous function when a/d approaches infinity.

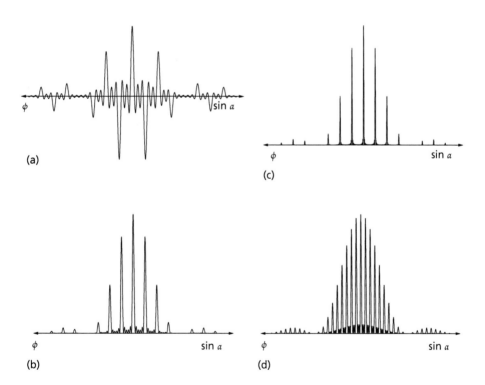

Fig. 13.17. (a) The amplitude distribution A_T for a grating with $N = 6$ slits and $a/d = 4$, and (b) the corresponding intensity distribution $(A_T)^2$. (c) The effect on the intensity distribution of increasing the number of slits to 20—the principal maxima are sharper and the subsidiary maxima are insignificant. (d) The effect of increasing a/d to 10 ($N = 4$)—the principal maxima approach closer together. Notice also the 10$^{\text{th}}$ missing orders. (Courtesy of Mr D.G. Wright.)

13.5 Abbe* theory of image formation

The formation of an image in a microscope may be considered to be carried out by
the objective lens in two stages. First, the lens collects some (not all) of the parallel
wavetrains resulting from the diffraction and interference of light across the grating into
a series of diffraction peaks (sometimes called 'spectra') situated in the back focal plane.
In this sense the objective lens performs a Fourier analysis of the light emanating from
the grating as shown diagrammatically in Fig 13.17(a). A screen placed here records
their positions and intensities (Fig 13.17(b)) or the diffraction peaks may be observed
by simply removing the microscope eyepiece and observing the back focal plane either
directly or with the aid of a close-focused telescope.[5]

The wavetrains emanating from the diffraction peaks then spread out and interfere to
form a set of interference fringes in the image plane—in short the objective lens then
performs a Fourier synthesis of the amplitude distribution of the light in the back focal
plane, the amplitudes of the zero, first, second, etc. diffraction peaks or spectra corres-
ponding to the zero, first, second etc. Fourier coefficients. And the greater the number of
terms included in the Fourier synthesis, i.e. the greater the number of diffracted beams
intercepted by the objective lens, the closer does the image correspond to the amplitude
distribution across the grating. Figure 13.18(a)–(d) shows the evolution of the image of
a diffraction grating with a 'square-top' waveform with $a/d = 4$ (as in Fig. 13.12(a))
synthesized with an increasing number of Fourier terms.

This two-stage image formation process may also be understood in simple terms by
making use of Abbe's 'classic' representation of image formation (Fig. 13.19). A plane
wave (Fraunhofer case) incident normally on the narrow-line grating \mathbf{G} is diffracted and
the diffracted beams are collected and focused by the objective lens \mathbf{L} to give a series
of diffraction peaks or spectra situated in the back focal plane of the objective lens.
Three such spectra are shown in Fig. 13.19, the zero order S_0 and the first orders S_1
and S_1' each side. The waves then 'spread out' and interfere in the image plane. The
wavefronts are shown on the right of Fig. 13.19; each (curved) line may be considered
to represent the 'crest' of the waves. Where all three intersect, at I_1, I_2 and I_3, we
have complete constructive interference and these interference maxima reproduce the
periodicity of the grating—in short the line-spacing of the grating is resolved in the
image plane. It is not a 'perfect' image because half-way in between I_1, I_2 and I_3 we see
that the two waves from S_1 and S_1' interfere constructively but that from S_0 interferes
destructively giving 'subsidiary' maxima of one-third the amplitude (and one-ninth the
intensity) of the maxima at I_1, I_2 and I_3. (Such subsidiary maxima are also evident in
Fig. 13.18(a)–(c).)

Figure 13.19 may also be used to see the effect of 'blocking off' one or more of
the waves emanating from S_0, S_1 and S_1'. If we 'block off', say S_1' then again we have
interference maxima at I_1, I_2 and I_3 and the amplitude distribution will be a simple

* Denotes biographical notes available in Appendix 3.

[5] Phase contrast and polarized-light microscopes often incorporate a *Bertrand lens* below the eyepiece
which when 'flipped in' allows observation of the back focal plane directly.

Fig. 13.18. The image of a diffraction grating with a 'square-top' waveform (as in Fig. 13.12(a)), $a/d = 4$ synthesized with an increasing number of Fourier terms (a)–(d). (Courtesy of Mr D.G. Wright.)

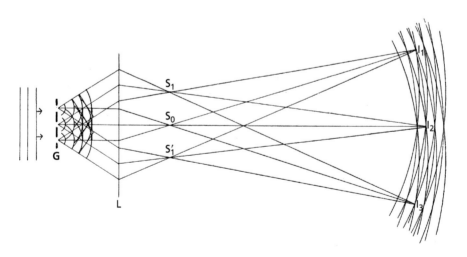

Fig. 13.19. Abbe's representation of image formation in a microscope. A plane wave (left) incident upon a narrow-slit grating G is diffracted and the sets of parallel diffracted beams are intercepted and focused by the objective lens **L** to give diffraction peaks (or spectra) S_0, S_1 and S'_1 situated in the back focal plane, the light from which interferes to give maxima I_1, I_2 and I_3 in the image plane (see text).

cosine curve, again reproducing the periodicity of the grating. If we 'block off' any two, say S_1 and S_1', then clearly there will be no interference in the image plane but just (for a narrow slit grating) a uniform distribution of light—the periodicity of the grating is not resolved. Finally (and this is an important case), if we 'block off' the zero order, S_0, then we have constructive interference (for the waves from S_1 and S_1') at I_1, I_2 and I_3 *and* half-way in between giving a (cosine curve) of periodicity *twice* that of the grating.

In order to demonstrate his theory of image formation (which was by no means immediately accepted, particularly by the English amateur microscopists), Abbe devised a special slide carrying diffraction gratings—the *Diffraktionsplatte*—together with a device enabling diaphragms to be inserted into the objective lens back focal plane to prevent (i.e. 'block off') some parts of the diffraction pattern from contributing to the image. This equipment was included in the Zeiss catalogue in 1878 and was available intermittently until the 1960s. Abbe's demonstrations have since been extended by Dr P.J. Evennett,[6] examples of which are shown in Figs 13.20 and 13.21.

Figure 13.20(a), left, shows a diffraction grating of line spacing $a = 16$ μm and line width $d = 4$ μm (as in Fig. 13.12(a)) imaged with an objective lens of numerical aperture, NA = 0.16. The corresponding diffraction pattern, Fig. 13.20(a) right, shows three diffraction orders S_1, S_2, S_3, and S_1', S_2', S_3', each side of the zero order, S_0. (The condition for plane wave illumination as in Fig. 13.19 is closely achieved in practice by closing down the aperture (condenser) diaphragm to a 'pinhole'; conversely, as the aperture diaphragm is opened out the diffraction spots broaden and overlap.)

Taking $\lambda = 0.55$ μm (green light) the corresponding $\sin \alpha$ angles (see Section 7.4, p. 180) are: 1st order: 1×0.55 μm/16 μm = 0.034; 2nd order: 2×0.55 μm/16 μm = 0.069; 3rd order: 3×0.55 μm/16 μm = 0.103, the 4th order is missing (because $a/d = 4$, see Fig. 13.12(b)), and the 5th order lies just outside the NA of the objective lens.

Figure 13.20(b)–(f) shows the effect on the image of the diffraction grating by 'blocking off' different diffraction spots. In Fig. 13.20(b) only S_1, S_0, and S_1' contribute to the image (as in Fig. 13.19): we see, as predicted, faint subsidiary maxima. In Fig. 13.20(c)

Fig. 13.20. The images of a diffraction grating (left) and the corresponding diffraction patterns (right) illustrating Abbe's theory of image formation (Fig. 13.19). In Fig. 13.20(a) seven diffraction orders, $S_3 \ldots S_3'$, contribute to the (sharp) image. In Fig. 13.20(b) only S_1, S_0, and S_1' contribute to the image and (as predicted) faint subsidiary maxima occur. In Fig. 13.20(c) only S_1 and S_0 contribute to the image and the grating is just resolved. In Fig. 13.20(d) it is not resolved. In Fig. 13.20(e) and (f) only S_1 and S_1' and S_2 and S_2' contribute to the image, leading to a halving and quartering of the image line spacings, respectively. The overlain (grey) bars represent the positions of the stops/apertures used to block off the various diffraction orders. (Courtesy of Dr P.J. Evennett.)

[6] P.J. Evennett (1996) *Abbe and the development of the modern microscope*, Proc. Roy. Micros. Soc. **31**, p. 283.

Fig. 13.21. (a) The image of a square cross-grid grating and the corresponding diffraction pattern. (b) and (c) The periodicity of the grating is only resolved in the direction normal to the line of diffraction spots which contribute to the image. (Courtesy of Dr P.J. Evennett.)

only S_1 and S_0 contribute to the image; the grating is just resolved and the lines are broad and 'fuzzy'—the intensity profile approaches that of a cosine curve. In Fig. 13.20(d) only one diffraction order (in this case S_0) passes the objective back focal plane and the grating is not resolved.

Figure 13.20(e) and (f) show two examples in which the periodicities or line spacings in the image **do not** correspond with the true periodicity or line spacing ($a = 16 \, \mu$m) of the grating. Figure 13.20(e), imaged with S_1 and S_1', shows a periodicity **twice** that of the grating (as predicted from Fig. 13.19) and Fig. 13.20(f), imaged with S_2 and S_2', shows a periodicity **four times** that of the grating. Clearly, unless the diffraction conditions are known, such periodicities can be confused with the true periodicity. This problem is of particular importance in the interpretation of high-resolution TEM images.

Fig. 13.21(a) shows the image of a square cross-grid grating and its associated diffraction pattern (see also Figs 7.1 and 7.3). When a slit diaphragm is inserted as in Figs 13.21(b) and (c), the periodicity is only resolved in the direction normal to the lines of the diffraction spots—and the spacing of the lines is of course inversely proportional to the spacings of the diffraction spots. We have, in short, the relationship between the lattice and the reciprocal lattice. How simple it all is!

14

The physical properties of crystals and their description by tensors

14.1 Introduction

In Section 4.4 we distinguished crystal properties that were direction-independent (scalars) and those that were direction-dependent (vectors).

Scalars and vectors are particular types of quantities that are known in general as **tensors**. There are other, more complicated types of tensors, which are of importance in crystal physics because they are able to describe properties such as conductivity that vary not only with the direction of the applied field or force but also with the orientation of the crystal. 'Everyday' examples of tensors (other than scalars or vectors) are not easy to provide but the following, perhaps rather contrived example, may illustrate the ideas involved.

Consider rainwater flowing down a corrugated roof, Fig. 14.1(a). The flow of water (a vector) is obviously down the roof; parallel to the corrugations and parallel also to a vector representing the inclined component of the gravitational field. The effective 'conductivity' may simply be regarded as the quantity that relates these two parallel vectors. Now consider a (badly made) roof in which the corrugations are set at an angle, Fig. 14.1(b). The water will partly run sideways, along the 'preferred direction' of the corrugations and will partly flow downwards over from one corrugation to the next. The overall water flow will lie in a direction somewhere between the lines of the corrugations and downwards. In this case the effective gravitational field vector and the water flow vector are not parallel and in estimating the effective conductivity of the roof, both the 'downwards' and 'sideways' components of the water flow need to be taken into account.

In crystals the 'corrugations' may be loosely regarded as preferred or easy directions of, say, heat or electrical flow. The conductivity of the crystal is the quantity that relates one vector (heat or electrical flow) to another vector (temperature gradient or electrical field). As indicated in the example here, these vectors may not, except in special cases, be parallel.

In this chapter, this analogy will be developed in order to explain second rank tensor properties (Section 14.2) and the relation between these properties and crystal symmetry (Neumann's principle Section 14.3). This is followed by the second rank tensors that describe stress and strain (Section 14.4). Then, in Section 14.5 the optical properties of crystals (the variation of refractive index with the directions of vibration and propagation of light) are described in terms of the indicatrix—not a tensor but closely related to a second rank tensor property. Finally, the third rank tensor property of piezoelectricity

The Basics of Crystallography and Diffraction. Fourth Edition. Christopher Hammond.
© Christopher Hammond 2015. Published in 2015 by Oxford University Press.

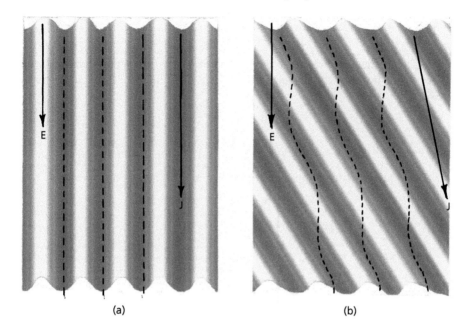

Fig. 14.1. 'Rainwater' illustration of a second rank tensor property: the grey/white lines show the corrugations of a roof and the dashed lines show the direction of water flow. In (a) the corrugations are parallel to **E**, the component of gravity down the roof and the water flow **J** is parallel to **E**. In (b) the corrugations are at an angle and the water flow **J** is partly along, and partly across, the corrugations.

and the fourth rank tensor property of elasticity are introduced in Sections 14.6 and 14.7, respectively. Throughout, the mathematics is kept to a minimum, but it is hoped that for a reader new to these topics the lack of mathematical rigour will be more than compensated by a clear appreciation of the basic principles—such that those books listed in Further Reading (e.g. the classic textbook *The Physical Properties of Crystals* by J.F. Nye, first published in 1957) can be tackled with greater confidence.

14.2 Second rank tensor properties

14.2.1 General expression for a second rank tensor relating two vectors

We develop the analogy of rainwater flowing down a 'badly made' corrugated roof. We first set up an axis system x_1 and x_2 (and x_3 in three dimensions). We use this terminology in preference to x, y, and z because we shall need to denote conductivities using suffixes—and number suffixes are much easier to follow than letter suffixes. $\mathbf{E} = E_1$ is a vector along x_1 representing the gravitational field and **J** is a vector representing the flow of water (Fig. 14.2(a)). We now resolve **J** into two components: J_1 along x_1 and J_2 along x_2. Now the 'conductivity' of the roof, symbol σ, may be expressed as two terms:

$$\sigma_{11} = J_1/E_1, \quad \sigma_{21} = J_2/E_1$$

where we adopt the convention that the **first** suffix in the conductivity term indicates the direction of flow and the **second** suffix indicates that the term is due to the vector $\mathbf{E} = E_1$.

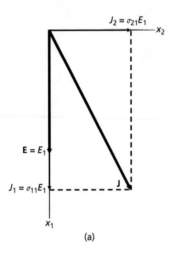

(a)

Fig. 14.2(a). Formal development of Fig. 14.1(b) showing the two components of \mathbf{J}, $J_1 = \sigma_{11}E_1$ and $J_2 = \sigma_{21}E_1$.

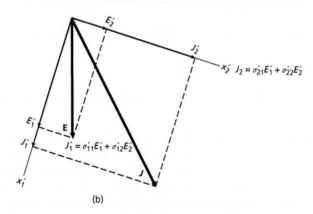

(b)

Fig. 14.2(b). Generalization of Fig. 14.2(a) showing another axis system x_1', x_2' where x_1' is **not** parallel to **E**. The flows of water J_1' and J_2' along the x_1' and x_2' axes are now the sums of two terms, but irrespective of the choice of axes they represent the **same** relationship between the vectors **E** and **J**.

We can generalize the analogy by choosing another axis system x_1', x_2' where \mathbf{E} is **not** parallel to the x_1'-axis (Fig. 14.2(b)). \mathbf{E} has a component E_1' along x_1' as before **plus** a component E_2' along x_2'. The flow of water J_1' along x_1' is now $\sigma_{11}'E_1'$ **plus** a contribution due to E_2', i.e. $\sigma_{12}'E_2'$. Hence, $J_1' = \sigma_{11}'E_1' + \sigma_{12}'E_2'$ and similarly $J_2' = \sigma_{21}'E_1' + \sigma_{22}'E_2'$. Notice (very important) that the conductivity terms are now also shown 'dashed'. This is to emphasize that their numerical values are different from those in Fig. 14.2(a) and depend upon our arbitrary choice of axis, x_1 and x_2, or x_1' and x_2'. In either case they describe the **same** relationship between \mathbf{E} and \mathbf{J}.

This is as far as our 'roof' analogy can go. But we can readily extend the analogy to a crystal where \mathbf{E} is a vector representing the electric field and \mathbf{J} is a vector representing the electrical current flow. These two vectors are now represented on an arbitrary axis system x_1, x_2, x_3:

$$\mathbf{E} = [E_1, E_2, E_3]; \qquad \mathbf{J} = [J_1, J_2, J_3]$$

There are now **three** components of J due to E_1:

$$J_1 = \sigma_{11}E_1, \qquad J_2 = \sigma_{21}E_1, \qquad J_3 = \sigma_{31}E_1$$

and similarly, three components of J due to E_2 and three due to E_3:

$$J_1 = \sigma_{12}E_2, \qquad J_2 = \sigma_{22}E_2, \qquad J_3 = \sigma_{32}E_2$$
$$J_1 = \sigma_{13}E_3, \qquad J_2 = \sigma_{23}E_3, \qquad J_3 = \sigma_{33}E_3$$

where, to repeat, the first suffix of the conductivity term refers to the direction of J and the second suffix refers to the component of E that gives rise to it.

We now gather all these terms together:

$$J_1 = \sigma_{11}E_1 + \sigma_{12}E_2 + \sigma_{13}E_3$$
$$J_2 = \sigma_{21}E_1 + \sigma_{22}E_2 + \sigma_{23}E_3$$
$$J_3 = \sigma_{31}E_1 + \sigma_{32}E_2 + \sigma_{33}E_3$$

or, in matrix form:

$$\begin{bmatrix} J_1 \\ J_2 \\ J_3 \end{bmatrix} = \begin{bmatrix} \sigma_{11} & \sigma_{12} & \sigma_{13} \\ \sigma_{21} & \sigma_{22} & \sigma_{23} \\ \sigma_{31} & \sigma_{32} & \sigma_{33} \end{bmatrix} \begin{bmatrix} E_1 \\ E_2 \\ E_3 \end{bmatrix}$$

Hence we see, in general, that the electrical conductivity of a crystal is specified by nine components or coefficients. This array of 9 (3×3) terms is called a **second rank tensor** and is enclosed in square brackets to show that it represents the crystal property of conductivity. We should note, however, that this does not constitute the formal definition of a tensor: the formal definition is expressed in terms of transformation equations of the components between different axis systems, as for example between the 'undashed' and 'dashed' components in Fig. 14.2(a) and (b).

There are several other second rank tensor properties that relate two vectors: for example thermal conductivity (temperature gradient and heat flow), permittivity (electric field and electric displacement), and magnetic susceptibility (magnetic field and intensity of magnetization).

The equations presented above may be written in more 'compact' forms, i.e.

$$\left.\begin{array}{l} J_1 = \sigma_{1j}E_j \ (j=1,2,3) \\ J_2 = \sigma_{2j}E_j \ (j=1,2,3) \\ J_3 = \sigma_{3j}E_j \ (j=1,2,3) \end{array}\right\} \quad \text{or } J_i = \sigma_{ij}E_j \ (i,j=1,2,3)$$

When written in these compact forms, we adopt what is known as the Einstein summation convention: that when a letter suffix appears twice in the same term (i.e. j in $\sigma_{ij}E_j$), then summation with respect to that suffix is automatically understood.

14.2.2 Simplification of second rank tensor equations: principal axes

The first simplification must be stated without proof. It is that the pairs of the off-diagonal terms are equal, i.e. $\sigma_{12} = \sigma_{21}$, $\sigma_{13} = \sigma_{31}$, $\sigma_{32} = \sigma_{23}$. The tensor is said to be **symmetrical** and the number of independent terms is reduced from nine to six. There follows another very important simplification. Recall our arbitrary choice of axes x_1, x_2, x_3. If we had chosen a different set of axes all six terms in the tensor would have had different numerical values—but of course the crystal property that they describe is unchanged. It is, however, possible (for symmetrical second rank tensors) to choose a particular set of axes—called the **principal axes** in which all the off-diagonal terms are zero. In this case we only have three terms σ_{11}, σ_{22}, σ_{33} that are abbreviated to σ_1, σ_2, σ_3 and are called the **principal conductivities**. When **E** is along a principal axis the current flow is also along that axis—there are no transverse components. Hence the equations reduce to the simple forms:

$$J_1 = \sigma_1 E_1, \qquad J_2 = \sigma_2 E_2, \qquad J_3 = \sigma_3 E_3$$

The principal axes (sometimes called eigenvectors) and the principal values (sometimes called eigenvalues) are worked out, using the stress tensor as an example, in Section 14.4.1.

14.2.3 Representation of second rank tensor properties: the representation quadric

In a practical experiment we may take, say, a slab of crystal, apply a field **E** perpendicular to its surface, and then measure the current flow **in the same direction** as **E**. But unless the field **E** was along a principal axis we would not be measuring the 'true' current flow **J** but only the component of **J** parallel to **E**. Call this $J_{//}$.

Figure 14.3 (drawn for two dimensions) shows **E** and **J** and their components referred to the principal axes x_1 and x_2. Now, by resolving J_1 and J_2 in the direction

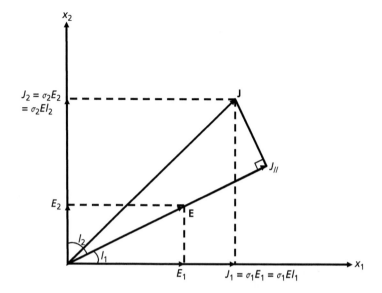

Fig. 14.3. E and J referred to principal axes x_1 and x_2. l_1 and l_2 are the direction cosines of E with respect to x_1 and x_2, respectively. $J_{//}$ is the component of J parallel to E.

of E we have $J_{//} = (J_1 l_1 + J_2 l_2)$ where l_1 and l_2 are the direction cosines of E. Then, substituting $J_1 = \sigma_1 E_1 = \sigma_1 E l_1$ and $J_2 = \sigma_2 E_2 = \sigma_2 E l_2$, we have $J_{//} = E(\sigma_1 l_1^2 + \sigma_2 l_2^2)$, or $J_{//}/E = (\sigma_1 l_1^2 + \sigma_2 l_2^2)$.

$J_{//}/E$ describes the conductivity in the direction of E. In three dimensions:

$$ J_{//}/E = \left(\sigma_1 l_1^2 + \sigma_2 l_2^2 + \sigma_3 l_3^2 \right) $$

This equation corresponds to the general equation for a quadric:

$$ \left(x_1^2 / a^2 + x_2^2 / b^2 + x_3^2 / c^2 \right) = 1 $$

which, for positive values of a, b, c, is an ellipsoid and a, b, c are the principal semi-axes. Substituting $x_1 = r l_1$, $x_2 = r l_2$, $x_3 = r l_3$ where r is the length of a radius vector and (as before) l_1, l_2, l_3 are the direction cosines of E to the principal axes we have:

$$ r^2 \left(l_1^2 / a^2 + l_2^2 / b^2 + l_3^2 / c^2 \right) = 1 $$

which, by comparison with the equation for $J_{//}/E$ we have a representation quadric that has principal semi-axes of length $1/\sqrt{\sigma_1}$, $1/\sqrt{\sigma_2}$, $1/\sqrt{\sigma_3}$ and the radius r in the direction of $E = \dfrac{1}{\sqrt{J_{//}/E}}$.

The representation quadric is drawn (in two dimensions) in Fig. 14.4. It also gives us the value of J/E:

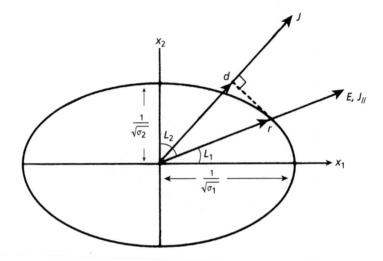

Fig. 14.4. Two-dimensional representation of the representation quadric for a second rank tensor property—(an ellipse, as shown, when the principal coefficients (or conductivities) are positive with semi-axes $1/\sqrt{\sigma_1}$ and $1/\sqrt{\sigma_2}$): $1/r = \sqrt{J_{//}/E}$ and $1/d = \sqrt{J/E}$.

$$J = \sqrt{J_1^2 + J_2^2} = \sqrt{(\sigma_1 E_1)^2 + (\sigma_2 E_2)^2} = E\sqrt{(\sigma_1 l_1)^2 + (\sigma_2 l_2)^2}$$

In three dimensions:

$$J/_E = \sqrt{(\sigma_1 l_1)^2 + (\sigma_2 l_2)^2 + (\sigma_3 l_3)^2}$$

By the radius normal property of the quadric (Exercise 14.1) this quantity is equal to $\sqrt{1/d}$ where d is the length of the normal to the tangent at E (Fig. 14.4). Hence, to summarize:

$$1/_r = \sqrt{J_{//}/_E} \quad \text{and} \quad 1/_d = \sqrt{J/_E}$$

In general, we may say that the reciprocal of the length of any radius vector of the representation quadric is equal to the square root of the conductivity in that direction. Finally, we note that the representation quadric is **centrosymmetrical**: all second rank tensors are centrosymmetrical as well as being symmetrical (see Section 14.2.2).

14.3 Neumann's principle

We now have to see how the principal axes of the representation quadric are related to the crystal axes. In order to do so we invoke **Neumann's principle** that may be stated: **the symmetry elements of any physical property of a crystal must include the symmetry of the point group of the crystal**. Note that the principle does not say 'the same as', just 'includes'—the physical property may, and often does, possess more symmetry than the point group.

We have already seen that second rank tensor properties are centrosymmetrical. This, as with X-ray diffraction (Friedel's law), reduces the number of point groups to 11 (see Table 9.1). Applying Neumann's principle to these in turn it turns out that the restrictions imposed by crystal symmetry are the same within each crystal system. In the cubic system (isotropic) the principal axes of the quadric (a sphere) can lie in any direction. In the tetragonal, hexagonal, and trigonal systems (uniaxial) we have a quadric of revolution; the x_3-axis is 'fixed' along the crystal z-axis and the x_1- and x_2-axes can lie anywhere in the x–y plane. In the orthorhombic system (biaxial) the principal axes of the quadric are all 'fixed' along the three crystal axes. In the monoclinic system (also biaxial) one axis of the quadric is 'fixed' along the y-(diad) axis and the other axes lie at variable angles to the x- and z-axes, i.e. the quadric can 'rotate' about the diad axis. For the triclinic system the principal axes of the quadric can lie at any angle to the crystal axes. The number of independent coefficients for each system is 1(cubic), 2(tetragonal, hexagonal, trigonal), 3(orthorhombic), 4(monoclinic) and 6(triclinic).

14.3.1 Pyroelectricity and ferroelectricity

Neumann's principle can be used to specify the ten polar point groups in which pyroelectricity and ferroelectricity occur (see Section 4.4).

The change in polarization in a crystal brought about by a small temperature change is specified by a vector **P** (a crystal property); the three coefficients or components of which are called the **pyroelectric coefficients**. **P** must conform to Neumann's principle and therefore a crystal showing the pyroelectric effect cannot exist in a crystal with a centre of symmetry. A further restriction is that the point group must include a direction that is **unique**, i.e. it is not repeated by any symmetry element. For point groups 2 and 2*mm* the unique axis is along the diad axis and similarly for point groups 3, 3*m*, 4, 4*mm*, 6, 6*mm*. In point group *m* the unique axis can lie anywhere in the mirror plane and in point group 1 it can lie anywhere. Hence we have a total of ten such groups that contain unique axes, which are rather confusingly called the **ten polar groups**: confusing because there are other point groups, e.g. 32, in which the diad axis is a polar, but not a unique, direction.

In ferroelectric crystals an applied electric field can reverse the direction of polarization. The name (nothing to do with iron) has arisen because ferroelectric crystals show a hysteresis loop akin to the ferromagnetic hysteresis loop in ferromagnetism. Why are not all pyroelectric crystals also ferroelectric? Possibly because the coercive field needed to reverse the polarization is so great that a breakdown of the crystal structure is first to occur.

14.4 Second rank tensors that describe stress and strain

14.4.1 The stress tensor: principal axes (eigenvectors) and principal values (eigenvalues)

Most, if not all, physical science students know about the distinction between tensile (normal) stress and shear stress. These can be expressed as the components of a **stress**

tensor, again a symmetrical second rank tensor that describes, in general, the state of stress in a crystal. It differs from the second rank tensors already described in that, since stress is imposed upon the crystal, it is not subject to Neumann's principle. The two may be distinguished by the terms matter tensor and field tensor, respectively.

Consider the 'body' in Fig. 14.5(a), drawn deliberately as an irregular 'lump' and subject to externally imposed forces as shown by the arrows. We need to specify the state of stress in the body. Consider, within the body, a 'unit cube' referred to arbitrary axes x_1, x_2, x_3. On each face there will be a force F_1, F_2, F_3 as shown at an arbitrary angle to the face and counterbalanced by equal and opposite forces on the rear faces (the condition for equilibrium). We now resolve these forces into normal and shear stress components (Fig. 14.5(b)). Note again the meaning of the suffixes: the first suffix refers to the **direction** and the second suffix refers to the **face** on which the stress acts. We have an array of nine components: three normal stresses and six shear stresses. In order to maintain equilibrium $\sigma_{13} = \sigma_{31}$, $\sigma_{23} = \sigma_{32}$, and $\sigma_{12} = \sigma_{21}$. Hence we have six independent components and the state of stress is described by a symmetrical second rank tensor.

We now show that, when referred to **principal axes**, the state of stress is described by three normal stresses on the faces of a cube. Figure 14.6 shows an (irregular) tetrahedron; the force P on the 'front' face ABC is counterbalanced by the normal and

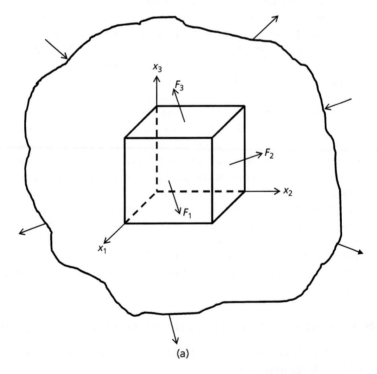

(a)

Fig. 14.5(a). Representation of a general state of stress on a body and the forces on the faces of a 'unit cube' (referred to arbitrary axes x_1, x_2, and x_3) and counterbalanced by equal and opposite forces in the rear faces (condition for equilibrium).

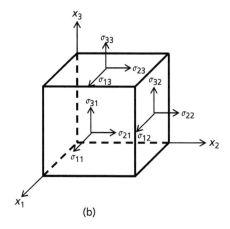

(b)

Fig. 14.5(b). Resolution of the forces acting upon the faces of the cube into normal and shear components. The **first** suffix indicates the direction of the stress and the **second** suffix indicates the face on which the stress acts.

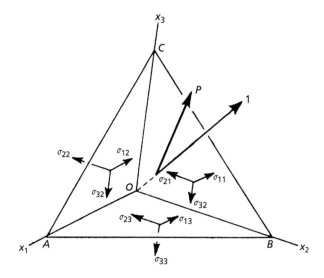

Fig. 14.6. Balance of forces in the 'front' and 'side' faces of an irregular tetrahedron (i.e. the cube in Fig. 14.5(b) with its four 'front' corners cut away). When the stress P is normal to the face ABC it constitutes one of the principal stresses of the stress tensor.

shear forces on the 'side' faces AOB, etc. When P is in the direction I, i.e. normal to face ABC it constitutes one of the three principal stresses and the direction cosines of I: $l_1 = COB/ABC$, $l_2 = AOC/ABC$, and $l_3 = AOB/ABC$ give the orientation of one principal axis with respect to the arbitrary axes x_1, x_2, x_3. The components of P along each of the axes are Pl_1, Pl_2, Pl_3. Resolving forces parallel to x_1 we have:

$$Pl_1 (ABC) = \sigma_{11} (COB) + \sigma_{12} (AOC) + \sigma_{13} (AOB)$$
$$Pl_1 = \sigma_{11}l_1 + \sigma_{12}l_2 + \sigma_{13}l_3$$

and similarly, resolving forces parallel to x_2 and x_3:

$$Pl_2 = \sigma_{21}l_1 + \sigma_{22}l_2 + \sigma_{23}l_3$$
$$Pl_3 = \sigma_{31}l_1 + \sigma_{32}l_2 + \sigma_{33}l_3$$

or

$$(\sigma_{11} - P)l_1 + \sigma_{12}l_2 + \sigma_{13}l_3 = 0$$
$$\sigma_{21}l_1 + (\sigma_{22} - P)l_2 + \sigma_{13}l_3 = 0$$
$$\sigma_{31}l_1 + \sigma_{32}l_2 + (\sigma_{33} - P)l_3 = 0$$

These three equations, written as a matrix, are

$$\begin{bmatrix} (\sigma_{11} - P) & \sigma_{12} & \sigma_{13} \\ \sigma_{21} & (\sigma_{22} - P) & \sigma_{23} \\ \sigma_{31} & \sigma_{32} & (\sigma_{33} - P) \end{bmatrix} \begin{bmatrix} l_1 \\ l_2 \\ l_3 \end{bmatrix} = 0$$

and the determinant of the matrix is zero.

The determinant is found from the sum of the three terms in any row or column multiplied by their co-factors (see Section 5.10). This gives a cubic equation in P, the three roots of which, P_1, P_2, P_3, are the principal stresses or eigenvalues. In order to determine the principal axes (eigenvectors), substitute each value of P, one at a time, to get the direction cosines l_1, l_2, l_3 for each principal stress: i.e. for P_1, l_1, l_2, $l_3 \to a_{11}$, a_{12}, a_{13}; for P_2, l_1, l_2, $l_3 \to a_{21}$, a_{22}, a_{23}; and for P_3, l_1, l_2, $l_3 \to a_{31}$, a_{32}, a_{33}. The array or matrix of nine direction cosines specifies the orientation of the principal axes with respect to the arbitrary axes (see Exercise 14.2).

As with the matter tensors described earlier in Section 14.2.3, the stress tensor may be represented by a quadric—the stress quadric: $\sigma_1 x_1^2 + \sigma_2 x_2^2 + \sigma_3 x_3^2 = 1$ where the lengths of the semi-axes are $1/\sqrt{\sigma_1}$, $1/\sqrt{\sigma_2}$, $1/\sqrt{\sigma_3}$. However, in many (if not most) situations one or more of the principal stresses will be negative (compressive stress) and so the surface is not an ellipsoid but a hyperboloid of one or two sheets—not nearly as visually useful. Finally, we may note that it is only in special circumstances that we would 'start off' with arbitrary axes and find the principal axes by calculation. Usually one or more of the principal axes are obvious (e.g. as in a 'tensile test') and given one principal axis the other two can be found by simple graphical constructions, e.g. Mohr's circle construction (see Nye, p. 43).

14.4.2 The strain tensor, Neumann's principle, and thermal expansion

As with the stress tensor, the strain tensor is symmetrical and a general state of strain, described by three normal and three shear components, can, when referred to the

principal axes, be described by three normal components. Just as before the components of strain, ϵ, are distinguished by their suffixes.

However, we must make a very important distinction between the 'engineering' strains, e (with which we may be familiar), and the components of ϵ of the strain tensor. Figure 14.7(a) shows a 'unit cube' deformed into a parallelogram. The simple, or engineering strain or shear is e_{12} (first suffix gives the direction, the second the face on which the strain occurs). However, this simple strain or shear consists of **two** components: strains of $1/2\,e_{12}$ and $1/2\,e_{21}$ (Fig. 14.7(b)) followed by strains of $1/2\,e_{12}$ and $-1/2\,e_{21}$ (Fig. 14.7(c)). The former (Fig. 14.7(b)) corresponds to the components of a symmetrical strain tensor, $1/2\,e_{12} = \epsilon_{12} = \epsilon_{21}$ and the latter (Fig. 14.7(c)) corresponds to a rotation (about the x_3-axis in this example). The result is quite general: all sets of strains can be resolved into strains ϵ (described by a symmetrical strain sensor) and a rotation ω (described by an antisymmetrical tensor).

In this example writing the strains in matrix form:

$$
\begin{bmatrix} 0 & e_{12} & 0 \\ 0 & 0 & 0 \\ 0 & 0 & 0 \end{bmatrix} = \begin{bmatrix} 0 & \tfrac{1}{2}\,e_{12} & 0 \\ \tfrac{1}{2}\,e_{21} & 0 & 0 \\ 0 & 0 & 0 \end{bmatrix} + \begin{bmatrix} 0 & \tfrac{1}{2}\,e_{12} & 0 \\ -\tfrac{1}{2}\,e_{21} & 0 & 0 \\ 0 & 0 & 0 \end{bmatrix}
$$

or in general: $e_{ij} \qquad = \qquad \varepsilon_{ij} \qquad + \qquad \omega_{ij}$

Note again that in the symmetrical strain tensor ε_{ij} the off-diagonal terms are equal and that in the antisymmetrical tensor ω_{ij} they are the reverse of each other.

When a crystal is strained as a result of an external directional influence (e.g. stress) the strain tensor does not conform to the crystal symmetry and (as with the stress tensor) Neumann's principle does not apply. However, when the external influence is non-directional (a scalar), the resulting strain in the crystal **does** conform to the crystal symmetry and Neumann's principle does apply. The important example is thermal expansion, a second rank tensor that relates temperature change (a scalar) to the second rank strain tensor: the coefficients of thermal expansion/contraction are directly

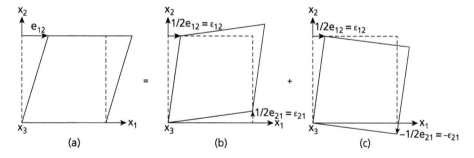

(a) (b) (c)

Fig. 14.7. A 'unit cube' deformed into a parallelogram (a) by the action of a simple strain or shear e_{12}. The resolution of this strain into strains (b) of $1/2\,e_{12}$ and $1/2\,e_{21}$ followed by strains (c) of $1/2\,e_{12}$ and $-1/2\,e_{21}$. The strains (b) correspond to the components $\varepsilon_{12} = \varepsilon_{21}$ of a symmetrical strain tensor and the strains (c) correspond to the components of an antisymmetrical tensor (a rotation), $\varepsilon_{12} = -\varepsilon_{21}$.

related to the coefficients of strain. In most situations the coefficients are all positive, so that the representation quadric is an ellipsoid. In some few cases (e.g. zinc at low temperatures) an expansion along the c-axis of the hexagonal unit cell is associated with a small contractions perpendicular to the c-axis: in this case the representation quadric is a single-surface hyperboloid.

Again, in accordance with Neumann's principle, the symmetry of the representation quadric must include the point group symmetry of the crystal (which again is the same for all point groups in a particular crystal system). Hence crystals are described as isotropic (cubic), uniaxial (hexagonal, trigonal, tetragonal), or biaxial (orthorhombic, monoclinic, triclinic) with respect to their thermal expansion properties.

14.4.3 Atomic displacement parameters (ADPs)

In order to describe precisely the atomic positions in a structure not only must the coordinates $x\, y\, z$ be specified but also the displacements arising from atomic disorder and the thermal displacements. These are in general not readily distinguished, are anisotropic, and are described by symmetric second rank tensors.

The temperature dependence of the thermal displacements is described by the Debye–Waller relation $\exp -2B\sin^2\theta/\lambda^2$ (see Section 13.3.4). B is known as the Debye–Waller or ADP and expresses the mean square atomic displacement \overline{U} through the relation $B = 8\pi^2\overline{U}^2$ (both \overline{U} and B measured in Å^2). In the anisotropic case B must be replaced by the six terms of a second rank tensor.

14.5 The optical properties of crystals

This topic was briefly introduced in Section 4.4 in which it was pointed out that, in the most general case (biaxial crystals), the optical properties are specified by three refractive indices, n_1, n_2, n_3.

Following Section 14.2.3, we now represent these as the semi-axes of a representation quadric, i.e. $\left(x_1^2/n_1^2 + x_2^2/n_2^2 + x_3^2/n_3^2\right) = 1$, which is an ellipsoid since the refractive indices n_1, n_2, and n_3 are all positive. It has a special name—the **indicatrix**—the properties and uses of which we discuss next.

However, it may appear from this equation that the refractive indices n_1, n_2, n_3 are the principal components of a second rank 'refractive index tensor'. This is not the case; refractive index is **not** a tensor property. However, the equation may be written in terms of the **relative dielectric impermeability**, B (the reciprocal of the relative dielectric constant K), which **is** a second rank tensor property, viz.

$$B_1 = {}^1\!/_{K_1} = {}^1\!/_{n_1^2}; \;\; B_2 = {}^1\!/_{K_2} = {}^1\!/_{n_2^2}; \;\; B_3 = {}^1\!/_{K_3} = {}^1\!/_{n_3^2}$$

where B_1, B_2, B_3 are the principal components of the tensor. Substituting these values in the previous equation and writing (as before) $x_1 = rl_1, x_2 = rl_2, x_3 = rl_3$ where r is a radius of the ellipsoid we have:

$$1\big/_{r^2} = \left(B_1 l_1^2 + B_2 l_2^2 + B_3 l_3^2\right)$$

an equation of the same form as that for the conductivity tensor (Section 14.2.3).

Hence, this representation quadric (of a second rank tensor property) and the indicatrix are identical and both must conform to Neumann's principle (Section 14.3). To repeat: for cubic crystals the indicatrix is a sphere, for hexagonal, trigonal, and tetragonal (uniaxial) crystals it is an ellipsoid of revolution about the x_3-axis and for orthorhombic, monoclinic, and triclinic (biaxial) crystals the principal axes of the ellipsoid bear the same relationships to the crystal axes as described before.

So much for the symmetry of the indicatrix. We are now going to use it to describe the passage of light through crystals. Two things need to be considered: (1) the direction of propagation and (2) the nature of the polarization of the light. In discussing polarization we are only concerned with the E-vector of the wave. Figure 14.8(a) shows a cross section (an ellipse) of the indicatrix for a uniaxial crystal. The refractive index in the direction of the optic axis is now denoted n_e (e standing for extraordinary wave or e-wave) and the refractive index in the circular cross section is denoted n_o (o standing for ordinary wave or o-wave). The significance of these denotions will shortly become plain. Note that in Fig. 14.8(a) $n_e > n_o$ (the 'long' direction of the ellipse is along the optic axis direction). This is the case for what are described as uniaxial positive crystals. When $n_e < n_o$ (the 'short' direction of the ellipse is along the optic axis direction) we have what are described as uniaxial negative crystals. Both are 'bad' terms; the distinction between 'positive' and 'negative' uniaxial crystals is quite arbitrary.

Suppose that a wave of plane (or linearly) polarized light incident perpendicular to the plane of the paper and with its plane of vibration (E-vector) at an angle to the optic axis as shown in Fig. 14.8(b). As it passes into the crystal the light **does not** remain

Optic axis

n_e

$\leftarrow n_o \rightarrow$

(a)

Fig. 14.8(a). A section of the indicatrix for a uniaxial (positive) crystal. The optic axis is shown as vertical and the corresponding long and short semi-axes of the ellipsoid of revolution are n_e and n_o, respectively.

(b)

Fig. 14.8(b). An incident wave of plane (or linearly) polarized light perpendicular to the plane of the paper and with its **E**-vector (or plane of vibration) shown by the double-headed arrow, at an angle to the optic axis. Also shown is the resolution of the incident plane-polarized wave into two plane-polarized components, one along the optic axis (*e*-wave) that propagates with a velocity c/n_e, and one perpendicular to the optic axis (*o*-wave) that propagates with a velocity c/n_o. At a depth t in the crystal the OPD between the waves is $t/\lambda(n_e - n_o)$. A helical wave (Fig. 14.8(c)) is equivalent to the combination of these two plane-polarized waves separated by an OPD.

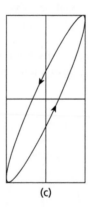

(c)

Fig. 14.8(c). On entering the crystal the wave is **no longer** plane polarized but the E-vector begins to sweep out a helix that (perpendicular to the direction of propagation) appears as an ellipse. The shape of the ellipse is not constant but continually changes as the light propagates further into the crystal.

plane polarized: the **E**-vector now begins to sweep out a helix along the direction of the propagation of the light, which, looking along the direction of propagation, appears as an ellipse (Fig. 14.8(c)). The shape of the ellipse is not constant: the ratio of its shorter to longer axes varies as we shall see next as the light propagates further and further into the crystal.

Now there is a simple way of understanding elliptically polarized light. It can be represented as the combination of two plane-polarized components: one with its E-vector along the optic axis (the e-wave) and the other with its E-vector perpendicular to the optic axis (the o-wave).

The resolution of the incident wave into these two components is shown in Fig. 14.8(b). The velocities of these two waves in the crystal are c/n_e and c/n_o (where c is the velocity of light *in vacuo*, or air). Their corresponding wavelengths are $\lambda_e = \lambda/n_e$ and $\lambda_o = \lambda/n_o$ (where λ is the wavelength of light *in vacuo*, or air). When they enter the crystal both waves are 'in step', but as they propagate through the crystal they become increasingly 'out of step' due to the velocity or wavelength difference. It is the combination of two 'out of step' plane polarized waves that is equivalent to a helical wave[1] and the amount to which they are 'out of step' determines the shape of the ellipse.

The amount by which the waves are 'out of step' is called the **optical path difference** (OPD). The OPD may be expressed in three ways. First, it may be expressed as a fraction (or multiple) of the wavelength λ (in air) that one wave is ahead of (or behind) the other. Consider the waves propagating through a crystal of thickness t. The number of extraordinary and ordinary waves in this thickness is $t/\lambda_e = tn_e/\lambda$ and $t/\lambda_o = tn_o/\lambda$, respectively. The difference, $t/\lambda(n_e - n_o)$, is the OPD expressed in terms of fractions (or multiples) of the wavelength. Second, multiplying by λ we have $t(n_e - n_o)$, i.e. the OPD expressed as a distance, and third, multiplying by $2\pi/\lambda$, we have $2\pi t/\lambda(n_e - n_o)$, i.e. the OPD expressed as a phase difference. The OPD is of supreme importance in polarized light microscopy because it determines the contrast and the interference colours that are observed.

Continuing on from Fig. 14.8(c), we now follow the changes in the form of ellipse as the light propagates through the crystal with an increasing OPD between the e-wave and o-wave. Figure 14.9 shows these changes for nine intervals of OPD from O to λ ($0°$ to $360°$ or 2π). Figure 14.9 is also drawn for the special case in which the **E**-vector of the incident plane polarized light is at an angle of $45°$ to the optic axis: in this case the amplitudes of the e-wave and o-wave are equal. In this case the ellipse expands into a circle at an OPD of $\lambda/4$, then contracts at an OPD of $\lambda/2$ to a plane polarized wave at $90°$ to the incident wave. Crystal plates cut with a thickness corresponding to an OPD of $\lambda/4$ or $\lambda/2$ therefore 'convert' plane polarized light to circularly polarized light and plane polarized light 'rotated' $90°$, respectively. The sequence is repeated: at an OPD of $3\lambda/4$ we again have circularly polarized light and at an OPD $= 1\lambda$ we have plane polarized light as at the beginning. Finally (very important), note the sense of rotation of the **E**-vector (corresponding to the sense of rotation of the helical wave) that changes from a left-handed to a right-handed screw (anti-clockwise (OPD $= \lambda/4$) and clockwise (OPD $= 3\lambda/4$) in Fig. 14.9).

Now, to conclude our discussion on the indicatrix, consider a wave incident at some angle to the optic axis (Fig. 14.10). Again, the wave is resolved into two components:

[1] A simple elegant demonstration of this is given in *Qualitative Polarised Light Microscopy* by P.C. Robinson and S. Bradbury, Oxford University Press (1992).

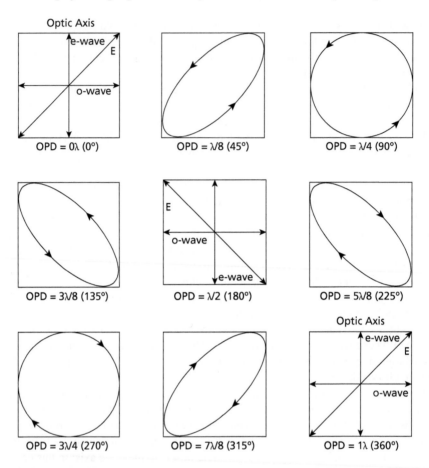

Fig. 14.9. The sequence of changes in the form of polarized light as it propagates through a crystal from an OPD = 0λ to an OPD = 1λ. Note that at OPDs of λ/4 and 3λ/4 we have left- and right-handed circularly polarized light, respectively and at an OPD = λ/2 plane polarized light vibrating at 90° to that of the incident light.

one (the o-wave) perpendicular to the optic axis as before and the other (the e-wave), in a plane containing the optic axis. The refractive index for this ray n'_e is given from the cross section of the indicatrix perpendicular to the direction of propagation and clearly has a value intermediate between n_e and n_o. We now see the distinction between the o-wave and e-wave: the o-wave has a fixed refractive index whereas that for the e-wave varies with direction of propagation of the light.

As the angle between the direction of propagation of the incident wave and the optic axis becomes smaller, n'_e becomes smaller—and when the wave travels along the optic axis $n'_e = n_o$, the OPD is zero and the polarization of the incident wave is unchanged. This simple analysis applies equally to biaxial crystals except that there are two optic

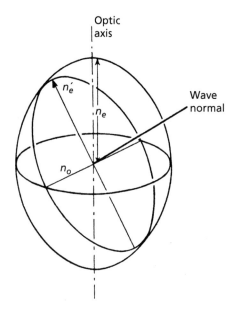

Fig. 14.10. A wave incident at an angle to the optic axis of a (positive) uniaxial crystal. The refractive indices for the e-wave and o-wave are given by the section of the ellipsoid perpendicular to the direction of propagation. The refractive index for the o-wave is constant but that for the e-wave, n'_e varies with the direction of propagation. $n'_e = n_e$ when the direction of propagation is perpendicular to the optic axis and $n'_e = n_o$ when the direction of propagation is along the optic axes.

axes (i.e. two circular cross sections of the indicatrix) and the refractive indices of both waves vary with direction. The geometry is more complicated but the principles are the same.

For cubic crystals, where the indicatrix is a sphere, the OPD for light travelling in any direction is zero. Cubic crystals are **optically isotropic**—viewed between crossed polars in the polarized light microscope they appear black or 'in extinction'. All other crystals are **optically anisotropic**. However, they may **appear** to be isotropic (i.e. in extinction) in the special case where the light is propagating along an optic axis.

14.6 Third rank tensors: piezoelectricity

A third rank tensor describes a crystal property that relates a second rank tensor and a first rank tensor (a vector), the most important example of which is piezoelectricity (or the piezoelectric effect). When a stress is applied to certain crystals (those, as we see next, which do not possess a centre of symmetry), they develop a polarization **P** (a vector).

Each component of **P**, P_1, P_2, P_3, is linearly related to all nine stress components, e.g. for P_1:

$$P_1 = d_{111}\sigma_{11} + d_{112}\sigma_{12} + d_{113}\sigma_{13}$$
$$+ d_{121}\sigma_{21} + d_{122}\sigma_{22} + d_{123}\sigma_{23}$$
$$+ d_{131}\sigma_{31} + d_{132}\sigma_{32} + d_{133}\sigma_{33}$$

where d_{111}, d_{112}, etc. are called the **piezoelectric moduli**. Note that the first suffix relates to the vector (P_1) and the second two suffixes relate to the stress tensor. Similarly, there are nine equations for P_2 and nine equations for P_3, giving $9 \times 3 = 27$ equations in all. The equations may be written in the compact form:

$$P_i = d_{ijk}\,\sigma_{jk} \quad (i,j,k = 1,2,3)$$

where again summation over the suffixes j and k is automatically understood.

Conversely, when an electric field **E** (a vector) is applied to such crystals they change shape. Again, each component of **E** (E_1, E_2, E_3) is linearly related to all nine strain components. The corresponding equations in the compact form are:

$$\epsilon_{jk} = d_{ijk}E_i \quad (i,j,k = 1,2,3)$$

These are called the direct and converse piezoelectric effects, respectively. It should be noted (not proved) that the piezoelectric moduli are the same in each case.

The number of independent coefficients is much reduced when the point group symmetry of the crystal is taken into account. Suppose a crystal with a centre of symmetry became polarized on the application of a stress. Suppose that the crystal plus stress is inverted through the centre of symmetry: the stress and crystal (both centrosymmetrical) will be unchanged but the polarization will be reversed. An impossible situation unless the polarization is zero.

The number of independent piezoelectric coefficients in the non-centrosymmetric point groups (see Table 3.1) may be determined by symmetry arguments, or more thoroughly by analytical methods: they vary from 18 (point group 1) to 0 (point group 432). Even though point group 432 is non-centrosymmetric it is the presence of the three tetrads that cancels out any polarization.

14.7 Fourth rank tensor properties: elasticity

Robert Hooke first recognized the linear relationship between (elastic) stress and strain that he expressed as 'ceiiinosssttuv'—an anagram for 'ut tensio, sic vis'. Tensile stress and strain are related by the Young modulus, E, and shear stress and strain by the shear modulus, G. Now we need to take into account that stress and strain are symmetrical second rank tensors. Each one of the nine stress components is related to nine strain components leading altogether to $3 \times 3 \times 3 \times 3 = 81$ terms in all. These are identified by four-number suffixes, e.g.

$$\sigma_{11} = c_{1111}\varepsilon_{11} + c_{1112}\varepsilon_{12} + c_{1113}\varepsilon_{13} + \cdots$$

$$\sigma_{12} = c_{1211}\varepsilon_{11} + c_{1212}\varepsilon_{12} + c_{1213}\varepsilon_{13} + \cdots \qquad \text{or, in compact notation}$$

$$\sigma_{ij} = c_{ijkl}\varepsilon_{kl}$$

The coefficients c_{ijkl} are called the elastic **stiffnesses**. The first two suffixes refer to the stress and the second two to the corresponding strain. The equations may be written 'the other way round', i.e.

$$\varepsilon_{11} = s_{1111}\sigma_{11} + s_{1112}\sigma_{12} + s_{1113}\sigma_{13} + \cdots \qquad \text{or, in compact notation}$$

$$\varepsilon_{ij} = s_{ijkl}\sigma_{kl}$$

where the coefficients s_{1111}, etc., are called (rather confusingly) the elastic **compliances**.

Since both stress and strain are symmetric tensors, each with six independent components, the number of independent elastic coefficients is reduced from 81 to $6 \times 6 = 36$; energy considerations further reduce the number to 21, which is the number for both point groups of triclinic crystals.

As before, considerations of crystal symmetry further reduce this number; but we will only state the results for hexagonal and cubic crystals and isotropic materials. (Notice straight away that cubic crystals are not isotropic with respect to their elastic properties.) But first we adopt a more concise (matrix) notation that avoids the use of four suffixes (this notation is also used for third rank tensor properties):

Tensor notation:	11	22	33	23,32	31,13	12,21
Matrix notation:	1	2	3	4	5	6

For hexagonal crystals there are five independent coefficients. Writing these in terms of compliances:

$$s_{11} = s_{22}, s_{12}, s_{13} = s_{23}, s_{33}, s_{44} = s_{55}, s_{66} = 2(s_{11} - s_{12})$$

For cubic crystals there are three independent coefficients:

$$s_{11} = s_{22} = s_{33}, s_{12} = s_{13} = s_{23}, s_{44} = s_{55} = s_{66}$$

For isotropic materials this reduces to two independent coefficients:

$$s_{11} = s_{22} = s_{33}, s_{12} = s_{13} = s_{23}, s_{44} = s_{55} = s_{66} = 2(s_{11} - s_{12})$$

For isotropic materials it is useful to relate the compliances to E, G, and ν (Poisson's ratio). Comparing equations:

$\varepsilon_1 = s_{11}\sigma_1 + s_{12}\sigma_2 + s_{13}\sigma_3$ and $\varepsilon_1 = 1/E\{\sigma_1 - \nu(\sigma_2 + \sigma_3)\}$ and similarly for ε_2 and ε_3.

$\varepsilon_4 = 2(s_{11} - s_{12})\sigma_4$ and $\varepsilon_4 = 1/G\ \varepsilon_4$ and similarly for ε_5 and ε_6.

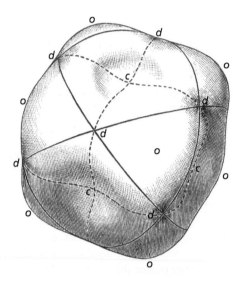

Fig. 14.11. Elasticity surface for iron pyrites, FeS$_2$. The radius in any direction, s'_{11}, is the reciprocal of the Young modulus.

Comparing coefficients we have: $s_{11} = 1/E$, $s_{12} = -v/E$, $2(s_{11} - s_{12}) = 1/G$ from which follows the relation: $2(1 + v)G = E$.

Unlike the representation quadric for symmetrical second rank tensor properties, there is no single surface that can completely represent the elastic behaviour of crystals. However, a single surface can represent, say, the variation of the Young modulus, E. Consider the compliance $s'_{//}$ with direction cosines $l_1 l_2 l_3$ to the principal axes. For cubic crystals:

$$s'_{//} \text{(the reciprocal of E in any direction)} = s_{11} - 2(s_{11} - s_{12} - 1/2s_{44})$$
$$\times \left(l_1^2 l_2^2 + l_2^2 l_3^2 + l_3^2 l_1^2 \right)$$

Figure 14.11 shows such an elasticity surface for iron pyrites, FeS$_2$, for which E is maximum in the <100> directions (labelled c) and minimum in the <111> directions (labelled o). For isotropic materials the surface is, of course, a sphere, radius $1/E$.

Exercises

14.1 With reference to Fig. 14.4, prove that the normal to the tangent at **E** is parallel to **J**.

14.2 A stress tensor referred to arbitrary axis x_1, x_2, x_3 has values:

$$\begin{bmatrix} 11 & \overline{6} & 2 \\ \overline{6} & 10 & 4 \\ 2 & 4 & 6 \end{bmatrix}$$

Find the values of the principal stresses (eigenvalues) and the direction cosines of the principal axes (eigenvectors).

14.3 As stated in Section 14.5, the condition for extinction in an optically anisotropic crystal is when the light propagates along an optic axis. Are there any other situations in which extinction can occur?

Appendix 1

Computer programs, models and model-building in crystallography

Computer programs now contribute a significant 'learning resource' in crystallography since, unlike the illustrations in books, images of crystal structures, lattices and space groups can be created and orientated at will such that a student can more readily appreciate the dispositions of atoms and symmetry elements. The same applies to the three-dimensional construction of the Ewald reflecting sphere and reciprocal lattice in providing an understanding of the geometry of diffraction. Moreover, some computer programs make use of anaglyphs such that the images appear to 'emerge' from the screen as three-dimensional objects when viewed with the appropriate left and right eye filter spectacles.

It might be supposed therefore that such computer programs have superseded models as a learning resource. I believe that this is not the case: what is missing is the *tangible* element, the enhancement of understanding brought about by the link between hand and eye. In short, models provide the basis for crystallographic *object lessons*. And by models I do not only mean the crystal-model building kits which are commercially available but the simpler ones which you can make yourself from cardboard or plastic spheres or even such everyday objects as clothes pegs, tennis balls or plant stems which provide the very first lessons in understanding symmetry.

Listed below are (1) a selection of computer programs which are of value in learning crystallography (though many include also much more advanced material) and (2) some suggestions for model-building and a list of suppliers of model-building kits.

A1.1 Computer programs in crystallography

The International Union for Crystallography (IUCr) maintains a website with links to many others. Its URL is:

(IUCR) http://www.iucr.org/

Useful links are to the American Crystallographic Association (ACA), the British Crystallographic Association (BCA), Crystallography World-Wide (CWW) and the European Crystallographic Association (ECA) whose URLs are:

(ACA) http://www.hwi.buffalo.edu/ACA/
(BCA) http://bca.cryst.bbk.ac.uk/BCA/index.html
(CWW) http://www.iucr.org/cww-top/crystal.index.html
(ECA) http://area.ba.cnr.it/eca/

The Science and Technology Facilities Council (STFC) Chemical Database Service (CDS) at Daresbury (<http://cds.dl.ac.uk/cds>), provides databases containing crystal structures data on organic, organometallic, inorganic compounds, and metals and alloys that can provide a basis for modelling studies on new materials. The major resources available are the Cambridge Structural Database (CSD), the Inorganic Crystal Structure Database (ICSD), CrystMet, and the Crystal Data Identification File (CDIF).

The following list provides brief descriptions of some programs currently available.

Ca.R.Ine Crystallography, version 3.1 and 4.0 by Cyrille Boudias and Daniel Monceau, distributed by Christine Leheutre, Uteam-Divergent, Centre de Transfert UTC—BP 50149, 60200 Compiegne, France, e-mail christine.leheutre@utc.fr.

CARINE CRYSTALLOGRAPHY version 3.1 is the preferred tool for teaching and research in materials science. It is used for simple rendering and for simulation and analysis. CaRIne 3.1 exposes in a clear way the relationships between different geometrical representations of crystals and brings them together in a user-friendly interface. It includes real lattice, interactive stereographic projections, X-ray diffraction on powders, reciprocal lattices in 3D and 2D, and (hkl) based calculation lists. A simplified version 3.1 for students, which includes 22 examples of use of the software, is given with version 3.1. CaRIne version 4.0 is a beta version delivered free with version 3.1. It is focused on 3D modelling of crystals, including surfaces and interfaces. CaRIne version 3.1 and 4.0 can read the unit cells created by each other. Version 3.1 is available for all versions of Windows, version 4.0 also runs under Linux.

Crystallographica © Oxford Cryosystems Ltd, 3 Blenheim Office Park, Long Hanborough, Oxford, OX29 8LN, UK, e-mail: info@oxcryo.com, follow @oxcryo.

CRYSTALLOGRAPHICA is a software toolkit aimed at scientists whose work involves crystallography and is now available free from the downloads page. Combining an easy-to-use Microsoft Windows interface with a powerful and flexible crystallographic engine, it can serve as a 'friendly crystallographer' for the non-specialist, but has more than enough power to satisfy the most demanding professional and for teaching. Functions include: crystal structure drawing with VRML export; X-ray and neutron powder pattern simulation; and weighted and unweighted reciprocal lattice display. It is also fully scriptable using the Pascal interpreter with an extensive library of crystallographic functions. Crystallographica runs under all 16- and 32-bit versions of Windows. Use with a 64-bit edition of Windows will require an emulator such as Microsoft's Windows Virtual PC.

CrystalMaker ® from CrystalMaker Software Ltd., Centre for Innovation and Enterprise, Begbroke Science Park, Woodstock Road, Begbroke, Oxford, OX5 1PF, UK, e-mail: info@crystalmaker.com.

CRYSTALMAKER is a program for building, displaying, and manipulating crystal and molecular structures, including real-time photo-realistic graphics and 'out-of-the-screen' three-dimensional displays. CrystalMaker ® allows visualization of massive structures (millions of atoms), with fast bond and polyhedra generation, and the display

of thermal ellipsoids. New features include the ability to seamlessly browse multiple structures in the same window and generate animations and movies (e.g. showing structural changes resulting from temperature or pressure). The program supports a wide range of data formats, exports high-resolution graphics in a wide range of formats, plus QuickTime movies and VR objects—and works with the 'SingleCrystal' and 'CrystalDiffract' diffraction software. SingleCrystal is a Mac and Windows application, designed to simulate X-ray, neutron, and transmission electron microscope (TEM) diffraction patterns of crystals, with tools for measurement, stereographic projections, display of observed diffraction images, and auto-indexing. CrystalDiffract is an application for measuring and characterizing X-ray and neutron powder diffraction patterns. The software includes tools for background subtraction and measurement, plus the ability to simulate diffraction patterns with interactive parameter adjustments. Multiple patterns can be viewed in the same window, with the ability to simulate multiphase mixtures.

CrystalMaker ® is available in two versions: for Mac (OS X 10.5 and later) and for Windows (Windows XP and later). Each version is a native application, developed from scratch for its respective operating system.

Crystals by David Watkin and Richard Cooper, Chemical Crystallography Department, University of Oxford, Oxford, OX1 3TA, UK, e-mail: david.watkin@chem.ox.ac.uk, richard.cooper@chem.ox.ac.uk.

CRYSTALS is available without charge for non-commercial use from the web server xtl.chem.ox.ac.uk. It provides all the basic and advanced facilities needed for crystal structure determination, plus a visualization window which continuously displays the structure and a graphics module for manipulating and drawing the structure. A 'guide' leads the user through all the stages of a normal crystal structure analysis and the manual is included as Postscript files and as browsable html. The distribution kit includes versions of SHELXS, SIR92 and Superflip (with the authors' permissions). Tutorials are included for a number of demonstration structures. A scripting interface enables the program to be customized as a teaching aid, and two extensible examples are included.

!Polynet, !Polysymm and !Stellate by Kate Crennell, 'Fortran Friends', The Lane, Chilton, Didcot, Oxon, OX11 OSE, UK. Website: <http://fortran.orpheusweb.co.uk/poly/prodp.htm.>

THESE POLYHEDRA PLOTTING PROGRAMS can run on RISCOS systems or on PCs under the Virtual Acorn emulator. They show the construction and symmetries of the Platonic and Archimedean solids and their duals and some of their stellations. For details see the website above, from which demonstration copies of the software may be freely downloaded. Free copies of plans to cut out and construct 3D models are also available to make space-filling models and examples of each of the crystal classes.

A1.2 Crystallography models, model-building and suppliers

The use of models in the teaching or exposition of crystallography has a long history; the first perhaps being the terracotta models made to accompany Rome di L'Isle's treatise

Cristallographie (1783). Later, comprehensive sets of models were made in wood by geological suppliers, particularly the German firm of Krantz: some were made in two halves linked by a rod to demonstrate twinning. Models of different space-filling poly-hedra were made by Alexander Herschel, brother of the rather more famous William Herschel, discoverer of the planet Uranus (which **he** called Georgium Sidus, after the King). More recently (1960s) the firm Cutrock Engineering made machined aluminium models based on forms shown in F.C. Phillips' book *Introduction to Crystallography*. Besides these, models in glass, paper, and cardboard have been made. Such models not only demonstrate crystal symmetry but have particular application in mineralogy as a practical substitute for crystals that are small, rare, or imperfectly formed. The Natural History Museum (NHM) in London has a substantial collection of such models, of which Figs A1.1–A1.3 show three examples[1].

Fig. A1.1. Shows glass models of five of the crystal systems with internal coloured wires indicating the crystal axes and some of the crystal forms in each system.

[1] This discussion is closely based upon Peter Tandy's article 'Crystallography and geometric modelling of minerals: a reflection on the models in the Natural History Museum, London', in *Crystallography News*, June 1999, published by the BCA. I am grateful to Peter Tandy and the NHM for permission to reproduce these figures.

Fig. A1.2. Is one of the most beautiful models in the Collection one of a set of 24 made by Nina Hosali in the 1920s. The exterior glass envelope shows faces of the {100} cubic form; within, fine silk threads of various colours over wire frames show other forms. Because of the construction method, faces of one form intersect with faces of another.

Fig. A1.3. Shows one of the most bizarre models in the Collection, made by William Barlow to illustrate his derivation of the 230 space groups. Of a total of 2~300 such models, only these survive. Barlow's workshop was said to represent a toy factory with hundreds of balls of different sizes and dolls' hands as shown here. They are arranged in groups of three (both left and right hands) around the body-diagonal (the triad axis) within a cubic frame. The cubic space group symmetries were achieved by stacking the frames together in their appropriate relationships.

Whilst such models may be an inspiration, the value of **making** models in under-standing symmetry, cannot be overstated. *Cardboard/paper models of polyhedra* are very easily constructed. 'Ready-made' templates are available on the BCA website and also in several **Tarquin Publications** (99 Hatfield Road, St Albans, AL1 4JL, UK; www.tarquinbooks.com), e.g. *Make Shapes*, books 1 and 2 by Jenkins and Wild; *Paper Polyhedra in Colour* by Jenkins and Bear; *Geodesic Domes* by van Loon and the classic text *Mathematical Models* by Cundy and Rollett. By making collections of polyhedra and glueing them together you can see, in a very direct way, those which are space-filling and those which are not. You can also represent twinned and multiply-twinned crystals. Figure A1.4 shows an interesting example. Each tetrahedron represents a ccp crystal, the faces of which correspond to the {111} twin planes. The twinning angle across each face is 70.53° and the total angle for the five twins is 70.53° × 5 = 352.65°, i.e. very close to 360°. Such a multiply-twinned ccp crystal shows 'pseudo-pentagonal' symmetry—not to be confused with the pentagonal symmetry of quasicrystals (see Section 4.9).

Plastic (polystyrene) sphere models of the ccp and hcp metal structures may be made by glueing spheres together to form close-packed hexagonal layers and then stacking them together in the ABCABC... and ABAB... sequences. The insertion of smaller (e.g. plasticine) spheres shows the sizes and distribution of the interstitial sites. The more complex stacking sequences (e.g. in SiC (Fig. 1.27) or of the oxygen atoms in Al_2O_3) can also be demonstrated. The bcc metal structure is best made by making layers of the closest-packed {110} planes (Fig. 1.9(b)) and stacking them in the ABAB... sequence, atoms in each layer being located at the 'saddle-points' of the layer below. Notice that the layer-separation is reduced if the layers are slipped into the recesses each side of the saddle-points, giving a more closely packed but no longer cubic structure.

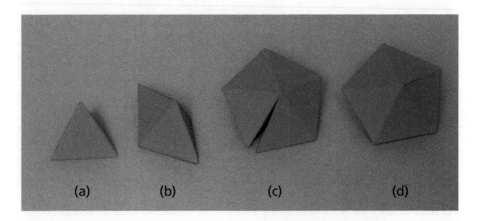

Fig. A1.4. (a) The faces of the tetrahedron correspond to the {111} twin planes in a ccp crystal; (b) a singly-twinned crystal; (c) a multiply-twinned crystal (five twins) showing pseudo-pentagonal symmetry. By a small distortion of the model the pentagonal symmetry of a quasicrystal (d) is obtained.

Childrens' wooden building bricks come in a variety of shapes—cubes, tetragonal, hexagonal and orthorhombic prisms—and (apart from building castles), may be used to represent the 32 point groups (monoclinic and triclinic models must be home-made). The plain models represent the holosymmetric point groups; those of lower symmetry may be made by cutting-off appropriate corners and edges (as shown in Fig. A1.5 for the three orthorhombic point groups) or by marking the faces (as shown in Fig. A1.6 for the five cubic point groups). Bricks may also be cut across a non-mirror plane and the two halves then rotated 180° to show the geometry of twinning. Clearly, many beautiful multiply-twinned crystals can be created in this way.

Model-building kits and complete models are obtainable from a number of suppliers of which the following is a selection.

Cochranes of Oxford Ltd., 29 Groves Yard, Shipton Road, Milton under Wychwood, Chipping Norton, Oxford, OX7 6JP, UK; http://www.cochranes.co.uk makes Orbit™ and Minit™ molecular model kits based on plastic atom centres colour-coded according to the element which are joined together with plastic tubes of the appropriate lengths. The scale of the Orbit™ system is 30 mm = 1 Å when used for organic and inorganic chemistry and typically 12–25 mm = 1 Å for crystal structures. For the smaller-scale Minit™ system the scale is typically 10–15 mm = 1 Å. Also available are larger-scale Unit™ molecular models in which the atoms are joined together with rigid aluminium or flexible vinyl tubes.

Miramodus Limited (formerly Beevers Miniature Models Unit), c/o School of Chemistry, West Mains Road, Edinburgh, EH9 3JJ; http://www.miramodus.com produces crystal models for display and teaching on the scale of 1 cm = 1 Å. Coloured PMMA balls are individually drilled and connected with stainless steel rods to give accurate and attractive models of any crystal or molecular structure. Models can be made to

Fig. A1.5. Models showing the symmetries of the three orthorhombic point groups: *mmm*, *mm2* and 222. Note the two enantiomorphic forms of point group 222.

Fig. A1.6. Models showing the symmetries of the five cubic point groups $m\bar{3}m$, $\bar{4}3m$, $m\bar{3}$, 432 and 23. Note the two enantiomorphic forms of point groups 432 and 23.

individual requirements, including mounting for gifts or presentations and are suitable for museum-quality displays. Rods and balls can also be supplied for self-assembly.

Spiring Enterprises Ltd., Billingshurst, West Sussex, RH14 9EZ, UK; http://www.molymod.com makes the Molymod™ molecular model system. The colour-coded atom centres are joined together with flexible mushroom-shaped links. A wide range of sets for building organic and inorganic molecular models is available as well as DNA (and RNA) model kits in which abstract-shaped parts are designed to represent the bases, sugar and phosphate groups of the double helix.

3-D Molecular Designs, 2223 North 72nd Street, Wauwatoza, WI 53213, USA; http://3DMolecularDesigns.com make molecular and crystal models in which the atoms or molecules are held together by internal magnets. These models simulate, for example, the hydrogen bonding and polarity of water molecules in the structures of liquid water and ice. A wide range of models of organic molecules are also available. The 'Nobel Prize' models include DNA, haemoglobin, and insulin.

Appendix 2

Polyhedra in crystallography

In this Appendix the crystallographic and geometrical properties of the polyhedra which are introduced and briefly described in this book are summarized. The word comes from the Greek meaning many (poly) faces or planes (hedra). The (simpler) polyhedra are simply described by their number of faces as indicated below. It is curious that only the hexahedron has a 'simple' name—the cube.

Polyhedron names

For the simpler polyhedra the prefix indicates the number of faces. Those of importance in crystallography are: tetra (4), penta (5), hexa (6), octa (8), deca (10), dodeca (12), icosa (20), icosidodeca (32). Further prefixes indicate the shapes of the faces themselves and/or their symmetry, e.g. pentagonal or rhombic dodecahedra (polyhedra with twelve pentagonal or rhombic (diamond shaped) faces respectively), or trapezorhombic dodecahedra (polyhedra in which the twelve rhombic faces are arranged in hexagonal symmetry). Further prefixes such as triakis or pentakis indicate that plane triangular or pentagonal faces are replaced by a group of three or five faces respectively. Thus a triakis tetrahedron has twelve (4 × 3) faces, and so on.

Truncation

Many polyhedra are described as 'truncated' which simply means 'cutting off the corners' of a polyhedron—and the truncated polyhedron thus created depends on 'how much' of the corners are cut off. As shown in Fig. A2.1(a)–(c) the truncated cube, cubeoctahedron and truncated octahedron are created by cutting increasing amounts off the cube corners. Alternatively, we could 'start off' with an octahedron and obtain the same polyhedra by again cutting off the corners as shown in Fig. A2.1(d)–(f). Truncation can also be applied to the newly-created corners and edges: the possibilities are almost endless and a whole series of polyhedra, with many faces of different shapes and sizes (and with rather long descriptive names) can be created. The process has its analogy in the beautiful patterns of faces which can develop in crystal growth.

Duality

Polyhedra of identical symmetry are said to be dual if the faces of one correspond to the corners (or vertices) of the other and vice versa. Thus the cube and octahedron are duals: the eight vertices of the cube correspond to the eight faces of the octahedron and, conversely, the six vertices of the octahedron correspond to the six faces of the cube.

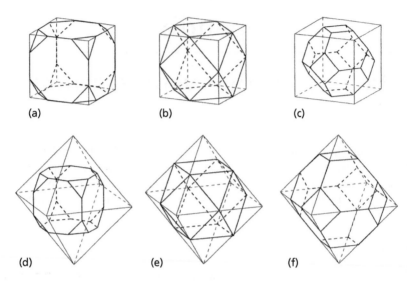

(a) (b) (c)

(d) (e) (f)

Fig. A2.1. Alternative truncation procedures to create a truncated cube, cubeoctahedron and truncated octahedron. (a) → (b) → (c) successive truncation of a cube; (d) → (e) → (f) successive truncation of an octahedron. (From *Polyhedra* by Peter R. Cromwell, Cambridge University Press, Cambridge, 1999.)

Similarly, the pentagonal dodecahedron and icosahedron are duals. The tetrahedron is *self-dual*: the four vertices of one correspond to the four faces of the other and vice versa.

Classification of polyhedra

There is an enormous number of polyhedra, including beautiful star-shapes with re-entrant faces. However, those of most importance in crystallography are the Platonic and Archimedean polyhedra, deltahedra, coordination and space-filling (Voronoi) polyhedra and Brillouin zones. These groupings are of course not mutually exclusive.

(A) The five perfect (regular) polyhedra or Platonic solids

These polyhedra (Fig. A2.2) each have identical faces and edges of equal length. The vertices and faces of each circumscribe and inscribe respectively the surface of a sphere.

The tetrahedron has cubic point group symmetry $\bar{4}3m$ and the octahedron and cube have the full (holosymmetric) cubic point group symmetry $4/m\bar{3}2/m \equiv m\bar{3}m$. The icosa-hedron and pentagonal dodecahedron both have point group symmetry $2/m\bar{3}\bar{5}$—the symmetry possessed by quasicrystals, buckminsterfullerene and some viruses. Of all the Platonic polyhedra, only the cube is space-filling.

These solids were known to the Ancient Greeks and were clearly identified and described by Theaetetus of Athens (d. 369 BC). They were considered by Plato to constitute the fundamental building-blocks of matter: fire (tetrahedron), solid (cube), air (octahedron), liquid (icosahedron) and the universe or cosmos itself (pentagonal

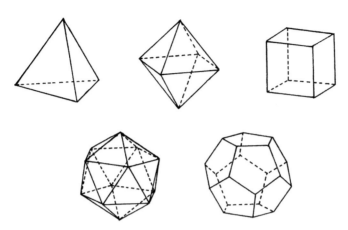

Fig. A2.2. The five perfect polyhedra or Platonic solids: tetrahedron, octahedron, cube, icosahedron, pentagonal dodecahedron.

dodecahedron). Hence, they are attributed to him, instead of (more properly) to Theaetetus.

The notion that the Platonic solids had a cosmic significance was taken up by Johannes Kepler in his first book *Mysterium Cosmographicum*, 1596, in which he attempted to show that the ratios of the diameters of the (circular) orbits of the then-known six planets corresponded with the ratios of the inscribed and exscribed diameters of the spheres of the five perfect solids. Thus the ratios of the diameters of the inscribed and exscribed spheres for an octahedron correspond to the orbits of Mercury and Venus; that for an icosahedron the orbits of Venus and Earth, and so on. In short he attempted to explain or model the orbits of the planets in purely geometrical terms. The model failed of course, even in Kepler's time, by observation and calculations of greater precision on the ellipticity of the planetary orbits.

(B) The thirteen semi-regular or Archimedean Polyhedra
In these polyhedra, the discovery of which is attributed to Archimedes, every face is a regular polygon as in the Platonic polyhedra but the faces may be of different kinds—two kinds in the case of ten of the polyhedra and three kinds in the remaining three. Around each vertex the faces are arranged in the same order, the edges are of equal length and all the polyhedra are inscribable in a sphere. A complete list is given in Table A2.1 and those of particular crystallographic interest are shown in Fig. A2.3.

We owe the (rather long) names of these polyhedra to Kepler, and as indicated, they may be obtained by truncation of the corresponding Platonic solids—either a simple truncation of the vertices as shown for example in Fig. A2.1, or a more complex truncation of the newly created vertices or edges. For example the truncated cubeoctahedron (also known as the great rhombicubeoctahedron) and the (small) rhombicubeoctahedron are obtained by truncating the cubeoctahedron in two different ways. The last two polyhedra, described by the curious word 'snub', are not derived simply by a process

Table A2.1 The Archimedean polyhedra

No.	Name	No. and shapes of faces	Point group symmetry
1	Truncated tetrahedron	4 hexag, 4 triang	$\bar{4}3m$
2	Truncated cube	8 triang, 6 oct	$m\bar{3}m$
3	Truncated octahedron	6 square, 8 hexag	$m\bar{3}m$
4	Truncated dodecahedron	20 triang, 12 decag	$2/_m\bar{3}\bar{5}$
5	Truncated icosahedron	12 pentag, 20 hexag	$2/_m\bar{3}\bar{5}$
6	Cuboctahedron	8 triang, 6 square	$m\bar{3}m$
7	Icosidodecahedron	20 triang, 12 pentag	$2/_m\bar{3}\bar{5}$
8	Truncated cubeoctahedron (great rhombicubeoctahedron)	12 square, 8 hexag, 6 oct	$m\bar{3}m$
9	(small) rhombicubeoctahedron	8 triang, 18 square	$m\bar{3}m$
10	Truncated icosidodecahedron	30 square, 20 hexag, 12 decag	$2/_m\bar{3}\bar{5}$
11	Rhombicosidodecahedron	20 triang, 30 square, 12 pentag	$2/_m\bar{3}\bar{5}$
12	Snub cubeoctahedron	32 triang, 6 square	432
13	Snub icosidodecahedron	80 triang, 12 pentag	235

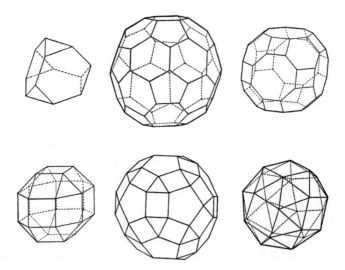

Fig. A2.3. Six of the thirteen Archimedean polyhedra, Nos. 1, 5, 8, 9, 11, 12 (see Table A2.1). Figure A2.1 shows Nos. 2, 3, 6. (From *Mathematical Models* by H.M. Cundy and A.P. Rollett, 2nd edn, Clarendon Press, Oxford, 1961 reprinted by Tarquin Publications, St Albans, 2006.)

of truncation, but also involve a 'twist' or rotation, which destroys mirror planes and centres of symmetry. The point groups of the Archimedean polyhedra are also of interest. Seven have crystallographic symmetry (cubic point groups $\bar{4}3m$, $m\bar{3}m$, and 432) and six have icosahedral symmetry (the two icosahedral point groups $2/m\bar{3}5$ and 235). Note again that the point groups 432 and 235 are enantiomorphous.

Of these thirteen polyhedra, only the truncated octahedron, also described by Lord Kelvin as a tetrakaidecahedron (the Voronoi polyhedron for the cubic I lattice) is space-filling (Fig. 3.9). The pattern of faces in the truncated icosahedron corresponds to that of C_{60} (see Section 1.11.6). The duals of the Archimedean polyhedra have the same symmetry. They are perhaps of less interest, except for the dual of the cubeoctahedron which is a rhombic dodecahedron (the space-filling Voronoi polyhedron for the cubic F lattice (Fig. 3.8(a))) and that of the icosidodecahedron which is a rhombic triacontahedron (Fig. A2.4[1]). The ratios of the lengths of the long and short diagonals of the rhombic or diamond-shaped faces are not the same: for the rhombic dodecahedron it is $\sqrt{2}:1$ and for the rhombic triacontahedron it is $\left(\sqrt{5}+1\right)\big/2:1$ or $1.618:1$—the Golden ratio (see Section 2.9).

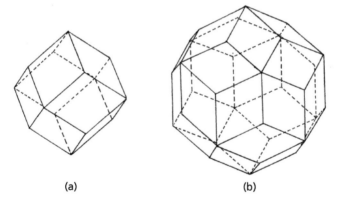

(a) (b)

Fig. A2.4. (a) Rhombic dodecahedron (dual of cubeoctahedron), and (b) rhombic triacontahedron (dual of icosidodecahedron). (a) is the space-filling polyhedron for the cubic F lattice and also the coordination polyhedron for the cubic I lattice.

(C) The eight deltahedra

In these polyhedra all the faces are equilateral triangles (Greek Δ). Three (the tetrahedron, octahedron and icosahedron) are also Platonic polyhedra. The remaining five have 6, 10, 12, 14 and 16 faces. Two, the triangular and pentagonal dipyramids with 6 and 10 faces, are shown in Fig. A2.5. Their duals are triangular and pentagonal prisms respectively.

[1] Only the rhombic dodecahedron was known to Archimedes, the rhombic triacontahedron was known to Kepler and the remaining 11 duals of the Archimedean polyhedra were discovered by the French mathematician E.C. Catalan in 1865.

 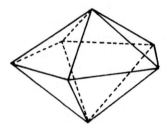

Fig. A2.5. Two of the eight deltahedra: the triangular and pentagonal dipyramids (point group symmetries $\bar{6}\,m2 = {}^3/_m\,m2$ and $\bar{10}\,m2 = {}^5/_m\,m2$ respectively).

(D) Coordination and space-filling polyhedra (Voronoi polyhedra or Dirichlet domains)

Coordination polyhedra, which represent the pattern of atoms or lattice points surrounding an atom or lattice point, should be clearly distinguished from the domains (Voronoi polyhedra or Dirichlet domains) which define the environment around an atom or lattice point and which are always space-filling (see Section 3.4). A vertex of a coordination polyhedron corresponds to the face of the corresponding domain, but the polyhedra are not necessarily duals.

They are duals, for example, in the cubic F lattice in which the coordination polyhedron is a cubeoctahedron (Fig. A2.1(b) and (e)) and the space-filling polyhedron is a rhombic dodecahedron (Fig. A2.4(a)) but not in the case of the cubic I lattice in which the coordination polyhedron is a rhombic dodecahedron[2] and the space-filling polyhedron is a truncated octahedron.

The procedure for constructing the space-filling Voronoi polyhedra for lattices is described in Section 3.4. For crystal structures consisting of only one kind of atom[3]—e.g. the simple cubic, bcc, fcc and hcp metal structures the procedure is the same. For structures consisting of more than one kind of atom the procedure is to construct the space-filling polyhedron around one kind of atom and its nearest neighbours *of the same kind*.[4] The atoms of other kinds are then shown as occupying the appropriate vertices of the space-filling polyhedron.

A few examples will make this clear. In the fcc NaCl and ZnS (zinc blende or sphalerite) structures, the space-filling polyhedron centred around a Cl or S anion is a rhombic dodecahedron (Fig. 3.8 (a)). In NaCl the Na cations, which occupy all six octahedral sites, are situated at the six vertices where four edges (or faces) meet (Fig. A2.6(a)). In sphalerite the Zn cations that occupy half the tetrahedral sites are situated at four of the eight vertices where three edges (or faces) meet (Fig. A2.6(b)). In the ZnS (wurtzite),

[2] The eight 'obtuse' vertices correspond to the lattice points at the corners of the cell and the six 'acute' vertices correspond to the lattice points at the centres of neighbouring cells as indicated in Fig. A2.4 (a)

[3] The term 'atom' encompasses charged species, i.e. both cations and anions.

[4] E.W. Gorter (1970) *Classification representation and prediction of crystal structures of ionic compounds.* J. of Solid State Chem. **1**, 279. See also R.J.D. Tilley, John Wiley & Sons, Ltd. (2006) *Crystals and Crystal Structures.*

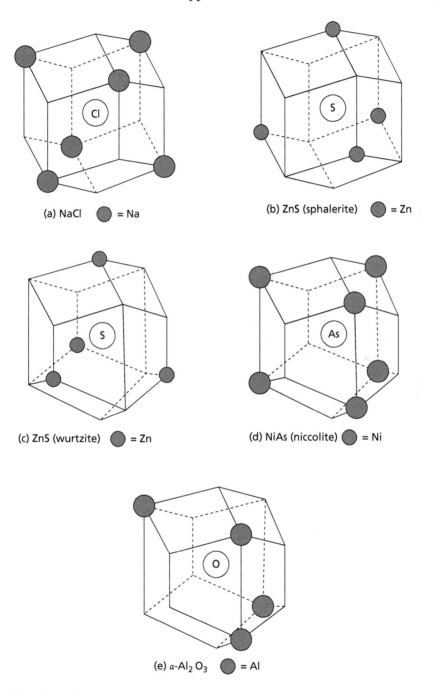

Fig. A2.6. Space-filling polyhedra for (a) the NaCl (rock salt) structure, (b) the ZnS (sphalerite) structure, (c) the ZnS (wurtzite) structure, (d) the NiAs (niccolite) structure, and (e) the α-Al_2O_3 (corundum) structure.

NiAs (niccolite), and α-Al_2O_3 (corundum), structures where the anions are (approximately) hexagonal close-packed, the anion-centred space-filling polyhedron is a trapezorhombic dodecahedron (Fig. 3.8(b)). In wurtzite the Zn cations are again situated at four of the eight vertices where three edges (or faces) meet (Fig. A2.6(c)); in niccolite the Ni cations are situated at all six vertices where four edges (or faces) meet (Fig. A2.6(d)); and in α-Al_2O_3 the Al cations are situated at four such vertices (Fig. A2.6(e)).

In more complex crystal structures, such as the spinels (Section 1.11.3, p. 36), two rhombic dodecahedra are needed to represent the ordered arrangement of the cations in the appropriate octahedral and tetrahedral interstitial sites.

Coordination polyhedra also determine the radius ratios of interstitial sites, e.g. between the atoms in metals as described in Section 1.6 or between close-packed oxygen anions. The data are summarized in Table A2.2 which includes the interstitial sites delineated by squares and triangles.

Table A2.3 lists the coordination and space-filling polyhedra for some simple structures.

(E) Brillouin zones

Brillouin zones are 'nests' of polyhedra in k-space (reciprocal space with the scale factor 2π). Their surfaces represent the directions and wavelengths (or wavenumbers) of free electrons in crystals which are Bragg-reflected by the lattice planes. The first, innermost Brillouin zone corresponds to the reflections from the planes of largest d-spacing, the next Brillouin zone corresponds to reflections from the planes of next largest spacing, and so on.

The first Brillouin zone is equivalent to the Voronoi polyhedron or Wigner–Seitz cell for the corresponding reciprocal lattice. For example, for the cubic F lattice, for which

Table A2.2 Coordination polyhedra for interstitial sites and radius ratios

Structure type	Co-ordination polyhedron around interstitial site	Radius ratio
Simple cubic	Cubic	0.732
	Square	0.414
Cubic close packed	Octahedral	0.414
	Square	0.414
	Tetrahedral	0.225
	Triangular	0.155
Hexagonal close packed	Octahedral	0.414
	Tetrahedral	0.225
	Triangular	0.155
Body centered cubic	'Distorted' octahedral	0.154
	'Distorted' tetrahedral	0.291
Simple hexagonal	Triangular prism	0.528
	Triangular	0.155

Table A2.3 Coordination and space-filling polyhedra

Structure/lattice type	Co-ordination polyhedron	Space-filling polyhedron
cubic P (simple cubic)	octahedron (square dipyramid)	cube (dual)
cubic F (fcc)	cubeoctahedron	rhombic dodecahedron (dual)
cubic I (bcc)	rhombic dodecahedron	truncated octahedron
hexagonal P (simple hexagonal)	hexagonal dipyramid	hexagonal prism (dual)
hexagonal close-packed (hcp)	hexagonal cubeoctahedron	trapezorhombic dodecahedron (dual)
NaCl (structure type)	octahedron, Na around Cl, or Cl around Na	rhombic dodecahedron, with Na or Cl at the 6 vertices where 4 edges meet, Fig. A2.6(a)
ZnS (zinc blende) (structure type)	tetrahedron, Zn around S, or S around Zn	rhombic dodecahedron with Zn at 4 of the 8 vertices where 3 edges meet, Fig. A2.6(b)
ZnS (wurtzite) (structure type)	tetrahedron, Zn around S, or S around Zn	trapezorhombic dodecahedron with Zn at 4 of the 8 vertices where 3 edges meet Fig. A2.6(c)
NiAs (niccolite) (structure type)	trigonal prism, Ni around As	trapezorhombic dodecahedron, with Ni atoms at the 6 vertices where 4 edges meet, Fig. A2.6(d)
α-Al$_2$O$_3$	tetrahedron (distorted), Al around O	trapezorhombic dodecahedron, with Al atoms at the 4 of the 6 vertices where 4 edges meet, Fig. A2.6(e)
CsCl (structure type)	cube, Cs around Cl, or Cl around Cs	cube, with Cs or Cl at the 8 vertices
CaF$_2$ and Li$_2$O (fluorite and anti-fluorite structure types)	cube, F, or Li around Ca, or O; tetrahedron, Ca, or O around F or Li	rhombic dodecahedron, with F or Li at the 8 vertices where 3 edges meet; cube, with Ca or O situated at 4 of the 8 vertices forming a tetrahedron

the corresponding reciprocal lattice is cubic I, it is a truncated octahedron and for the cubic I lattice, for which the corresponding reciprocal lattice is cubic F, it is a rhombic dodecahedron.

The constructional procedure for Brillouin zones may be illustrated for a cubic P (simple cubic) lattice with a cell edge length a and for which the corresponding reciprocal (k-space) lattice is also cubic P with a cell edge length $2\pi/a$ (Fig. A2.7).

Consider electrons travelling in the \mathbf{x}^*-\mathbf{y}^* plane reflected from lattice planes parallel to the \mathbf{z}^* axis. In order of decreasing d_{hkl}-spacing (or increasing \mathbf{d}^*_{hkl} spacing) these are 100, 110, 200, 210, ... Consider an electron with wavevector \mathbf{k} as indicated:

$$k = |\mathbf{k}| = \frac{2\pi}{\lambda}.$$

For reflection from the 100 planes, the corresponding value of k is found by substituting for λ in Bragg's law

i.e.
$$k = \frac{2\pi}{\lambda} = \frac{2\pi}{2a\sin\theta} = \frac{2\pi}{a}\cdot\frac{1}{2\sin\theta}$$

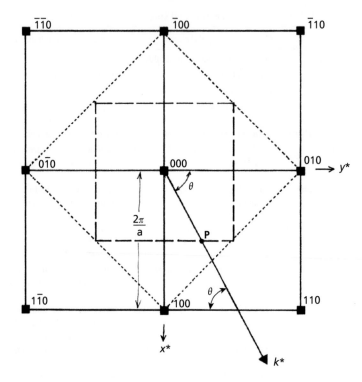

Fig. A2.7. Construction for the first Brillouin zone for a cubic *P* lattice (dashed lines). θ is shown for reflection from the (100) plane. The second Brillouin zone is shown by the dotted lines.

which value is indicated by point P. The locus of P, for all values of θ, and for reflections from the 010, etc. planes, gives rise to the outline of the first Brillouin zone, as indicated by the dashed lines. In three dimensions the Brillouin zone is a cube.

The second Brillouin zone arises from reflections from the 110 planes and, proceeding as before, the corresponding values of **k** are indicated by the dotted lines. In three dimensions this second Brillouin zone is a rhombic dodecahedron (like the first Brillouin zone for the cubic *I* reciprocal lattice).

Appendix 3

Biographical notes on crystallographers and scientists mentioned in the text

Ernst Abbe 1840–1905

Abbe's father was a mill worker in Eisenach, Germany, and it was his employers who provided the scholarships which enabled Abbe to proceed from High School to study physics at the Universities of Jena and Göttingen. In 1863 Abbe achieved his ambition of becoming a lecturer at Jena, but his career may be said to commence in 1866 when he began his long and fruitful collaboration with Carl Zeiss. As a theoretician he established the relation between the aperture and the limit of resolution of a lens (the equation is engraved on his memorial at Jena); as an optical designer he invented the apochromatic objective lens (made possible with the new glasses developed by the Schott glassworks); and as an industrialist he and Zeiss developed systematic microscope production on a large scale.

In 1876 he was made a partner in the Carl Zeiss firm and became the sole owner following the death of Zeiss in 1888. In 1891 he created the Carl Zeiss Foundation whose charter was used in Prussia as a model for progressive social legislation.

George Biddell Airy 1801–92

Airy's life was characterized, to a remarkable degree, by a determination to succeed and meticulous organization: he made a record of everything that he wrote, with dates, and discarded not the merest note or receipt. His entry to Cambridge in 1819, which was largely facilitated by a wealthy uncle in Colchester, provided him with the opportunity to break away from a modest parental background and thence his career developed rapidly. He was appointed Lucasian Professor of Mathematics in 1826, Plumian Professor of Astronomy in 1828 and Astronomer Royal in 1835, a post he filled for 46 years and which entirely fitted his temperament. He largely re-equipped the Greenwich Observatory and transformed it into a highly efficient Institution—but at a price: it provided no centre for the training of innovative scientific minds and the discoveries of John Herschel and John Couch Adams were made elsewhere. Airy was, in effect, the prototype of the modern government scientist and played a major role in the development of institutional science in Victorian England. He designed (with the lawyer, Edmund Beckett Denison) the clock for the new Palace of Westminister—it had to be redesigned

in 1852 when it was discovered that the interior walls had been made too small—but its continued functioning almost without interruption for over 140 years is a measure of their achievement.

Airy calculated the magnitude of the image formed by a telescope objective of a star—a point source of light—and thereby laid the foundations for the theory of the resolving power of optical instruments. He was knighted in 1872, having refused that honour three times previously on the grounds that he couldn't afford the fees.

William Thomas Astbury 1898–1961

Astbury was born in Longton in Staffordshire, England, one of the 'five towns' of the pottery manufacturing industry in what was then called the 'Black Country'. His family was poor, but Astbury's success in winning scholarships provided him with an excellent early education and entry into Jesus College, Cambridge, which would otherwise have been beyond his reach. His initial studies in mathematics and physics were interrupted by the First World War, during which time he served with the X-ray unit of the Royal Army Medical Corps. On his return to Cambridge, through the influence of Arthur Hutchinson, Demonstrator in Mineralogy, he studied in addition chemistry, mineralogy and the newly emerging science of X-ray crystallography pioneered by the Braggs. Arthur Hutchinson indeed was the 'source' of many of the crystals first studied by the Braggs which he 'borrowed' (without permission) from the Mineralogical Museum.

In 1921 Astbury joined W.H. Bragg's newly formed research group at University College, London and in 1923 moved with the group to the Royal Institution where (together with Kathleen Yardley) he developed the first space group tables to help determine, from X-ray evidence, to which of the 230 space groups a crystal belongs. W. H. Bragg thought them unnecessary—all that was needed was common sense. Astbury replied that not everyone has common sense. However, the major direction of his life's work, in macromolecular structures, came about as a result of a request from Bragg in 1926 that he take some X-ray photographs of fibres which Bragg wished, if possible, to use in one of his Royal Institution lectures. This was typical of Bragg's subtle way of directing research—to simply ask for help on some particular topic. Astbury was so successful in applying X-ray techniques to the study of cellulose, silk and wool fibres that in 1928 Bragg recommended him for the post of Lecturer in Textile Physics and Director of the Textile Physics Research Laboratory at Leeds University which, particularly through the work of J. B. Speakman, was pre-eminent in the study of the physical and chemical properties of fibres. Here he made rapid progress in distinguishing the fully-extended chain structures characteristic of cellulose and silk (the simplest polypeptide protein structure) and the much more complex protein structure, keratin—the basic material not only of wool and hair fibres but also of nails, horn and quills. He showed, in wool, that the keratin polypeptide chains occurred in two forms, a folded form, which he called α-keratin and an extended form, β-keratin, a discovery of enormous scientific and technological importance in understanding the extensibility of wool fibres and in the processing of woollen fabrics. He disseminated this information in a series of University Extension lectures in 1932 and which formed the basis of his book *Fundamentals of Fibre Structure* (Oxford University Press, 1933).

Astbury proposed a model for α-keratin in which the amino acid peptide links were folded in the form of hexagons, parallel to the chain axis and perpendicular to the cross-linking side arms. This model was superseded in 1951 by Pauling's α-helix model but it did nevertheless represent the first attempt to model a protein chain structure in which specific linkages held the polypeptide chains in a characteristic conformation.

In 1935, together with a research student, Florence Ogilvy Bell, Astbury began to study the structure of nucleic acids and in 1938 together they published the first X-ray diffraction photograph (or fibre diagram) of a nucleic acid and attempted, unsuccessfully, to interpret its structure.

In 1945 Astbury was appointed to the new Chair of Biomolecular Structure at Leeds which he occupied until his early death in 1961.

As a pioneer it was perhaps inevitable that Astbury should see his own work super-seded. His great legacy is his early stimulation to so many others in the field of molecular biology.

William Barlow 1845–1934

Following an inheritance from his father, Barlow, one of the last great English amateurs in science, took up the study of crystallography in his early thirties and attempted to relate the properties of crystals to the packing arrangements of their constituent atoms. He collaborated with W. J. Pope at Cambridge and guessed (correctly) the structure of the alkali halides. His mathematical ability and developing interest in symmetry enabled him to extend the methods of Bravais and Sohncke to find the 'symmetrical groupings to fit the forms and compositions of a variety of different substances' but he did not publish his derivation of the 230 space groups until 1894, three years after Fedorov and Schönflies, of whose work he almost certainly had no prior knowledge.

John Desmond Bernal 1901–1971

Bernal was born in Nenagh, County Tipperary. His father, a prosperous farmer, came from a branch of Sephardic Jews who had fled from Spain to Ireland to escape the Inqui-sition but which had yet converted to Catholicism in the nineteenth century. His mother, an American citizen, but also Irish by birth, was undoubtedly a formative influence in Bernal's early life. As a young boy, Bernal's scientific interests were first awakened on reading Faraday's *The Chemical History of a Candle* (which led, in turn, to him carrying out his own chemical experiments).

At the age of 10, Bernal and his younger brother, Kevin, were sent to Hodder, a preparatory school for Stonyhurst in England and then to Stonyhurst itself. Stonyhurst was, and is, a leading Catholic school, but Bernal recalled 'I learned nothing there but the joys of prison life'. Wisely, his parents removed him to Bedford School, where Bernal was free to pursue his scientific interests and from where he won a scholarship to Emmanuel College Cambridge to read mathematics.

Bernal rapidly immersed himself in what might be called the post-war revolution-ary zeal and mores of Cambridge: he became a member of 'The Heretics' (an elite gathering that opposed the then religious orthodoxy of university life), abandoned his

Catholic faith, and became a convert to Marxism. His bohemian appearance, evident intellectual ability, and encyclopaedic knowledge earned him the flattering nickname 'Sage' (conferred on him by Dora Grey, one of his many lovers). It came as a shock that he did not (as a result of the social and intellectual distractions of his life) achieve a first class in Part 1 of the Mathematics Tripos: a happy result of which led him to transfer to Natural Sciences and to encounter, through the benign teaching of Arthur Hutchinson, the subject of crystallography. Here he attempted, entirely unaided, to describe the 230 space groups mathematically, a task that so impressed Hutchinson that he recommended to William Bragg that Bernal should be accepted as a research student at the Davy–Faraday Laboratory in the Royal Institution.

At the R.I. Bragg set Bernal the task of determining the structure of graphite. At first, Bernal did not find experimental work to his liking and mentioned to Bragg his mathematical paper. A bond between these two very different men was established when Bragg confessed that he had not even looked at it! Bernal went on to study metallic compounds, wrote a systematic account of X-ray rotation methods, and devised 'Bernal' charts to assist in the interpretation of X-ray rotation photographs.

In 1927 Bernal returned to Cambridge as Lecturer in Structural Crystallography (and later Assistant Director of Research in the Cavendish Laboratory). The years up to 1937 (when he was appointed Professor of Physics at Birkbeck College, London and elected FRS in the same year) were scientifically the most creative of his life. His personal magnetism and intellectual brilliance inspired a succession of able young men and women who passed through his laboratory (physically, a leaky four-roomed but centrally located hut). He and his students, most notably Isodore Fankuchen, Dorothy Crowfoot, Helen Megaw, Nora Wooster, and (later) Max Perutz were (along with W.T. Astbury at Leeds) the first to apply crystallographic techniques to organic molecules (sterols, pepsin, vitamins, enzymes, and the tobacco mosaic virus). Indeed, Bernal and Astbury were the pioneers of the New Science of Molecular Biology.

But Bernal's intellect was not confined to crystallography. In 1929 he published *The World, the Flesh and the Devil: An Enquiry into the Three Enemies of the Rational Soul*, which expressed his optimism about the social benefits of science and that Arthur C. Clark called 'the most brilliant attempt at scientific prediction ever made' (perhaps Aldous Huxley's *Brave New World*, published a few years later, and that expresses the opposite view, was written with Bernal's book in mind). In 1934, C.P. Snow (who attempted, unsuccessfully, to get Bernal elected to a Fellowship at Emmanuel College) based his first novel *The Search* on Bernal and his work.

Bernal's appointment at Birkbeck College was soon interrupted by the outbreak of the Second World War. But his Marxist politics and espousement of the Soviet system (as expressed in his 1939 book *The Social Function of Science*) were no obstacles to his commitment to the war effort. He joined the Ministry of Home Security to assess the effects of enemy bombing and from 1942 he and Solly Zuckerman served as scientific advisors to Lord Louis Mountbatten. Bernal's detailed mapping of the Normandy beaches was a significant contribution to the success of the D-Day landings.

Bernal's stock was at its zenith in the years following the war. He established the Biomaterials Research Laboratory at Birkbeck (inconveniently housed in two bomb-damaged houses) and again attracted researchers of the calibre of Aaron Klug, Andrew

Booth and (later) Rosalind Franklin. In 1946 he was awarded the Royal Medal of the Royal Society and together with Olga Kennard began the systematic recording of crystallographic data that developed into the Cambridge Crystallographic Data Centre. But increasingly he became a hard-line supporter of the Soviet system at a time when its failures and oppression were clearly evident to his contemporaries. He became a founder of the Soviet-sponsored World Peace Council and was, as C.P. Snow describes 'as being in continuous motion to socialist capitals'. Not even the Lysenko affair affected his admiration for Stalin and in 1953 he was awarded the Stalin Peace Prize.

The last major act of Bernal's scientific career was the Royal Society Bakerian Lecture that he presented in 1962 on 'The Structure of Liquids'. Here, he called into question the whole definition of what 'crystalline' meant. In his opinion biomolecular studies had 'broken formal crystallography, shattered it completely'. How delighted he would have been to see the discovery of quasicrystals!

In 1963 Bernal's strong constitution began to give way: he was unable to walk for more than a few miles on the Annual Easter Aldermaston March and in the summer suffered his first stroke. A further stroke in 1965 left him increasingly incapacitated, unable to speak, and more and more cut off from the world. Towards the end of his life his young daughter Jane would visit to tell him about her day at school: perhaps he understood her but by this time he was incapable of outward expression.

It is not easy to summarize what must have been the conflicting passions in the life of Desmond Bernal. Perhaps the last words should be those of Francis Crick: 'Bernal was the first genius he had met who had consideration for the feelings of others'.

Johannes Martin Bijvoet 1892–1980

Bijvoet was born and brought up in an old waterside house in the centre of Amsterdam. He recalls that his wish to become a chemist arose from the excellent teaching at his secondary school, where the emphasis was on understanding rather than rôte learning. As a result (and also having passed the then rigorous examinations in Latin and Greek required for university entrance) he entered Amsterdam University in the years preceding the First World War. Here the emphasis was on routine analytical chemistry, which gave him much less satisfaction.

During the war, Bijvoet, like all young men of his age, was conscripted for the military service, but since the Netherlands was a non-combatant nation, it was simply a period of enforced idleness which Bijvoet used 'to study quietly thermodynamics and statistical mechanics'—an indication perhaps of his serious attitude to life.

Bijvoet's move to the area of crystallography arose from the excitement aroused by Bragg's model of NaCl, a model which was by no means accepted by the professorial staff at the University, including his own supervisor, as an adequate representation of the structure. However, these objections had a happy consequence in that it was decided that X-ray investigations should be undertaken in which Bijvoet would have a leading role.

However, despite sound progress made in the solution of crystal structures, the 'chemists' remained unconvinced. Bijvoet recalls a 'painful memory' of a visit by W. H. Bragg in which in despair at the persistent hackneyed arguments against the results of X-ray analysis, Bragg 'raised his arms up to heaven'. Even after Bijvoet's

appointment in 1928 as lecturer in crystallography and thermodynamics he records that he 'passed through a time of struggle for the *raison d'etre* of X-ray analysis'.

In 1939 Bijvoet was appointed to the Chair of Chemistry at the van't Hoff Laboratory of the University of Utrecht, a post which the previous holder, Ernst Cohen, was forced, as a Jew, to vacate and who, despite his support for Germany following the First World War, was later to be killed by the Nazis.

It was at Utrecht, in the years 1949–1951, that Bijvoet and his colleagues, made their most outstanding contribution to X-ray crystallography: the use of anomalous scattering to determine absolutely (and not relatively) the right and left handed forms of enantiomorphic crystals—a fulfilment, in a sense, of van't Hoff's work of 1874 on tetrahedrally bound carbon in organic crystals. The experimental work, carried out with crystals of sodium rubidium tartrate, was of great difficulty. Exposures of over 100 hours were needed, using an improvised X-ray tube with a zirconium target and a 'freakish' pump which had to be watched continuously.

Bijvoet was closely involved in the establishment of the International Union of Crystallography in 1946 and became President in 1951. He retired to Winterswijk, a rural district of the Netherlands, in 1962.

William Henry Bragg 1862–1942

W.H. Bragg was born in Westward, Cumberland. At the age of seven, following his mother's death, he went to live first with an Uncle in Market Harborough, Leicestershire and then to a boarding school in the Isle of Man. In 1881, he won a scholarship at Trinity College, Cambridge to read mathematics; he graduated as 'Third Wrangler' in Parts I and II of the mathematical tripos in 1884, then a first class in Part III in 1885 On the basis of this remarkable mathematical achievement in the same year he was appointed Professor of Mathematics and Physics at the University of Adelaide. The story goes that some friends pointed out that he did not know any physics. 'Ah', said Bragg 'but the boat takes ten weeks to reach Australia, plenty of time to read up on the subject.'

For nearly twenty years, Bragg immersed himself in the affairs of the University, developed his clear teaching style but did little serious research. It was only in 1903, at the age of 41, that he identified a research area suited to his particular gifts—the problems of radioactive emission, the ionization of gases and the nature of X-rays which he considered to be 'corpuscular' in nature. His rapidly growing reputation resulted in election to the Royal Society in 1907 and, in the same year, Bragg was offered the Cavendish Chair of Physics in the University of Leeds, a post which he took up in 1909.

The contrast between Leeds and Adelaide could not have been more great: his Australian wife, Gwendoline, was particularly appalled by Leeds, the dirt and smoke, the rows of back-to-back houses and the poor rickety children. It was no better in the University; Bragg was restricted by the syllabus, found the students inattentive and also found himself in a losing battle with his corpuscular theory of X-rays and γ-rays. However, he derived support from Arthur Smithells, Professor of Chemistry, from Ernest Rutherford at Manchester and not least from the emerging scientific abilities of his

elder son, Lawrence. Gwen immersed herself in social work for poor children and for the family there were weekends spent in the beautiful surrounding Dales countryside.

In June and July 1912, Laue's two papers announcing the diffraction of X-rays were presented at meetings of the Bavarian Academy of Sciences. They were published in late August, but even before then word 'got around'; it did so to the Braggs during a family holiday at Cloughton, a village on the Yorkshire coast, by way of a letter from Lars Vegard, a former colleague, who enclosed a photograph which he had obtained from Laue.[1] Laue's photograph clearly showed the wave nature of X-rays but both father and son were reluctant to abandon the corpuscular theory and on their return to Leeds, attempted to interpret the patterns in terms of corpuscles travelling along 'avenues' between the atoms. This work resulted in a letter to *Nature* in October 1912, authored solely by W.H. Bragg. It represents the last attempt to 'save' the corpuscular theory. Thereafter, the wave theory was unquestioned.

Bragg's significant achievement at this time was the design and construction in the Leeds physics workshops in the winter of 1912–1913 of the X-ray spectrometer and it was this instrument that enabled him and Lawrence to make the first determinations of crystal structures, for which work they were jointly awarded the Nobel Prize in 1915.

Bragg records how he and Lawrence worked together in the physics laboratory far into the night and it is near impossible to disentangle the relative contributions of father and son. Certainly Lawrence's deep physical insight and his exploitation of the law of reflection that bears his name was of primary importance. But set against this is Bragg's foresight, expressed in the Bakerian lecture to the Royal Society in 1915, that the series of spectra (reflections) from a given set of crystal planes represented a series of harmonic terms which could be analysed by Fourier methods—a notion of profound importance in crystallography which awaited some ten years before it could be exploited in crystal structure determination.

In 1915 Bragg accepted the Quain Chair of Physics at University College London but only took up the post in 1918, following secondment for research at the Admiralty during the course of the War. At University College he assembled and encouraged a remarkable team of researchers—W.T. Astbury, G. Shearer, A. Muller, K. Yardley—and made the first attacks on the structure of organic crystals (by tacit agreement Lawrence, then at Manchester, worked on inorganic, mainly mineral and silicate, crystals).

In 1919 Bragg was invited to give the Christmas lectures 'to a Juvenile Auditory' at the Royal Institution, continuing a long-standing tradition that had begun with Michael Faraday. The lectures, subsequently published as 'The World of Sound', show his powers of simple exposition and his affection for young people (he dedicated the book 'to Peggy, Gwendy and Phyllis who discussed with me so many things in this book as we walked to school in the mornings').

[1] In 2013 The University of Leeds co-sponsored a plaque at 'Whin Brow', the house in Cloughton where the Braggs' collaboration began. It was unveiled by Mr Charles Bragg, great grandson of WHB and reads 'At this house in August 1912, W.H. Bragg, Professor of Physics in the University of Leeds and his son, W.L. Bragg, later Sir William and Sir Lawrence, first discussed M. Laue's discovery of X-ray diffraction. The Braggs' collaboration led to the joint award of the Nobel Prize for Physics in 1915 and the new science of X-ray crystallography'.

It was perhaps as a result of these lectures that Bragg was elected in 1923 to the headship of the Royal Institution and the title, *inter alia* of Fullerian Professor of Chemistry. Bragg was by now the distinguished icon of British science, he was elected to the Order of Merit in 1931, yet in spite of his heavy administrative duties continued to attract to the Institution younger scientists such as J.D. Bernal who maintained Britain's leading role in X-ray crystallography. His book *The Universe of Light* (1933) was based on his 1931 Royal Institution Christmas lectures.

In 1935, at the age of 73, Bragg was elected to the Presidency of the Royal Society, a post he held until 1940. It was a difficult time for him, the gathering clouds of war and the administrative duties taxed his strength. His daughter, Gwendolen, records seeing on one occasion how wearily he looked up from his papers. 'Daddy, need you work so hard' she cried, and he answered simply 'I must dear, I'm always afraid they'll find out how little I know.'

William Lawrence Bragg 1890–1971

W. L. Bragg, the elder son of W. H. Bragg and Gwendoline Bragg (née Todd) was born in Adelaide, Australia. He attended St Peter's Collegiate School there and enrolled at Adelaide University at the early age of 15, achieving a first class degree in mathematics before coming to England with his family in 1909. He immediately entered Trinity College, Cambridge and within a year again achieved a first class in Part I of the Mathematics Tripos. For Part II he made the momentous decision to switch from Mathematics to Natural Sciences—a decision perhaps prompted by his father. He now began to acquire his knowledge of optics which was to be of such seminal importance in his later discoveries, from C. T. R. Wilson, his most influential teacher (whose lectures were the best, and whose delivery was the worst!).

In June 1912, Bragg graduated with a first class degree but before then, from late 1911 he had already begun research at the Cavendish Laboratory under the supervision of Professor J. J. Thomson. It was a hopeless time for him, there was very little apparatus available, virtually no workshop facilities and he made no progress with his research project (on ionic mobility) at all. However, all this was to shortly change. In order, perhaps, to escape from the frustrations of the Cavendish Laboratory, in August 1912 he joined his family on their summer holiday where he heard from his father the first news of Laue's discovery of the scattering of X-rays by crystals which clearly showed the wave nature of X-rays and where, as he records in his unpublished autobiography, 'I realised that my golden opportunity for research had come.'

The first 'breakthrough' was made by Bragg following his return to Cambridge in October. He noticed that the spots were elliptical in shape—and this was just the shape one would expect if the X-rays were effectively *reflected* from sheets of atoms in the crystal. Moreover he realized, following the ideas of William Pope, Professor of Chemistry at Cambridge, that if the atoms in Laue's zinc blende crystals were packed in the form of a face-centred cubic, rather than a simple cubic pattern, he was able to explain the positions of all the spots in the Laue photographs. Bragg announced his interpretation in a paper 'The diffraction of short electromagnetic waves by a crystal'

read to the Cambridge Philosophical Society on 11th November 1912 and which included for the first time the Bragg law in embryonic form $n\lambda = 2d\cos\theta$ (where θ is the complement of θ subsequently used).

How was it that a twenty-two year old student succeeded where all the 'best brains' of Europe had failed? Genius certainly, but perhaps a genius whose mind was not encumbered with complicated theory and which rather drew on simple optical ideas. There is a story recounted by R.W. James, an early close colleague. At a meeting of English and German delegates at Leipzig in 1931 it did not go unnoticed by the German delegates that, in comparison with their own rigorous education, there was a distinct lack of training in mathematics and physics among the English. 'Tell me', confided one of James's German colleagues, 'how does Bragg discover things? He doesn't know anything.'

In comparison to the Cavendish Laboratory, his father's laboratory at Leeds was much better equipped, in particular the availability of an X-ray spectrometer, with which father and son carried out their joint researches in the spring, summer and autumn of 1913. The notebook in which they recorded their observations is now held in the Brotherton Library at Leeds University and in its closely-pencilled pages one can recognize the data which resulted in a stream of papers in which the structures of ZnS, NaCl, KCl, FeS_2, sulphur, diamond, $CaCO_3$ and the calcite series of crystals, the spinels and CaF_2 were all wholly or partially determined and for which father and son were jointly awarded the Nobel Prize in 1915.[2]

At the outbreak of war, Bragg enlisted with the Royal Artillery and was posted to France in August 1915. It was from here, in the closing weeks of 1915, that the news reached him both of the Nobel Prize and also the death of his younger brother, Bob, in the Gallipoli landings.

In 1919 Bragg was appointed to the Langworthy Chair of Physics at Manchester, a post vacated by Ernest Rutherford. It was a tough assignment: the mainly ex-servicemen students had little time for authority and in Bragg's lectures little short of hooliganism prevailed. However, by 1921 this difficult transitional phase was over: he was elected Fellow of the Royal Society, became married, solved the structure of α-quartz and in 1923 broke what he later called the 'sound barrier' in X-ray analysis in his analysis of the structure of aragonite, in which he made explicit use of space group theory for the first time. He also pioneered the use of Fourier methods in the analysis of diopside and beryl and studied the structures of metals and alloys.

At this time Bragg and his father made a tacit agreement to divide their work on crystal structure determination: Bragg would work on inorganic structures and the father, then at University College, London, would work on organic structures. Many years were to elapse before Bragg became involved, as Cavendish Professor, in organic structures, but as early as 1935, cognisant of the work of Astbury at Leeds, he realized that the structures produced by living matter were the most interesting field of all.

[2] *The Bragg Notebook: A Commentary and Interpretation.*
https://library.leeds.ac.uk/special-collections/research-spotlight/780/braggnotebook Written by Christopher Hammond; edited by Diana Joyce, Jane Saunders and Catherine Robson; website designed and developed by Matt Taylor and Tom Grahame.

In 1937 he left Manchester to become Director of the National Physical Labora-
tory (NPL)—a socially but perhaps not a scientifically, prestigious appointment. Bragg
(and certainly his wife) never overcame their disparaging view of life in the north of
England, the (then) black and smoky Manchester in particular. He mentioned that 'the
students are rather rough diamonds, as indeed they must be when one considers their
circumstances'.

The appointment at NPL was short-lived. Even before Bragg took it up, Rutherford
died and in 1938 Bragg was offered, and accepted, the Cavendish Chair of Experimen-
tal Physics at Cambridge. Almost immediately the Second World War intervened and
Bragg was primarily involved in war work such that his tenure only commenced in
earnest when the war was over. Even so, Bragg pursued the analogies between light and
X-ray diffraction in the analysis of crystal structures (his so-called 'flys' eye' and 'X-ray
microscope' techniques). But most significant of all was his support, from 1939, of the
work of Max Perutz, an Austrian refugee who had been working with J. D. Bernal on the
structure of haemoglobin. It seemed an entirely impossible task; haemoglobin clearly
had a far more complicated structure than anything then known but it is a measure of
Bragg's willingness to take chances that, as he said 'some intuition led me to think that
this line of research must be pursued'. Bragg's support led, in 1947 to the establishment
of the Medical Research Council Unit on the molecular structures of biological systems
and ultimately to the solution of the structure of haemoglobin and myoglobin in 1960
by Perutz and Kendrew and the structure of DNA in 1953 by Watson and Crick. There
is no doubt that this Nobel prize-winning work owed its origin to Bragg's willingness
'to take chances'.

Bragg resigned the Cavendish Chair in 1953 to take up the post of Director and
Fullerian Professor in the Royal Institution. Since the days of his father, the Institute
had declined to a point where almost no research was done at all. Bragg set about a
major programme of reform; he assembled what was to become a prestigious research
group on biological structures and in addition to the long-standing Christmas lectures
revived the programme of afternoon schools' lectures, which were far more enthusi-
astically received than his university lectures. When asked why (as with his father) he
had such an easy rapport with school children his explanation was simple 'we enjoy
the same things'. He records that the period 1960–1966 was 'the happiest time of my
life': a family atmosphere prevailed at the Royal Institution and the only thing lacking,
for Bragg, was a garden. In 1966, the year of his retirement, Bragg was awarded the
Copley Medal of the Royal Society—its highest award—and a year later was made a
Companion of Honour. But the anticipated peace of his retirement was soon to be shat-
tered when he received the first draft of Watson's book *The Double Helix*, which he
himself had encouraged Watson to write many years previously and which now, in view
of its scurrilous remarks (even on Bragg himself) provoked widespread hostility. Bragg
had always had a high regard for Watson's imaginative and brilliant ideas (whereas
Crick did all the talking!); perhaps he also felt an affinity for Watson, who, like himself,
had arrived in England very much as an outsider. But Bragg recognized the value of this
personal record and on condition of the removal of some of the most offending passages
agreed to write a supportive foreword.

Auguste Bravais 1811–63

Bravais was born in Annonay in France, studied mathematics in Paris and became a naval cadet in Toulon, which enabled him to participate in explorations to Algeria and the North Cape. He was appointed Professor of Physics at the École Polytechnique in 1845, and his interest in the relationships between the external forms and internal structures of crystals led to the derivation of the fourteen space lattices in 1848, published in his famous *Mémoire sur les systèmes formés par des points distribués régulièrement sur un plan ou dans l'espace* (1850)—work which was partly based on Frankenheim's fifteen 'nodal' lattice types. However, constant application to a wide range of studies which aroused his curiosity led, in 1857, to a breakdown of his health.

Martin Julian Buerger 1903–86

Buerger was born in Detroit, USA, and studied mining engineering at the Massachusetts Institute of Technology, specializing after graduation in research in mineralogy. It was during this time that he attended, in 1927, a course of lectures on X-ray diffraction given by W. L. Bragg, as a result of which Buerger realized that in order to understand the chemical and physical properties of crystals it was necessary to know their structures. Hence began his life-long work in X-ray crystallography, not only in the solving of crystal structures but in the invention of the equi-inclination Weissenberg camera, the precession camera and the writing of several well-known textbooks, the earliest of which *X-Ray Crystallography* is, like Bragg's *The Crystalline State*: Volume 1, *A General Survey*, a classic which is still current today.

Buerger was closely associated with the founding of the International Union of Crystallography and was a member of the Commission for the International Tables from its establishment in 1948.

Francis Harry Compton Crick 1916–2004

Crick was born in the village of Weston Favell, close to the shoe manufacturing town of Northampton where his father owned a factory. He recalls that, in his middle-class boyhood, it was the gift from his parents of the ten volumes of Arthur Mee's *Childrens' Encyclopaedia* that stimulated his interest in the world. In this great and once popular work, all the articles on science ('Wonder'), art, literature, and engineering ('How it Works') are jumbled up together. Crick read it all avidly and kept the ten volumes throughout his life.

Crick was educated first at Northampton Grammar School (in Miss Holding's class who 'made everything interesting'), then on a scholarship at Mill Hill School in North London. In 1934 he enrolled on the Physics course at University College London; the teaching did not stimulate him and in 1937 he obtained a 'good' second class degree. However, he became a graduate student under the autocratic Professor Andrade who set him the 'dullest PhD project imaginable' (on the viscosity of water above 100°C). Fortunately, the war intervened, during which time his apparatus was (happily) destroyed

by bombing. During and after the war he worked in the Admiralty Research Laboratory but, having been inspired by Erwin Schrödinger's book, *What is Life?* decided that research on the biophysics of the cell was what he really wanted to do. This decision fulfilled the requirement of what he later called *The Gossip Test*: 'What you are really interested in you gossip about'. Hence, in 1947, he obtained a studentship at Strangeways Research Laboratory and in 1949 joined the recently formed Medical Research Council Unit at the Cavendish Laboratory in Cambridge to research for a PhD on protein structure under Max Perutz and John Kendrew. Crick rapidly assimilated both the theoretical aspects and experimental techniques of X-ray crystallography and made a major contribution (together with W. Cochran and V. Vand) in his derivation of the Fourier transform of a set of atoms arranged in a helical pattern. Thus was set the groundwork for his fruitful collaboration with Jim Watson that led to the model of the structure of DNA in 1953. The story of the 'race to DNA'—of Crick and Watson's uneasy relationship with Maurice Wilkins and Rosalind Franklin at Kings College London, of the rivalry between Lawrence Bragg and Linus Pauling—has been widely told. For Crick it led to a Fellowship at Brooklyn Polytechnic with David Harker working on protein structure, but the collaboration was not a fruitful one; neither Harker's fixed working routine, nor the environment of downtown Brooklyn (in stark contrast to Cambridge) was amenable to him and he returned to the MRC unit in 1954. His work now broadened to address the problem of the role of DNA (and RNA) in protein synthesis and to an unravelling of the genetic code—the subject of his 1962 Nobel Prize lecture. Increasing fame brought increasing honours—Fellowship of the Royal Society, a Visiting Fellowship at Harvard and, in 1960, Fellowship of the newly-established Churchill College in Cambridge—which he resigned a year later because, as a committed atheist, he objected to the proposed building of a chapel.

In 1977 Crick was offered, and accepted, a Chair at the Salk Institute in Southern California, thus ending his long association with the MRC unit at Cambridge. By now Molecular Biology, and indeed Technology, was firmly established and Crick turned to what he considered to be the central problem of existence: namely the nature of consciousness itself and the neurology of the brain.

This led to a second long collaboration—with Christof Koch whom he first met in 1981 when Koch was a graduate student at Tübingen—and which culminated in his 1994 book *The Astonishing Hypothesis—The Scientific Search for the Soul* in which he suggests that the part of the brain called the claustrum is the key structure in producing consciousness. Crick's ideas brought him into conflict with those of Roger Penrose (author of *The Emperor's New Mind*), but only history will be able to judge this last phase of his life's work.

Crick refused a knighthood but accepted, in 1991, the Order of Merit—an honour personally bestowed by the Sovereign herself.

Peter Joseph William Debye 1884–1966

Debye was born in Maastricht in the Netherlands, was educated at Aachen and Munich and held a number of Chairs in Physics before becoming Director of the Kaiser

Wilhelm Institute for Theoretical Physics 1935–40. He then emigrated to the USA in 1940, becoming Professor of Chemistry at Cornell University.

Debye was primarily a theoretician concerned with the application of physical methods to molecular structure and is best known for his work of specific heat, electrolysis and the electron diffraction of gases. He was a member of staff in the Institute of Experimental Physics at the University of Munich at the time that Laue carried out his famous X-ray diffraction experiments. His development of a technique and camera (with P. H. Scherrer) for the investigation of X-ray diffraction from powders was made at the University of Göttingen in 1915.

Paul Peter Ewald 1890–1985

Ewald, a posthumous child, was brought up by his mother, Clara Philippson Ewald, a renowned portrait painter; they travelled widely and Ewald learned English and French at an early age. He graduated from the Köningliches Victoriagymnasium in Potsdam in 1905 and spent a short time studying chemistry at Cambridge before returning to the University of Göttingen in 1906. Ewald's interests then moved from chemistry to mathematics; he transferred to the University of Munich to attend the courses of lectures given by Sommerfeld on hydrodynamics.

The story of Ewald's thesis topic is given in Section 8.1 and it is clear that Ewald deserves greater recognition for his contribution to the interpretation of X-ray diffraction patterns than he has hitherto been accorded—unlike Laue or the Braggs he was not awarded a Nobel Prize. Yet his doctor's thesis of 1912, which he discussed with Laue, contained within it the basis of the 'reciprocal lattice and reflecting sphere construction' analysis of the geometry of (X-ray) diffraction which is equivalent to Bragg's law. Indeed, it was only after reading Laue, Friedrich and Knipping's published papers that Ewald realized the relevance of his own approach and, in particular, the applicability of the formula in his thesis which he had brought to the attention of Laue but which in fact he (Laue) had not used.

During the First World War, Ewald became a field X-ray technician in the northern Russian Front but had sufficient time to publish, from 1916 onwards, a series of papers on crystal optics and dynamical theory. In 1921 he became first Professor of Theoretical Physics in the University of Stuttgart and then Rector in 1932. It was an inauspicious time: only one year after his inaugural address, in which he pleaded for the pursuit of social and political harmony, he was forced, by a Nazi Party decree, to resign his Rectorship because his wife was Jewish and he part-Jewish himself. The increasing persecution and corruption of the objective academic ideals of German science by the Nazi Party forced him in 1937 to leave Germany for England with the support of W. L. Bragg. He lived in Cambridge with his family in a financially precarious situation until 1939 when he became Professor of Theoretical Physics in the Queens University in Belfast. In 1949 he became Professor of Physics at the Polytechnic Institute of Brooklyn, and although he remained in the USA for the rest of his life he never gave up his British citizenship.

The International Union of Crystallography was founded in 1947 largely through Ewald's initiative, with W. L. Bragg as President and he himself as Vice-President. It was, in a sense, a fulfilment of those ideals for which he had pleaded fifteen years previously.

Evgraph Stepanovich Fedorov 1853–1919

Son of an army engineer, Fedorov attended military school in Kiev and became, in turn, a combat officer and member of the revolutionary underground. His first ideas on the derivation of the 230 space groups were contained in his book The *Elements of Configurations*, started in 1879, when he was 26 years old but not published until 1885. His complete derivation of the space groups was circulated in 1890 to his friends (including Schönflies) as a series of preprints but was not published until 1891, shortly before Schönflies' independent derivation.

Jean Baptiste Joseph Fourier 1768–1830

Fourier, son of poor parents who died when he was nine, attended the military school in his home town of Auxerre, France, where his great mathematical ability was first recognized. As a young man he was caught up in the Revolution; he was imprisoned both by Robespierre (being released only as a result of Robespierre's execution in 1794) and, subsequently, by the counter regime, ironically as a supporter of Robespierre.

Fourier's administrative and political talents were recognized by Napoleon who made Fourier secretary of the newly formed Institut d'Egypte, but after Napoleon's fall he found himself out of favour with Louis XVIII and was not elected a member of the Academie Française until 1827.

Fourier's main achievements lie in his development of the mathematical techniques applied to the diffusion of heat and (what are now called) Fourier series and Fourier integrals. He also made substantial contributions to the theory of equations, linear inequalities, and probability.

Frederick Charles Frank 1911–1998

Frank was born in Durban, South Africa. His parents were English and the family returned to England, to Abbots Farm in Suffolk, when Frank was just ten weeks old. His early schooling was at the village of Denham Abbots, then Ipswich Grammar School from which he won, in 1929, an Open Scholarship to Lincoln College, Oxford, to read chemistry. Having achieved a first class degree in 1933 he switched to research in engineering, gaining the DPhil degree in 1937, and then went on to do two years' research in physics at the Kaiser Wilhelm Institute für Physik in Berlin. It was this wide educational background (and fluency in German) which was to be of such value to him in his later wide-ranging research from the physics and properties of crystals to geophysics and continental drift.

At the outbreak of World War II, Frank joined the Chemical Defence Research Establishement at Porton Down in Wiltshire—but in view of his experience was 'head-hunted' by R. V. Jones to join him in Scientific Intelligence in the Air Ministry. Then followed a remarkably fruitful six-years' partnership, vividly described in Jones' book *Most Secret War* (1978) and in which he pays tribute to Frank's acute powers of observation and interpretation—perhaps the most spectacular being his identification, near Bruneval on the French coast, of the paraboloid antenna of the much-sought Würtzburg 53 cm radar and which was captured intact in a British parachute raid in 1942.

Immediately following the war, Frank was involved in secretly recording conversations among ten eminent German scientists detained at Farm Hall, near Cambridge. The transcripts, of enormous historical interest, were eventually released in 1992 and published, with notes and an introduction by Frank, as *Operation Epsilon: The Farm Hall Transcripts* (1993).

In 1946, Professor Nevill Mott, Director of the H. H. Wills Physics Laboratory in the University of Bristol invited Frank to examine problems associated with the deformation and growth of crystals. Here Frank was spectacularly successful and contributed largely to the subject now known as physical metallurgy. He showed how, during deformation, dislocations (the line-defects in crystals) could be continuously generated (Frank–Read sources); he identified different types (Frank partial and Frank sessile dislocations) and showed how crystal growth could be explained in terms of (screw) dislocations intersecting growth surfaces. From here he moved to polymeric materials and described the chain-folded structures of polymer crystals and the linear defects, for which he coined the term *disclinations*, which occur in single crystals. These physical intuitions extended to geophysics; island arcs for example being simply explained by analogy with the shape of a dimple in a squashed ping-pong (table tennis) ball.

Frank was elected Fellow of the Royal Society in 1954, was its Vice-President 1967–69 and received the Copley Medal, the Society's premier award, in 1994. He became Head of the Physics Department at Bristol in 1969 and was knighted soon after his retirement in 1977.

Frank was a man with a strong sense of duty and social conscience, which extended from his tireless help to young research students to his membership of the 'Pugwash' movement, which strove to alert nations to the dangers of nuclear weapons proliferation.

Moritz Ludwig Frankenheim 1801–69

Frankenheim was born and educated in Brunswick and the University of Berlin, where he was appointed lecturer in 1826. He was subsequently appointed Professor of Physics at Breslau, a post which he held until 1866. It was in a work of 1835 that he showed there could be only fifteen 'nodal', i.e. space lattice types, which much later (1856) he corrected to fourteen configurations, 8 years after Bravais' derivation of the fourteen space lattices.

Rosalind Elsie Franklin 1920–58

Rosalind Franklin was born in London into a prosperous upper middle class Jewish family. She was one of five children (three brothers and a younger sister) and followed a Franklin family tradition in being educated at St. Paul's Girls' School where she was a Foundation Scholar. By the age of 15 she had no doubts that she wanted to be a scientist (which was a new departure away from her cultural background) and in 1938 won an Exhibition (a scholarship) to Newnham College, Cambridge. Her studies were affected by the outbreak of war when many of the university staff were seconded to the war effort and in 1941 she graduated, to her disappointment and against expectations, with a second, rather than a first class degree. However, she was awarded a Research Scholarship at Newnham College which she held for a year before being appointed as a Research Officer at the British Coal Utilization Research Council—a youthful organization where she established her reputation as a research scientist and for her work there was awarded the PhD degree by Cambridge University in 1945. In 1947 she was appointed Chercheur in the Laboratoire Centrale des Services Chimique de l'Etat in Paris and it was here that she developed X-ray diffraction techniques in the analysis of carbon structures—work which has partly laid the foundations to a whole new materials technology. In 1951 she decided it was time to return to England and on the basis of her expertise in a difficult area of X-ray analysis was appointed to a Newall and Turner Research Fellowship in the Medical Research Council Unit at King's College, London to work on DNA—the elucidation of the structure of which was widely recognized as being a matter of paramount importance. It has, indeed, been represented as a race between competing laboratories and individuals, but if so it was a race which could not be won by one group alone but rather with the inputs from various individuals. Franklin's great contribution was her meticulous experimental work, and cautious, unspeculative but well-founded deductions. Together with a research student, Raymond Gosling, who, under M. H. F. Wilkins' direction had obtained diffraction patterns of DNA which showed a high degree of crystallinity, she demonstrated the effect of varying humidity on the crystal structure of DNA and in 1951 showed that the B-form (that which occurs under conditions of high humidity) had a structure consistent with a helical conformation and that the bases must lie on the inside, rather than the outside, of the sugar phosphate chains. Her photograph of DNA, which was crucial in demonstrating the correctness of Crick and Watson's model in 1953, was taken with Gosling over the weekend of 1 May 1952.

In the last five years of her life, before her untimely death from cancer, she worked in the Crystallography Laboratory at Birkbeck College London, where she discovered the structure of the tobacco mosaic virus.

Joseph Fraunhofer 1787–1826

Fraunhofer was the eleventh child of a poor master glazer in Straubing, Germany, who, after the early death of his parents and an unsatisfactory apprenticeship, joined the optical instrument workshop of a scientific instrument maker in Munich in 1806. Here his versatility as experimental scientist, optical instrument technician and glassmaker laid

the foundations of the pre-eminence of the German optical instrument industry which was to be consolidated with such success by Otto Schott, Carl Zeiss and Ernst Abbe later in the century. It was a result of his investigations into the optical properties of glass that Fraunhofer discovered that the solar spectrum was crossed by many fine dark lines (some of which William Wollaston had first noticed in 1802), thereby laying the foundations of spectrometry. Following the work of Thomas Young he made the first quantitative study of diffraction and found the reciprocal relationship between the separation of the lines in diffraction gratings and the dispersion of light. In 1823, three years before his death from tuberculosis, he was appointed to the post of director of the Physics Museum of the Bavarian Academy of Sciences.

Augustin Jean Fresnel 1788–1827

Fresnel was born in Broglie, France, and grew up in a stern Jansenist home environment. In 1804 he entered the Ecole Polytechnique in Paris, his intended career being in civil engineering; from there he entered the Ecole des Ponts et Chaussees, graduating in 1809. His first big job as a civil engineer was the construction of the Imperial Highway to link France and Italy through the Alpine Pass of Col Montgenevre.

Fresnel's interests in optics and the nature of light appear to have begun in about 1814, but were stimulated in 1815 when he was suspended from his engineering work because of his opposition to Napoleon's return to France from exile in Elba—a period of enforced leisure during which time Fresnel was under police surveillance. However, in 1823 he become a member of the French Academy of Sciences and in 1824 was appointed to the lighthouse commission in France and developed the echelon lens which bears his name.

Like Thomas Young, but working entirely independently, Fresnel advocated and elaborated on the wave theory of light, though it is not known to what extent he was familiar with the ideas of Christiaan Huygens. Like Young, he introduced the principle of interference, drawn from analogy with his work on acoustics, publishing the results of his experiments on the diffraction and interference of light in 1815.

Fresnel collaborated closely with François Arago in studies on the polarization phenomena of light and followed Young in the realization that light waves must be transverse rather than longitudinal. For most of his short life Fresnel suffered severe ill-health; he was awarded the Rumford Medal of the Royal Society shortly before his early death.

Georges Friedel 1865–1933

Georges Friedel was one of a long line of distinguished scientists—his great-grandfather had been Dean of the Faculty of Sciences at Strasbourg University, his father Charles, son Edmond, and grandson Jacques, all made significant contributions to chemistry and mineralogy.

Georges was born in Mulhouse, but the family left their Alsace home for Paris prior to the Franco–Prussian war where they had an apartment in the building of the School of

Mines—the proximity of which was to be a great influence during Georges' boyhood. In 1885 he was placed first in the competitive examination of the École polytechnique and three years later both married and moved to the École Nationale des Mines in St Etienne where he was subsequently appointed Professor, and then Director, in the years before the First World War. After the war, it was a great joy for him to be able to return to his beloved Alsace as Chairman of the Institute of Geological Sciences in the newly-reopened French University of Strasbourg.

Friedel made significant contributions to what he called the mesomorphic states of matter (nematic and smectic phases, such as exist in liquid crystals) that exist between the amorphous and crystalline states. The Law that bears his name (the 11 centrosymmetries that can be determined by X-rays) was worked out in 1913 (i.e. soon after the discovery of X-ray diffraction) and stemmed from his early work in the symmetries of Bravais lattices.

Walter Friedrich 1883–1968

Walter Friedrich was born in Magdeburg in Germany. His introduction to X-rays commenced at a remarkably early stage in his life: whilst still a schoolboy at the Stephaneum Gymnasium in Aschersleben he took X-ray radiographs for the local hospital, using apparatus that had been given to him by his father. He studied physics, first at Geneva and then in Röntgen's Institute of Experimental Physics in Munich, obtaining his PhD degree in 1911. It was as Sommerfeld's assistant that he carried out, with Paul Knipping, the first successful X-ray diffraction photographs in April 1912. Sommerfeld was reluctant to let Friedrich carry out an experiment that he did not think would work. However, urged on by Laue and with the assistance of Paul Knipping, the experiments were carried out in a cellar in Sommerfeld's Institute while Sommerfeld was away. The first attempts were unsuccessful since the photographic plates were placed either parallel to the primary beam or behind the crystal. The first successful experiments (using a crystal of copper sulphate pentahydrate) showed a clear, albeit blurry, pattern of spots; then using a crystal of zinc blende orientated with a <100> direction along the primary beam, a clear pattern was observed (see the front of this book, p. xvi). Further experiments followed with the crystal in different orientations. On seeing these results Sommerfeld 'had a complete turnaround' and thence gave Friedrich and Knipping all the help they needed. (There is an uncanny resemblance here to W.L. Bragg's early discouragement of Francis Crick and then, realizing that another laboratory may win the race to DNA, giving Watson and Crick all his support.)

These experiments were to be Friedrich's only excursion into X-ray diffraction: in 1914 he joined the University Clinic in Freiburg and worked on problems of radiation therapy and the prevention of radiation injuries. In 1922 he was appointed Professor of Medical Physics in Berlin and laid the foundations of radium therapy and dosimetry. After the Second World War he became a senior figure in science in the German Democratic Republic, first as Rector of the Humbolt University in Berlin (1949–1951) and then as President of the Academy of Sciences.

Richard Buckminster Fuller 1895–1983

Richard Buckminster Fuller was born in Milton, Massachusetts, into a patrician New England family—his great-great-great grandfather was the Massachusetts delegate at the Federal Constitutional Assembly that created the USA. He broke from family tradition by failing to graduate from Harvard and was packed off to work in a cotton mill in Canada. In 1917 he enlisted with the United States Navy which first provided him with experience of engineering design and organization. The death of his young daughter from influenza in 1922 was a turning-point in his life; he developed an obsession with the role of damp and bad housing on public health and founded a company to develop his 'Stockade Building System'—a method of cheaply building walls from cement and wood shavings. However, the company collapsed in 1927. Thereafter followed a strange and crucial period of his life, 'the year of silence' when he refused to speak to anyone, not excepting his wife, but consumed books and articles on mathematics, science, engineering and architecture with a compulsive and almost destructive inner energy. It was a time when the inventiveness of his mind took root—an inventiveness subsequently expressed in a welter of ideas, proposals and projections throughout the rest of his life which it is difficult, if not impossible, to assess. Certainly he was far ahead of his time in his advocacy of a 'world design science' to avoid ecological catastrophe and the invention of the geodesic dome was a major engineering design achievement; however these must be set against so much else of his work and ideas which now seem merely fanciful and which he would tirelessly express in presentations and lectures of up to four hours at a stretch!

The initial outcome of the year of silence was '4-D' (meaning four-dimensional thinking) in time as well as space. It was the first of many terms or words coined by Buckminster Fuller to express ideas or concepts for which the English language was then deficient; the most well-known of which are *Synergy*, or *Dymaxion*, the latter a word not invented by Buckminster Fuller himself but by the public relations department of a Chicago department store (the 'spin doctors' of the time) and made up of the words he most frequently used—'dynamic', 'maximum' and 'ion'. It was Buckminster Fuller's favourite adjective which he applied to light-weight, easily erectable houses, towers (stacks of Dymaxion houses) and, most famous of all, the Dymaxion car of 1933–4, a hoped-for breakthrough in automobile design which would lift the industry out of the depression.

The geodesic dome was largely the outcome of Buckminster Fuller's work with the Design Class of Black Mountain Art School, North Carolina, from 1948. The Patent was applied for in 1951 and granted in 1954; it represented the last turning-point in his life from failed entrepreneur to visionary designer. The basic structural design has been applied to simple tent-like structures to radomes and, following the success of the dome covering the Ford Motor Company's building in Dearborn, Michigan, has become an accepted architectural form. The use of his name to describe the newly-discovered carbon structures was made after his death.

René-Just Haüy 1743–1822

The son of a poor weaver, Haüy received a classical and theological education at the College de Navarre in Paris and was ordained in 1770. His *Essai d'une théorie sur la structure des cristaux* of 1784 laid the foundation for the mathematical theory of crystal structure. In 1793 he proposed that there were six 'primary forms'—a parallelpiped, rhombic dodecahedron, hexagonal dipyramid, right hexagonal prism, octahedron and tetrahedron. In the *Traité de Minéralogie* of 1801 these were further divided, which led to the notion of 'molécules intégrantes'. Haüy survived the Revolution and was made Honorary Canon of Notre Dame in 1802.

Carl Heinrich Hermann 1898–1961

Formal structure theory, following the derivation of the 230 space groups by Fedorov, Schönflies and Barlow, remained dormant even in the early years of crystal structure analysis by X-rays, largely because of the inconvenient and difficult notation then used. Hermann's great contribution (carried out initially independently of Mauguin) was to simplify the notation for the symmetry elements and space groups, making the theory much more accessible. Hermann was also instrumental in the preparation of the *Strukterberichte (Structure Reports)* from 1925 to 1937. He was a member of the Society of Friends and after the Second World War (during which he was jailed for listening to BBC broadcasts) he was appointed Professor of Crystallography at Marburg, a post he held until his death.

Norman Fordyce McKerron Henry 1909–83

N. F. M. Henry was born at Grangemouth, Stirlingshire and was brought up in Aberdeen where his father was a master at the Grammar School. He read geography at Aberdeen University and in 1934 entered the newly founded Department of Mineralogy and Petrology at Cambridge University and St. John's College which was to become, to a great extent, the focus of his life.

Henry's research interests lay in the fields of X-ray crystallography and reflected light microscopy and, although he published few papers, the two books *The Interpretation of X-ray Diffraction Photographs* and the *Microscopic Study of Opaque Minerals* which he co-authored have been instrumental in the wide dissemination of these subjects. Indeed, he and his colleagues introduced the subject of crystallography to a greater number of students than at any other university. As a teacher Henry presented an image of formality and dryness, but even undergraduates could recognize the humour and generosity that lay underneath. He conducted tutorials two at a time, passing from group to group (with perhaps time for a cigarette in between), setting problems and expecting the solutions to be ready at the next visitation: a simple and most effective technique.

Henry played a major role in the work of the International Union of Crystallography where his knowledge of European languages and cultures did much to establish an atmosphere of goodwill and collaboration. He edited (with Kathleen Lonsdale) the first

volume of the *International Tables for X-ray Crystallography*, a task entirely fitted to his rigorous standards and painstaking scholarship.

In 1960 he was elected a Fellow of St. John's College, was made Steward (1961–9) in recognition of his knowledge of food and wine and was elected Praelector or 'college-father' in 1971 where again his support and concern for his younger colleagues found full expression.

Johann Friedrich Christian Hessel 1796–1872

Hessel was born and educated at Nuremberg but spent most of his professional life as Professor of Mineralogy and Mining Technology at Marburg. He was the first to show (in 1830) that only two-, three-, four- and six-fold axes of symmetry can occur in crystals and that considerations of symmetry lead to the thirty-two crystal classes. However, his work was unrecognized by his contemporaries and remained so until long after his death.

Dorothy Mary Crowfoot Hodgkin 1910–1994

Dorothy Crowfoot was born in Cairo, the daughter of expatriate parents of the British Educational Service in Egypt and the Sudan. At the outbreak of the First World War, she and her younger sisters were evacuated to England, first to Nettleham in Lincolnshire, their mother's childhood home, and then to Beccles in Norfolk. There she attended a PNEU (Parents' National Educational Union) class held in the Rectory in which she first encountered chemistry and by the age of 10 had set up her own attic laboratory using chemicals freely purchased (before the days of 'Health and Safety') from the local chemist. Her passion for chemistry, which was to continue unabated in her life, was further encouraged at the progressive Sir John Leman School (where, with her friend, Norah Pusey, she was allowed to attend the boys' chemistry classes) and by her reading Sir William Bragg's book *Concerning the Nature of Things*.

In 1928 she was admitted to Somerville College, Oxford to read for a degree in chemistry and in the final (Honours) year opted to carry out a research project under the supervision of Marcus ('Tiny') Powell, the leading crystallographer at Oxford at the time. In 1932 Dorothy graduated with First Class Honours in Chemistry—the first woman to do so—a remarkable achievement considering the restrictions then endured by university women. (The prestigious Alembic Club (for chemists) neither admitted women to membership, nor even allowed them to attend its meetings.)

Dorothy's project left her in no doubt that it was in the field of X-ray crystallography that she wished to work: a wish fulfilled when, following a recommendation by one of her examiners, she joined J.D. Bernal's group soon after graduation. There she took up the structures of the sterols and together with Bernal showed that crystals of the enzyme pepsin only gave clear X-ray reflections when photographed in the 'wet' condition. She recalls 'my years in Cambridge were rich with new discoveries'. Then came the offer of a two year research fellowship at her old college—possibly leading to a full fellowship. Bernal, magnanimous as ever, urged her to accept—not only were 'jobs'

in crystallography far and few between, but at Oxford she would be able to establish herself as an independent scientist.

At Oxford, Dorothy's laboratory consisted of wholly inadequate accommodation in the 'Abbot's Kitchen' of the University Museum that she occupied for much of her working life. (Many years later she upbraided a student who wanted to work in the United States 'because of the better facilities' by pointing out that good research was not dependent on 'good facilities'.)

In 1934, Robert Robinson, Professor of Organic Chemistry, presented her with crystals of the protein insulin that had been prepared by Boots Pure Drug Company. It was the beginning of a life-long search to determine its structure. Dorothy pioneered the technique of isomorphous replacement (comparing Zn- and Cd-bearing crystals) and was the first to fully exploit the Patterson technique—but at this time, before the advent of computers, the structure proved to be intractable. And so she turned to other structures. Her first major success was her analysis, together with her first research student, Harry Carlisle, of the structure of cholesterol iodide. It was, in 1943, the first complete three-dimensional analysis of a complex organic molecule. This led to her work on penicillin—a project of international importance—the structure of which, in collaboration with Charles Bunn of the Imperial Chemical Laboratory at Northwich was completed in 1945. The model which she made, showing the three-dimensional electron density map built up in sheets of Perspex, is preserved in the Museum of the History of Science at Oxford.

Then in 1948 she began to study the structure of vitamin B_{12}, a molecule intermediate in size between penicillin and a protein. And apart from its medical importance she was encouraged by the fact that the crystals prepared by the Glaxo Laboratory naturally contained the heavy Co atom. The solution of the structure of vitamin B_{12} in 1955 (which Lawrence Bragg described as 'breaking the sound barrier') was very much a team effort involving colleagues at Oxford, UCLA, and in particular, access to the computing facilities at UCLA.

It was her analysis of the structures of penicillin and vitamin B_{12} that led to a succession of awards and honours—election to the Royal Society in 1947, the Royal Medal in 1956, the Copley Medal in 1976, the (unshared) Nobel Prize for Chemistry in 1964 and the award of the Order of Merit in 1965—the second woman after Florence Nightingale to be so honoured. But, as David Sayre, a colleague and pioneer of direct methods records, Dorothy was acutely disappointed that she was not the first to solve the structure of a protein. Indeed, she returned to insulin in 1966 and continued to elucidate its structure, at an ever increasing resolution, until a few years before her death.

Dorothy Hodgkin's single minded determination as a scientist contrasts strongly with her gentle, reticent personality. Her students and colleagues were drawn into her relaxed, perhaps bohemian home lifestyle, with her husband Thomas Hodgkin whom she married in 1937, and her three children. Inevitably, as an ambassador for British Science, she was drawn into international science and peace movements—she opposed the Vietnam War and nuclear weapons and was awarded the Lenin Peace Prize in 1987. But she probably failed to recognize the darker aspects of those political systems that sought her recognition.

In summary, it may be said that Dorothy Hodgkins' work represents the supreme embodiment of that stage of crystal structure determination between the trial and error methods of such as the Braggs and Bernal and the computer-based direct methods of later generations.

Robert Hooke 1635–1703

Hooke attracted the attention of Robert Boyle at Oxford and it was through his mechanical skill that he made a success of Boyle's air pump. He became a Fellow of the Royal Society and held the post of 'Curator of Experiments' from 1662 until the end of his life. It was in this capacity that Hooke was solicited by the Council of the Royal Society to prosecute his microscopical observations in order to publish them and was also charged to bring in at every meeting one microscopical observation at least. Hooke more then fulfilled this onerous obligation and, in doing so, caused the great capability of the microscope to be realized in England. The fruit of his work, *Micrographia, or Some Physiological Descriptions of Minute Bodies made by Magnifying Glasses, with Observations and Inquiries Thereupon*, was printed in 1665. In his book the word 'cell', so important in biology, was first applied to describe the porous structure of cork.

Albert Wallace Hull 1880–1966

Albert W. Hull was born in Connecticut and studied Classics at Yale University. However, he became increasingly interested in physics and after a period teaching modern languages, returned to Yale and completed, as it were, his educational transition by the award of a doctorate in 1909. After a further period of teaching (this time physics), he joined the General Electric Research Laboratory at Schenectady where he discovered (independently of Debye and Scherrer) the powder diffraction technique. Hull's later achievements were in the fields of electronics (principally the invention of the Thyratron) and what is now called materials (principally the development of metal–glass vacuum seals). He was elected President of the American Physical Society in 1942.

Arthur Hutchinson 1866–1937

Hutchinson was born in London, the only child of parents from (what is now) Cumbria. He attended Clifton College in Bristol from which he won a scholarship to Christ's College, Cambridge and achieved first class in both parts of the Natural Sciences Tripos (1886 and 1888). From there he carried out research at the University of Würtzburg under Emil Fischer and W.C. Röntgen and was awarded the PhD degree in 1891. On his return to Cambridge he was made a Fellow of Pembroke College and in 1895 was appointed Demonstrator in Mineralogy. Then, for no less than 28 years, he taught, single-handedly, the whole of mineralogy in the Natural Sciences Tripos and was only appointed Professor in 1927 following the long moribund professorship of W.J. Lewis. In 1928 he was elected Master of Pembroke College, a post that he held until shortly before his death.

Hutchinson made few contributions to research (he identified the mineral stokesite) but his greatest contribution to science was the inspiration and support that he gave to younger colleagues. He provided W.L. Bragg with the crystals that the Braggs used in their earliest X-ray experiments (including a particularly fine specimen of diamond) despite strict orders from Lewis that no mineral should leave the collection at Cambridge. W.L. Bragg recalled,

I shall never forget Hutchinson's kindness in organizing a black market in minerals for a callow young student. I got all my first specimens and all my advice from him and I am afraid that Professor Lewis never discovered the source of my supply.

Hutchinson's teaching in crystallography inspired both W.T. Astbury and J.D. Bernal and both owe the beginnings of their subsequent careers to letters of recommendation from Hutchinson to W.H. Bragg at the Royal Institution.

Christiaan Huygens 1629–95

Huygens was born in The Hague, studied at the University of Leiden and Breda and became one of the founding members of the French Academy of Sciences. He made important contributions to dynamics and showed that a pendulum which describes a cycloid (rather than the arc of a circle) is exactly isochronous and succeeded in inventing and constructing a pendulum clock based on this principle. He also made major contributions to observational astronomy and was the first to describe Saturn's rings correctly.

Huygens' greatest achievement was his development of the wave theory of light described in his *Traite de Lumiere* (1690). The notion of action at a distance was abhorrent to him; in his view space is pervaded by particles (the ether), light being the effect of the disturbance of the particles—that of one particle being transmitted to the next and so on. The net effect is a wave spreading out from the point of origin, each point in the wavefront being the source of secondary wavelets. Huygens was thereby able to explain refraction and reflection and to predict (correctly) that the velocity of light decreased in passing from a rarer to a denser medium—whereas Newton's corpuscular theory required the denser medium to attract the corpuscles of light and therefore increase its velocity. However, because the waves were conceived as being longitudinal, rather than transverse, he could not satisfactorily explain double refraction.

The problem of double refraction constituted one of Newton's objections to the wave theory of Huygens for 'Pressions and Motions', as he says in *Opticks*, 'propagated from a shining body through a uniform medium must on all sides be alike', whereas experiment had shown that the 'Rays of Light have different properties in their different sides'—a clear presentiment of the idea of polarization. The other, even more serious objection, was that without the concept of interference, Huygens' wave theory could not explain the most obvious character of a light beam, namely rectilinear propagation. It was not until the work of Fresnel and Young at the beginning of the nineteenth century that the wave theory of light was finally established.

Johannes Kepler 1571–1630

The turmoil of Kepler's life is in strong contrast to his belief in the existence of a mathematical harmony underlying the Universe, the search for which was the guiding inspiration of his life's work. He was born in Würtemburg in Germany and in 1594 was appointed teacher of mathematics at the seminary at Gratz. Here he encountered the work of Nicolaus Copernicus and published his first book *(Mysterium Cosmographicum*, 1596), in which he tried to show that the orbits of the planets were determined by the nestling, one within the other, of Plato's five regular solids—cube, tetrahedron, dodecahedron, icosahedron and octahedron. Forced, as a Lutheran, to leave Gratz, he joined Tycho Brahe in Prague under the patronage of the emperor Rudolph II, a fruitful but uneasy collaboration. Here Kepler discovered the first two of the three laws of planetary motion which were to be of such importance to Newton, wrote a major work on optics, discovered the second new star visible to the naked eye since antiquity and prepared his astronomical tables, against a background of the social unrest of the Thirty Years' War and continual lack of financial support from the emperor.

In 1611 civil war broke out in Prague, the emperor was forced to abdicate and Kepler's young son and wife died. He moved to Linz, remarried in 1613, and had to face his mother's trial for witchcraft in Würtemburg. At Linz he published *Harmonices Mundi* (1619) which contains his third law.

On completion of the astronomical (Rudolphine) tables, Kepler moved to Sagan in Silesia under the patronage of Imperial General Wallenstein. He died of a fever at Ratisbon in the unsuccessful quest of 12 000 florins owed to him by the emperor.

Seishi Kikuchi 1902–1974

Seishi Kikuchi's mother, Tatsu, and father, Dairoku, were both from distinguished families of scholars in Japan. Dairoku (1855–1917), a member of the Mitsukuri family, was particularly eminent during the Meiji period (the period of modernization in Japan) and became President of both Tokyo and Kyoto Imperial Universities and Minister of Education. More particularly, he was educated in England at University College School and St John's College, Cambridge from which he graduated in the Mathematical Tripos in 1877—the first Japanese to do so.

Kikuchi's home environment was a particularly happy one; he attended the First High School in Tokyo and entered the Physics Department of the University of Tokyo in 1923—his interest in physics being almost certainly inspired by his elder brother, Taiji, who had graduated in physics and who went on to study radioactivity under Ernest Rutherford at the Cavendish Laboratory. Taiji's promising career was tragically cut short when he died in Cambridge 'of an infectious disease'—an enormous shock to the Kikuchi family.

On graduating, Kikuchi commenced research in radioactivity, continuing his beloved brother's footsteps. His first paper, published in 1927, reached Rutherford, who, to Kikuchi's delight, wrote to commend this contribution to science from the younger brother of his former pupil.

Then the direction of his research materials changed. The papers by de Broglie and Schrödinger on the wave nature of electrons, and the early experiments of Davisson and Germer in the United States, made him anxious to carry out electron diffraction experiments of his own. In this he was strongly supported by Shoji Nishikawa and in 1928 entered Nishikawa's laboratory in the Institute of Physical and Chemical Research to embark on this new project. After some false starts, and possibly influenced by G.P. Thomson's work on thin polycrystalline metal thin films, Kikuchi, aided by Nishikawa and I. Sumoto was able to prepare thin (\sim0.1 μm) flakes of (muscovite) mica that yielded the very first single crystal electron diffraction 'spot' patterns. Sumoto recalls that in the first experiment Kikuchi was so impatient to see the faint emerging pattern that he removed the plate from the fixer before it was properly fixed and washed away the emulsion in the heat of the dark-room lamp! Thicker specimens showed the characteristic dark and light lines that Kikuchi called the P-pattern because the lines occurred in Pairs. All this work was published in six papers in 1928.

Following this success, Kikuchi left Japan to study quantum mechanics in Germany, first in Max Born's Institute in Göttingen and then Werner Heisenberg's Institute in Leipzig, returning to Japan in 1931 to marry Taeko Kawada. In 1932 he received the Meldenhall Prize from the Imperial Academy and in 1934 was appointed Professor of Physics at Osaka University that he established as an active centre for nuclear physics. The work of the Laboratory was severely curtailed during and immediately after the Second World War. However, in order to help re-establish the Laboratory, in 1950 he went to the United States, first to Cornell University and then to the Radiation Laboratory in the University of California, Berkeley, returning to Osaka in 1952, where he presided over the construction of a new cyclotron that was completed in 1955. Then he became successively Director of the Institute for Nuclear Study, University of Tokyo and (in 1959) the third Director of the Japan Atomic Energy Research Institute. His last public position was as President of the Tokyo Science University (1966–1970).

Kikuchi was a man of wide abilities and versatile tastes: he was skilled as a golfer, as a pianist and as a cultivator of roses (an enthusiasm shared with W.L. Bragg). He retained a generous and broad-minded spirit throughout his life.

Alexander Isaakovich Kitaigorodskii 1914–1985

Kitaigorodskii was born in Moscow into a professional family background: his father was a distinguished chemical engineer and one of the creators of devitrified glasses. In his early education at home Kitaigorodskii became fluent in French (the then polite language of the Russian aristocracy and professional classes) and learnt German and some English. He also became a skilled, indeed prize-winning dancer. In short, his childhood seems to have been little affected by the political and social turmoil of the times.

On leaving secondary school at the age of sixteen he worked in the laboratory of a metallurgical (steelmaking) plant in the Urals and there found both the time and opportunity to acquire a knowledge of physics and mathematics at sufficient depth not only to allow him to enter the Physics Department of Moscow University directly into the third year but also contribute to the teaching of students. His biographers assert that he

achieved this knowledge 'completely by himself',[3] but it is difficult to believe that he had no teacher nor mentor to encourage him. One thinks of the near-parallel case of Harry Brearley, the discoverer of stainless steel, working in the laboratory of the Sheffield steelmaking firm of Thomas Firth and Sons and becoming a skilled analyst through the encouragement of the Head of the Laboratory.

On graduating in 1935, Kitaigorodskii commenced research on the X-ray diffraction of amino acids in the Biophysics Division of the Institute of Experimental Medicine, gaining his PhD degree in 1939. During the second world war he was Head of the Physics Department of an armaments plant in the City of Ufa in the Urals. In 1944 he returned to Moscow to work in the Institute of Organic Chemistry of the Soviet Academy of Sciences.

In these years Kitaigorodskii developed his ideas concerning the packing arrangements of organic molecules in crystals, the first fruit of which was the subject of his DSc thesis *Arrangement of molecules in crystals of organic compounds* which he defended in the Institute of Physical Problems. This famous Institute of the Soviet Academy of Sciences has a remarkable history. Its founder was Peter Kapitza, a Soviet citizen who had yet been long resident in England (and hence eligible for election as a Fellow of the Royal Society) and who in 1930 had been appointed as Head of the Royal Society's Mond Laboratory at Cambridge. On a visit to his home country in 1934, Kapitza was 'detained' by the Authorities, but in order that his work could continue, the equipment at Cambridge was purchased by the Soviet Government to equip the new and prestigious Institute. From 1946 (the year of Kitaigorodskii's DSc thesis) until 1955, Kapitza was under house arrest for refusing to take part in nuclear weapons development.

In 1954, Kitaigorodskii moved to the Institute of Organoelement Compounds where he remained until his death although he was never elected to a University professorship.

Kitaigorodskii's work on the geometry of the packing of organic molecules and his elucidation of the space groups for closest, limitingly and permissible close packing became widely recognized through the English translation of his book *Organic Chemical Crystallography* (1961). It is in his later book *Molecular Crystals and Molecules* (1973) that the limitations of the geometrical approach became apparent: Kitaigorodskii assumed that molecules were rigid objects, undeformed by crystal forces and that variations in bond lengths were spurious. This of course is not the case, but at the time such variations were within the experimental error of the equipment then available to him.

Towards the end of his life, Kitaigorodskii was allowed to travel abroad and amazed his hosts not only by his science but by his convivial and outgoing nature—his dancing skills developed into water-skiing and ice-skating! But despite being awarded the D. I. Mendeleev (1949) and E. S. Fedorov (1967) prizes of the Soviet Academy of Sciences, he was never himself elected an Academician—doubtless a consequence of his irreverent attitude even to the Academy itself. Of his life's work he should have the last word. Not long before his death he was asked which of his achievements in science

[3] Yu. T. Struchkov and E. I. Fedin (1993) *Alexander Kitaigorodsky his life and scientific activity*, Acta Chimica Hungarica **130** (2), 159.

he considered the most important. 'I've shown', he replied 'that a molecule is a body. One can take it, one can hit with it—it has mass, volume, hardness. I followed the ideas of Democritus.'

Paul Knipping 1883–1935

Paul Knipping, the son of a doctor, was born in Neuwied-on-Rhine. He studied physics first at Heidelberg and then, like his exact contemporary, Walter Friedrich, in Röntgen's Institute of Experimental Physics in Munich. Having just completed his PhD degree in 1912 he volunteered to assist Friedrich in carrying out the 'Laue' X-ray diffraction experiments in April 1912. Thereafter their careers diverged: after working at the Siemens Laboratories in Berlin and military service in World War 1 he transferred to the Kaiser Wilhelm Institute für Physikalische und Elektrochemie in Berlin to work with Fritz Haber. In 1929 he proposed an Institute for X-ray Physics and Techniques at the Technical University in Darmstadt (inaugurated in 1933) and continued there as Professor until his untimely death as a result of a motorcycle accident.

Max von Laue 1879–1960

Max Laue (the title 'von' appears after his father was raised to the hereditary nobility in 1914) was born in Pfaffendorf near Koblenz. He was educated at the University of Berlin and, as a pupil of Max Planck, obtained his doctorate in 1903 for a thesis on the interference of light between plane parallel plates.

In 1909 he became *Privatdozent* in the Institute for Theoretical Physics in the University of Munich under Arnold Sommerfeld. He was an early 'convert' to Einstein's special theory of relativity and wrote the first monograph on the subject.

Laue's intuition (that the regular arrangement of atoms in a crystal might give rise to an interference effect if the waves travelling through were of a wavelength of the same order as the atom or 'resonator' spacing) almost certainly stems from a meeting with P.P. Ewald in January 1912. Ewald was completing his doctorate thesis and wished to discuss some of his conclusions with Laue. Laue encountered strong disbelief amongst his colleagues of a significant outcome of such a diffraction experiment on the grounds that the thermal vibration of the atoms would obscure any diffraction maxima. However, he persevered, and with the assistance of Walter Friedrich (an assistant of Sommerfeld) and Paul Knipping (who had just finished his thesis under Röntgen), the famous experiment on a copper sulphate crystal which 'happened to be in the laboratory' was carried out.

Laue recalls in his *Autobiography* the very time and place when the idea for a mathematical explanation of the spots in the photograph taken by Friedrich and Knipping came to him: 'I was plunged in deep thought as I walked home along Leopoldstrasse just after Friedrich showed me this picture. Not far from my own apartment at Bismarckstrasse 22, just in front of the house at Siegfriedstrasse 10, the idea for a mathematical explanation of the phenomena came to me I had to re-formulate Schwerd's theory of diffraction by an optical grating so that it would be valid, if iterated, also for a cross-grating. I needed only to write down the same equation for

a third time, corresponding to the triple periodicity of the space lattice, in order to explain the new discovery.'

The new discovery, and Laue's analysis (in terms of a three-dimensional diffraction grating), which Einstein referred to as one of the most beautiful in physics, led to the award of the Nobel Prize for Physics in 1914. But unlike the Braggs, who were awarded the Nobel Prize in 1915, Laue did not pursue crystal structure determination since his main interests lay towards the 'great general principles' of Physics.

In 1919 Laue returned to the University of Berlin to be near his mentor, Max Planck, and in 1921 was elected a member of the Prussian Academy of Sciences. He protested against the Nazi Party's persecution of Einstein: at his opening address on 18 September 1933 to the Physics Congress held in Würtzburg he drew the parallel between Galileo, champion of the Copernican World View, and Einstein, the founder of relativity theory; just as Galileo had won out against the church's prohibition so also would Einstein win out against the proscriptions of National Socialism. However, Laue was unable to stem the tide of Nazi infiltration into German Science; he was increasingly 'sidelined', his advice and judgement were no longer asked for and in 1943 he took early retirement from teaching and moved away from Berlin to Württemberg-Hohenzollern. After the war he was briefly interned in England and until his death took a major role in the rebuilding of German science and its institutions.

Kathleen Lonsdale 1903–71

Kathleen Yardley was born in Newbridge, S. Ireland, the youngest in a family of ten children. In 1908 the Yardley family moved to England in view of the unsettled state in Ireland. As a schoolgirl Kathleen showed her remarkable aptitude for physics and mathematics; at the age of 16 she won a scholarship to Bedford College for Women and in 1922 came top of the class list, as a result of which the examiner, W. H. Bragg, offered her a place in his research team first at University College and then at the Royal Institution—a team which then consisted of some of the brightest stars in X-ray crystallography. In 1927 she married Thomas Lonsdale and moved with him to Leeds when he took up a post in the Textile Department. It was a short but formative period in her life: she proved that the benzene molecule was planar, established her position as a leading X-ray crystallographer and also showed that it was possible to combine the roles of a scientist and a wife and mother. Her recollections of Leeds—the food bargains which could be obtained (and can still be obtained today) in the closing-time auctions at the Leeds market—are evocative. In 1929 she returned to London as W. H. Bragg's research assistant and remained with him at the Royal Institution until his death in 1942. Her early upbringing was strict Baptist, but by degrees she became a member of the Society of Friends, a vegetarian and a pacifist. These latter ideals led to her imprisonment for a short time in 1943 in Holloway Prison which was another positive formative influence in her life.

Kathleen Lonsdale (and Marjory Stephenson) were the first women to be elected Fellows of the Royal Society (22 March 1945). It was only subsequently that she was appointed to a university post, and became involved in teaching, first as Reader (1946)

and then as Head (1949) of the Department of Crystallography at University College, London. However, perhaps her most enduring achievement was her leading role in the preparation of the *International Tables for X-ray Crystallography* as General Editor 1948–63 and as joint editor (with N. F. M. Henry) of the first volume, published in 1952.

To the end of her life her social conscience and concern for what is now called women's liberation was unwavering. She wrote in 1957 the Penguin Special *Is Peace Possible?* and in 1971 the broadsheet 'Women Scientists—Why so Few?'

Charles Mauguin 1878–58

Mauguin's early career expectation was that of a teacher in a teacher's training college, but he quickly developed an interest in mathematics and natural philosophy and commenced his scientific work in the field of organic chemistry. His interest in crystallography probably stems from a course of lectures given by Pierre Curie which he attended in 1905. Mauguin was one of a small group of crystallographers who, in 1933, undertook the publication of the International Tables, and it is his symbolism of the 230 space groups (worked out in collaboration with C. H. Hermann) which is now in almost universal use.

Helen Dick Megaw (1907–2002)

Helen Megaw was born in Dublin in 1907 into a distinguished and well-connected Northern Irish family—her father was a Leading Ulster politician. The family moved to Belfast in 1921 just before the partition of Ireland, but Helen continued her education in England—at Roedean School in Brighton. It was here that her interest in science, and crystallography in particular, was stimulated by her reading the Braggs' book *X-ray and Crystal Structure*.

She briefly returned to Northern Ireland to study at the Queen's University, Belfast, but in 1926 returned to England with a scholarship to read natural sciences at Girton College, Cambridge (then an all-female college). By a happy chance one of her chosen subjects was mineralogy and she achieved a first class in Part I of the Tripos examination. But in Part II (physics) she failed to achieve a first class. This was the second happy chance because it effectively debarred her from beginning research in the Cavendish Laboratory; instead she went to see Arthur Hutchinson of the Department of Mineralogy (who many years previously had provided the Braggs with crystals) and through his good offices began research with J.D. Bernal. Her subject was the structure of ice and Megaw identified the 'oscillating' hydrogen bonding between oxygen atoms. For this work 'Megaw Island' in the Antarctic was named after her.

But in 1934, the year she was awarded her PhD, the country was in a state of depression and jobs in research were virtually non-existent. So, from 1935 to 1943 she taught physics at high-flying girls' schools in Bedford and Bradford.

However, in 1943 she joined Philips Research Laboratories at Mitcham and by another happy chance began to study the structure of barium titanate (samples of which had been sent to Philips from the USA by mistake). This was the beginning of her life's

work on the perovskite structures which she continued briefly with Bernal at Birkbeck College (1945) and finally at the Cavendish Laboratory and Girton College (1946 until her retirement in 1972).

In 2000, at the age of 93, Helen Megaw received an Honorary Degree of the Queen's University.

William Hallowes Miller 1801–80

Miller was educated in St. John's College, Cambridge, where, in 1829, he became a Fellow. In 1839 he published *A Treatise on Crystallography*, in which he made the fundamental assertion that crystallographic reference axes should be parallel to possible crystal edges. His system of indexing was based on a 'parametral' plane making intercepts *a*, *b* and c on such axes [i.e. (111)]; a plane making intercepts a/h, b/k and c/l was assigned the (integral) indices (*hkl*). Although the algebraic advantages of this system were immediately apparent to his contemporaries (and were quickly adopted), their full significance was not fully appreciated until Bragg's and Ewald's interpretation of X-ray diffraction.

Franz Ernst Neumann 1798–1895

Neumann was born in Joachimsthal (now Jáchymov in the Czech Republic) and was brought up by his grandparents (his mother, a divorced Countess, being unable to marry his father who was from a lower social class). He showed an early aptitude for mathematics and attended the Berlin Gymnasium after which, at the age of 16, he volunteered to serve in the Prussian Army. He was wounded in the Battle of Ligny (the forerunner to Waterloo) in 1815.

Neumann's contributions to mineralogy and crystallography, which represent a relatively small part of his life's work, arose through his collaboration with his friend and mentor C. S. Weiss. He introduced and developed the stereographic projection in the study of crystals and extended Weiss' work on zones in publications of 1823 and 1830. He was appointed curator of the Mineral Cabinet at the University of Berlin in 1823.

Although throughout his long life Neumann made many original contributions to optics, heat and electrodynamics (for which he was awarded the Copley Medal of the Royal Society in 1887), he was chiefly remembered by his contemporaries as an inspiring teacher. He was also a firm Prussian patriot and supported Bismarck's policy for the unification of Germany.

Isaac Newton 1642–1727

'Nature was to him an open book, whose letters he could read without effort.'

A. Einstein. Foreword to Newton's Opticks

Isaac Newton was born, a premature and posthumous child, at Woolsthorpe Manor in Lincolnshire. He was educated at the King's School, Grantham, and was admitted

to Trinity College, Cambridge, in 1661. It is while he was an undergraduate that he seems to have made his first acquaintance with optics; he read the works of Kepler and Descartes and from about 1663 was involved in the construction and performance of telescopes and the observation of lunar crowns and haloes. In the summer of 1665 he went home to Woolsthorpe Manor to escape the plague and remained there until April 1667, except for a three-month visit to Cambridge from March 1666. It was during this time that Newton says of himself 'I was in the prime of my age for invention and minded Mathematicks and Philosophy more than at any time since.' These two years of intellectual achievement—the formulation of the binomial theorem and the differential and integral calculus, the theory of colours and of universal gravitational attraction—together with the two years (1685–86) during which he composed the *Principia*, have never been surpassed.

Newton's first paper to the Royal Society (published in 1672), in which he showed that ordinary white light is a mixture of rays of every variety of colour, received such a hostile reception from Robert Hooke that it made Newton cautious of publishing his work; the *Principia* or the *Mathematical Principles of Natural Philosophy* itself was not published until 1687 and then largely due to the encouragement of Edmund Halley. It immediately brought to Newton enormous prestige, a place in society and public affairs but thereafter little more science.

The *Opticks* was not published until 1704, a year after the death of Robert Hooke, Newton's most pertinacious antagonist. Further editions came out in 1717, 1721 and 1730, the last being 'corrected by the Author's own hand'. In this great work Newton states that his 'design in this book is not to explain the Properties of Light by Hypothesis but to propose and prove them by Reason and Experiments'. However, there are many comments suggestive of a deeper penetration into the nature of light than exact knowledge could then achieve, including the notion of interference, for in attempting to explain the colours of thin films Newton describes the passage of light as being made up of 'alternate fits of easy Reflection and easy Transmission'.

The greatest interest in the *Opticks* lies in the *Queries*, in which Newton speculates on the nature of light. The question form may have been used to allay criticism that he was departing from his dictum 'hypotheses non fingo' (I frame no hypotheses), but all are couched in the negative 'Do not Bodies act upon Light at a distance?' A negatively phrased question suggests or implies a positive answer; 'do *not* bodies act upon light at a distance?' Is not Newton implying that the answer must be in the affirmative?

Shoji Nishikawa 1884–1952

Nishikawa was born in Hachioji near Tokyo where his father was a prosperous and well-established merchant of silk textiles. However, he was brought up and educated in Tokyo itself, entering the Faculty of Science at Tokyo University and completing the undergraduate degree course in Physics in 1910. As a young man he was shy and modest (personal characteristics which were to remain with him for most of his life), but of his scientific ability there was no doubt. It was therefore a natural progression for

him to continue at the University as a postgraduate student where he commenced a research project in the area of radioactivity. His change of direction into the area of X-ray crystallography arose from a visit, either by chance, or perhaps in view of his scientific insight, to the laboratory of Torahiko Terada in 1913. Terada, like the Braggs in England, on hearing the news of the discovery of X-ray diffraction by Laue, Friedrich and Knipping, immediately began experiments on single crystals of rock-salt and other minerals. His unique contribution at the time was to visually record the movements, as the crystal was rotated, of the diffraction spots on a fluorescent screen which he interpreted (like W. L. Bragg) as arising from reflections from the crystal planes. There can be little doubt that he arrived at this conclusion independently since it appears unlikely, given the problems of communication with Japan, that he would have known of W. L Bragg's November 1912 Cambridge Philosophical Society paper in which the Bragg law in the embryonic form $\lambda = 2d \cos\theta$ was first announced (θ being the complement of the θ angle subsequently used). In the event Nishikawa, encouraged by Terada, commenced X-ray diffraction studies, principally on fibrous materials, asbestos and gypsum, and published the earliest X-ray fibre patterns. This work was extended to lamellar materials, talc and mica, thin sheets of rolled and annealed metals and finely-powdered materials, rock-salt, quartz and corundum. Again, independently of W. L. Bragg, he determined the crystal structure of spinel in 1915.

In 1917 the Institute of Physical and Chemical Research of the University of Tokyo was established, one of the first acts of which was to send Nishikawa first to join R. W. G. Wyckoff at Cornell University and then, after the end of the war, to join W. H. Bragg at University College, London. On his return to Tokyo in 1920 he organized, and led, the Nishikawa Laboratory, the work of which was seriously interrupted by an earthquake in 1923.

In 1924 Nishikawa married Kiku Ayai, was appointed Professor of Physics and gave largely inaudible (but otherwise excellent) lectures to undergraduates. As with Terada it might be said that he suffered from the then lack of communication from abroad; he was forced to abandon his nearly completed structural analysis on aragonite and α-quartz on finding that W. L. Bragg (1924) and W. H. Bragg and R. E. Gibbs (1925), respectively, had already solved these structures. However, following the discovery of electron diffraction in 1927 by G. P. Thomson and A. Reid in England and Davisson and Germer in the USA, he was instrumental in his guidance to his pupil, Seishi Kikuchi, in the discovery of what are now known as Kikuchi lines. Nishikawa's last statesmanlike act, long after his retirement, was to secure for Japan membership of the International Union of Crystallography.

Louis Pasteur 1822–95

Pasteur is perhaps best remembered for his work in microbiology and immunology and, in particular, the practical applications—the discovery of vaccines, the treatments for rabies and anthrax, silkworm disease, etc. His work in crystallography was confined to the early years of his scientific career. Beginning in about 1847 (shortly after completing his dissertation in physics), he began a series of investigations into the relations between

optical activity, crystalline structure and chemical composition in organic compounds. His guiding principle was that optical activity was somehow associated with life; that whereas enantiomorphic crystals or molecules produced in the laboratory, or of mineral origin, occurred in equal quantities of left- and right-handed forms, those of organic origin—from plants or animals—were always of one form. The origin of life it seemed was bound up with an original asymmetric chemical synthesis. Hence Pasteur believed that molecular asymmetry was an essential characteristic of the chemistry (and biology) of life.

Pasteur's demonstration that the optically inactive (or racemic) paratartaric acid was composed of equal amounts of two optically active forms of opposite senses was made on crystals of sodium-ammonium paratartrate. In using these, and in separating out the two forms, he was perhaps fortunate, as in no other compounds is the relationship between crystal structure and molecular asymmetry so straightforward; also, the distinctive 'hemihedral' crystalline forms of these compounds occur only under certain conditions of crystallization. From this he was led to study asparagine and its derivatives (aspartic acid and the aspartates, malic acid and the malates).

Arthur Lindo Patterson 1902–1966

Patterson was born in Nelson, New Zealand (also the birthplace of Ernest Rutherford). His family emigrated to Canada and he spent his boyhood partly in Montreal and partly in England where he 'received a very sound education at Tonbridge School'. On his return to Canada he enrolled for a physics degree at McGill University (still 'haunted' by Rutherford who had left McGill in 1907)—but in 1923 became a casualty of the British/Colonial 'one-shot' degree classification system—he was placed in the second class (BSc Honours). How (he later recalled) 'might I live this down'. But the system works (or worked) both ways: Patterson's evident ability, and personal recommendations, led in 1924 to the award of a two-year McGill Travelling Fellowship. He was accepted to work under W.H. Bragg in the Davy–Faraday Laboratory at the Royal Institution where he made his own X-ray apparatus—the DC current interrupters for which 'wailed like banshees on the lab roof' (and occasionally blew up)—and worked on space groups with, among others, William Astbury and Kathleen Yardley. He then spent a year, under a Canadian National Research Council Fellowship, at the Kaiser-Wilhelm Institute in Berlin where the excitement of each week was the Physical Colloquium with Laue as Chairman and Planck, Einstein, and Nernst sitting on the front row! He worked on the determination of the particle size in cellulose by X-ray diffraction techniques and on the theory of particle size broadening. But more important, here began his 'obsession with the notion that something had to be learned about structural analysis from Fourier theory'.

In 1927, Patterson returned to McGill to complete his PhD and then took a succession of research posts in the United States, in particular the Massachusetts Institute of Technology where he met the mathematician, Norbert Weiner and through whose influence was born in 1934 the Patterson function—which Patterson himself modestly referred to as the 'F^2 series'. It was and still is, a most important tool in crystal structure

determination, although its full practical implementation in summing the Fourier series took many years of development.

From 1936 to 1949 Patterson joined the faculty at Bryn Mawr Women's College in Philadelphia—an Institution remarkable for its tradition that the faculty, and not the administration, governed its research and teaching programmes. Finally he left Bryn Mawr in 1949 to start an X-ray structure analysis group at the Institute of Cancer Research in Philadelphia.

Lindo Patterson was remembered by his contemporaries as a man who 'possessed the rare combination of a keen mind, a lively humour, and a gentle disposition'. His sudden death from a brain haemorrhage was a great loss to the wider crystallographic community.

Linus Carl Pauling 1901–1994

Pauling was born in Portland, Oregon, USA. His childhood, following the death of his father, was a difficult one, but he was encouraged by an early school teacher, Pauline Geballe, who recognized his clearly emerging scientific abilities. Years afterwards, as the recipient of two Nobel prizes, he revisited, at a school reunion, his old teacher who had been such an influence on him.

After graduating from Oregon Agricultural College with high honours, Pauling enrolled not in one of the well-established 'Ivy League' Universities of the USA but in the fledgling California Institute of Technology at Pasadena: 'Caltech' was to be Pauling's Alma Mater for another forty years. His advisor was Roscoe Dickinson who had introduced X-ray diffraction techniques from England and Pauling determined the structures of microcline and labradorite, gaining his PhD degree in 1925. Thence the nature of the chemical bond dominated his interest: he realized early on that the newly emerging field of quantum mechanics was relevant to the problems with which he was concerned. From 1926 Pauling made extended visits to Europe to make contact with the leading figures of European science: Arnold Sommerfeld in Munich, Niels Bohr in Copenhagen and W.L. Bragg in Manchester. It was during this time that he developed his 'rules' for the solution of crystal structures, first enunciated in a paper of 1928 and which were eventually to be encapsulated in his great book *The Nature of the Chemical Bond and the Structure of Molecules and Crystals*, which was first published in 1939. Pauling's presentation of his rules, as solely the fruit of his own chemical insight, was the source of friction with W.L. Bragg whose analyses of the structures of minerals, silicates in particular, involved, in effect, an application of the first three rules and whose work and contribution received scant acknowledgement. It was not the only such occasion: Pauling's undoubted brilliance (and perhaps his own feeling of superiority and/or lack of recognition of the contributions of others) led to difficult relations with many of his fellow scientists.

In 1932, armed with the first of a series of large grants from the Rockefeller Foundation, Pauling abandoned the analysis of inorganic compounds and entered the field of biology and medicine—a major phase in his life's work that was ultimately to lead to the 'α-helix' model for the structure of proteins. Pauling's initial work was on haemoglobin and led, from 1936, to his long collaboration with Robert Corey.

During the war years 1940–1945, Pauling virtually abandoned scientific work and became involved with political and social issues which initially stemmed from his outrage at the treatment of Japanese-American citizens and the vilification of his doctor—an outspoken communist whose treatment in 1941 for a serious illness certainly saved Pauling's life. After the war, Pauling became a strong supporter of the Peace Movement, an advocate of rapprochement with the Soviet Union and an opposer of nuclear weapons, activities which during the McCarthy era stigmatized him as a security risk and which led to the withdrawal of his passport in 1952.

The story of Pauling's discovery of the α-helix protein structure is a curious one. He recalls that during a visit to Oxford as a visiting Professor in 1948 he was laid up in bed with a bad cold. Sick of reading detective stories he drew pictures of amino-acid molecules on pieces of paper, folding the papers in such a way as to achieve the correct bond angles and then fitting them together to form a helix—a helix with the unexpected property that the successive turns of the helix did not correspond with an integral number of amino acid residues. On his return to Caltech he set a senior fellow, Herman Branson, with the problem of finding possible helical structures and Branson, largely independently, came up with two solutions—the α and γ helices. The discovery of the α-helix was formally announced at a congress in Stockholm in 1951. It did not go down well at first with the English scientists—Astbury, Perutz and Bragg—because it contravened their innate feeling that the turns of the helix should correspond with an integral number of amino acid residues.

In 1952, W. T. Astbury organized a Discussion Meeting on the Structure of Proteins at the Royal Society, to which of course Pauling was invited as a principal speaker. But without a passport, he was unable to attend and a colleague, Edward Hughes, had to deputize on his behalf.

The great opus of Pauling's scientific work was recognized with the award of his first Nobel Prize for Chemistry in 1954. His second Nobel Prize, for Peace, was announced on 10th October 1963, the same day as the announcement of the (partial) Test Ban Treaty for which Pauling had campaigned so hard.

William Jackson Pope 1870–1939

Pope's major scientific work was in the field of organic chemistry. He was educated at Finsbury Technical College and the City and Guilds of London Institute in Kensington; in 1908 he was appointed Professor of Chemistry at Cambridge at the early age of 38. His crystallographic work extends over the period 1906–10 when, in collaboration with W. Barlow, he developed models of crystal structures based on the close-packing of ions or atoms of different sizes—models which were to prove so valuable to the Braggs in their analyses of crystal structures.

Jean Baptiste Louis Rome de L'Isle 1736–90

Rome de L'Isle, the son of a cavalry officer, was born in Gray in France. In 1756 he joined the Royal Corps of Artillery and Engineering and travelled to Pondicherry in

the French East Indies. When Pondicherry was seized by the English in 1761 he was taken prisoner and transported to China where he remained until 1764 when he returned to France. He first studied mineralogy and in 1767 he was asked to prepare a catalogue of a collection of mineral 'curiosities'. In doing so he followed the method of Linnaeus (the great Swedish botanist) in emphasising the importance of crystalline form in minera-logical description. In 1772 he published *Essai de cristallographie* and in 1783 his major work *Cristallographie*. Here he identified 450 'crystalline forms' and characterized the cube, regular octahedron, parallelepiped, rhomboidal octahedron and dodecahedron. In the course of measuring the angles between the faces of terracotta (clay) crystal models using a contact goniometer, he recognized the constancy of interfacial angles.

Rome de L' Isles' great contribution, through his experimental work, was to establish crystallography as the firm basis for the study of minerals although he did not follow (and was critical of) Hauy's notion of 'molecules integrantes'.

Wilhelm Conrad Röntgen 1845–1923

Röntgen was born in Lennep, a small town near Remscheid in the Rhine province of Germany. His father was a prosperous cloth merchant and both parents came from established Lutheran Rhineland families. When Röntgen was aged three, the family moved to Appeldoorn in Holland. During his boyhood Röntgen showed no special aptitudes and his education was interrupted as a result of being expelled from school for refusing to name a fellow pupil who had been involved in some misdemeanour. He entered Utrecht Technical School (rather than University) in 1862 and then went on to study mechanical engineering at Zurich Polytechnic, gaining the PhD degree in 1869.

By this time Röntgen's academic ability, and moreover his outstanding practical experimental skills, were evident. In 1871 he became assistant to Professor August Kundt in the University of Würtzburg and moved with him to Strasbourg before being himself appointed in 1879 to the Chair of Physics at the University of Giessen—a post he held until 1888 despite the offer of Chairs at the prestigious Universities of Jena and Utrecht. However, in 1889 he returned to Würtzburg as Professor of Physics: perhaps the atmosphere of that old Bavarian town suited his own reticent personality—for Röntgen shunned all public engagements. He worked almost entirely alone in the laboratory such that there should be no disturbance to his concentration and developed acute powers of observation which undoubtedly led to his discovery of X-rays on 8th November 1895.

Röntgen was experimenting with discharge (cathode ray) tubes and the action of the rays in generating fluorescence in crystals and luminescence in gases—hardly a 'hot topic' since such tubes had been developed by William Crookes some twenty years earlier. Röntgen had enclosed the tube in a light-tight cardboard box and noticed that every time the induction coil discharged through the tube, a crystal of barium platinocyanide, lying on a nearby table, gave a flash of fluorescence. Röntgen, ever cautious, did not immediately announce his discovery, but in a feverish bout of work found that the rays (which, not knowing their nature, he called X-rays), originated from where the

cathode rays fell on the glass walls of the tube; that they were neither refracted, reflected nor diffracted (from optical gratings), were more heavily absorbed by denser elements and blackened photographic plates. On 22nd December he took the famous radiograph of his wife's hand and on 28th December communicated his observations to the Physical and Medical Society of Würtzburg which he followed up by sending friends and colleagues reprints of his paper as New Year gifts. On 5th January 1896 the news of these amazing rays which could reveal the living skeleton broke around the World and on 15th January, Röntgen received a Royal Summons from the young Kaiser Wilhelm II to demonstrate his discovery at the Royal Palace in Berlin (a summons which, of course, he could not refuse), following which he was immediately awarded the Prussian Order of the Crown, Second Class. Further honours followed. In 1896, jointly with Lenard, he was awarded the Rumford Medal of the Royal Society and in 1901 the first Nobel Prize for Physics; at the presentation ceremony Röntgen declined to give the expected lecture.

Röntgen's discovery is an example of serendipity in science as well as a testament to his observational and experimental skills. For X-rays were being generated long before 1895. As early as 1879 Crookes complained of fogged photographic plates and Goodspeed and Jennings in Philadelphia in 1890 noted a 'peculiar' blackening of photographic plates after a demonstration with a Crookes discharge tube. Finally, we should record that Röntgen's other great achievement was to detect and measure the (very weak) magnetic effects, predicted by Maxwell's theory, when a dielectric is moved between charged condenser plates.

Alexei Vasilievich Shubnikov 1887–1970

Shubnikov's childhood in Moscow was one of great poverty, a family of six children being raised by his mother following the early death of his father. In 1906 Shubnikov attended a 'popular' course in crystallography given by G.V. Wulff and following his enrolment in 1908 at the State University, became Wulff's devoted pupil and assistant. At this time the seeds of the Russian Revolution were already being sown: in 1911 Wulff, Shubnikov and many of the academic staff resigned in protest against the reactionary policy of the Minister (of Education) Lev A. Kasso and transferred to the Shaniavskii People's University from which Shubnikov graduated in 1912. He was seriously wounded in 1914 and spent the remaining war years on non-combat duty as a chemist, working during his free time on geometrical crystallography and corresponding with Fedorov. In 1918 he returned to the People's University as Wulff's assistant but the political circumstances of the time—the isolation from Western Europe and the almost complete lack of equipment—made it impossible to do any serious research work. However, by 1927 the situation had eased and Shubnikov was 'sent' to Germany and Norway to gain first hand experience of X-ray analysis and where he met many of the leading figures of German science. He recalls asking Laue to show him the laboratory where he (Laue) investigated crystal structures and being amazed to find that Laue did not have a laboratory at all!

In 1925 Wulff died and Shubnikov took up a position in the Mineralogical Museum of the Soviet Academy of Science in Leningrad where he set up an X-ray diffraction

facility and also studied piezoelectricity in quartz, work which entailed the full use of his great experimental skills. This work was severely interrupted in 1934 when the government transferred the Academy to Moscow and again in 1941 when the crystallography laboratory was evacuated to the Urals. It was during this time that Shubnikov developed the idea of 'antisymmetry' (surely an unsatisfactory and potentially misleading word) which rather means the unification of opposite concepts – plus and minus, black and white, etc., initially an abstract field of crystallography which has later found wide application in the study of magnetic and other properties of crystals.

John William Strutt, third Baron Rayleigh 1842–1919

Like his near contemporary the geologist and microscopist Henry Clifton Sorby, Lord Rayleigh (as he became when he succeeded to his father's title in 1873) had the leisure and financial resources to devote his life to science. His early life was dogged by ill health; in 1865 he graduated in mathematics at Cambridge and remained there until 1872 when he went to Egypt for a period of convalescence. It was not, however, a period of idleness; he wrote his classic book *The Theory of Sound* on a houseboat on the Nile.

On his return to England he built his own private laboratory and carried out most of his work there except for a short period (1879–1884) when he was Cavendish Professor of Experimental Physics at Cambridge.

Rayleigh's main contributions to science were in optics and acoustics; his work on the scattering of light by small particles led to the explanation of the blue colour of the sky and his work on the density of gases led to the discovery of argon.

Paul Hermann Scherrer 1890–1969

Scherrer was born at St. Gall, Switzerland. During his education his interests moved away from commercial topics towards natural history, first to botany and then to mathematics and physics. In 1913 he entered the University of Göttingen and became a member of Debye's research group. The stimulus for the investigation of polycrystalline specimens appears to have come from Debye although the construction of the cylindrical camera was due to Scherrer. With it, Debye and Scherrer took the first X-ray powder photographs (of lithium fluoride) in 1915. A very similar method was devised in 1917 by Albert W. Hull at the General Electric Laboratories, USA.

In later years Scherrer was active in the foundation of the European Centre for Nuclear Research (CERN). He was also prolific teacher and brought to his lectures a considerable element of showmanship.

Arthur Schönflies 1853–1928

Schönflies was born in Landsberg an der Warte, now in Poland. He studied mathematics at Berlin and became successively high school teacher, Associate Professor of Mathematics (at Göttingen), Professor (at Königsberg) and Rector at the University at Frankfurt. He extended the work of Sohncke on periodic discrete groups by taking into

consideration rotation–reflection and inversion axes of symmetry, adding another 165 to Sohncke's sixty-five groups. The work was completed in his book *Kristall-systeme und Kristallstruktur*, published in Leipzig in 1891, a few months after Fedorov's paper.

William Bradford Shockley 1910–89

William Shockley was born in London to American parents and moved to California (with them) in 1913. He was educated at the California Institute of Technology and Massachusetts Institute of Technology and, after war service with the U.S. Navy's Submarine Warfare Operations Research Group, became a supervisor in the semiconductor research group at Bell Telephone Laboratories in New York. There, in 1947, together with John Bardeen and Walter Brattain, he discovered the principle of the point contact transistor and then went on to develop the junction transistor. This may be said to be the starting point of the electronic transformation of the latter part of the twentieth century, which was recognized by the award of the Nobel prize for Physics, jointly with Bardeen and Brattain, in 1956.

Shockley then left 'Bell Labs' to set up the Shockley Semiconductor Corporation in California and from 1963 became Professor of Engineering at Stanford University.

The latter part of his life was clouded by controversy because of his uncompromisingly held views on dysgenics, i.e. retrogressive evolution through the excessive reproduction of the genetically disadvantaged, and for which, as one example of his increasing isolation, the University of Leeds refused to confirm his nomination for an Honorary Doctorate.

Leonhard Sohncke 1842–97

Sohncke was born in Halle, Germany, where his father was Professor of Mathematics at the University. Like his father he followed an academic career, being appointed to a succession of professorships in physics, firstly at Halle and at the end of his career at the Technische Hochschule in Munich. Sohncke considered the possible arrays of points which have identical environments but not necessarily in the same orientation (as in the definition of Bravais lattices), and arrived at sixty-five of the possible 230 space groups. He published his findings in 1867, while he was professor of Physics at the Technische Hochschule in Karlsruhe, and used cigar-box models by way of illustrations.

Nicolaus Steno (Niels Stensen) 1638–86

Stensen (Steno being the Latinized form of his name) was born in Copenhagen where his father, a goldsmith, kept a shop close to the famous Round Tower. He studied medicine at the University of Copenhagen and carried out anatomical researches in Florence, ultimately being appointed Royal Anatomist in Denmark in 1672.

Stensen combined his medical and anatomical career with an interest in minerals, metals, the refraction of light, telescopes and microscopes which presumably stemmed both from boyhood associations in his father's shop and also from his later travels in

Italy where he became acquainted with the works of Kepler and Galileo. He published in Florence in 1669 the *Prodromum* (i.e. an advance notice) *to a treatise on solid bodies naturally enclosed in other solids*—a work in which he outlined the principles of geology. This was translated into English by Henry Oldenberg, Secretary of the Royal Society, and re-published in London in 1671. In this work Stensen shows that crystals grow by external deposition on faces already laid down and not (as was then supposed by analogy with plants) by the absorption of fluids and the deposition of material from within. The *Law of the Constancy of Interfacial Angles*, for which he is best known, is illustrated in just one figure and stated in two sentences. In the latter part of his life Stensen's strongly held religious beliefs led him to renounce scientific research.

Gustav Heinrich Johann Apollon Tammann 1861–1938

Tammann can be regarded as a pioneer physical metallurgist in the light of his work in metallography and the study of solid-state chemical reactions. He was born in Yamburg (now Kingisepp) in Russia and, after a career as a lecturer at Charlottenburg and Leipzig, was chosen to head the newly formed Institute at Göttingen in 1903. It was here that he began his studies on metallic compounds, the crystal structures and mechanical properties of metals and alloys, and the general problems of the mechanisms of plastic deformation and work hardening in terms of crystallographic slip and crystalline rearrangements. He continued to work on these problems until shortly before his death, applying his ideas at the same time to the flow of ice.

Georgy Fedoseevich Voronoi 1868–1908

Voronoi was born in Zhuravka, Russia, into an academic family—his father was Superintendent (a sort of Government Inspector) of the Gymnasiums in towns in Southern Ukraine. His mathematical ability was apparent from an early age; he attended the Gymnasium at Priluka, graduated in mathematics in the University of St. Petersburg in 1889 and was appointed Professor of Pure Mathematics in the University of Warsaw at the early age of 26. Voronoi's main interests were in the Theory of Numbers, for which he planned a series of memoirs—a series cut short by his early death. It was in the second of these that he established the possible methods of filling n-dimensional Euclidean space with identical non-intersecting convex polyhedra (Voronoi polyhedra or parallelepipeds), all of which possessed contiguous boundaries—an analysis which extended the work of his near contemporary, Fedorov.

James Dewey Watson 1928–

James Watson was born and brought up in Chicago's south side, the son of parents who, although poor, managed to retain their middle class status throughout the years of the depression. But they loved learning 'my family had no money but lots of books', which Watson devoured from an early age, especially science ones 'full of facts'. His other keen interest was bird-watching with his father, which might have ultimately led to a

career as an ornithologist. Despite the family circumstances Watson was fortunate in his schooling: from the ages of 5–13 he attended the Horace Mann Grammar School, and then the South Shore High School. During these years the makings of an infant prodigy became evident; in 1942 he was a contestant on a popular radio show called 'The Quiz Kids'—but only survived three broadcasts on account of his lack of knowledge of Shakespeare and the Old Testament.

At the age of 15, under the enlightened early admissions policy of the President, Robert Hutchins, he enrolled in the University of Chicago to read Zoology. He was, perhaps, not the ideal student: he studiously distained from taking any notes in class and avoided social contacts with his fellow students.

However, in 1945 he came across Erwin Schrödinger's book '*What is Life?*' —which inspired him and which, by a remarkable collusion of events, was also inspiring his future joint Nobel Prize winners on the other side of the Atlantic. A final year course in physiological genetics was further instrumental in leading him away from ornithology to genetics and, after graduation in 1947, the award of a scholarship at the University of Indiana at Bloomington. He wasted no time; following a course on viruses given by Salvador Luria, he boldly offered himself as a research student and, despite his inexperience, Luria accepted him to work on bacteriophages. It was a turning point in Watson's career: membership of Luria's 'phage group' brought him into contact with leading geneticists, most notably Max Delbrück, a refugee from Germany who had studied at Niels Bohr's laboratory in Copenhagen. Delbrück's biological work was, in a large part, the inspiration that lay at the root of '*What is Life?*'

Following the PhD in 1950, Watson was awarded a National Research Council grant to study biochemistry in the laboratory of Herman Kalckar in Copenhagen. This move was not a success but did provide him with the opportunity of attending, an International Meeting in Naples at which Maurice Wilkins presented his first X-ray diffraction photographs of DNA. This to Watson was a revelation; clearly X-ray diffraction studies were the way forward. Wilkins' laboratory was the obvious place to go, but Wilkins shied away. Undeterred, Watson was able, with the backing of Luria and John Kendrew, to transfer his scholarship to Cambridge to work with Francis Crick. Crick and Watson complemented each other extraordinarily well and it was their close collaboration that led to the discovery of the double helix structure of DNA in 1953. Thereafter, except for a short period in Cambridge in 1955–1956, their careers diverged. Following a short period as Senior Research Fellow at Caltech, Watson was appointed to the Faculty of the Biology Department at Harvard. To say that he was unpopular, with his tradition-bound colleagues, would be an understatement: he is said to have 'radiated contempt in all directions'. But set against this must be his achievement in the publication (1965) of *The Molecular Biology of the Gene*.

Then in 1968 comes the last major epoch in Watson's career: his appointment as Director of the Cold Spring Harbor Laboratory (CSHL), his marriage and the publication of *The Double Helix*. The reception of this book has been told elsewhere but what is remarkable and what gives the book its great appeal, is that it reads as the story of a young man describing near-contemporary events and not as the recollection of events some 15 years in the past.

As Director, and subsequently President, of CSHL, Watson was enormously successful in transforming a minor laboratory into one of world-class significance. He was, of course, crowned with honours—the National Medal for Science, the Copley Medal of the Royal Society and (in 2002) an Honorary KBE from the Queen. But his later years were marred (as with William Shockley) by his views on eugenics, enforcing both his retirement and (in 2007) the cancellation of a UK tour to publicize his book *Avoid Boring People: Lessons from a Life in Science*.

Leonhard Weber 1883–1968

Weber was awarded a doctorate at the University of Fribourg in 1917 for a thesis on the refractive indices of biaxial crystals. He was curator of the mineralogical collection in Zurich and then Professor of Mineralogy at the University of Fribourg. The 4-index system for lattice directions, which now bears his name, was published in 1922. However, his most significant contributions to crystallography were his derivation and illustration, in 1925, of the 230 space groups (in which Bravais lattice points are represented by point group polyhedra) and his geometrical derivation, in 1929, of the 80 two-colour (2-sided) plane groups.

Christian Samuel Weiss 1780–1856

After studying medicine at Leipzig, Weiss switched to chemistry and physics, becoming Professor of Physics there in 1808. In 1810 he was appointed Professor of Physics at the newly established and prestigious University of Berlin, a post which he held until his death. Weiss developed the idea of crystal axes, from which he was able to distinguish crystal systems by the ways in which crystal faces were related to such axes. He also formulated the concept of a zone: originally conceived as a direction of prominent crystal growth, the term was defined as a collection of crystal faces parallel to a line—the zone axis; from this concept the zone law was derived.

Maurice Hugh Frederick Wilkins 1916–2004

Wilkins was born in Pongaroa, New Zealand where his father, a doctor, hoped that the opportunities for the practice of preventive medicine would be so much greater than in his native Dublin. These hopes were not fulfilled and the family returned first to Dublin, then to London and finally settled in Birmingham. Here, in 1929, Wilkins enrolled at the prestigious King Edward's High School: at home he set up his own workshop, made model aeroplanes and telescopes and his favourite reading matter was 'The Modern Boy'. In 1935 he won a place at St John's College, Cambridge, to read physics but graduated with 'only' a lower second class degree, which destroyed his expectations for 'a cloistered college existence'. Perhaps it was a piece of good luck: his tutor, Mark Oliphant, was appointed Professor of Physics at the University of Birmingham and suggested that Wilkins should join John Randall (later to be the inventor, with Henry Boot, of the cavity magnetron) to carry out research on the luminescence of solids. Thus began

Wilkins' lifelong but uneasy association with Randall: they were men of opposed but curiously complementary temperaments. On gaining his PhD, Wilkins joined Oliphant's atomic physics 'bomb' team that included Rudolf Peierls and Otto Frisch, both refugees from Nazi Germany. In 1944 Wilkins moved to Berkeley to take part in the Manhattan project: it was an episode in his career, following the atomic bombing of Hiroshima and Nagasaki, which Wilkins always felt uncomfortable about. However, it was after reading Erwin Schrödinger's book *What is Life?* that Wilkins decided that his future lay in the biological and genetic applications of physics: Schrödinger's notion that a gene was of the nature of an aperiodic crystal attracted him enormously. In fact *What is Life?* made the same profound impression upon him as his exact contemporary, Francis Crick.

In 1945 Randall was appointed Professor of Physics at the University of St Andrews and invited Wilkins to join him there. But St Andrews could not contain a man of Randall's ambition and D'Arcy Wentworth Thomson, the most distinguished Professor there, author of *On Growth and Form*, had little to contribute to the subject of genetics. In 1946 Randall was appointed Head of the Department of Physics at Kings College, London and subsequently Director of the Medical Research Council's Biophysics Research Unit with Wilkins as his deputy. In both respects Randall was enormously successful; both he and Wilkins built up a strong research team, including a large proportion of women researchers. Wilkins developed polarized light microscope techniques to study DNA but only took up X-ray diffraction studies following the gift of a phial of carefully prepared DNA from Rudolf Signer in Berne. Together with a research student, Raymond Gosling, he obtained sharp diffraction patterns of DNA in the moist state and it was as a result of their initial success that led Randall to appoint Rosalind Franklin to continue the work, with Gosling, on the Signer material. The ensuing unhappy relationship between Franklin and Wilkins; has been the subject of innumerable books and articles: Wilkins' own biography *The Third Man of the Double Helix*, published the year before his death, is largely an apologia for this episode in his life.

Following the publication of the historic papers 'Molecular Structure of Nucleic Acids' in *Nature* (25 April 1953) and Franklin's move to Birkbeck College, Wilkins continued to refine the 'Watson–Crick' model of DNA and confirmed the essential correctness. Thereafter, his life took on a more tranquil existence: the bouts of depression that had suffered in his student and later days no longer returned, in 1959 he married again, became a member of the Campaign for Nuclear Disarmament and, as a Nobel Laureate, in 1969, became first President of the British Society for Social Responsibility in Science.

Georg (Yuri Viktorovich) Wulff 1863–1925

Georg Wulff (the name he used in his German-language publications) was born in Nezhin in the Ukraine. He published his first papers (on the morphology and physical properties of crystals) in 1883 and 1884 whilst still a student at Warsaw University. In 1907 he was appointed Professor of Crystallography at Moscow University, a post which he held until his death. He proposed what has become known as the Wulff net

(the stereographic projection of a sphere orientated with the polar 'north-south' axis in the plane of projection rather than perpendicular to it) in 1909.

Wulff's contribution to X-ray crystallography was his demonstration of the geometrical equivalence of Laue's diffraction and W. L. Bragg's reflection interpretation of X-ray diffraction. The equation which he derived in 1913[4] is equivalent to that which W. L. Bragg presented in the previous year;[5] in both cases the equation is expressed in the form $n\lambda = 2d\cos\Theta$ where $\Theta = (90 - \theta)$. On this basis some writers link Wulff's name with Bragg's—the Bragg-Wulff equation. However, the present form of the equation was first given in a later joint paper by W. H. and W. L. Bragg[6] and it was of course the Braggs who made the first great leap forward in crystal structure analysis.

Ralph Walter Graystone Wyckoff 1897–1994

Ralph Wyckoff had a restless career: during his long life he was employed in no less than eleven laboratories in the United States (including periods of unemployment in between). He was born in Geneva in New York State into a family of Dutch migrants who had first arrived in America in 1634. At school, although left-handed, he was forced to write with his right hand, but rather than making him autistic he became ambidextrous that, he recalls, was a great advantage in experimental work. He attended Hobart College and in graduating moved to Cornell University in the hope of fulfilling his boyhood dream of becoming an astronomer. But jobs in astronomy were far and few between. Happily, he came under the influence of Shoji Nishikawa who was then spending two years in the Physics Department and began research in X-ray diffraction, gaining his PhD in 1919. In strong contrast to British crystallographers (who studied single crystals), Wyckoff largely made use of the Laue and powder diffraction techniques. In 1922 he published *The Analytical Expression of the Theory of Space Groups*—a prelude to the description of the 230 space groups in the *International Tables for X-ray Crystallography*. Thence, his interests moved towards medical research. At the Rockefeller Institute for Medical Research he developed a vaccine against typhus fever and pioneered electron microscope techniques for imaging the influenza virus, work that first led (in 1946) to an appointment at the Institute for Health in Bethesda and then as Scientific Attaché to the American Embassy in London. Wyckoff felt very much 'at home' in Europe and found, on his return to Bethesda, that his irregular working habits conflicted with the demands of the bureaucracy there. He resigned in 1960 and ended his career as Professor of Physics at the University of Arizona.

Thomas Young 1773–1829

Young was an infant prodigy who could read at the age of 2 and at the age of 14 knew Latin, Greek, French and Italian, and had begun to study several ancient languages— Hebrew, Chaldean, Syriac and the like—linguistic knowledge that he put to use in his

[4] Phys. Zeitschrift (15 Mar. 1913) Vol. 14, p. 217.
[5] Proc. Cam. Phil. Soc. (11 Nov. 1912) Vol. 17, pp. 43–57.
[6] Proc. Roy. Soc. A (17 April 1913) Vol. 88, pp. 428–438.

contribution to the translation of the text of the Rosetta stone. Young intended to take up a medical career and commenced his education in 1793 at St. Bartholomew's Hospital, moving from there to Edinburgh, Göttingen and Cambridge. He was elected a Fellow of the Royal Society as early as 1794 for his explanation on how the ciliary muscles of the eye change the shape (and focal length) of the lens.

Young's great contribution to science was his revival in 1800–04 of Huygens' wave theory of light, to which he added the concept of interference to explain rectilinear propagation and the diffraction phenomena at single or double slits or pinholes and in which he was closely shadowed by the work of Fresnel in France. In England Young was strongly attacked because he questioned the hegemony of Newton, but his work quickly won acceptance in the USA and Europe. Young was also first to suggest (in a letter to François Arago in 1816) that light might be propagated as a transverse wave, thus accounting for polarization and meeting the main objection of Newton. Regrettably, Young did not live to witness Foucault's demonstration, in 1850, that the speed of light was less in water than in air—the crucial experiment which distinguished between the corpuscular and wave theories.

Young was also active in many other fields. His name comes down to us through his work on the elastic behaviour of solids as the constant of proportionality (E) in Hooke's law, called Young's (or latterly Young) modulus.

Appendix 4

Some useful crystallographic relationships

In Sections A4.1–A4.3 the rather lengthy expressions for the triclinic system are omitted.

A4.1 Interplanar spacings, d_{hkl} (see Section 6.5.4)

Orthorhombic:

$$\frac{1}{d_{hkl}^2} = \frac{h^2}{a^2} + \frac{k^2}{b^2} + \frac{l^2}{c^2}.$$

(Tetragonal: $a = b$; cubic: $a = b = c$.)

Hexagonal:

$$\frac{1}{d_{hkl}^2} = \frac{4}{3}\left(\frac{h^2 + hk + k^2}{a^2}\right) + \frac{l^2}{c^2}.$$

Rhombohedral:

$$\frac{1}{d_{hkl}^2} = \frac{(h^2 + k^2 + l^2)\sin^2\alpha + 2(hk + kl + hl)(\cos^2\alpha - \cos\alpha)}{a^2(1 - 3\cos^2\alpha + 2\cos^3\alpha)}.$$

Monoclinic:

$$\frac{1}{d_{hkl}^2} = \frac{1}{\sin^2\beta}\left(\frac{h^2}{a^2} + \frac{k^2\sin^2\beta}{b^2} + \frac{l^2}{c^2} - \frac{2hl\cos\beta}{ac}\right).$$

A4.2 Interplanar angles, ρ, between planes $(h_1 k_1 l_1)$ of spacing d_1 and $(h_2 k_2 l_2)$ of spacing d_2 (see Section 6.5.5)

Orthorhombic:

$$\cos\rho = \frac{(h_1 h_2/a^2) + (k_1 k_2/b^2) + (l_1 l_2/c^2)}{\sqrt{\{[(h_1^2/a^2) + (k_1^2/b^2) + (l_1^2/c^2)][(h_2^2/a^2) + (k_2^2/b^2) + (l_2^2/c^2)]\}}}.$$

(Tetragonal: $a = b$; cubic: $a = b = c$.)

Hexagonal:

$$\cos \rho = \frac{h_1 h_2 + k_1 k_2 + \frac{1}{2}(h_1 k_2 + h_2 k_1) + (3a^2/4c^2)l_1 l_2}{\sqrt{\{[(h_1^2 + k_1^2 + h_1 k_1 + (3a^2/4c^2)l_1^2][h_2^2 + k_2^2 + h_2 k_2 + (3a^2/4c^2)l_2^2]\}}}.$$

Rhombohedral:

$$\cos\rho = (a^4 d_1 d_2/V^2)[\sin^2 \alpha (h_1 h_2 + k_1 k_2 + l_1 l_2)$$
$$+ (\cos^2 \alpha - \cos \alpha)(k_1 l_2 + k_2 l_1 + l_1 h_2 + l_2 h_1 + h_1 k_2 + h_2 k_1)].$$

Monoclinic:

$$\cos \rho = \frac{d_1 d_2}{\sin^2 \beta} \left[\frac{h_1 h_2}{a^2} + \frac{k_1 k_2 \sin^2 \beta}{b^2} + \frac{l_1 l_2}{c^2} - \frac{(l_1 h_2 + l_2 h_1)\cos \beta}{ac} \right].$$

A4.3 Volumes of unit cells (see Section 6.5.6)

Orthorhombic:

$$V = abc.$$

(Tetragonal: $\alpha = b$; cubic: $\alpha = b = c$.)
Hexagonal:

$$V = (\sqrt{3}/2)a^2 c = 0.866a^2 c.$$

Rhombohedral:

$$V = a^3 \sqrt{(1 - 3\cos^2 \alpha + 2\cos^3 \alpha)}.$$

Monoclinic:

$$V = abc \sin \beta.$$

A4.4 Angles between planes in cubic crystals (variants between 100 and 221) (see Section 6.5.5)

100	100	0.00	90.00		
	110	45.00	90.00		
	111	54.74			
	210	26.56	63.43	90.00	
	211	35.26	65.90		
	221	48.19	70.53		
110	110	0.00	60.00	90.00	
	111	35.26	90.00		
	210	18.43	50.77	71.56	
	211	30.00	54.74	73.22	90.00
	221	19.47	45.00	76.37	90.00

111	111	0.00	70.53				
	210	39.23	75.04				
	211	19.47	61.87	90.00			
	221	15.79	54.74	78.90			
210	210	0.00	36.87	53.13	66.42	78.46	90.00
	211	24.09	43.09	56.79	79.48	90.00	
	221	26.56	41.81	53.40	63.43	72.65	90.00
211	211	0.00	33.56	48.19	60.00	70.53	80.40
	221	17.72	35.26	47.12	65.90	74.21	82.18

A4.5 Relationships between zones and planes (see Section 6.5.3)

For a plane (hkl) which lies in a zone $[uvw]$:

$$hu + kv + lw = 0.$$

For a plane $(hkil)$ which lies in a zone $[UVTW]$:

$$hU + kV + iT + lW = 0.$$

For a lattice point with coordinates $[uvw]$ which lies in the nth (hkl) plane from the origin:

$$hu + kv + lw = n.$$

For a plane (hkl) which lies in the; zones $[u_1v_1w_1]$ and $[u_2v_2w_2]$:

$$h = (v_1w_2 - v_2w_1); \quad k = (w_1u_2 - w_2u_l); \quad l = (u_1v_2 - u_2v_1).$$

For a zone $[uvw]$ which contains the planes $(h_1k_1l_1)$ and $(h_2k_2l_2)$:

$$u = (k_1l_2 - k_2l_1); \quad v = (l_1h_2 - l_2h_1); \quad w = (h_1k_2 - h_2k_1).$$

Appendix 5

A simple introduction to vectors and complex numbers and their use in crystallography

A5.1 Scalars and vectors

In science—and, of course, everyday life—we are concerned with different types of quantities as well as quantities of the same type but of different size or magnitude. In many cases the quantities are independent of direction—mass for example, or volume or temperature. Such quantities, which can be represented by points on scales, are known as **scalars** and are denoted algebraically in familiar italic type—m for mass, t for time, x and y for unknown (scalar) quantities and so on. In many other cases, quantities are direction dependent, or, in other words, in order to specify the quantity completely the direction as well as the magnitude of the quantity needs to be described. Such quantities are called **vectors** and are denoted algebraically by printing in bold type—**a, b, x, y,** etc. In handwriting wavy underlining is substituted for bold type, i.e. $\underset{\sim}{a}, \underset{\sim}{b}, \underset{\sim}{x}, \underset{\sim}{y}$, etc.

Velocity is an everyday example of a vector quantity—we need to specify not only the speed of an object but the direction in which it is travelling. A motor car travelling 50 km h^{-1} due north has the same speed as a motor car travelling 50 km h^{-1} due east, but their velocities are different—one has a velocity of 50 km h^{-1} *north* and the other has a velocity of 50 km h^{-1} *east*. The velocities may be represented as vectors by arrows— one pointing north and one pointing east. The lengths of the arrows or vectors gives the speed—the greater the speed the longer the arrows. In the above case (50 km h^{-1}), the arrows are of equal length.

Heat and electrical current flow, electrical and magnetic fields are examples of vector quantities of interest in crystal physics, but in geometrical crystallography we are invariably concerned with vectors which denote displacements: for example the displacement from one corner of a unit cell to the next, or the displacement from one atom to another. Crystal axes or directions are examples of such displacement vector quantities. In Fig. 5.1 (p. 136) the directions and lengths of the edges of the unit cell are all denoted by the vectors **a**, **b**, and **c**. In the special case of cubic crystals (Fig. 3.1, p. 87) these vectors are all at 90° (orthogonal) and are also the same length, but they are *not* identical vectors because they 'point' in different directions.

The 'length' of a vector is called its **modulus**. In the case of the motor car the modulus of the velocity is the speed. The modulus of a vector is written in two ways, either in italic type, a (denoting that it is a scalar quantity), or with the heavy type vector symbol bracketed between straight vertical lines thus: |**a**|. Hence |**a**| $= a$.

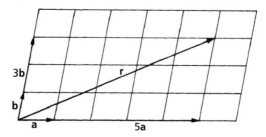

Fig. A5.1.

In Fig. 3.1 the moduli a, b, and c of the unit cell vectors for the fourteen Bravais lattices are indicated. In the case of cubic (or rhombohedral) crystals, the unit cell edge lengths, or moduli, are all identical and are represented (for all three axes) by a. Hence $|a| = |b| = |c| = a$ (as in Fig. 3.1).

A5.2 Resolving vectors into components and vector addition

Consider the (displacement) vector r drawn in Fig. A5.1 which is outlined in a (two-dimensional) lattice with the unit cell vectors a and b. Note that all these vectors have a common origin.

The vector r may be resolved into its components in the directions of the unit cell vectors a and b. These components, $5a$ and $3b$, are drawn in Fig. A5.1. Added together they give r, i.e. $r = 5a + 3b$.

In vector analysis in general, cubic or Cartesian axes are chosen, and the unit cell vectors are labelled i, j, k rather than a, b, c. Also, the moduli of i, j, k are unity and they are known as **unit vectors**. The general symbol for a unit vector is n, i.e. $|n| = 1$.

A5.3 Multiplying vectors—scalar products and vector products

There are two ways in which vectors can be multiplied, namely, the scalar product and the vector product, the general significance of which go beyond the scope of this book. We will consider only the multiplication of displacement vectors.

The scalar (or 'dot') product $a \cdot b$ may be thought of as 'representing' the product of the moduli of the vectors resolved in the direction of one or other of the vectors. For example (Fig. A5.2), if the angle between the vectors, drawn from a common origin, is θ, the resolved component of the modulus of b along the a direction is $b \cos \theta$. Hence $a \cdot b = ab \cos \theta$. The same result applies if a is resolved along the direction of b.

If one of the vectors is a unit vector—say in Fig. A5.2 $a = n$—then the scalar product $n \cdot b = b \cos \theta$ (since $|n| = 1$): it is simply the modulus of b resolved along the direction of the unit vector n. The scalar product may therefore be thought of as a 'shorthand' means of resolving the vectors along different directions.

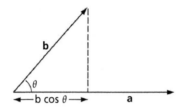

Fig. A5.2.

Clearly, the maximum value of the scalar product occurs when the vectors are parallel ($\theta = 0°$) and the minimum value, zero, occurs when they are at right angles ($\theta = 90°$). The scalar product of a vector multiplied by itself is simply the modulus squared, i.e. $\mathbf{a} \cdot \mathbf{a} = |a||a| = a^2$.

The vector (or 'cross') product $\mathbf{a} \times \mathbf{b}$ may be thought of as 'representing' the area outlined by two vectors. In Fig. A5.3 the area outlined by the vectors \mathbf{a} and \mathbf{b} is $ab \sin \theta$. But there is more: the area may represent a crystal plane or surface, the **orientation** of which also needs to be specified. The orientation of the surface may be specified by the direction normal to it, i.e. perpendicular to the page. Let the unit vector along this direction be \mathbf{n}; then $\mathbf{a} \times \mathbf{b} = (ab \sin \theta)\mathbf{n}$.

In general, the vector product is expressed as $\mathbf{a} \times \mathbf{b} = \mathbf{c}$, where \mathbf{c} is perpendicular to \mathbf{a} and \mathbf{b} and $|c| = ab \sin \theta$. The largest value of \mathbf{c} occurs when \mathbf{a} and \mathbf{b} are at right angles and the smallest value, zero, occurs when they are parallel. Finally, the vector product involves the use of a sign convention: the vector \mathbf{c} or \mathbf{n} is perpendicular to the plane defined by \mathbf{a} and \mathbf{b} but can have one of two directions, i.e. with reference to Fig. A5.3, it can be 'up' from the page or 'down'. One vector must have a + sign and the opposite must have a − sign. The sign convention is as follows. Consider in Fig. A5.3 an (ordinary) right-handed screw coming out of the page (i.e. parallel to \mathbf{c} or \mathbf{n}) through the common origin of \mathbf{a} and \mathbf{b}. If we 'screw' from \mathbf{a} to \mathbf{b} the screw will move 'upwards' and this upwards direction is the direction or sense of \mathbf{c}. If we screw in the opposite direction, from \mathbf{b} to \mathbf{a}, the screw will move downwards. Hence, the order of the multiplication of vectors is important and leads to the **anticommutative law** for the vector product:

$$\mathbf{a} \times \mathbf{b} = -(\mathbf{b} \times \mathbf{a}).$$

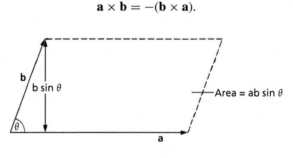

Fig. A5.3.

A5.4 Multiplication of vectors in terms of components

Calculations are generally made easier if vectors are expressed in terms of their components on a Cartesian axis system with unit vectors, $\mathbf{i}, \mathbf{j}, \mathbf{k}$ at right angles. In this case $\mathbf{i} \cdot \mathbf{i} = 1, \mathbf{i} \cdot \mathbf{j} = 0, \mathbf{i} \times \mathbf{i} = 0, \mathbf{i} \times \mathbf{j} = \mathbf{k}$, and similarly for the other combinations of \mathbf{i}, \mathbf{j} and \mathbf{k}.

Writing

$$\mathbf{a} = (a_1\mathbf{i} + a_2\mathbf{j} + a_3\mathbf{k}) \quad \text{and} \quad \mathbf{b} = (b_1\mathbf{i} + b_2\mathbf{j} + b_3\mathbf{k}).$$

The scalar product

$$\begin{aligned}
\mathbf{a} \cdot \mathbf{b} &= (a_1\mathbf{i} + a_2\mathbf{j} + a_3\mathbf{k}) \cdot (b_1\mathbf{i} + b_2\mathbf{j} + b_3\mathbf{k}) \\
&= a_1b_1\mathbf{i} \cdot \mathbf{i} + a_1b_2\mathbf{i} \cdot \mathbf{j} + \cdots \\
&= (a_1b_1 + a_2b_2 + a_3b_3)
\end{aligned}$$

(all the other terms cancel because $\mathbf{i} \cdot \mathbf{i} = 1$ and $\mathbf{i} \cdot \mathbf{j} = 0$, etc.)

The vector product

$$\begin{aligned}
\mathbf{a} \times \mathbf{b} &= (a_1\mathbf{i} + a_2\mathbf{j} + a_3\mathbf{k}) \times (b_1\mathbf{i} + b_2\mathbf{j} + b_3\mathbf{k}) \\
&= a_1b_1(\mathbf{i} \times \mathbf{i}) + a_1b_2(\mathbf{i} \times \mathbf{j}) + \cdots \\
&= (a_2b_3 - b_2a_3)\mathbf{i} + (a_3b_1 - b_3a_1)\mathbf{j} + (a_1b_2 - b_1a_2)\mathbf{k}
\end{aligned}$$

(all the other terms cancel because $\mathbf{i} \times \mathbf{i} = 0$ and $\mathbf{i} \times \mathbf{j} = \mathbf{k}$, etc.).

The vector product may be expressed in matrix form as a determinant:

$$\mathbf{a} \times \mathbf{b} = \begin{vmatrix} \mathbf{i} & \mathbf{j} & \mathbf{k} \\ a_1 & a_2 & a_3 \\ b_1 & b_2 & b_3 \end{vmatrix}.$$

A5.5 Volumes of unit cells in vector notation

The volume of a unit cell defined by vectors \mathbf{a}, \mathbf{b}, and \mathbf{c} is 'the area of the base times the height'.

The 'area of the base' is $(\mathbf{a} \times \mathbf{b}) = (ab \sin \theta)\mathbf{n}$. The 'height' is the component of \mathbf{c} resolved in the direction of the normal \mathbf{n} (Fig. A5.4), i.e. the height $= \mathbf{n} \cdot \mathbf{c} = c \cos \phi$. Hence the volume $V = (\mathbf{a} \times \mathbf{b}) \cdot \mathbf{c}$. It is a scalar quantity. The vectors $\mathbf{a}, \mathbf{b}, \mathbf{c}$ in this equation may be 'cycled' but the sequence must be maintained, i.e.

$$V = (\mathbf{a} \times \mathbf{b}) \cdot \mathbf{c} = \mathbf{a} \cdot (\mathbf{b} \times \mathbf{c}) = (\mathbf{c} \times \mathbf{a}) \cdot \mathbf{b}, \text{ etc.}$$

If the sequence is reversed, i.e. $(\mathbf{b} \times \mathbf{a}) \cdot \mathbf{c}$, then a negative volume results. In practice this can occur if a left-handed set of axes are substituted, inadvertently or otherwise, for a right-handed set.

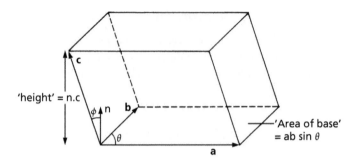

<div align="center">Fig. A5.4.</div>

The reciprocal unit cell \mathbf{a}^*, \mathbf{b}^*, \mathbf{c}^* can be derived from the 'direct' unit cell \mathbf{a}, \mathbf{b}, \mathbf{c} in a similar way as outlined in Section 6.5.6. Note that the relationships

$$\mathbf{a}^* = \frac{\mathbf{b} \times \mathbf{c}}{V} \text{ etc.}$$

are valid for the unit cells for any axis system.

A5.6 Representing vectors by complex numbers: the Argand diagram

The sum of a number of vectors $\mathbf{a} + \mathbf{b} + \mathbf{c} + \cdots$ may be simply obtained graphically by adding them 'top to tail' as shown in Fig. A5.5. In the case where the vectors represent the atomic scattering factors, f, for the atoms in the unit cell, the sum is the structure factor F_{hkl}.

However, although graphical procedures are very instructive in that they clearly show the physical principles involved, they are also very clumsy and do not readily lend themselves to analytical procedures. We need therefore an alternative method of representing vectors and the method we adopt is that of representing vectors as complex numbers.

A complex number, z, is a number which has two parts of the form

$$z = x + iy$$

where x and y are real numbers and $i = \sqrt{-1}$. The first part, x, is called the real part of the complex number and the second part, iy, is called the imaginary part. These two parts can be represented graphically on what is known as an Argand diagram—a graph in which the horizontal axis (abscissa) represents x, the real part and the vertical axis (ordinate) represents iy, the imaginary part (Fig. A5.6).

The complex number z is represented on the Argand diagram by the point \mathbf{P}:(x, y) and the length OP $= r = \sqrt{x^2 + y^2}$ is called the **modulus** of the complex number z (or simply 'mod z' or $|z|$). Hence (from Fig. A5.6) the complex number z can be written in the form:

Fig. A5.5.

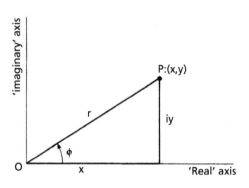

Fig. A5.6.

$$z = r \cos \phi + i r \sin \phi$$

where r (the modulus) and ϕ (the angle) are known as polar coordinates. This expression can further be written in exponential form (Euler's formula); $\cos \phi + i \sin \phi = \exp i\phi$ and $\cos \phi - i \sin \phi = \exp -i\phi$, i.e. $z = r \exp i\phi$.

This representation of complex numbers in exponential form by means of the polar coordinates r and ϕ allows us to add, subtract and carry out other mathematical operations upon them without having to resort to rulers and graph paper.

For example, the addition of complex numbers, $z_1 + z_2 + \cdots$ is simply given by:

$$z_1 + z_2 + \cdots = r_1 \exp i\phi_1 + r_2 \exp i\phi_2,$$

or:

$$z_1 + z_2 + \ldots = r_1 \cos \phi_1 + r_2 \cos \phi_2 + \cdots + i(r_1 \sin \phi_1 + r_2 \sin \phi_2 + \cdots)$$

(sum of 'horizontal' components) + i(sum of 'vertical' components).

This is the procedure used to find the structure factor F_{hkl}, the sum of all the atomic scattering factors, f, for the atoms in the unit cell.

The complex conjugate of a complex number, symbol z^* is one which has the same modulus r, but a negative angle ϕ, i.e.

$$z^* = r\cos\phi - ir\sin\phi = r\exp-i\phi.$$

The complex conjugate can be used, for example to determine the modulus r of a complex number, i.e.

$$z \cdot z^* = r\exp i\phi \cdot r\exp-i\phi = r^2.$$

Appendix 6

Systematic absences (extinctions) in X-ray diffraction and double diffraction in electron diffraction patterns

A6.1 Systematic absences

In X-ray diffraction (and also, except for the complication of double diffraction, in electron diffraction) the intensities of reflections from certain planes with Laue indices *hkl* are zero. Such reflections are said to be *forbidden* or *extinguished* or, better, *systematically absent* since they arise from the centring of the unit cell and /or the presence of translational symmetry elements—glide planes and screw axes. The identification of systematic absences is very useful since it provides the 'first step' in crystal structure determination.

The subject may be best introduced by considering the diffraction of light from a wide slit diffraction grating (Sections 7.4 and 13.3). In the diffraction of light extinctions or 'missing orders' occur for certain ratios of the slit spacing *a* and the slit width *d*. For example, referring to Fig. 7.7 or 13.7, if the slit width *d* is set equal to one-third of the slit spacing *a*, then the zero order minimum of the central peak from a single slit coincides with the angle of the third order diffracted peaks $n = \pm 3$. These peaks (and subsequently the sixth, ninth etc. order peaks) are therefore of zero intensity (i.e. extinguished). Other sets of systematic absences may be derived for other combinations of *a* and *d*—the most important being that when $d = \frac{1}{2}a$ (see Exercise 7.2).

Systematic absences in crystals may be derived in two ways: either by a consideration of the geometrical consequences arising from the use of centred, rather than primitive, unit cells and the presence of glide planes and screw axes, or by application of the structure factor equation. The former approach is the proper one, since systematic absences arise solely as a consequence of the architecture of crystals—but the latter approach is a useful means of gaining practice and familiarity with the structure factor equation. We will start with both approaches for an fcc crystal.

Consider Fig. A6.1 which shows the unit cell vectors **a, b, c** for the conventional fcc unit cell and the unit cell vectors **A, B, C** for (one variant) of the primitive rhombohedral unit cell. The origin is drawn at the front left-hand corner to help make clear the shape of the rhombohedral cell.

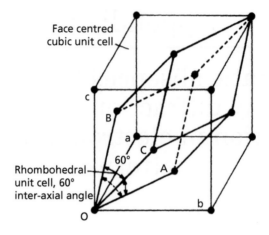

Fig. A6.1.

Writing down the transformation equations for unit cell vectors or axes (see Section 5.8):

$$\mathbf{A} = \tfrac{1}{2}\mathbf{a} + \tfrac{1}{2}\mathbf{b} + 0\mathbf{c}$$
$$\mathbf{B} = \tfrac{1}{2}\mathbf{a} + 0\mathbf{b} + \tfrac{1}{2}\mathbf{c}$$
$$\mathbf{C} = 0\mathbf{a} + \tfrac{1}{2}\mathbf{b} + \tfrac{1}{2}\mathbf{c}.$$

Now, as shown in Section 5.8, Miller indices transform in the same way as unit cell vectors. Hence if (HKL) and (hkl) are the Miller indices of a plane referred to the primitive and fcc unit cells, respectively:

$$H = \tfrac{1}{2}h + \tfrac{1}{2}k + 0l \qquad 2H = h + k$$
$$K = \tfrac{1}{2}h + 0k + \tfrac{1}{2}l \quad \text{or} \quad 2K = h + l$$
$$L = 0h + \tfrac{1}{2}k + \tfrac{1}{2}l \qquad 2L = k + l.$$

Now H, K, L are all integers, odd or even; hence $2H, 2K, 2L$ are all *even* integers, from which it follows that $(h + k)$, $(h + l)$, $(k + l)$ are also all even integers.

An even integer is either the sum of two even integers, or two odd integers, but not mixed. Hence, for each identity, h, k, l are either all odd or all even integers. In other words the lattice planes from which reflections occur are those for which h, k, l are all even or all odd integers. When h, k, l are mixed (some odd, some even) the Miller index (hkl) describes a set of planes which are not lattice planes since they do not all pass through lattice points.

Now let us derive the same result using the structure factor equation applied to an fcc crystal with an identical atom (atomic scattering factor f) at each of the four lattice points in the cell. The $(u_n v_n w_n)$ fractional coordinate values of these atoms are:

$$(000), \left(\tfrac{1}{2}\tfrac{1}{2}0\right), \left(\tfrac{1}{2}0\tfrac{1}{2}\right), \left(0\tfrac{1}{2}\tfrac{1}{2}\right).$$

Hence,

$$F_{hkl} = f \exp 2\pi i(h0 + k0 + l0) + f \exp 2\pi i \left(h\tfrac{1}{2} + k\tfrac{1}{2} + l0\right)$$
$$+ f \exp 2\pi i \left(h\tfrac{1}{2} + k0 + l\tfrac{1}{2}\right) + f \exp 2\pi i \left(h0 + k\tfrac{1}{2} + l\tfrac{1}{2}\right).$$

Simplifying, and remembering that since the cell is centrosymmetric the sine components in the exponents vanish:

$$F_{hkl} = f(1 + \cos \pi(h+k) + \cos \pi(h+l) + \cos \pi(k+l)).$$

Now, $\cos\pi$ (even integer) $= +1$ and $\cos\pi$ (odd integer) $= -1$. The value of F_{hkl} will depend on whether h, k, l are all even, all odd or mixed. It is easily seen that if they are all even or all odd $F_{hkl} = 4f$ and if they are mixed $F_{hkl} = 0$.

The conditions for systematic absences from body- and base-centred lattices can be derived in the same ways: either by referring the lattice to a primitive unit cell or by applying the structure factor equation. For example, for a body-centred cubic crystal with identical atoms at (000) and $\left(\tfrac{1}{2}\tfrac{1}{2}\tfrac{1}{2}\right)$

$$F_{hkl} = f \exp 2\pi i(h0 + k0 + l0) + f \exp 2\pi i \left(h\tfrac{1}{2} + k\tfrac{1}{2} + l\tfrac{1}{2}\right)$$
$$= f(1 + \cos \pi(h+k+l)).$$

Hence when $(h+k+l)$ = even integer $F_{hkl} = 2f$ and when $(h+k+l)$ = odd integer $F_{hkl} = 0$.

Systematic absences for glide planes and screw axes may be derived in the same way; an example of each will suffice to show the principles involved.

Figure A6.2 shows a (monoclinic) crystal with a c-glide plane in the xz plane or (010) plane (the plane of the paper). The action of the c-glide is to translate any atom, atomic scattering factor f, with fractional coordinates (uvw) to $\left(u\bar{v}\left(\tfrac{1}{2}+w\right)\right)$. Substituting

Fig. A6.2.

these two values in the structure factor equation:

$$F_{hkl} = f \exp 2\pi i (hu + kv + lw) + f \exp 2\pi i \left(hu + k\bar{v} + l \left(\tfrac{1}{2} + w \right) \right).$$

Consider the $h0l$ reflections ($k = 0$), i.e. the reflections in the zone whose axis is perpendicular to the glide plane (010)

$$F_{h0l} = f \exp 2\pi i (hu + lw) + f \exp 2\pi i (hu + lw) \exp 2\pi i \tfrac{1}{2} l$$
$$= f \exp 2\pi i (hu + lw)[1 + \exp \pi il].$$

When $l =$ odd integer, $\exp \pi il = -1$ and $F_{h0l} = 0$. This then gives the condition for systematic absences for the $h0l$ reflections; systematic absences occur when l equals an odd integer and, correspondingly, reflections occur when l equals an even integer. Clearly, for a-glide in the (010) plane reflections occur when h equals an even integer.

Figure A6.3 shows a (monoclinic) crystal with a screw diad axis, 2_1 along the y-axis, the action of which is to translate an atom with fractional coordinates (uvw) to $\left(\bar{u} \left(\tfrac{1}{2} + v \right) \bar{w} \right)$. Proceeding as before:

$$F_{hkl} = f \exp 2\pi i (hu + kv + lw) + f \exp 2\pi i \left(h\bar{u} + k \left(\tfrac{1}{2} + v \right) + l\bar{w} \right)$$

consider the $0k0$ reflections ($h = l = 0$) i.e. reflections arising from planes perpendicular to the screw axis.

$$F_{0k0} = f \exp 2\pi i (kv) + f \exp 2\pi i \left(k\tfrac{1}{2} + kv \right)$$
$$= f \exp 2\pi i (kv)[1 + \exp \pi ik].$$

When $k =$ odd integer, $\exp \pi ik = -1$ and $F_{0k0} = 0$ and reflections only occur from $0k0$ planes when k equals an even integer. This result can also be seen intuitively: the action of the screw diad axis repeats the atomic layers at spacings equal to half the **b** lattice repeat distance.

Operation of screw diad

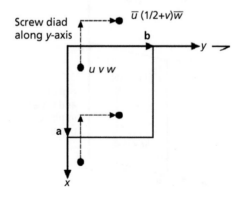

Fig. A6.3.

Table A6.1 lists the extinction criteria for lattices and translational symmetry elements. n and d are the symbols for diagonal glide planes where the translations are along two axes in the plane, each of which are half the unit cell repeat distance for an n glide plane and a quarter of the unit cell repeat distance for a d glide plane (d stands for 'diamond glide'). In summary, Table A6.1 shows that lattices affect all reflections, glide planes affect only reflections which lie in the zone whose axis is perpendicular to the glide plane and screw axes affect only the reflections from planes perpendicular to the screw axis.

Table A6.2 lists the reflecting planes in cubic P, I, F and diamond-cubic crystals in order of decreasing d_{hkl}-spacing. For cubic crystals $d_{hkl} = a/\sqrt{N}$ where $a =$ the lattice parameter and $N = (h^2 + k^2 + l^2)$. The conditions for reflection for a diamond-cubic crystal are h, k, l are all odd or all even integers (since a diamond-cubic crystal has an fcc lattice) with the additional condition that $(h + k + l)$ is either an odd integer or an integer which is an even multiple of 2 (see Exercise 9.5, p. 251).

A6.2 Double diffraction

In electron diffraction the presence of face- or body-centring lattice points also gives rise to systematically absent reflections just as in the case of X-ray diffraction (the first group in Table A6.1). However, reflections which are systematically absent in X-ray diffraction as a result of the presence of translational symmetry elements (glide planes and screw axes, Table A6.1) may, and usually do, occur in electron diffraction patterns. This is known as double diffraction and occurs because the intensities of the diffracted beams may be comparable with that of the direct or undiffracted beam—a consequence of the dynamical interactions between the direct and diffracted beams. The effect, in terms of the geometry of the electron diffraction pattern, is that a strong diffracted beam can behave, as it were, as the direct beam and the whole pattern is in effect shifted so as to be centred about the diffracted beam. In the case of electron diffraction patterns from centred lattices all the spots are coincident and no new ones arise; but in most other cases new ones arise in positions in which the diffraction spots should be systematically absent. A simple example—that of electron diffraction from the hcp structure—will make this clear.

In the hcp structure the relevant translational symmetry element is the screw hexad 6_3 (Fig. 4.10) which describes the symmetry of the sequence of A and B layer atoms (Fig. 1.5(b)). Consider the conditions for constructive interference for Bragg reflections from the A layers of atoms, i.e. the (0001) planes with interplanar spacing $d_{0001} = c$. When the path difference (PD) is 1λ (constructive interference), the path difference between the interleaving A and B layers of atoms, interplanar spacing $d_{0002} = c/2$ is $\frac{1}{2}\lambda$, which is the condition for *destructive* interference; hence the 0001 reflection is (systematically) absent. Second-order reflection from the (0001) planes (PD $= 2\lambda$) corresponds to the first-order reflection from the (0002) planes (PD $= 1\lambda$). Continuing in this way it turns out that $000l$ reflections where l is odd are systematically absent, as shown in Table A6.1. The same result may be found by applying the structure factor equation to the (0001), (0002), (0003), etc. planes, as in Example 4, Section 9.2: when l is odd the atomic scattering factors for the A and B layer atoms are equal and opposite.

Table A6.1 Extinction criteria for lattices and symmetry elements

Lattice or symmetry element type	Symbol	Class of reflections	Condition for presence
Lattice type:		hkl	
primitive	P		none
body-centred	I		$h+k+l=2n$
centred on the C face	C		$h+k \quad=2n$
centred on the A face	A		$\quad k+l=2n$
centred on the B face	B		$h \quad +l=2n$
centred on all faces	F		h,k,l all $=n$(odd) or all $=2n$(even)
rhombohedral, obverse	R		$-h+k+l=3n$
rhombohedral, reverse	R		$h-k+l=3n$
Glide plane ∥ (001)	a	$hk0$	$h \quad=2n$
	b		$\quad k \quad=2n$
	n		$h+k \quad=2n$
	d		$h+k \quad=4n$
Glide plane ∥ (100)	b	$0kl$	$k \quad=2n$
	c		$l=2n$
	n		$k+l=2n$
	d		$k+l=4n$
Glide plane ∥ (010)	a	$h0l$	$h \quad=2n$
	c		$l=2n$
	n		$h+ \quad l=2n$
	d		$h+ \quad l=4n$
Glide plane ∥ (1$\bar{1}$0)	c	hhl	$l=2n$
	b		$h \quad=2n$
	n		$h+k \quad=2n$
	d		$2h+ \quad l=4n$
Screw axis ∥ c	$2_1,4_2,6_3$	$00l$	$l=2n$
	$3_1,3_2,6_2,6_4$		$l=3n$
	$4_1,4_3$		$l=4n$
	$6_1,6_5$		$l=6n$
Screw axis ∥ a	$2_1,4_2$	$h00$	$h \quad=2n$
	$4_1,4_3$		$h \quad=4n$
Screw axis ∥ b	$2_1,4_2$	$0k0$	$k \quad=2n$
	$4_1,4_3$		$k \quad=4n$
Screw axis ∥ [110]	2_1	$hh0$	$h \quad=2n$

(n = odd integer, $2n$ = even integer etc.)

Table A6.2 Reflecting planes ($\sqrt{}$) in cubic *P*, *I*, *F* and diamond-cubic crystals.

N	hkl	cubic P	cubic I	cubic F	Diamond cubic
1	100	$\sqrt{}$			
2	110	$\sqrt{}$	$\sqrt{}$		
3	111	$\sqrt{}$		$\sqrt{}$	$\sqrt{}$
4	200	$\sqrt{}$	$\sqrt{}$	$\sqrt{}$	
5	210	$\sqrt{}$			
6	211	$\sqrt{}$	$\sqrt{}$		
7	—				
8	220	$\sqrt{}$	$\sqrt{}$	$\sqrt{}$	$\sqrt{}$
9	300/221	$\sqrt{}$			
10	310	$\sqrt{}$	$\sqrt{}$		
11	311	$\sqrt{}$		$\sqrt{}$	$\sqrt{}$
12	222	$\sqrt{}$	$\sqrt{}$	$\sqrt{}$	
13	320	$\sqrt{}$			
14	321	$\sqrt{}$	$\sqrt{}$		
15	—				
16	400	$\sqrt{}$	$\sqrt{}$	$\sqrt{}$	$\sqrt{}$
17	410/322	$\sqrt{}$			
18	330/411	$\sqrt{}$	$\sqrt{}$		
19	331	$\sqrt{}$		$\sqrt{}$	$\sqrt{}$
20	420	$\sqrt{}$	$\sqrt{}$	$\sqrt{}$	

Note 1. There are no squares of three numbers which add up to 7, 15 etc. This means that there are 'gaps' in the sequence of reflections for cubic *P* crystals, i.e. 'line 7', 'line 15' etc. are missing.
Note 2. Cubic *I* crystals show a uniform sequence of reflections (with no 'gaps' as in cubic *P* crystals).
Note 3. Cubic *F* crystals show a characteristic sequence of lines: 'two-together', 'one-on-its-own', 'two-together' etc.

Fig. A6.4.

Now let us draw a commonly observed hcp electron diffraction pattern from the
[$1\bar{2}10$] zone (Fig. A6.4), showing the systematically absent reflections 0001, 000$\bar{1}$,
0003, 000$\bar{3}$ etc. in the row through the origin.[1] Now suppose that the (strong) 10$\bar{1}$0
reflected beam acts as it were as the direct beam: the whole pattern will, in effect,
be shifted from the origin 0000 to 10$\bar{1}$0 as shown by the arrow and spots will now
appear in the systematically absent positions. The same result will be obtained for
other choices of strong diffracted beams—as can easily be seen by making a tracing
paper overlay in Fig. A6.4 and shifting the origin to the positions of different diffracted
spots.

Double diffraction can also occur in situations in which the electron diffraction spots
of two (or more) crystals are present in the pattern. In such cases the double-diffracted
spots which appear may correspond to no *d*-spacing in either crystal—and are probably
the biggest cause of the frustrated electron microscopists' 'unindexable' patterns. Fig-
ure A6.5 indicates the geometry involved. Here, two (high order) patterns are shown

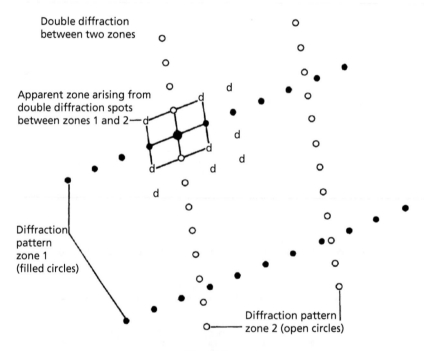

Fig. A6.5.

[1] Systematic absences also occur in every *third* row (not shown in Fig. A6.4). These arise because the A-
layer and B-layer atoms are in special equivalent positions in the hexagonal unit cell (see Section 4.6). This
leads to the conditions for *hkil*: if $h - k = 3n$, $l = 2n$, i.e. in the third row the 30$\bar{3}$1,30$\bar{3}$3, etc. reflections
are systematically absent. These conditions are given in the table in Fig. 4.14, space group P6$_3$/*mmc*, for the
special equivalent positions for hcp structures denoted by the Wyckoff letters *c* and *d*.

consisting of widely separated rows (indicated by filled and open circles). The strong spots in the rows through the origin may act in effect as the direct beam, generating additional double diffraction spots in the positions marked by the letter d. The pattern *appears* to be from a relatively simple low-order zone—but of course it is not, and is either unindexable or (worse) may be indexed incorrectly. An example of the latter case is that of the electron diffraction patterns from twin-related plates or laths of ferrite (bcc) which occur in the microstructures of some structural steels.

Figure A6.6 shows the spots (indicated by filled and open circles) for two bcc [110] zones twin-related in the (112) plane. Double diffraction between these twin-related zones gives rise to the spots marked **d** (i.e. $\mathbf{d}^*_{110} + \mathbf{d}^*_{110_r} = \mathbf{d}_d$ where $(110)_T$ is the twin-spot of (110)). The presence of these additional spots indicates a pattern in which the spots outline an 'elongated' hexagon, as indicated, which is precisely the same shape as for an fcc $\langle 110 \rangle$ zone—the four bcc {110} spots closer to the centre spot being identified as fcc {111} spots and the two double-diffraction spots further away being identified as fcc {200} spots. Since in steels d_{110} (ferrite) is closely similar to d_{111} (austenite), then this pattern may be misinterpreted as a $\langle 110 \rangle$ austenite zone—leading to the erroneous conclusion that this phase is present at the ferrite twin boundary.

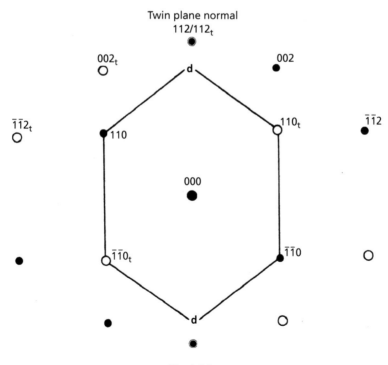

Fig. A6.6.

A note on double diffraction in X-ray diffraction
Double diffraction can also occur in X-ray diffraction when two (or more) sets of crystal planes are simultaneously at the Bragg diffracting condition (i.e. the reciprocal lattice points lie simultaneously on the Ewald reflecting sphere). The effect was first described by M. Renninger in 1937 and can lead to violations of the systematic extinction criteria listed in Table A6.1.

Appendix 7

Group theory in crystallography

From our reading of Chapter 4 in this book, we understand the term **point group** of a crystal simply to denote the 'collection and arrangement' of symmetry elements that together describe its symmetry and distinguish it from all the other 31 point groups. Similarly, we understand the term **space group** of a crystal to denote the distribution and arrangements of symmetry elements in space that describe its three-dimensional pattern and distinguish it from all the other 229 space groups.

But the word **group** has a precise and much deeper meaning. We now need to make clear the distinction between the terms **symmetry element** and **symmetry operation**. A symmetry operation is a movement of body (a molecule or crystal) in which, after the operation has been carried out, is in an **equivalent configuration**: that is to say that it is indistinguishable from the original configuration, though not necessarily identical to it. For example, rotation of a pencil by 60° about the pencil-lead direction or axis results in an equivalent configuration—and of course, repeating the process a total of six times returns us to the original configuration. The symmetry element (the hexad axis) is the geometrical line about which these symmetry operations are carried out. The same applies to all the other symmetry elements that we have learned about: their existence is demonstrated (as with the pencil) by showing that the appropriate symmetry operations exist.

Group theory in crystallography is concerned with the number and distribution of symmetry operations in a crystal and the relationships between them. By its means, the 32 point groups and the 230 space groups can be formally identified and enumerated. In the words of Vainshtein (1981) 'Group theory ... enables one to reveal the deeper meaning and regularities of the symmetry concept itself'. It also has important applications in the analysis of the symmetry properties of atoms and molecules, as well as crystals. However, largely because of the mathematical formalism and abstract terms in which group theory is inevitably expressed, the subject is generally by-passed in most books on crystallography—either because of the 'difficulty' of the subject or because it is only needed in such particular applications as those mentioned earlier. Another practical hindrance for the crystallographer is that the Schönflies' notation for point groups and space groups is generally used rather than the Hermann–Mauguin notation (see *Note* and Table A7.4).

Here, we will describe group theory as it applies to crystallographic point group symmetry in the very simplest terms, with only a brief mention of space group symmetry. Readers with a sufficient mathematical background and who wish to pursue the subject further will find the books listed in Further Reading useful.

A7.1 Group theory: axioms and multiplication tables

We begin by reconsidering the symmetry operations between the pair of right and left hands shown in Fig. 4.5. We see that a reflection (symbol m) in Fig. 4.5(b) plus a rotation of 180° (a diad axis, symbol 2, perpendicular to the mirror plane) is equivalent to a centre of symmetry or inversion monad axis, symbol $\bar{1}$ as shown in Fig. 4.5(a). Writing this algebraically we may put $m.2 = \bar{1}$ where the dot means the product or combination of one symmetry operation (m) followed by another (2) giving rise to a centre of symmetry ($\bar{1}$). Conversely, we may write $\bar{1}.2 = m$. Furthermore, the order in which the symmetry operations are carried out gives the same result, e.g. $m.2 = 2.m = \bar{1}$, etc. (This is only the case for the so-called Abelian groups, named after the Norwegian mathematician, Niels Henrik Abel 1802–1829.) Finally, if we find the product or combination of a mirror symmetry operation with itself, we 'get back where we started'; e.g. the operation of a reflection m (right to left hand) followed by another reflection m in the same plane returns us back to a right hand. This combination of symmetry operations is equivalent to the operation of a **monad** axis of symmetry, symbol 1 (see Section 2.3, p. 59), i.e. a rotation of 360° (about any axis) that brings the crystal into coincidence with itself. Writing this algebraically: $m.m = 1$ and similarly $2.2 = 1$ and $\bar{1}.\bar{1} = 1$.

In group theory the symmetry operations m, 2, $\bar{1}$, and 1 are called a **set of elements** of the group and they fulfil the first requirement or **first axiom**, of group theory, namely that **the combination or product of any two elements (including an element with itself) is also an element of the set**. As we see, these four symmetry operations do indeed accord with this axiom. They are also in accord with the **second axiom** of group theory, namely **the combination of elements is associative**. This means that if we combine three elements, the sequence of combination gives the same result, e.g. for m, 2 and $\bar{1}$, $(m.2).\bar{1} = \bar{1}.\bar{1} = 1$, and $m.(2.\bar{1}) = m.m = 1$.

The **third axiom** introduces a new idea. It may be stated as follows: **the set of elements contains a single element known as the identity, which has the property that, when it is combined with any other element, the result is that same element**. Here, the identity is the monad axis, symbol 1[1]. Clearly, in the previous example, $1.m = m$ etc. and thus the third axiom is fulfilled. Finally, the **fourth axiom** may be stated: **for each element in the set there exists another unique element called its inverse, such that the combination of the element and its inverse is the identity**. In this example, each of the symmetry elements in the set is its own inverse: we may consider the combination $m.m = 1$ to mean, as indicated earlier 'reflection one way followed by reflection back again', or $2.2 = 1$ to mean 'a rotation of 180° clockwise followed by a rotation of 180° anti-clockwise' (which is equivalent of course to two clockwise or anti-clockwise rotations).

The combinations, or products, of a set of elements in a group can be set out in the form of a matrix array, known as a **multiplication table** or 'Cayley's Square' (after the English mathematician Arthur Cayley 1821–1895) as shown in Fig. A7.1(a).

[1] The identity is also given the symbol E (German: Einheit)

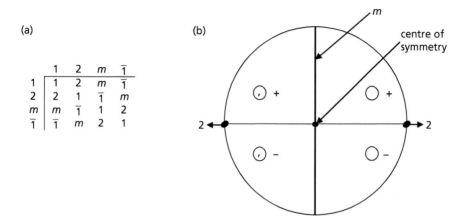

Fig. A7.1. (a) The multiplication table for monoclinic point group 2/m and (b) the corresponding stereographic projection. Note that, as with space group symbols, the symbols ◯ and ⊙ are related by a reflection or a centre of symmetry.

Notice that all the combinations of symmetry elements in the table contain the original four symmetry elements, i.e. the first axiom is fulfilled. This multiplication table (said to be of order 4 since there are four elements in the set), represents monoclinic point group symmetry 2/m that is also represented by a stereogram (see Section 12.5.1) in Fig. A7.1(b).

We can now write down the multiplication tables of monoclinic point group 2 (elements 1 and 2) and monoclinic point group m (elements 1 and m), both are of order 2.

	1	2
1	1	2
2	2	1

(a)

	1	m
1	1	m
m	m	1

(b)

Fig. A7.2. Multiplication tables for (a) point group 2 and (b) point group m.

These two groups are called **subgroups** of group 2/m; conversely group 2/m is the **supergroup** of these subgroups. A considerable part of the application of group theory to crystallography consists of first devising simple groups and then finding supergroups based upon them by adding the appropriate symmetry operations. In this way, complicated groups are constructed from simple ones. We can proceed in this way to write down the multiplication tables of all the 32 point groups—which are of higher order of course—the greater the number of the symmetry operations present. However, we shall only do this for four further examples, orthorhombic point groups 222 and $mm2$, trigonal point group 32, and tetragonal point group 4.

For point group 222 the symmetry operations are 1, 2_x, 2_y, 2_z; the subscripts referring to the crystallographic axis along which the corresponding diad axis lies.

(a)

(b)

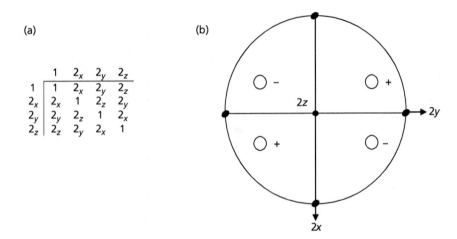

Fig. A7.3. **(a)** The multiplication table for orthorhombic point group 222 and **(b)** the corresponding stereographic projection.

Notice that a rotation of 180° about x followed by a rotation of 180° about y equals a rotation of 180° about z, i.e. $2_x.2_y = 2_z$, and so on.

For point group $mm2$ the symmetry operations are 1, m_x, m_y, and 2_z.

(a)

(b)

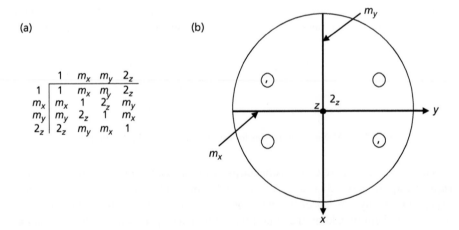

Fig. A7.4. **(a)** The multiplication table for orthorhombic point group $mm2$ and **(b)** the corresponding stereographic projection.

Notice that a reflection in a plane perpendicular to x followed by a reflection in a plane perpendicular to y equals a rotation of 180° about z, i.e. $m_x.m_y = 2_z$. Indeed, we encountered this equivalence in our derivation of the ten plane point groups (Section 2.3)

where we found (Fig. 2.3(3)) that the operation of two perpendicular mirror lines 'automatically' generates a diad axis of symmetry along their intersection.

Notice also the similarity between the multiplication tables for point groups 222 and $mm2$, i.e. $1 \equiv 1$, $m_x \equiv 2_x$, $m_y \equiv 2_y$, $2_z \equiv 2_z$. These two groups are said to be **isomorphic** (meaning 'equal form') since there is a unique one-to-one correspondence between the elements of the two groups. The groups in Fig. A7.2 are also isomorphic.

For point group 32 (quartz) the symmetry operations are $1, 3, 3^2, 2_x, 2_y, 2_u$. These are indicated on the stereographic projection in Fig. A7.5(b). Here, we have to distinguish a symmetry operation and its inverse: 3 represents an anti-clockwise rotation of $120°$ and 3^2 a clockwise rotation of $120°$ (equivalent to an anti-clockwise rotation of $240°$), i.e. $3.3^2 = 1$ (fourth axiom) and $3.3 = 3^2$. (The symbol 3^2 should not be confused with the symbol 3_2 that denotes a screw triad axis.) Also, unlike our previous examples, the order of combination is important (non-Abelian). For example, $3.2_x = 2_y$ and $2_x.3 = 2_u$; i.e. depending on the order of combination, the operation of an anti-clockwise rotation of $120°$ about the z-axis combined with a rotation of $180°$ about the x-axis gives a rotation of $180°$ either about the y-axis or the u-axis. The multiplication table is given in Fig. A7.5(a), the order of combination being 'column times row'.

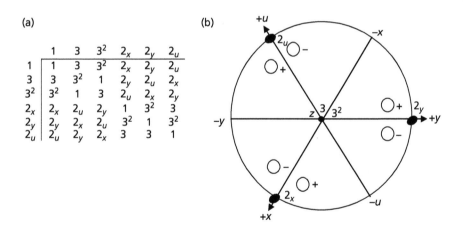

Fig. A7.5. (a) The multiplication table for trigonal point group 32 and (b) the corresponding stereographic projection.

For point group 4, the symmetry operations (about the z-axis) are $1, 4$ ($90°$ anti-clockwise), 4^2 ($180°$ anti-or-clockwise $= 2$), and 4^3 ($270°$ anti-clockwise or $90°$ clockwise). These operations are shown in the multiplication table and stereographic projection in Fig. A7.6(a) and (b).

Notice that for each symmetry operation there is an inverse, i.e. $4^{-1} = 4^3$, $(4^2)^{-1} = 4^2$ and $(4^3)^{-1} = 4$. We may check that the **second axiom** (the associative law) holds, e.g. $4.(4^2.4^3) = 4.4 = 4^2$ and $(4.4^2).4^3 = 4^3.4^3 = 4^2$

(a)

	1	4	4^2	4^3
1	1	4	4^2	4^3
4^2	4	4^2	4^3	1
4	4^2	4^3	1	4
4^3	4^3	1	4	4^2

(b)

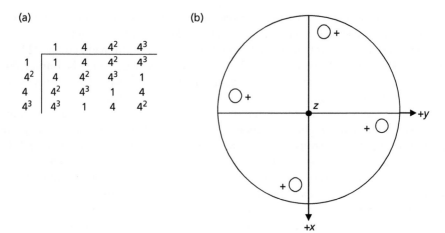

Fig. A7.6. (a) The multiplication table for tetragonal point group 4 and (b) the corresponding stereographic projection.

A note on nomenclature for symmetry operations

The operations of rotation about a symmetry axis, both rotation axes and screw axes, are sometimes called **proper operations**, or operations of the first sort because they do not result in a change of hand of the asymmetrical unit. Conversely, the operations of inversion, reflection, and glide-reflection, are called **improper operations**, or operations of the **second sort** since they do result in a change of hand. This distinction is important in the classification of point groups (see Table 3.1) and space groups.

A7.2 Matrix representation of symmetry operations: character tables

Matrices were used (see Section 5.8) to describe the transformations of plane indices and zone axis symbols from one axis system to another. Here, we use matrices to describe symmetry operations and their combinations: in effect we substitute the algebraic process of matrix multiplication for the geometrical process of successively applying symmetry operations. This in turn leads to a new (matrix) way of representing point groups. Note that in what follows we refer the matrices to axes x, y, and z.

The identity (monad symmetry operation), symbol 1 or E

Suppose we have a point with coordinates x, y, z. The identity transforms the point 'into itself', i.e. $x \rightarrow x$, $y \rightarrow y$, $z \rightarrow z$. In matrix notation this is:

$$\begin{bmatrix} 1 & 0 & 0 \\ 0 & 1 & 0 \\ 0 & 0 & 1 \end{bmatrix} \begin{bmatrix} x \\ y \\ z \end{bmatrix} = \begin{bmatrix} x \\ y \\ z \end{bmatrix}$$

Reflection m_x (i.e. in the y–z plane, perpendicular to the x-axis)
Here, a point (x, y, z) becomes (\bar{x}, y, z). In matrix notation this is:

$$\begin{bmatrix} \bar{1} & 0 & 0 \\ 0 & 1 & 0 \\ 0 & 0 & 1 \end{bmatrix} \begin{bmatrix} x \\ y \\ z \end{bmatrix} = \begin{bmatrix} \bar{x} \\ y \\ z \end{bmatrix} \qquad \text{and similarly, for reflections } m_y \text{ and } m_z.$$

Inversion, symbol $\bar{1}$
Here, a point (x, y, z) becomes $(\bar{x}, \bar{y}, \bar{z})$. In matrix notation this is:

$$\begin{bmatrix} \bar{1} & 0 & 0 \\ 0 & \bar{1} & 0 \\ 0 & 0 & \bar{1} \end{bmatrix} \begin{bmatrix} x \\ y \\ z \end{bmatrix} = \begin{bmatrix} \bar{x} \\ \bar{y} \\ \bar{z} \end{bmatrix}$$

Rotation operations
First, consider a general rotation of $\phi°$ anti-clockwise about the z-axis (Fig. A7.7). A point (x_1, y_1, z_1) becomes a point (x_2, y_2, z_2) (or vector $\mathbf{r_1}$ becomes vector $\mathbf{r_2}$). These are related by the equations:

$$x_2 = x_1 \cos \phi - y_1 \sin \phi + 0z_1$$
$$y_2 = x_1 \sin \phi + y_1 \cos \phi + 0z_1$$
$$z_2 = 0 + 0 + 1z_1$$

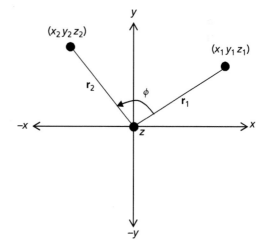

Fig. A7.7. Coordinate transformation of a point following an anticlockwise rotation, angle [Greek] phi.

(I leave it to the reader to work out these equations. *Hint*: consider the x_1 and y_1 components of vector $\mathbf{r_1}$ separately, rotate both anti-clockwise and find x_2 and y_2 from the sums.)

In matrix notation this is:

$$\begin{bmatrix} \cos\phi & -\sin\phi & 0 \\ \sin\phi & \cos\phi & 0 \\ 0 & 0 & 1 \end{bmatrix} \begin{bmatrix} x_1 \\ y_1 \\ z_1 \end{bmatrix} = \begin{bmatrix} x_2 \\ y_2 \\ z_2 \end{bmatrix} \qquad \text{(anti-clockwise rotation)}$$

and:

$$\begin{bmatrix} \cos\phi & \sin\phi & 0 \\ -\sin\phi & \cos\phi & 0 \\ 0 & 0 & 1 \end{bmatrix} \begin{bmatrix} x_1 \\ y_1 \\ z_1 \end{bmatrix} = \begin{bmatrix} x_2 \\ y_2 \\ z_2 \end{bmatrix} \qquad \text{(clockwise rotation)}$$

Notice (a) that these matrices are the **transposes** of each other (i.e. rows and columns interchanged) and (b) that the **inverse** of each matrix (see Section 5.10) is simply equal to the transpose. They are said to be **conjugate**.

For $\phi = 180°$ the anti-clockwise and clockwise matrices are identical.

$$\begin{bmatrix} \bar{1} & 0 & 0 \\ 0 & \bar{1} & 0 \\ 0 & 0 & 1 \end{bmatrix} \begin{bmatrix} x_1 \\ y_1 \\ z_1 \end{bmatrix} = \begin{bmatrix} x_2 \\ y_2 \\ z_2 \end{bmatrix} \qquad \begin{bmatrix} \bar{1} & 0 & 0 \\ 0 & \bar{1} & 0 \\ 0 & 0 & 1 \end{bmatrix} \begin{bmatrix} x_1 \\ y_1 \\ z_1 \end{bmatrix} = \begin{bmatrix} x_2 \\ y_2 \\ z_2 \end{bmatrix}$$

For $\phi = 90°$ the matrices are:

$$\begin{bmatrix} 0 & \bar{1} & 0 \\ 1 & 0 & 0 \\ 0 & 0 & 1 \end{bmatrix} \begin{bmatrix} x_1 \\ y_1 \\ z_1 \end{bmatrix} = \begin{bmatrix} x_2 \\ y_2 \\ z_2 \end{bmatrix} \qquad \phi = 90° \text{ anti-clockwise about } z\text{-axis}$$

$$\begin{bmatrix} 0 & 1 & 0 \\ \bar{1} & 0 & 0 \\ 0 & 0 & 1 \end{bmatrix} \begin{bmatrix} x_1 \\ y_1 \\ z_1 \end{bmatrix} = \begin{bmatrix} x_2 \\ y_2 \\ z_2 \end{bmatrix} \qquad \phi = 90° \text{ clockwise about } z\text{-axis}$$

Combinations of symmetry operations; here we give two examples:
For the rotation and reflection operations 2_z and m_z (see Fig. A7.1)

For the reflection operations m_x and m_y (see Fig. A7.4)

$$\begin{bmatrix} \bar{1} & 0 & 0 \\ 0 & \bar{1} & 0 \\ 0 & 0 & 1 \end{bmatrix} \cdot \begin{bmatrix} 1 & 0 & 0 \\ 0 & 1 & 0 \\ 0 & 0 & \bar{1} \end{bmatrix} = \begin{bmatrix} \bar{1} & 0 & 0 \\ 0 & \bar{1} & 0 \\ 0 & 0 & \bar{1} \end{bmatrix} \qquad \begin{bmatrix} \bar{1} & 0 & 0 \\ 0 & 1 & 0 \\ 0 & 0 & 1 \end{bmatrix} \cdot \begin{bmatrix} 1 & 0 & 0 \\ 0 & \bar{1} & 0 \\ 0 & 0 & 1 \end{bmatrix} = \begin{bmatrix} \bar{1} & 0 & 0 \\ 0 & \bar{1} & 0 \\ 0 & 0 & 1 \end{bmatrix}$$

$$\quad 2_z \qquad \cdot \qquad m_z \qquad = \qquad \bar{1} \qquad\qquad m_x \qquad \cdot \qquad m_y \qquad = \qquad 2_z$$

Matrix representation of point groups
Consider for example point group $2/m$ (Fig. A7.1). We now represent the symmetry operations 1, 2_z, m_z, and $\bar{1}$ by the matrices:

$$\begin{bmatrix} 1 & 0 & 0 \\ 0 & 1 & 0 \\ 0 & 0 & 1 \end{bmatrix} \quad \begin{bmatrix} \bar{1} & 0 & 0 \\ 0 & \bar{1} & 0 \\ 0 & 0 & 1 \end{bmatrix} \quad \begin{bmatrix} 1 & 0 & 0 \\ 0 & 1 & 0 \\ 0 & 0 & \bar{1} \end{bmatrix} \quad \begin{bmatrix} \bar{1} & 0 & 0 \\ 0 & \bar{1} & 0 \\ 0 & 0 & \bar{1} \end{bmatrix}$$
$$\quad 1 \qquad\qquad 2_z \qquad\qquad m_z \qquad\qquad \bar{1}$$

and for point group $mm2$ (Fig. A7.4), we now represent the symmetry operations 1, m_x, m_y, and 2_z by the matrices:

$$\begin{bmatrix} 1 & 0 & 0 \\ 0 & 1 & 0 \\ 0 & 0 & 1 \end{bmatrix} \quad \begin{bmatrix} \bar{1} & 0 & 0 \\ 0 & 1 & 0 \\ 0 & 0 & 1 \end{bmatrix} \quad \begin{bmatrix} 1 & 0 & 0 \\ 0 & \bar{1} & 0 \\ 0 & 0 & 1 \end{bmatrix} \quad \begin{bmatrix} \bar{1} & 0 & 0 \\ 0 & \bar{1} & 0 \\ 0 & 0 & 1 \end{bmatrix}$$
$$\quad 1 \qquad\qquad m_x \qquad\qquad m_y \qquad\qquad 2_z$$

Each matrix corresponds to a single operation in the group, which may be summarized by considering their operations on a point $x\,y\,z$. For example, for point group $mm2$ above:

1 (or E)	m_x	m_y	2_z
No change in x, y or z	$x \to -x$ No change in y or z	$y \to -y$ No change in x or z	$x \to -x$ $y \to -y$ No change in z

The results may be set out in the form of a table.

$mm2$	1 (or E)	m_x	m_y	2_z
x	1	-1	1	-1
y	1	1	-1	-1
z	1	1	1	1
xy	1	-1	-1	1

Also included in the table is the effect of the symmetry operations on the product function xy (simply multiply x and y in each of the columns here). As can be seen, xy transforms differently to x, y, and z. If we do the same for the product functions xz, yz, x^2, y^2, and z^2, we find that we simply repeat the combinations already in the table. For example, yz transforms like y and xz transforms like x. Hence, in summary,

Table A7.1 Character table for point group $mm2$[2].

$mm2$	1(E)	m_x	m_y	2_z		
A_1	1	1	1	1	z	x^2, y^2, z^2
A_2	1	-1	-1	1	R_z	xy
B_1	1	-1	1	-1	x, R_y	xz
B_2	1	1	-1	-1	y, R_x	yz

[2] This is the character table for the water molecule. Its use in describing the atomic vibrations of the water molecule may be found (e.g.) in Hargittai & Hargittai (1995) and Ladd (2014).

there are four ways in which the functions transform in point group $mm2$. They are tabulated in the form of a **character table** (not to be confused with a multiplication table—Section A7.1).

The four rows A_1, A_2, B_1, and B_2 are called **representations** of the point group $mm2$ and R_z, R_y, and R_x are **axial vectors**. However, at this point our introduction to character tables must stop, leaving a reader to pursue these topics in the books listed in Further Reading.

A7.3 Symmorphic and enantiomorphous space groups

The 230 space groups are derived by combining the 32 point group symmetries with the appropriate Bravais lattices and then taking into account the additional space groups arising from the existence of the translational symmetry elements—glide planes and screw axes. It is, in essence, the same procedure by which we derived the 17 plane groups (Sections 2.4 and 2.5) but is obviously more complicated. In group theory, the procedure is to combine the multiplication table of each point group with the appropriate lattice translations. Consider the general symbol for a direction (Section 5.7):

$$\mathbf{r}_{uvw} = u\mathbf{a} + v\mathbf{b} + w\mathbf{c}$$

For an 'infinite' crystal the integers u, v, w take any value from $-\infty$ to $+\infty$ and specify the coordinates of the origins of every unit cell. Each set of such integers u, v, w constitute a group, i.e. they are in accord with the four axioms stated here except (1) the rule of combination is **addition** and (2) the identity is zero. From this, multiplication tables may be constructed but the mathematical procedures involved (which begin so simply!) are too sophisticated to be covered in detail in this book. Instead we derive the 73 **symmorphic** space groups by combining the appropriate point groups and Bravais lattices. For example, the five cubic point groups (23, 432, $\bar{4}3m$, $m\bar{3}$, and $m\bar{3}m$) combined with the three (P, I, and F) Bravais lattices give rise to 15 symmorphic space groups. In other systems there is more than one way in which the point group and Bravais lattice can be combined. For example, the trigonal point group $3m$ combined with the hexagonal P lattice gives rise to two space groups, $P3m1$ and $P31m$ (recall the two possible plane groups, $p31m$ and $p3m1$ in Table 2.6). In the orthorhombic system point group $mm2$ may be combined with the C-centred lattice with the diad axis *either* along the C-axis (space group $Cmm2$) **or** along the a- (equivalently b-) axis (space group $Amm2$, equivalently $Bmm2$). The 73 symmorphic space groups are listed in Table A7.2. As can be seen, they are almost equally divided between the three groups of point groups (see Table 3.1).

The 24 symmorphic space groups based on the 11 centrosymmetric point groups correspond to the 24 Patterson space groups (Section 13.3.2) and include the eleven Laue groups (Table 9.1) with the appropriate/permissible cell centrings.

The space groups may also be divided into those based on:

1. The eleven non-centrosymmetric enantiomorphous point groups. These space groups, total 65, which were first worked out by Sohncke in 1867 (Section 4.6),

Table A7.2 The 73 symmorphic space groups (based on Burns and Glazer, 2013).

		Space groups—based on point groups:		
System	Bravais lattice	Non-centrosymmetric enantiomorphous (total 24)	Non-centrosymmetric non-enantiomorphous (total 25)	Centrosymmetric non-enantiomorphous (total 24)
Cubic	P	P23, P432	$P\bar{4}3m$	$Pm\bar{3}, Pm\bar{3}m$
	I	I23, I432	$I\bar{4}3m$	$Im\bar{3}, Im\bar{3}m$
	F	F23, F432	$F\bar{4}3m$	$Fm\bar{3}, Fm\bar{3}m$
Tetragonal	P	P4, P422	$P\bar{4}, P4mm, P\bar{4}2m, P\bar{4}m2$	P4/m, P4/mmm
	I	I4, I422	$I\bar{4}, I4mm, I\bar{4}2m, I\bar{4}m2$	I4/m, I4/mmm
Orthorhombic	P	P222	Pmm2	Pmmm
	I	I222	Imm2	Immm
	C(A)	C222	Cmm2, Amm2	Cmmm
	F	F222	Fmm2	Fmmm
Trigonal	R	R3, R32	R3m	$R\bar{3}, R\bar{3}m$
	P	P3, P312, P321	P3m1, P31m	$P\bar{3}, P\bar{3}1m, P\bar{3}m1$
Hexagonal	P	P6, P622	$P\bar{6}, P6mm, P\bar{6}m2, P\bar{6}2m$	P6/m, P6/mmm
Monoclinic	P	P2	Pm	P2/m
	C	C2	Cm	C2/m
Triclinic	P	P1		$P\bar{1}$

are sometimes called the axial space groups and are important in that proteins and nucleic acids (which do not possess mirror planes or inversion axes of symmetry) crystallize in these space groups. They include the eleven enantiomorphous pairs (Table A7.3).

2. The ten non-centrosymmetric non-enantiomorphous point groups, total 73 space groups; the same number (coincidence?) as the symmorphic space groups.

3. The eleven centrosymmetric (non-enantiomorphous) point groups, total 92 space groups.

Table A7.3 The eleven pairs of enantiomorphous space groups.

$P3_1$ and $P3_2$	$P4_1$ and $P4_3$	$P6_1$ and $P6_5$	$P6_2$ and $P6_4$
$P3_112$ and $P3_212$	$P3_121$ and $P3_221$	$P4_122$ and $P4_322$	$P4_12_12$ and $P4_32_12$
$P6_122$ and $P6_522$	$P6_222$ and $P6_422$	$P4_132$ and $P4_332$	

A note on Schönflies' notation for symmetry elements, point groups, and space groups
Schönflies treated the crystallographic symmetries as mathematical 'groups' of operations. He called a group containing one rotation operation and its powers a **cyclic group**, symbol C. For example, the operations of a four-fold axis, symbol C_4 are the rotations $1, \pi/2, \pi, 3\pi/2$ which are all 'powers' of the fundamental rotation $\pi/2$. A mirror (reflection) operation, symbol σ, **normal** to the rotation axis he symbolized with a superscript h. A mirror operation **parallel** to the rotation axis he symbolized with the superscript v.

As noted in Section 4.3 (see p. 104), Schönflies used rotation-reflection or alternating axes (rather than rotation-inversion (or simply inversion/axes)) symbolized by S (from Sphenoidisch). Next, we have the groups with normal diad axe(s) symbolized by D (from Diëdergruppe). Those 'D' groups that also include a plane of symmetry normal to the rotation axis are symbolized, as before, by the superscript h and those with a diagonal plane of symmetry are symbolized by the superscript d. In the cubic system the symbols T and O are used to represent tetrahedral and octahedral symmetry, respectively.

The Schönflies symbol may also include the letter i denoting a centre of symmetry, so C_i is a synonym for S_2 $(=\bar{1})$ and C_3^i is a synonym for S_6 $(=\bar{3})$. We may also note that the $\bar{5}$ inversion axis that describes icosahedral (non-crystallographic) symmetry (Section 4.9) corresponds to Schönflies' S_{10} rotation-reflection axis.

The corresponding Schönflies and Hermann–Mauguin point group symbols are set out in Table A7.4 (arranged in the same form as Table 3.1).

Table A7.4 Hermann–Mauguin and Schönflies point group symbols.

	Non-centrosymmetric point groups		Centrosymmetric point groups (non-enantiomorphous)
System	Enantiomorphous	Non-enantiomorphous	
Cubic	$T(23), O(432)$	$T_d(\bar{4}3m)$	$T_h(m3), O_h(m\bar{3}m)$
Tetragonal	$C_4(4), D_4(422)$	$S_4(\bar{4}), C_4^v(4mm), D_2^d(\bar{4}2m)$	$C_4^h\,(4/m), D_4^h\,(4/mmm)$
Orthorhombic	$D_2(222)$	$C_2^v(mm2)$	$D_2^h(mmm)$
Trigonal	$C_3(3), D_3(32)$	$C_3^v\,(3m)$	$S_6(\bar{3}), D_3^d(\bar{3}m)$
Hexagonal	$C_6(6), D_6(622)$	$C_3^h(\bar{6}), C_6^v(6mm), D_3^h(\bar{6}m2)$	$C_6^h(6/m), D_6^h(6/mmm)$
Monoclinic	$C_2(2)$	$C_s, C_1^h(m)$	$C_2^h(2/m)$
Triclinic	$C_1(1)$		$C_i(\bar{1})$

Clearly, the Hermann–Mauguin notation is to be far preferred since it contains the essential point group symmetry information. This is even more the case for Schönflies' space group notation in which the space groups, with the same point group, are simply numbered sequentially (presumably in the order in which he derived them). For example, the five space groups with point group 23 (Nos. 195–199 in the International Tables) are denoted T^1(P23), T^2(F23), T^3(I23), T^4(P2$_1$3), and T^5(I2$_1$3).

Answers to exercises

Chapter 1

1.1 Figure E1.1 shows a tetrahedron of height h and edge-length equal to the interatomic spacing a. The 'B layer' atom lies above the centre of the triangle formed by three 'A layer' atoms. The distance AC is $\frac{2}{3}a\cos 30 = a/\sqrt{3}$. Hence from a right-angled triangle ABC $h^2 = a^2 - \frac{1}{3}a^2$ whence $h = \sqrt{(2/3)}a$. Hence $c/a = 2h/a = 2\sqrt{(2/3)} = 1.633$. The centre of the tetrahedron above the triangular base (denoted by X in Fig. E1.1), is obtained by projecting a bisector from any edge, e.g. AB to X. $\cos x = h/a$ and $BX = a/2.\frac{1}{\cos x} = a/2\,a/h = a^2/2h$ and since $a = \sqrt{3/2}\,h$, $BX = \frac{3h^2}{2.2h} = \frac{3}{4}h$ or $CX = 1/4\,h$: i.e. the centre of a tetrahedron is at $1/4$ of its height. To determine the packing density, PD, consider the **primitive** hexagonal unit cell that contains two atoms (see Fig. 9.6). The volume of the cell $V = 4r^2\frac{\sqrt{3}}{2}$ (area of face). $2\frac{\sqrt{2}}{3}2r$ (height $2h$) and the volume occupied by the atoms, $V_a = \frac{24}{3}\pi r^3$. Hence, the packing density, $PD = \pi/3\sqrt{2} = 0.7405$ as it is for the ccp structure.

1.2 As can be seen in Figure 1.13(d), the tetrahedron of four A atoms is not regular but has four shorter edges of length $a\sqrt{3}/2$ and two longer edges of length a. However, unlike the octahedral interstitial site all four A atoms are equidistant from the interstitial X atom at the centre of the tetrahedron. The geometry is best appreciated with reference to a plan view (Fig. E1.2) showing two bcc cells and one tetrahedral interstitial site which lies, by symmetry, at height $\frac{1}{4}a$ in the common cell face. The distance from the centre of the tetrahedron to any corner is $\sqrt{(a/2)^2 + (a/4)^2} = a\sqrt{5}/4 = r_X + r_A$. The closest

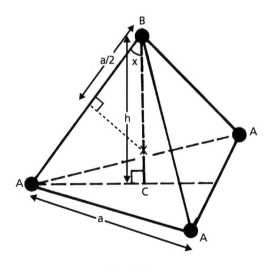

Fig. E1.1.

distance between A atoms is along the direction of the body-diagonal of the cube, i.e.
$2r_A = 1/2\sqrt{3}a$. Hence $r_X + r_A = a\sqrt{5}/4 = r_A\sqrt{5/3} = 1.291r_A$ whence $r_X = 0.291r_A$.

1.3 The interstitial site is at the centre of a triangular prism of height and side-length $2r_A$
(Fig. E1.3). The interstitial atom X at height r_A is equidistant from all six corners of the
prism. Hence, proceeding as in Exercise 1.1, AC $= \frac{2}{3}2r_A \cos 30° = r_A2/\sqrt{3}$, and AX
$= \sqrt{4/3r_A^2 + r_A^2} = r_A\sqrt{7/3}$.
Hence AX $= r_X + r_A = 1.528r_A$ whence $r_X = 0.528r_A$.

Fig. E1.2.

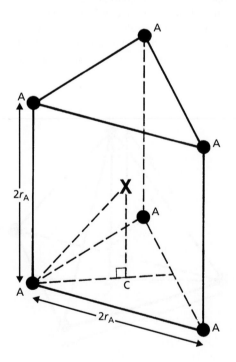

Fig. E1.3.

1.4 (a) In the ccp structure there are four sets of close-packed planes and six close-packed directions and in the hcp structure only one set of close-packed planes and three close-packed directions.

(b) In the bcc structure there are six sets of closest-packed planes and four close-packed directions.

1.5 In the ccp structure each close-packed plane contains three close-packed directions, i.e. three slip systems share a common close-packed plane. As there are four sets of close-packed planes in the ccp structure there are therefore twelve slip systems. In the bcc structure each closest-packed plane contains two close-packed directions, i.e. two slip systems share a common closest-packed plane. As there are six sets of closest-packed planes in the bcc structure there are therefore also twelve slip systems.

1.6 The atomic coordinates are (000) (for all the A-layer' atoms situated at the corners of the unit cell), and either $\left(\frac{1}{3}\frac{2}{3}\frac{1}{2}\right)$ or $\left(\frac{2}{3}\frac{1}{3}\frac{1}{2}\right)$ for the 'B layer' or the 'C layer' atom within the cell, depending upon the arbitrary choice made between these two possible (B or C) sites.

1.7 For B6 (carborundum II) the stacking sequence is ABCACBA... giving a six-layer unit cell repeat distance. In the Frank notation this is △ △△ ▽ ▽ ▽ ..., i.e. an inversion every three layers (rather than an inversion every layer in B4 and an inversion every two layers in B5).

For B7 (carborundum I) the stacking sequence is ABCBACABACBCACB... giving a fifteen-layer unit cell repeat distance. In the Frank notation this is △△▽▽▽△△▽▽▽△△▽▽▽ ..., i.e. inversions every two and three layers alternately.

1.8 Fault an A layer to a B layer as in Fig. 1.20(a)–(b). Then fault not the C layer in Fig. 1.20(b) but the A layer above which becomes a B layer; all the layers above are re-labelled B → C and C → A as before. Again, fault the next A layer above to become a B layer and repeat as before. The stacking sequence is clearly CBCB ..., i.e. the hcp stacking sequence.

1.11 The sequence is: △△▽▽△△▽▽ (a four-layer repeat).

1.13 Both models represent the body-centred cubic structure. Figure 1.43(a) shows the primitive unit cell and Fig. 1.43(b) (a rhombic dodecahedron) shows the conventional bcc cell to which the body-centring atoms of the six neighbouring cells have been added to the cube faces. Notice that in both cases the models have diamond-shaped faces in which the ratio of the long to short diagonals is $\sqrt{2} : 1$ (see Exercise 3.4).

Chapter 2

2.1 There are twice as many mirror lines in p3m1 and the triads lie at the intersections of all of them; in p31m half the triads are not intersected by mirror lines.

2.2 The lattice type is p3m1 (refer to your drawings or Question 2.1).

2.3 The lattice type is p2—i.e. the pattern has identical symmetry to Fig. 2.7; it is based on the oblique plane lattice, not a rectangular plane lattice (primitive or centred) because there are no mirror lines of symmetry present.

2.4 The lattice type is pg. The glide lines run vertically through the back heels of the black men and the raised hands of the white men.

2.5 A 4' axis with vertical and horizontal m' lines and diagonal m lines (point group 4'mm').

Fig. E2.8.

2.6 The one-dimensional lattice types (Fig. 2.24) are:
 (a) *p*112 (b) *p*112 (c) *p*112
 (d) *p*112 (e) *pmm*2 (f) *pmm*2
 (g) *p*1*a*1 (h) *p*111 (i) *p*112.
2.7 Figure 2.24(b) is counterchange plane group *p*2*g′g′*.
2.8 There are three such counterchange patterns based upon *pma*2 (*p*2*mg*) as shown in Fig. E2.8.

Chapter 3

3.1 (a) is orthorhombic *C*; (b) is not a Bravais lattice because the points *do not* all have an identical environment [see the two-dimensional examples, Figs 2.2(b) and (c)]; (c) is orthorhombic *P* (two primitive cells are drawn together).
3.2 (a) is orthorhombic *P* (relocate the origin at $x'y'z'$); (b) is orthorhombic *I* (relocate the origin at $0\frac{1}{2}0$); (c) is orthorhombic *C* (relocate the origin at $00z'$; the motif is two atoms, one of each type).
3.3 For $\alpha > 90°$ it is a 'distorted' truncated octahedron (i.e. rectangular, rather than square faces and unequal-sided hexagonal faces) and for $\alpha < 90°$ it is a 'distorted' rhombic dodecahedron (i.e. with 12 faces of unequal-shaped diamonds). See Fig. 3.7(d) and (e).
3.4 For cubic *I* the ratio is $\sqrt{2} : 1$ (see also Exercise 1.13) and for cubic *F* the ratio is $\sqrt{3} : 1$ (for cubic *P* it is, of course, 1:1).

Chapter 4

4.2

Structure	Bravais lattice	Motif
NaCl	Cubic F	(Na and Cl)
CsCl	Cubic P	(Cs and Cl)
Diamond	Cubic F	(C and C)
ZnS (sphalerite)	Cubic F	(Zn and S)
ZnS (wurtzite)	Hexagonal P	(2Zn and 2S)
Li_2O (or CaF_2)	Cubic F	(2Li and 1O)
$CaTiO_3$	Cubic P	(1Ca, 1Ti, and 3O).

NB: in ferroelectric crystals the distortions which occur at the Curie temperature and below change the Bravais lattice from Cubic P to Tetragonal P to Orthorhombic P and to Rhombohedral with decreasing temperature.

4.3 The space group of Fig. 1.33(a) is $P3_121$ and the space group of Fig. 1.33(b) is $P6_422$. Notice that in α-quartz a group of six silicon atoms is arranged around an **irregular** hexagon: the two pairs of three atoms are related by concentric 3_1 (right-handed) screw axes. But we can also see that the six atoms in the irregular hexagon correspond very closely with the symmetry of a 6_4 (left-handed) screw axis—which becomes a 'perfect' 6_4 axis in β-quartz. The irregular 6_4 axis in α-quartz is called an **imperfect** or **structural** helix—and it is the presence of such a helix (rather than the 3_1 helices) that determines the right- or left-handed (dextra- or laevo-) rotating optical properties of α-quartz.

Chapter 5

5.1 In fcc crystals the slip planes are $\{111\}$, slip directions are $\langle110\rangle$.
In bcc crystals the slip planes are $\{110\}$, slip directions are $\langle111\rangle$.

5.2 $[23\bar{1}]$, i.e. $2.1 + 3.1 + \bar{1}.5 = 0$.
(321), i.e. $3.1 + 2.\bar{1} + 1.\bar{1} = 0$.

5.3 Triclinic, none; monoclinic, (010) and $[010]$; tetragonal, (001) and $[001]$; $(hk0)$ and $[hk0]$.

5.4 $(4\bar{1}\bar{1})$; $(2\bar{3}1)$; $[\bar{1}\bar{2}11\bar{4}]$; $[3\bar{3}1]$ (or, their opposites, e.g. $(\bar{4}11)$, etc.).

5.5 (a) $d_{101} = 375$ pm; $d_{100} = 452$ pm; $d_{111} = 302$ pm; $d_{202} = 188$ pm.
(b) $\alpha = 48.1°$; $\beta = 53.5°$; $\gamma = 63.4°$.
(c) $p = 61.8°$; $q = 58.0°$; $r = 45.3°$.
α, β, γ are not identical to p, q, r because the (111) plane normal is not parallel to the $[111]$ direction in orthorhombic crystals. These directions are only parallel in the case of cubic crystals.

5.6 Writing planes as row matrices (Section 5.8);

$$\text{Cubic } I : (HKL) \leftrightarrows (hkl) \begin{pmatrix} \frac{1}{2} & \frac{\bar{1}}{2} & \frac{1}{2} \\ \frac{1}{2} & \frac{1}{2} & \frac{\bar{1}}{2} \\ \frac{\bar{1}}{2} & \frac{1}{2} & \frac{1}{2} \end{pmatrix}$$

$$\text{Cubic } F : (HKL) \leftrightarrows (hkl) \begin{pmatrix} \frac{1}{2} & 0 & \frac{1}{2} \\ \frac{1}{2} & \frac{1}{2} & 0 \\ 0 & \frac{1}{2} & \frac{1}{2} \end{pmatrix}.$$

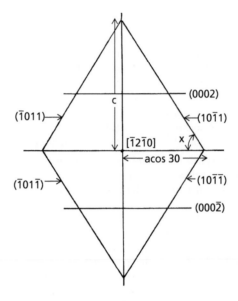

Fig. E5.9.

The answers you obtain will depend upon your particular choices of **A, B, C** and **a, b, c**—but they will be the same form as those given above. The reverse transformations for planes are given by the inverse of the matrix in each case. Note that each matrix and its inverse also gives the transformation for direction symbols $[UVW] \leftrightarrows [uvw]$ when these are written as column matrices (see for example, Section 5.8). Finally, check that your determinants are positive ($\frac{1}{2}$ for Cubic I, $\frac{1}{4}$, for Cubic F in the examples above). A *negative* value means that you have inadvertently chosen a left-handed axial system to describe one or other of the unit cells.

5.7 $[001] = [0001]$; $[010] = [\bar{1}2\bar{1}0]$; $[210] = [30\bar{3}0] = [10\bar{1}0]$; $[110] = [11\bar{2}0]$.

5.8 The direction $[UVTW]$ is $[4\bar{3}\bar{1}\bar{1}]$ and the plane $(hkil)$ is $(1\bar{4}3\bar{1})$ (or their opposites).

5.9 Figure E5.9 shows traces of planes (0002), $(10\bar{1}1)$, etc. lying in the $[\bar{1}2\bar{1}0]$ zone axis (normal to the page). We require $\tan x = \tan 60° = c/a \cos 30°$ i.e. $c/a = \tan 60. \cos 30 = 3/2 = 1.5$.

Chapter 6

6.1 These reciprocal lattice sections all consist of hexagonal arrays of reciprocal lattice points, those closest to the origin representing the $\{10\bar{1}0\}$ planes (hexagonal lattice), $\{110\}$ planes (cubic I lattice) and $\{220\}$ planes (cubic F lattice).

6.2 Check your answers with the equations given in Appendix 4.

6.3 Yes.

6.4 $r^*_{uvw} = \left| \frac{1}{\mathbf{r}_{uvw}} \right| = 0.149 \text{Å}^{-1}$

Chapter 7

7.1 The unit cells of the net are 'skew', like that for the monoclinic crystal (Fig. 6.4(c)) or the net (Fig. 7.1(d)). The (reciprocal lattice) unit cells of each twin orientation indicated by circles

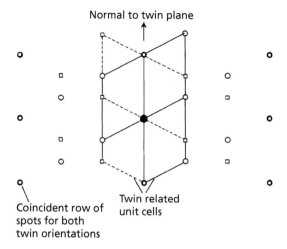

Normal to twin plane

Coincident row of spots for both twin orientations

Twin related unit cells

Fig. E7.1.

and squares are shown in Fig. E7.1. The vertical row of spots through the origin and perpendicular to the twin plane is common to both twin orientations and the spots are unstreaked. Similarly, in every third row spots are common to both twin orientations and are also unstreaked. This arises because one-in-three of the points in the net have the same position, or are *coincident* in both twin orientations. Streaking, or the extension of the reciprocal lattice points normal to the twin plane, increases as the twin thickness decreases, *except* for those reciprocal lattice points common to both orientations.

7.2 The angles at which the second, fourth etc. maxima occur, i.e. $2(\lambda/a)$, $4(\lambda/a)$, etc., coincide with the angles at which the zero order, first order etc. minima occur, i.e. $(1\lambda/d)$, $2(\lambda/d)$. Hence alternate peaks or maxima in the diffraction pattern are missing.

Chapter 8

8.1 α_n, the angle between the direct and diffracted beams, is equal to 2θ (see Fig. E8.1); the planes are inclined to the diffraction grating such that they make equal angles with the incident and reflected beams. Hence $d_{hkl} = a\cos\theta$. Substituting for θ and d_{hkl} in the equation $n\lambda = a\sin\alpha_n$ gives $n\lambda = (d_{hkl}/\cos\theta)\sin 2\theta = (d_{hkl}/\cos\theta)2\cos\theta\sin\theta = 2d_{hkl}\sin\theta$—i.e. Bragg's law.

8.2 $d_{110} = 0.2027$ nm; $d_{200} = 0.1433$ nm; and $d_{211} = 0.1170$ nm. Planes whose d-spacings are smaller than $\lambda/2 (= 0.1146$ nm) do not satisfy Bragg's law since the maximum possible value of θ is $90°$.

8.3 Let the angle Ψ and distance OB be as shown in Fig. E8.3. Then $AB = OB\sin(\theta + \Psi)$ and $BC = OB\sin(\theta - \Psi)$. Summing and expanding the sine terms:

$$(AB + BC) = OB\sin\theta\cos\Psi + OB\cos\theta\sin\Psi + OB\sin\theta\cos\Psi - OB\cos\theta\sin\Psi.$$

Fig. E8.1.

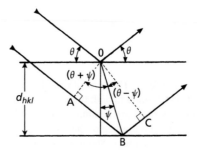

Fig. E8.3.

Hence, cancelling and substituting OB cos $\Psi = d_{hkl}$:

$$(AB + BC) = 2OB \sin\theta \cos\Psi = 2d_{hkl}\sin\theta;$$

i.e. the path difference is independent of the value of Ψ, ie is independent of the atomic arrangement in a plane.

Chapter 9

9.1 Bragg's law is satisfied for the planes $(2\bar{2}0),(3\bar{1}0),(400),(310),(220)$ (in the $hk0$ section); $(4\bar{1}1),(4\bar{1}\bar{1}),(301),(411),(41\bar{1})$ (in the $hk1$ and $hk\bar{1}$ sections) and $(3\bar{1}2)$, $(3\bar{1}\bar{2}),(202),(20\bar{2}),(312),(31\bar{2})$ (in the $hk2$ and $hk\bar{2}$ sections). The (larger) reflecting sphere does not extend beyond the $hk2$ and $hk\bar{2}$ sections.

9.2 Bragg's law is satisfied for the planes $(420),(4\bar{2}0),(220)$ and $(2\bar{2}0)$.

9.3 Let v be the root mean square velocity of the neutrons. Hence, the kinetic energy $\frac{1}{2}mv^2 = 3/2kT$. The wavelength λ is given by de Broglie's equation $\lambda = h/mv = h/\sqrt{3mkT}$. Substituting the values of h, m, k and $T = 373$ K gives $\lambda = 1.31 \times 10^{-10}m = 1.31$Å.

9.4 The reflection condition for the cubic F lattice is h, k, l are all odd or all even and $F_{hkl} = 4f$ (Appendix 6). For f we substitute the values for Zn and S, using the exponential form since

ZnS is non-centrosymmetric (point group $\bar{4}3m$):

$$F_{hkl} = 4\left(f_{Zn} \exp 2\pi i\,(h0 + k0 + l0) + f_S \exp 2\pi i\left(h/4 + k/4 + l/4\right)\right)$$
$$= 4\left(f_{Zn} + f_S \exp \pi i\frac{(h + k + l)}{2}\right).$$

Hence, when
$$(h + k + l) = 4n, \qquad F_{hkl} = 4\,(f_{Zn} + f_S)$$
$$(h + k + l) = 4n + 2, \quad F_{hkl} = 4\,(f_{Zn} - f_S)$$
$$(h + k + l) = 2n + 1, \quad F_{hkl} = 4\,(f_{Zn} \pm if_S).$$

The third pair of equations, (h, k and l all odd) give identical values of I_{hkl} except under conditions of anomalous scattering, see Figs 9.8 and 9.9.

9.5 For the origin at the centre of symmetry the atomic coordinates are $\pm\,(1/8\ 1/8\ 1/8)$, hence

$$F_{hkl} = 4 \cdot 2f_c \cos 2\pi\,\left(h/8 + k/8 + l/8\right) = 8f_c \cos \pi/4\,(h + k + l)$$

which gives the condition that when

$$(h + k + l) = 4n + 2, \quad F_{hkl} = 0$$

in addition to the condition for a cubic F lattice that $F_{hkl} = 0$ when h, k, l are mixed (see Table A6.2).

9.6 For f we substitute the values for f_{Cl} at (000) and f_{Na} at $(1/2\ 0\ 0)$ (equivalent to $(0\ 1/2\ 0)$ or $(0\ 0\ 1/2\,)$).

$$F_{hkl} = 4\left(f_{Cl} \exp 2\pi i\,(h0 + k0 + l0) + f_{Na} \exp 2\pi i\left(h1/2 + l0 + k0\right)\right)$$
$$= 4\,(f_{Cl} + f_{Na} \exp \pi ih).$$

Hence when h, k, l all odd $F_{hkl} = 4(f_{Cl} - f_{Na})$
 when h, k, l all even $F_{hkl} = 4(f_{Cl} + f_{Na})$.

9.7 Simply measure the distances, r, of the $hk1$ and $hk2$ layer lines from the $hk0$ layer line. Figure E9.7 shows the Ewald sphere. The $hk0$, $hk1$, $hk2$, etc. reflections lie in the layer lines, spacing c^* as indicated, where $\frac{1}{|c^*|} = c$, the lattice repeat distance.
From Fig. E9.7, $\tan 2\theta = r/30\,mm$ and $\sin 2\theta = \left|c^*\right|\frac{1}{1/\lambda} = \lambda/c$

For the $hk1$ layer line r (corrected for scale factor) = 9.1 mm, whence $c = 5.31$Å.
For the $hk2$ layer line r (corrected for scale factor) = 21.3 mm, whence $c/2 = 2.66$Å, $c = 5.32$Å.

9.8 This is part of the calculation that in 1913 gave W.L. Bragg 'the greatest thrill'. We do not need to know the atomic coordinates as such since x immediately gives us the phase angle between the beams scattered from the layers of Fe and S atoms (see Fig. 9.2(b) and Section 9.2, p.214).

$Z_{Fe} = 26$ and $Z_S = 16$.
For $x = 1/4$ the phase angles are $2\pi/4$, $4\pi/4$, etc. hence:
$(F_{1st\ order})^2 = (26 + 2.16.\cos 2\pi/4)^2 = 676$, 676 (1) = 676
$(F_{2nd\ order})^2 = (26 + 2.16.\cos 4\pi/4)^2 = 36$, 36 (0.2) = 7.2
$(F_{3rd\ order})^2 = (26 + 2.16.\cos 6\pi/4) = 676$, 676 (0.07) = 47.32

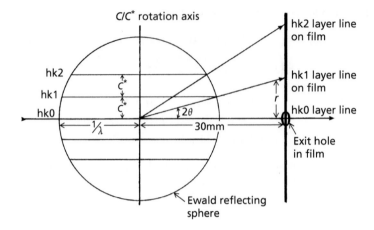

Fig. E9.7.

etc. and similarly for $x = 1/5$, phase angles $2\pi/5$, $4\pi/5$, etc. and $x = 1/6$, phase angles $2\pi/6$, $4\pi/6$, etc. In summary, the calculations show:

	1st	2nd	3rd	4th	5th
$x = 1/4$	100	1.1	7	14.9	1.0
$x = 1/5$	100	0	0	3	2.6
$x = 1/6$	100	1.1	0.14	0.17	1.0

The value of $x = 1/5$ gives intensity values that closely match W.L. Bragg's observed values. It was the first parameter or atomic position to be determined from the intensities of the X-ray reflections.

9.9　The estimates, which do not account for the (small) energy level differences in the L and M shells, are simply based on the differences in energy levels between the K, L, and M shells. For example, $(W_K - W_L) = \left(hc/\lambda_K - hc/\lambda_L\right) = hc/\lambda_{ka}$, whence $\left(1/\lambda_K - 1/\lambda_{Ka}\right) = 1/\lambda_L$, $\lambda_L = 13.23$ Å. Similarly, $\lambda_M = 174.8$ Å and $\lambda_{La} = 14.31$ Å.

Chapter 10

10.4　The rings in both Figs 10.11(a) and (b) have the same diameters and hence show the same sequence of d-spacings. The diameters of the rings represent a 4θ (*not* 2θ) angle, hence $\tan 2\theta$ = radius of ring/specimen film distance, from which, and using Bragg's law, the d-spacings can be determined. Referring to the table, the sequence of indices is (110), (040), (130) (the first three strong reflections in the zero layer-line in Fig. 10.11(b)), (111), (131) (the weaker reflections in the first layer-lines above and below) and (022), (112) (the very faint reflections in the second layer lines).

Since the reflections in the zero layer-line are all of type $\{hk0\}$, they have a common zone axis [001] which is parallel to the tensile axis and the direction of the polymer chains.

10.5 The 2θ angles and d-spacings of the peaks in (a) should first be checked to confirm that they correspond to the 110, 200 and 211 reflections for α-Fe (a $= 2.866$Å). The half-height peak widths (measurable only to a limited accuracy) give $\beta (= \beta_{obs} - \beta_{inst})$ values ≈ 0.004, 0.006 and 0.008 respectively which give residual stress values ($\sigma = {}^{1}\!/_{2}E\beta \cot \theta$) in the range $1.10 \sim 0.92$ GPa. These are considerably greater than the yield stress of mild steel indicating that a large component of the broadening arises from a reduction in crystallite size.

10.6 The β values for the peaks are determined in the same way as for Exercise 10.5. Substituting these values into the Scherrer equation $\beta = \lambda/t \cos \theta$ gives grain size (t) values in the range 35–50 nm. Notice also the small peak shifts; these arise from variations in the α lattice parameters as a result of a change in vanadium concentration following precipitation of the vanadium-rich β phase.

Chapter 11

11.1 The equation to be derived follows from the discussion in Section 11.1. Since $eV = \frac{1}{2}mv^{2}$, then $\sqrt{2meV} = mv$. Substituting in de Broglie's equation gives $\lambda = h/\sqrt{2meV}$. The numerical values obtained by substituting the m, e and V values are 10 kV: 12.2 pm (12.2 pm); 100 kV: 3.88 pm (3.70 pm); 1 MV: 1.22 pm (0.87 pm). The relativistically corrected values are in brackets, from which it is seen that the differences between the values are insignificant at 10 kV, significant at 100 kV and very significant at 1 MV.

11.2 The zone axis of the pattern is $\langle 112 \rangle$. The particular variant obtained depends upon the variants of the indices chosen to index the pattern consistently.

11.3 This problem is essentially the same as that for the simple cubic crystal as shown in Figs 11.1 and 11.2, except that the cubic P reciprocal lattice is replaced by the cubic F reciprocal lattice (for a cubic I crystal—see Section 6.4). In the ZOLZ $0kl$ and FOLZ $1kl$ the reflecting planes are those for which $(h + k + l) =$ an even number (see Fig. 6.8(a)).

11.4 The diffraction pattern is from $Fe_{3}C$ (cementite) and is indexed in Fig. E11.4. Notice that the pattern is *not* rectangular, but slightly skew. Since the structure is orthorhombic the order of the indices hkl may not be changed; only their signs may be changed, e.g. the allowed permutations of 103 are: $\bar{1}03,\ 10\bar{3}, 103, \bar{1}0\bar{3}$.

Fig. E11.4.

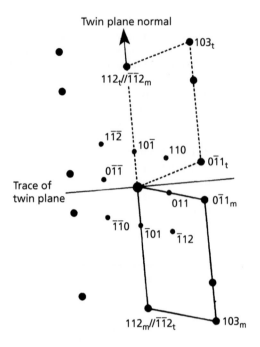

Fig. E11.6.

11.5　In Fig. 11.6 the camera constant $\lambda L = 27 \text{Å}$ mm.

11.6　The diffraction patterns may be indexed as shown in Fig. E11.6. The 'matrix' α-Fe orientation is outlined by solid lines and the twin orientation by dashed lines. The trace of the twin plane and the twin plane normal are indicated (compare with the twin-related unit cells in the optical diffraction pattern as shown in the answer to Exercise 7.1). The reflections for matrix (m) and twin (t) orientations are indexed such that (e.g.) $0\bar{1}1_t$ is related to $0\bar{1}1_m$ by reflection in the twin plane. The Fe_3C reflections may be indexed as shown, giving the orientation relationship between the α-Fe (matrix) and Fe_3C as:

$$(011)_{Fe_3C} \| (0\bar{1}1)_{\alpha-Fe} \quad \text{and} \quad (\bar{1}01)_{Fe_3C} \| (112)_{\alpha-Fe} .$$

For the zone axes, obtained by cross multiplication (see Section 5.6.2):

$$[\bar{1}1\bar{1}]_{Fe_3C} \| [3\bar{1}\bar{1}]_{\alpha-Fe} .$$

11.7　Figure E11.7(a) shows a [001] projection of the diamond-cubic structure and the traces (dashed lines) of the (220) planes, the atoms in successive planes being denoted by the ● and ■ symbols; separation $= \sqrt{3}/4$ (0.566 nm) = 0.245 nm. Figure E11.7(b) shows the $\langle 110 \rangle$ projection and the 'dumbbell' (circled) pattern of atoms; spacing (in projection) $= 1/4(0.566nm) = 0.142$ nm.

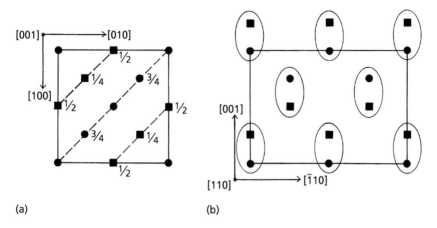

Fig. E11.7.

Chapter 12

12.1 Figure E12.1 shows a 'vertical' section with the axis of the cone of elliptical cross-section marked OS as in Fig. 12.3(a) and at angle x to the N–S direction. The tangent to the circle (heavy line), which is parallel to the small circle on the sphere, is also tilted at this angle x to the axis of the cone since the angle at the centre of the circle is $2x$ (Euclid, Book III, Proposition 20). Since the two circular sections must, by symmetry, be inclined at equal and opposite angles to the axis of the cone OS, then the other circular section, indicated by the other heavy line, is parallel to the plane of the stereographic projection.

12.2 The α, β and γ angles of the (212) plane to the x, y and z axes are 48° ,71° and 48°, hence the direction cosines cos α, cos β and cos γ are 0.67, 0.33 and 0.67. The ratios between these direction cosines are equal to the ratios between the Miller indices, i.e. 2:1:2. The α, β and γ angles of the ($\bar{3}\bar{3}2$) plane are $-50°$, $-50°$ and 65°; hence the direction cosines are $-0.64, -0.64, 0.42$ which are equal to the ratios $\bar{3}:\bar{3}:2$.

12.3 For plane A, using the Wulff net with its axis A ... A along the x-axis, draw in the small circle 36.5° from the +x-axis; this small circle represents the locus of all planes which lie 36.5° from the +x-axis. Now rotate the Wulff net 90° so that its axis A ... A lies along the y-axis and draw in the small circle 57.5° from the +y-axis. The two points of intersection of the small circles both in the northern and southern hemispheres (i.e. above and below the plane of projection) show the two possible positions of plane A. Finally, draw in the small circle (the 'line of latitude') at 74.5° from the +z-axis. The three small circles all intersect at the point in the northern hemisphere and hence give the position of plane A (see Fig. E12.3).

 The position of plane B is found by the same procedure except that a−sign indicates that the small circle is drawn from the−(minus)-axis.

 To find the Miller indices of planes A and B, write down the direction cosines of the angles to the x, y and z-axes respectively and express these as the ratios of whole numbers, i.e.

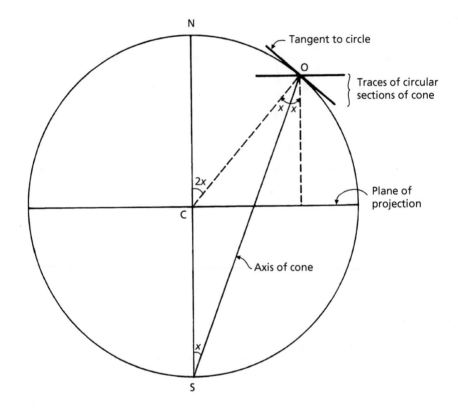

Fig. E12.1.

Plane A	+36.5°	+57.5°	+74.5°
direction cosines:	+0.804 :	+0.537 :	+0.267
ratios:	3 :	2 :	1

Hence the Miller index of plane A is (321).
Similarly, the Miller index of plane B is ($\bar{2}1\bar{2}$).
The angles between the planes are found by rotating the Wulff net until each pair of planes lie in the same great circle.
Plane A and plane B : 121°
Plane A and ($\bar{1}\bar{1}1$) : 126°
Plane B and (101) : 90°
Plane B and ($\bar{1}01$) : 19°

12.4 Figure E12.4 (p.496) shows the stereographic projections using space group symbolism (Section 4.6). Point group $\bar{2} \equiv m$ is shown for two orientations of mirror plane/inversion diad axis. The + and − symbols indicate that the poles are in the 'northern' and 'southern' hemispheres, respectively and the presence/absence of the, (comma) symbols indicate that the poles are related by a mirror plane (reflection) or centre of symmetry.

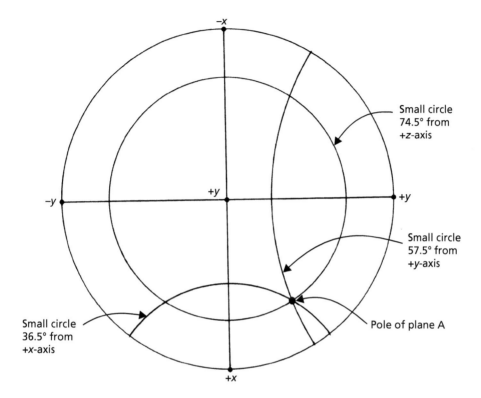

Fig. E12.3.

Chapter 14

14.1 Since $J_1 = \sigma_1 E_1 = \sigma_1 l_1 E$, $J_2 = \sigma_2 E_2 = \sigma_2 l_2 E$, $J_3 = \sigma_3 E_3 = \sigma_3 l_3 E$, the direction cosines of J are in the ratios $\sigma_1 l_1 : \sigma_2 l_2 : \sigma_3 l_3$. Re-writing the equation for a quadric: $\sigma_1 x_1^2 + \sigma_2 x_2^2 + \sigma_3 x_3^2 = 1$ (where $\sigma_1 = {}^{1}/_{a^2}$ etc.). The coordinates of a point P on the quadric are $r l_1$, $r l_2$, $r l_3$ (Fig. 14.4 where $OP = r$ is in the direction of E). The tangent plane at P is:

$$\sigma_1 r l_1 x_1 + \sigma_2 r l_2 x_2 + \sigma_3 r l_3 x_3 = 1$$

Hence, the normal at P has direction cosines in the ratios $\sigma_1 l_1 : \sigma_2 l_2 : \sigma_3 l_3$ and hence is parallel to J.

14.2 The principal stresses are $P_1 = 3$, $P_2 = 6$, $P_3 = 18$. This answer may be checked by substituting these values, one at a time, in the general equation and then checking that the determinant (see Section 5.10) is zero. For example, for $P_1 = 3$:

$$\begin{vmatrix} (11-3) & \bar{6} & 2 \\ \bar{6} & (10-3) & 4 \\ 2 & 4 & (6-3) \end{vmatrix} = 0 \qquad \text{and similarly for } P_2 = 6 \text{ and } P_3 = 18$$

The array of direction cosines relating the arbitrary axes to the principal axes is:

$$\begin{bmatrix} 1/3 & 2/3 & 2/3 \\ \bar{2}/3 & \bar{1}/3 & 2/3 \\ 2/3 & \bar{2}/3 & 1/3 \end{bmatrix}$$

14.3 Yes. When a crystal is of a thickness in which the OPD = 1λ, 2λ, etc. the polarization of the emerging light is identical to that of the incident light (see Fig. 14.9) and the crystal is in extinction. However, this situation only applies to monochromatic light. In general, n_e and n_o, and hence $(n_e - n_o)$ for a crystal vary with wavelength. Hence, for white light there will only be one wavelength for which the OPD will be exactly 1λ, 2λ, etc. and other wavelengths will not be in extinction. For example, the '1λ' or 'sensitive tint' plate used in polarized light microscopy gives extinction for green-yellow light, the blue and red wavelengths are not in extinction and give rise to a deep purple colour—the 'sensitive tint'.

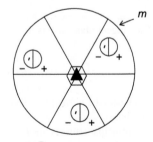

(a) $\bar{2} \equiv$ perp mirror plane (m) shown for two orientations of the 2 axis. (Left) perp. to plane of paper; (Right) in plane of paper

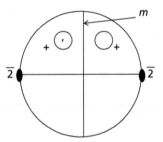

(c) $\bar{3} = 3$ + centre of symmetry (b) $\bar{6} = 3$ + perp. mirror plane

Fig. E12.4.

Further Reading

Introductory books on crystallography

Bhadeshia, H. (1987). *Worked Examples in the Geometry of Crystals*. The Institute of Materials, London.

Glazer, A. M. (1987). *The Structures of Crystals*. Adam Hilger, Bristol.

Glazer, A. M. (2016). *Crystallography: A Very Short Introduction*, Oxford University Press.

Holden, A. and Morrison, P. (1982). *Crystals and Crystal Growing*. MIT Press.

Kennon, N. F. (1978). *Patterns in Crystals*. John Wiley, New York.

Steadman, R. (1982). *Crystallography*. Van Nostrand Reinhold, Wokingham.

Windle, A. (1977). *A First Course in Crystallography*. G. Bell and Sons Ltd, London.

Books on two-dimensional patterns and symmetry

Grünbaum, B. and Shephard, G. C. (1989). *Tilings and Patterns: An Introduction*. W. H. Freeman, New York.

Hann, M. A. and Thomson, G. M. (1992). *The Geometry of Regular Repeating Patterns*. The Textile Institute, Manchester.

Jackson, L. (2008) *From Atoms to Patterns: Crystal Structure Designs from the 1951 Festival of Britain*. Richard Dennis Publications, The Old Chapel, Shepton Beauchamp, Somerset, UK.

Jones, Owen (1856). *The Grammar of Ornament*. Day and Son, London, reprinted (1986) by Studio Editions, London.

MacGillavry, C. H. (1976). *Symmetry Aspects of M. C. Escher's Periodic Drawings*. Published for the I.U.Cr. by Bohn, Scheltema and Holkema, Utrecht. Re-issued as *Fantasy and Symmetry: the Periodic Drawings of M. C. Escher*. Abrams, New York.

Shubnikov, A. V. and Koptsik, V. A. (1974). *Symmetry in Science and Art* (English translation, ed. David Harker). Plenum Press, New York and London.

Books mainly concerned with crystal structure and symmetry

Buerger, M. J. (1970). *Contemporary Crystallography*. McGraw-Hill, New York.

Buerger, M. J. (1971). *Introduction to Crystal Geometry*. McGraw-Hill, New York.

Burns, G. and Glazer, M. (2013). *Space Groups for Solid State Physicists* (3rd edn). Academic Press (imprint of Elsevier), Waltham MA, USA and Kidlington, Oxford.

Evans, R. C. (1964). *An Introduction to Crystal Chemistry* (2nd edn). Cambridge University Press, Cambridge.

Janot, C. (2012). *Quasicrystals: A Primer* (2nd edn). Oxford Classic Texts in the Physical Sciences, Oxford University Press, Oxford.

McKie, D. and McKie, C. (1986). *Essentials of Crystallography*. Blackwell Scientific Publications, Oxford.

Phillips, F. C. (1963). *Introduction to Crystallography* (3rd edn). Longmans, London.

Tilley, J.D. (2006). *Crystals and Crystal Structures*. John Wiley & Sons, Ltd, Chichester.

Vainshtein, B.K., Fridkin, V.M., and Indenbom, V.L. (1982). *Modern Crystallography II: Structure of Crystals*. Springer-Verlag, Berlin, Heidelberg and New York.

Books which cover, generally in more detail, structural crystallography, symmetry, X-ray and electron diffraction techniques from the standpoint of the chemist, geologist, materials scientist or physicist

Allen, S. M. and Thomas, E. L. (1999). *The Structures of Materials*. John Wiley, Chichester and New York.

Bacon, G. E. (1962). *Neutron Diffraction* (2nd edn.). Oxford University Press, Oxford.

Barrett, C. S. and Massalski, T. B. (1980). *The Structure of Metals* (3rd edn). Pergamon, Oxford.

Bragg, W.L. and Claringbull, G.F. (1965). *Crystal Structures of Minerals: The Crystalline State* (Vol. IV). G. Bell and Sons Ltd, London.

Brown, P. J. and Forsyth, J. B. (1973). *The Crystal Structure of Solids*. Arnold, London.

Bunn, C. W. (1967). *Chemical Crystallography* (2nd edn). Oxford University Press, Oxford.

Dent-Glasser, L. S. (1977). *Crystallography and its Applications*. Van Nostrand Reinhold, New York.

Giacovazzo, C. (ed.) (2011). *Fundamentals of Crystallography* (3rd edn). IUCr Texts on Crystallography 15. International Union of Crystallography, Oxford Science Publications, Oxford University Press, Oxford.

Glusker, J. P., Lewis, M. and Rossi, M. (1994). *Crystal Structure Analysis for Chemists and Biologists*. VCH Publishers, New York.

Hahn, T. (ed.) (1985), 3rd edn (1993). Brief Teaching Edition of Volume A of the *International Tables for Crystallography*. Kluwer Academic Publishers, Dordrecht, Boston, and London.

Hargittai, M. and Hargittai, I. (2008). *Symmetry Through the Eyes of a Chemist* (3rd edn). Springer, New York and London.

Kelly, A., Groves, G. W. and Kidd, P. (2000), revised edn. *Crystallography and Crystal Defects*. John Wiley, Chichester and New York.

Klein, C. and Hurlbut, C. S., Jr. (1993). *Manual of Mineralogy* (21st edn). John Wiley, New York.

McKie, D. and McKie, C. (1974). *Crystalline Solids*. Nelson London.

Putnis, A. (1992). *An Introduction to Mineral Sciences*. Cambridge University Press, Cambridge.

Smith, J. V. (1982). *Geometrical and Structural Crystallography*. John Wiley, New York.

Verma, A. R. and Srivastava, O. N. (1982). *Crystallography for Solid State Physics*. Wiley Eastern, New Delhi.

Wells, A. F. (1984). *Structural Inorganic Chemistry* (5th edn). Oxford Science Publications, Oxford.

Whittaker, E. J. W. (1981). *Crystallography for Earth Science (and other solid state) Students*. Pergamon, Oxford

Books on optics and the diffraction of light

Harburn, G., Taylor, C. A. and Welberry, T. R. (1983). *Atlas of Optical Transforms*. Bell & Hyman Ltd. London (first published in 1975 by G. Bell and Sons Ltd).

Hecht, E. (1998). *Optics* (3rd edn). Addison-Wesley, London.

Jenkins, F. A. and White, H. E. (1976). *Fundamentals of Optics* (4th edn). McGraw-Hill, New York.

Lipson, A., Lipson S. G. and Lipson H. (2011). *Optical Physics*. Cambridge University Press, Cambridge, London, New York.

Robinson, P. C. and Bradbury, S. (1992). *Qualitative Polarized Light Microscopy*. Royal Microscopical Society Handbook No. 09. Oxford University Press/Royal Microscopical Society.

Smith, F. G. and Thomson, J. H. (1988). *Optics*. John Wiley and Sons Ltd., Chichester, New York, Brisbane, Toronto, Singapore.

Steward, E. G. (1989). *Fourier Optics*. Ellis Horwood Ltd., Chichester.

Taylor, C. A. (1978). *Images: a Unified View of Diffraction and Image Formation with all Kinds of Radiation*. Wykeham Publications, London and Basingstoke.

Wood, E. A. (1977). *Crystals and Light* (2nd edn). Dover Publications, New York.

Books on X-ray diffraction and X-ray crystallography

Azaroff, L. V. (1968). *Elements of X-ray Crystallography*. McGraw-Hill, New York.

Bijvoet, J. M., Kolkmeyer, N. H., MacGillavry, C. H. (1951). *X-ray Analysis of Crystals*. Butterworths Scientific Publications, London.

Blow, D. (2002). *Outline of Crystallography for Biologists*. Oxford University Press, Oxford.

Bragg, W. H. and Bragg, W. L. (1966). *The Crystalline State*, Vol. 1. *A General Survey* by W. L. Bragg. G. Bell and Sons Ltd, London.

Buerger, M. J. (1942). *X-ray Crystallography*. John Wiley, New York.

Clegg, W. (1998). *Crystal Structure Determination*. Oxford University Press, Oxford, New York and Tokyo.

Clegg, W., editor (2009). *Crystal Structure Analysis: Principles and Practice* (2nd edn). IUCr Texts on Crystallography 13. International Union of Crystallography, Oxford Science Publications, Oxford University Press, Oxford.

Cullity, B. D. and Stock, S. R. (2001). *Elements of X-ray Diffraction* (3rd edn). Prentice-Hall, Upper Saddle River, New Jersey.

Glusker, J.P. and Trueblood, K.N. (2010). *Crystal Structure Analysis: A Primer* (3rd edn). IUCr Texts on Crystallography 14. International Union of Crystallography, Oxford Science Publications, Oxford University Press, Oxford.

Henry, N. F. M., Lipson, H. and Wooster, W. A. (1961). *The Interpretation of X-ray Diffraction Photographs* (2nd edn). Macmillan, New York.

Ladd, M. F. C. and Palmer, R. A. (2003). *Structure Determination by X-ray Crystallography*. 4th edn. Kluwer-Plenum Press, New York.

Lipson, H. and Cochran, W. (1968). *The Determination of Crystal Structures: The Crystalline State, Vol. III*. G. Bell and Sons Ltd., London.

Rhodes, G. (1993). *Crystallography Made Crystal Clear*. Academic Press, New York.

Woolfson, M. M. (1997). *An Introduction to X-ray Crystallography* (2nd edn). Cambridge University Press, Cambridge.

Books on electron diffraction techniques

Brandon, D. and Kaplan, W. D. (2001). *Microstructural Characterisation of Materials*. John Wiley and Sons, Chichester and New York.

Champness, P. M. (2001). *Electron Diffraction in the Electron Microscope*. Royal Microscopical Society Handbook No. 48, Bios Scientific Publishers/Royal Microscopical Society, Oxford.

Dyson, D. J., Andrews, K. W. and Keown, S. R. (1971). *Interpretation of Electron Diffraction Patterns* (2nd edn). Hilger and Watts, London.

Edington, J. W. (1975). *Electron Diffraction in the Electron Microscope. Monograph* 2. Practical Electron Microscopy in Materials Science. Philips Technical Library, Macmillan, London.

Goodhew, P. J., Humphreys, F. J. and Beanland, T. F. (2001). *Electron Microscopy and Analysis* (3rd edn). Taylor and Francis, London and New York.

Loretto, M. H. (1994). *Electron Beam Analysis of Materials* (2nd edn). Chapman and Hall, London.

Thomas, G. and Goringe, M. J. (1979). *Transmission Electron Microscopy of Materials*. Wiley, Chichester and New York. Reprinted by Tech books, 4012 Williamsburg Court, Fairfax, Virginia 22032.

Williams, D. B. and Carter, C. B. (1996). *Transmission Electron Microscopy: A Textbook for Materials Science*. Plenum Press, New York and London.

Zhou, X., Hovmöller, S., and Oleynikov, P. (2011). *Electron Crystallography: Electron Microscopy and Electron Diffraction*. IUCr Texts on Crystallography 16. International Union of Crystallography, Oxford Science Publications, Oxford University Press, Oxford.

Books of Reference

The *International Tables for Crystallography* replace the earlier *International Tables for X-ray Crystallography*. The volumes, a joint IUCr-Wiley publication are planned and maintained by the IUCr's Commission on International Tables and are available in print form and on-line. The most recent on-line revisions are indicated by [brackets].

Volume A: *Space Group Symmetry*. Hahn, T. (ed.) (1983, corrected reprint of 5th edn, 2005). [2006]. Brief Teaching edn (1985, 5th edn, 2005)

Volume B: *Reciprocal Space*. Shmueli, U. (ed.) (1993, 3rd edn, 2008). [2010]

Volume A1: *Symmetry Relations between Space Groups*. Wondratschrek, H. and Müller, U. (eds) (2004, 2nd edn, 2010). [2011]

Volume C: *Mathematical, Physical and Chemical Tables*. Wilson, A.J.C. and Prince, E. (eds) (1992, 3rd edn, 2004). [2006]

Volume D: *Physical Properties of Crystals*. Authier, A. (ed.) (2003, corrected reprint 2010). [2013]

Volume E: *Sub-Periodic Groups*. Kopský, V. and Litvin, D.B. (eds) (2002, 2nd edn, 2010). [2010]

Volume F: *Crystallography of Biological Macromolecules*. Rossmann, M.G. and Arnold, E. (eds) (2001 2nd edn, 2012). [2012]

Volume G: *Definition and Exchange of Crystallographic Data*. Hall, S.R. and McMahon, B. (eds) (2005, corrected reprint 2010).

The *X-Ray Powder Diffraction File* is published by the International Centre for Diffraction Data (ICDD) (the Joint Committee for Powder Diffraction Standards, JCPDS), 12 Campus Boulevard, Newtown Square, PA 19073-3273, USA. It is updated as sets are added at approximately yearly intervals, together with the following manuals.

Alphabetical Indexes: Inorganic Phases
Hanawalt Search Manual: Inorganic Phases
Organic and Organometallic Phases Search Manual
Indexes and Search Manual: Frequently Encountered Phases.

Books on, or including, the stereographic projection

Barrett, C. S. and Massalski, T. B. (1980). *The Structure of Metals* (3rd edn). Pergamon, Oxford.

Cullity, B. D. and Stock, S. R. (2001). *Elements of X-ray Diffraction* (3rd edn). Prentice-Hall, Upper Saddle River, New Jersey.

Hammond, C. (1993). *Stereographic Techniques*, Module 9.17 in *Procedures for Transmission Electron Microscopy*, eds Robards, A. and Wilson, A. J. John Wiley, Chichester and New York.

Johari, O. and Thomas, G. (1969). *The Stereographic Projection and its Applications.* Techniques for Metals Research, Vol. 2A. Interscience Publishers, New York.

Krawitz, A. D. (2001) *Introduction to Diffraction in Materials Science and Engineering.* John Wiley and Sons, New York.

Moon, J. R. (1978). *Worked Examples in Stereographic Projections.* Institute of Metallurgists Monograph No. 2. The Institute of Metals, London.

Books on the physical properties of crystals

Lovett, D.R. (1999). *Tensor Properties of Crystals* (2nd edn). Institute of Physics Publishing, Bristol and Philadelphia.

Nye, J.F. (1985). *Physical Properties of Crystals: Their Representation by Tensors and Matrices.* Oxford University Press, Oxford.

Wooster, W.A. (1973). *Tensors and Group Theory for the Physical Properties of Crystals.* Clarendon Press, Oxford.

Books on or including group theory

Buerger, M.J. (1956). *Elementary Crystallography.* John Wiley & Sons, Inc., New York and Chapman Hall Ltd., London.

Cotton, F.A. (1971). *Chemical Applications of Group Theory.* John Wiley & Sons Inc., New York.

Ladd, M. (2014). *Symmetry of Crystals and Molecules.* Oxford University Press, Oxford.

Vainshtein, B.K. (1981). *Modern Crystallography I: Symmetry of Crystals, Methods of Structural Crystallography.* Springer-Verlag, Berlin, Heidelberg and New York.

Wooster, W.A. (1973). *Tensors and Group Theory for the Physical Properties of Crystals.* Clarendon Press, Oxford.

Books of general or historical interest

Aste, T. and Weaire, D. (2000). *The Pursuit of Perfect Packing.* Institute of Physics Publishing, Bristol.

The authors recount the story of the problem of the densest packing of identical spheres and many others which have to do with packing things together—from atoms in crystals,

soap bubbles in foams to the structure of the Giant's Causeway; together with brief accounts of the lives of the scientists who devoted themselves to these problems.

Authier, A. (2013). *Early Days of X-ray Crystallography*. Oxford University Press, Oxford.
Authier's aim in writing this book has been 'simply to outline the successive steps that our understanding of the inner structures of crystals went through, and to tell the story of the early days of X-ray crystallography, putting the events into the context of their time'. In this endeavour he has succeeded brilliantly.

Bragg, W. H. and Bragg, W. L. (1966). *The Crystalline State*, Vol. 1. *A General Survey* by W. L. Bragg. G. Bell and Sons Ltd, London.
A book which shows the clarity and incisiveness of Bragg's scientific method. It is so well written that it remained in print, substantially unrevised from its first publication in 1933, for nearly forty years. Even today very little is 'out of date' and it contains much material which is not adequately covered at the same level elsewhere.

Bragg, W. L. (1975). *The Development of X-ray Analysis*, edited by Phillips, D.C. and Lipson, H. G. Bell and Sons, London.
A book published four years after Bragg's death; a summary in effect of his life's work, written in a simple and engaging fashion.

Bunn, C. (1964). *Crystals, their Role in Nature and in Science*. Academic Press, New York and London.
A book very much in the same category as A. F. Wells (1968) (see below) but with an even wider range of subject matter (including speculations on crystals and the origin of life) to appeal to the general scientific reader.

Crick, F. (1989). *What Mad Pursuit: A Personal View of Scientific Discovery*. Weidenfeld and Nicolson, London. Reprinted by Penguin Books, London, 1990.
A book which should be read together with J. D. Watson's *The Double Helix*; written much later, it provides in retrospect a more balanced, but no less personal, account of the discovery of the structure of DNA and its aftermath.

Cundy, H. M. and Rollett, A. P. (1961). *Mathematical Models* (2nd edn). Oxford University Press, Oxford. Reprinted by Tarquin Publications, St. Albans (2006).
This book not only includes clear descriptions of polyhedra and how to construct them but also describes mechanical and models for logic and computing—a compendium of information not to be found elsewhere.

Ewald, P. P. and numerous crystallographers (1962). *Fifty Years of X-ray Diffraction*. Published for the International Union of Crystallography by N. V. A. Oosthoek's Uitgevers-maatschappij, Utrecht.
An invaluable source-book which describes the beginnings of X-ray diffraction, the growing field and development of schools of crystallography throughout the world, and the personal reminiscences and memoirs of those who participated. One of the strongest impressions is that of an international community of scientists which transcended the barriers of nationality, politics and war.

Hall, K.T. (2014). *The Man in the Monkeynut Coat* Oxford University Press, Oxford.
The story of William Astbury who pioneered the science of molecular biology and yet whose early X-ray diffraction studies on the structure of DNA (preceding those of Wilkins and Franklin) have been largely overlooked.

Hargittai, M. and Hargittai, I. 3rd edn. (2008). *Symmetry Through the Eyes of a Chemist.* Springer, New York and London.

A book which transcends the usual subject barriers and which draws illustrations and analogies from a wide range of sources. It also bridges the gap between a popular book on patterns and symmetry and a textbook on 'the third dimension' in chemistry.

Lima de Faria, J. (Editor) (1990). *An Historical Atlas of Crystallography.* Kluwer Academic Publishers, Dordrecht.

A very comprehensive historical survey of the development of geometrical, physical and chemical crystallography and crystal structure determination—topics which are covered in most histories of science in a fragmentary manner. Special features of the book are the beautifully presented 'time maps' of crystallography, the title pages of important works and a bibliography on the history of crystallography.

Lipson, H. (1970). *Crystals and X-rays.* Wykeham Publications (London) Ltd.

An excellent introduction to X-ray diffraction, largely treated historically and with reference to the diffraction of light and microscopy. Crystal structure analysis is also described in a very clear and simple manner.

Newton, I. (1979). *Opticks* or *A Treatise of the Reflections, Refractions, Inflections and Colours of Light.* Dover Publications, New York, based on the Fourth Edition, London, 1730. Published by G. Bell and Sons Ltd. in 1931.

First published in 1704, Newton's *Opticks* is one of the most readable of all the great classics of physical science; in simple language Newton describes his famous experiments with colours, lenses, reflection and diffraction of light, the colours of rainbows and soap bubbles. This (Dover) edition has a foreword by Albert Einstein, a preface by I. Bernard Cohen and an introduction by Sir Edmund Whittaker.

Thomas, J.M. and Phillips D. (1990). *Selections and Reflections: The Legacy of Sir Lawrence Bragg.* Science Reviews Ltd., Middlesex.

The great value of this book is that, in drawing together brief reminiscences of family members, students and colleagues over many years it presents a fully-rounded portrait of Lawrence Bragg, the man. Each contribution adds its own flavour. Moreover, in the second half of the book, in which are gathered together some of Bragg's papers, letters, and Royal Institution schools' lectures (most not now readily accessible) we hear him, as it were, speaking for himself.

Thompson, D'Arcy Wentworth (1942). *On Growth and Form: A New Edition.* Cambridge University Press. Reprinted by Dover Publications, New York 1992.

A classic book, first published in 1917, which provides in many ways a unique perspective on the evolution of biological structures and processes in terms of their physical and mathematical aspects. It is also a major work of literature. An abridged edition, edited by J. T. Bonner, was first issued by Cambridge University Press in 1961.

Watson, J. D. (1968). *The Double Helix: A Personal Account of the Discovery of the Structure of DNA.* With a Foreword by Sir Lawrence Bragg. Weidenfeld & Nicolson, London, and Penguin Books, Harmondsworth (1970).

A book which has become notorious for its lop-sided treatment of the other scientists in the UK and USA whose work also contributed to the discovery of the structure of deoxyribonucleic acid (DNA)—but which should not detract from Watson and Crick's

great achievement in recognizing that the helices ran in opposite directions and the crucial idea of base-pairing.

Wells, A. F. (1968). *The Third Dimension in Chemistry*. Clarendon Press, Oxford.
A very scholarly work, with a much wider range of subject matter than its title suggests: it includes elegant descriptions of polygons and plane nets, polyhedra and the symmetry of crystals from a broad scientific and historical framework.

Weyl, Hermann (1952). *Symmetry*. Princeton University Press, Princeton, NJ.
A book in the same mould as *On Growth and Form*, Weyl expresses symmetry as *harmony of proportions* and shows that it is one of the ideas by which mankind throughout the ages has tried to create order, beauty and perfection in the world.

IUCr Commission on Crystallographic Teaching: Teaching Pamphlets

The aim of this series of pamphlets, edited by C. A. Taylor and first published in 1981, is to produce a collection of short statements, each dealing with a specific topic at a specific level. They are listed below and may be downloaded from the IUCr website.

First Series

1 *A Non-Mathematical Introduction to X-ray Diffraction. C. A. Taylor*
2 *An Introduction to the Scope, Potential and Applications of X-ray Analysis. M. Laing*
3 *Introduction to the Calculation of Structure Factors. S. C. Wallwork*
4 *The Reciprocal Lattice. A. Authier*
5 *Close-Packed Structures. P. Krishna and D. Pandey*
6 *Pourquoi les Groupes de Symétrie en Cristallographie. D. Weigel*
7 *Crystal Structure Analysis using the 'Superposition' – and 'Complementary' – Structures. L. Höhne and L. Kutchabsky*
8 *Anomalous Dispersion of X-rays in Crystallography. S. Caticha-Ellis*
9 *Rotation Matrices and Translation Vectors in Crystallography. S. Hovmöller*
10 *Metric Tensor and Symmetry Operations in Crystallography. G. Rigault*

Second Series

11 *The Stereographic Projection. E. J. W. Whittaker*
12 *Projections of Cubic Crystals. I. O. Angell and M. Moore*
13 *Symmetry. L. S. Dent Glasser*
14 *Space Group Patterns. W. M. Meier*
15 *Elementary X-ray Diffraction for Biologists. J.P. Glusker*
16 *The Study of Metals and Alloys by X-ray Powder Diffraction Methods. H. Lipson*
17 *An Introduction to Direct Methods. The Most Important Phase Relationships and their Application in Solving the Phase Problem. H. Schenk*
18 *An Introduction to Crystal Physics. E. Hartmann*
19 *Introduction to Neutron Powder Diffractometry. E. Arzi*

Third Series

20 *Crystals – A Handbook for School Teachers. E. A. Wood*
21 *Crystal Packing. A. Gavezzotti and H. Flack*
22 *Matrices, Mappings and Crystallographic Symmetry. H. Wondratschek*
23 *Teaching Crystallographic and Magnetic point group symmetry using three-dimensional rendered visualisations. M. De Graef*

Index

Note: Illustrations are indicated by *italic* page numbers, footnotes are indicated by the letter 'n'